Lecture Notes in Geoinformation and Cartography

Series Editors: William Cartwright, Georg Gartner, Liqiu Meng, Michael P. Peterson

Georg Gartner · Karl Rehrl (Eds.)

Location Based Services and TeleCartography II

From Sensor Fusion to Context Models

5th International Conference on Location Based Services and TeleCartography 2008, Salzburg

Springer

Editors

Univ. Prof. Dr. Georg Gartner
Institute of Geoinformation and Cartography
Vienna University of Technology
Erzherzog-Johannplatz 1
A-1040 Vienna
Austria
georg.gartner@tuwien.ac.at

Karl Rehrl
Salzburg Research
Forschungsgesellschaft mbH
Mobile & Web-based
Information Systems
Jakob Haringer Str. 5/III
A-5020 Salzburg
Austria
karl.rehrl@salzburgresearch.at

ISBN: 978-3-642-09950-2 e-ISBN: 978-3-540-87393-8

DOI 10.1007/978-3-540-87393-8

Lecture Notes in Geoinformation and Cartography ISSN: 1863-2246

Softcover reprint of the hardcover 1st edition 2009

Cover illustration: Salzburg Research

Cover design: deblik, Berlin

Printed on acid-free paper

9 8 7 6 5 4 3 2 1

springer.com

Preface

This book provides a general overview of research activities related to location-based services. These activities have emerged over the last years, especially around issues of positioning, sensor fusion, spatial modelling, cartographic communication as well as in the fields of ubiquitious cartography, geo-pervasive services, user-centered modelling and geo-wiki activities. The innovative and contemporary character of these topics has led to a great variety of interdisciplinary contributions, from academia to business, from computer science to geodesy. Topics cover an enormous range with heterogenous relationships to the main book issues. Whilst contemporary cartography aims at looking at new and efficient ways for communicating spatial information, the development and availability of technologies like mobile networking, mobile devices or short-range sensors lead to interesting new possibilities for achieving this aim. By trying to make use of available technologies, cartography and a variety of related disciplines look specifically at user-centered and context-aware system development, as well as new forms of supporting wayfinding and navigation systems.

The contributions are a selection of full reviewed papers submitted to the 5th Conference on Location Based Services and TeleCartography in Salzburg November 2008, jointly organized by Salzburg Research and Vienna University of Technology, Research Group Cartography.

The production of this book would not have been possible without the professional and formidable work of Manuela Schmidt and Felix Ortag. The editors are grateful for their help as well as for the help of Sven Leitinger.

Georg Gartner, Vienna, Austria
Karl Rehrl, Salzburg, Austria

Table of Contents

List of Reviewers

Ahas, Rein
Department of Geography
University of Tartu
Tartu, Estonia

Bortenschlager, Manfred
Mobile and Web-based Information Systems
Salzburg Research
Salzburg, Austria

Cartwright, William
School of Mathematical and Geospatial Science
RMIT University
Melbourne, Australia

Clarke, Keith C.
Department of Geography
University of California
Santa Barbara, CA, USA

Döllner, Jürgen
Hasso-Plattner-Institute
University of Potsdam
Potsdam, Germany

Eichhorn, Andreas
Geodätisches Institut
Technische Universität Darmstadt
Darmstadt, Germany

Frank, Andrew
Institute of Geoinformation and Cartography
Vienna University of Technology
Vienna, Austria

Gartner, Georg
Institute of Geoinformation and Cartography
Vienna University of Technology
Vienna, Austria

Grejner-Brzezinska, Dorota A.
Department of Civil and Environmental Engineering and Geodetic Science
The Ohio State University
Columbus, OH, USA

Karimi, Hassan
Department of Information Science and Telecommunications
University of Pittsburgh
Pittsburgh, PA, USA

Kranz, Matthias
Institute of Communications and Navigation
German Aerospace Center
Oberpfaffenhofen-Wessling, Germany

Krüger, Antonio
Institute of Geoinformatics
University of Muenster
Münster, Germany

Meng, Liqiu
Department of Cartography
Technical University of Munich
Munich, Germany

Merminod, Bertrand
Institute of Geomatics
Swiss Federal Institute of Technology
Lausanne, Switzerland

Millonig, Alexandra
Institute of Geoinformation and Cartography
Vienna University of Technology
Vienna, Austria

Mok, Esmond
Department of Land Surveying and Geo-Informatics
The Hong Kong Polytechnic University
Hong Kong SAR, China

Morita, Takashi
Department of Civil and Environmental Engineering
Hosei University
Tokyo, Japan

Peterson, Michael
Department of Geography and Geology
University of Nebraska at Omaha
Omaha, NE, USA

Ratti, Carlo
Massachusetts Institute of Technology
Department of Urban Studies and Planning and Media Lab
Cambridge, MA, USA

Rehrl, Karl
Mobile and Web-based Information Systems
Salzburg Research
Salzburg, Austria

Retscher, Guenther
Institute of Geodesy and Geophysics
Vienna University of Technology
Vienna, Austria

Rizos, Chris
School of Surveying and Spatial Information Systems
University of New South Wales
Sydney, Australia

Sarjakoski, Tapani
Department of Geoinformatics and Cartography
Finnish Geodetic Institute
Masala, Finland

Schmalstieg, Dieter
Institute of Computer Graphics and Vision
Graz University of Technology
Graz, Austria

Schmidt, Albrecht
Pervasive Computing und User Interface Engineering
University of Duisburg-Essen
Essen, Germany

Strobl, Josef
Department of Geography and Geoinformation
University of Salzburg
Salzburg, Austria

Verbree, Edward
Section GIS-technology
Delft University of Technology
Delft, The Netherlands

Virrantaus, Kirsi
Department of Surveying, Cartography and Geoinformatics
Helsinki University of Technology
Helsinki, Finland

Winter, Stephan
Department of Geomatics
The University of Melbourne
Melbourne, Australia

Zipf, Alexander
Department of Geography
University of Bonn
Bonn, Germany

List of Contributors

Ahas, Rein
Department of Geography
University of Tartu
Tartu, Estonia
rein.ahas@ut.ee

Beinat, Euro
Department of Geography and Geology
Salzburg University
Salzburg, Austria
euro.beinat@sbg.ac.at

Bulens, Jandirk
Centre for Geo-Information
Wageningen University and Research Centre
Wageningen, The Netherlands
jandirk.bulens@wur.nl

Cartwright, William
School of Mathematical and Geospatial Science
RMIT University
Melbourne, Australia
william.cartwright@rmit.edu.au

Chaouali, Marcel
Institute of Cartography and Geoinformatics
Leibniz University of Hannover
Hannover, Germany
marcel.chaouali@gmx.de

Cheong, Joon Wayn
School of Surveying and Spatial Information Systems
University of New South Wales
Sydney, Australia
joon.cheong@student.unsw.edu.au

Dal Fiore, Filippo
GIS-Science
Austrian Academy of Sciences
Salzburg, Austria
filippo.dalfiore@oeaw.ac.at

De Meer, Hermann
Computer Networks and Computer Communication
University of Passau
Passau, Germany
demeer@fim.uni-passau.de

De Sabbata, Stefano
Department of Mathematics and Computer Science
University of Udine
Udine, Italy
stefano.desabbata@gmail.com

Delikostidis, Ioannis
Department of Geo-Information Processing
International Institute for Geo-Information Science and Earth Observation
Enschede, The Netherlands
delikostid13285@itc.nl

Dempster, Andrew G.
School of Surveying and Spatial Information Systems
University of New South Wales
Sydney, Australia
a.dempster@unsw.edu.au

Dias, Eduardo
Department of Spatial Economics
Vrije Universiteit Amsterdam
Amsterdam, The Netherlands
edias@feweb.vu.nl

Döllner, Jürgen
Hasso-Plattner-Institute
University of Potsdam
Potsdam, Germany
doellner@hpi.uni-potsdam.de

Elias, Birgit
Institute of Cartography and Geoinformatics
Leibniz University of Hannover
Hannover, Germany
birgit.elias@ikg.uni-hannover.de

Fu, Qing
Institute of Geodesy and Geophysics
Vienna University of Technology
Vienna, Austria
fu@mail.tuwien.ac.at

Gartner, Georg
Institute of Geoinformation and Cartography
Vienna University of Technology
Vienna, Austria
georg.gartner@tuwien.ac.at

Goossen, Martin
Landscape Centre
Wageningen University and Research Centre
Wageningen, The Netherlands
martin.goossen@wur.nl

Grejner-Brzezinska, Dorota A.
Department of Civil and Environmental Engineering and Geodetic Science
The Ohio State University
Columbus, OH, USA
grejner-brzezins.1@osu.edu

Haber, Peter
Information Technology and Systems Management,
University of Applied Sciences Salzburg
Salzburg, Austria
peter.haber@fh-salzburg.ac.at

Hadjiefthymiades, Stathes
Department of Informatics and Telecommunications
National and Kapodistrian University of Athens
Athens, Greece
shadj@di.uoa.gr

Hiller, Wolfgang
Institute of Geoinformation and Cartography
Vienna University of Technology
Vienna, Austria
hillerwolfgang@hotmail.com

Houyou, Amine M.
Computer Networks and Computer Communication
University of Passau
Passau, Germany
houyou@fim.uni-passau.de

Huang, Haosheng
Institute of Geoinformation and Cartography
Vienna University of Technology
Vienna, Austria
haosheng.huang@tuwien.ac.at

Itäranta, Joonas
Nokia Research Center
Nokia Corporation
Tampere, Finland
joonas.itaerantar@nokia.com

Järv, Olle
Department of Geography
University of Tartu
Tartu, Estonia
olle.jarv@ut.ee

Jobst, Markus
Hasso-Plattner-Institute
University of Potsdam
Potsdam, Germany
markus@jobstmedia.at

Karanikolas, Nikolaos
School of Urban-Regional Planning and Development Engineering
Faculty of Engineering
Thessaloniki, Greece
karanik@topo.auth.gr

Kettunen, Pyry
Department of Geoinformatics and Cartography
Finnish Geodetic Institute
Masala, Finland
pyry.kettunena@fgi.fi

Kleijer, Frank
OTB Research Institute for Housing, Urban and Mobility Studies
Delft University of Technology
Delft, The Netherlands
frankkleijer@yahoo.com

Kohlmann, Manuel
Information Technology and Systems Management
University of Applied Sciences Salzburg
Salzburg, Austria
manuel.kohlmann@fh-salzburg.ac.at

Kolomvatsos, Kostas
Department of Informatics and Telecommunications
National and Kapodistrian University of Athens
Athens, Greece
kostasks@di.uoa.gr

Kouroupetroglou, Georgios
Department of Informatics and Telecommunications
National and Kapodistrian University of Athens
Athens, Greece
koupe@di.uoa.gr

Kranzer, Simon
Information Technology and Systems Management
University of Applied Sciences Salzburg
Salzburg, Austria
simon.kranzer@fh-salzburg.ac.at

Laakso, Katri
Devices R&D
Nokia Corporation
Tampere, Finland
katri.laakso@nokia.com

Li, Binghao
School of Surveying and Spatial Information Systems
University of New South Wales
Sydney, Australia
binghao.li@unsw.edu.au

Ligtenberg, Arend
Centre for Geo-Information
Wageningen University and Research Centre
Wageningen, The Netherlands
arend.ligtenberg@wur.nl

Lippautz, Michael
Information Technology and Systems Management
University of Applied Sciences Salzburg
Salzburg, Austria
michael.lippautz@fh-salzburg.ac.at

Liu, Gang-Jun
School of Mathematical and Geospatial Sciences
RMIT University
Melbourne, Australia
gang-jun.liu@rmit.edu.au

Meng, Liqiu
Department of Cartography
Technical University of Munich
Munich, Germany
liqiu.meng@bv.tum.de

Millonig, Alexandra
Institute of Geoinformation and Cartography
Vienna University of Technology
Vienna, Austria
alexandra.millonig@tuwien.ac.at

Mizzaro, Stefano
Department of Mathematics and Computer Science
University of Udine
Udine, Italy
mizzaro@dimi.uniud.it

Moafipoor, Shahram
Department of Civil and Environmental Engineering and Geodetic Science
The Ohio State University
Columbus, OH, USA
shahram@mail.cfm.ohio-state.edu

Mustafa, Masria
Department of Cartography
Technical University of Munich
Munich, Germany
masria.mustafa@tu-muenchen.de

Nivala, Annu-Maaria
Department of Geoinformatics and Cartography
Finnish Geodetic Institute
Masala, Finland
annu-maaria.nivala@fgi.fi

Nomikos, Vangelis
Department of Informatics and Telecommunications
National and Kapodistrian University of Athens
Athens, Greece
grad0917@di.uoa.gr

Odijk, Dennis
Department of Earth Observation and Space Systems
Delft University of Technology
Delft, The Netherlands
d.odijk@tudelft.nl

Paelke, Volker
Institute of Cartography and Geoinformatics
Leibniz University of Hannover
Hannover, Germany
volker.paelke@ikg.uni-hannover.de

Panov, Ilija
Institute for Geoinformatics
University of Münster
Münster, Germany
ilijapanov@gmx.de

Papageorgas, Panagiotis
Department of Informatics and Telecommunications
National and Kapodistrian University of Athens
Athens, Greece
ppapag@teipir.gr

Papataxiarhis, Vassilis
Department of Informatics and Telecommunications
National and Kapodistrian University of Athens
Athens, Greece
vpap@di.uoa.gr

Piralli, Irene
Information Technology and Systems Management
University of Applied Sciences Salzburg
Salzburg, Austria
irene.piralli@fh-salzburg.ac.at

Raubal, Martin
Department of Geography
University of California
Santa Barbara, CA, USA
raubal@geog.ucsb.edu

Rauchenschwandtner, Renate
Information Technology and Systems Management
University of Applied Sciences Salzburg
Salzburg, Austria
rrauchenschwandtner.its2005@fh-salzburg.ac.at

Rizos, Chris
School of Surveying and Spatial Information Systems
University of New South Wales
Sydney, Australia
c.rizos@unsw.edu.au

Retscher, Günther
Institute of Geodesy and Geophysics
Vienna University of Technology
Vienna, Austria
gretsch@pop.tuwien.ac.at

Riga, Vivi
Department of Informatics and Telecommunications
National and Kapodistrian University of Athens
Athens, Greece
std02073@di.uoa.gr

Roustanis, Themistocles
School of Urban-Regional Planning and Development Engineering
Faculty of Engineering
Thessaloniki, Greece
throu@topo.auth.gr

Saluveer, Erki
Department of Geography
University of Tartu
Tartu, Estonia
erk@ut.ee

Sarjakoski, Tiina
Department of Geoinformatics and Cartography
Finnish Geodetic Institute
Masala, Finland
tiina.sarjakoski@fgi.fi

Sekkas, Odysseas
Department of Informatics and Telecommunications
National and Kapodistrian University of Athens
Athens, Greece
sekkas@di.uoa.gr

Silm, Siiri
Department of Geography
University of Tartu
Tartu, Estonia
siiri@ut.ee

Sternig, Christian
Information Technology and Systems Management
University of Applied Sciences Salzburg
Salzburg, Austria
christian.sternig@fh-salzburg.ac.at

Stylianidis, Efstratios
School of Urban-Regional Planning and Development Engineering
Faculty of Engineering
Thessaloniki, Greece
sstyl@auth.gr

Toth, Charles K.
Department of Civil and Environmental Engineering and Geodetic Science
The Ohio State University
Columbus, OH, USA
toth.2@osu.edu

Tsetsos, Vassileios
Department of Informatics and Telecommunications
National and Kapodistrian University of Athens
Athens, Greece
b.tsetsos@di.uoa.gr

van der Spek, Stefan
Faculty of Architecture, Urbanism and Building Sciences
Delft University of Technology
Delft, The Netherlands
s.c.vanderSpek@tudelft.nl

van Elzakker, Corné P. J. M.
Department of Geo-Information Processing
International Institute for Geo-Information Science and Earth Observation
Enschede, The Netherlands
elzakker@itc.nl

van Lammeren, Ron
Centre for Geo-Information
Wageningen University and Research Centre
Wageningen, The Netherlands
ron.vanlammeren@wur.nl

Vassena, Luca
Department of Mathematics and Computer Science
University of Udine
Udine, Italy
vassena@dimi.uniud.it

Verbree, Edward
OTB Research Institute for Housing, Urban and Mobility Studies
Delft University of Technology
Delft, The Netherlands
e.verbree@tudelft.nl

Vourakis, Stelios
Department of Informatics and Telecommunications
National and Kapodistrian University of Athens
Athens, Greece
ea04504@di.uoa.gr

Wu, Falin
School of Mathematical and Geospatial Sciences
RMIT University
Melbourne, Australia
falin.wu@rmit.edu.au

Xouris, Vasileios
Department of Informatics and Telecommunications
National and Kapodistrian University of Athens
Athens, Greece
std02193@di.uoa.gr

Zhang, Kefei
School of Mathematical and Geospatial Sciences
RMIT University
Melbourne, Australia
kefei.zhang@rmit.edu.au

Zhang, Meng
Department of Cartography
Technical University of Munich
Munich, Germany
meng.zhang@bv.tum.de

Zhu, Ming
School of Mathematical and Geospatial Sciences
RMIT University
Melbourne, Australia
s3139786@student.rmit.edu.au

1 LBS and TeleCartography II: About the book

Georg Gartner

Department of Geoinformation and Cartography
Vienna University of Technology, Austria

Abstract

This book is based on a series of symposia on Location Based Services and Telecartography that have been held since 2002 at the Vienna University of Technology, the Polytechnical University HongKong and Salzburg Research. The meetings themselves were a response to technological developments in miniaturizing devices for telecommunication, computing and display and an increased interest in both incorporating cartographic presentations on such mobile devices and developing services that are specific to a particular location. The broad variety of disciplines involved in this research and the differences in approaching the basic problems are probably typical for a developing field of interdisciplinary research. However, some main areas of research and development in the emerging area of LBS and Telecartography can now be identified. The contributions to this book are selected from the full papers of the 5th Symposium on LBS and TeleCartography and reflect the main areas of interest: positioning, modelling and awareness, visualisation and cartographic communication and application development. It follows the book "Location Based Services and TeleCartography" (Gartner et al. 2007), which reflects a selection of papers of the first symposia.

1.1 A Series of Symposia on LBS and TeleCartography

Research and business activities in the field of applying cartographic presentations on mobile devices (TeleCartography) and developing innovative services, where the location of a mobile device becomes a "variable" of an information system (LBS), have increased internationally since the mass-market availability of mobile devices, satellite positioning and telecommunication infrastructure. The Cartography Department of the Vienna University of Technology has been interested in these developments from the early stages, which have led to various research projects, student activities, multidisciplinary cooperations and the organization of a series of symposia on LBS and TeleCartography in Vienna. The success of the first symposium in 2002 (Kelnhofer et al. 2002) and the increasing activities in various disciplines related to LBS and TeleCartography were the major impetus for setting up a second symposium in 2004 (Gartner 2004), a third symposium in 2005 (Gartner 2005), a fourth symposium in 2007 in Hong Kong (Gartner & Mok 2007) and a fifth symposium in 2008 in Salzburg, being meant as forum for bringing together experts from academia and business as well as representatives of different disciplines.

The symposia on Location Based Services (LBS) and TeleCartography have been organized by the Institute of Cartography and Geo-Media Techniques (IKGeoM)[1] of the Vienna University of Technology (TU) in close cooperation with local organizers (namely Prof. Esmond Mok of the Department of Land Surveying and Geoinformatics of the Hong Kong Polytechnic University for the 4th Symposium and DI Karl Rehrl of Salzburg Research for the 5th Symposium), the Commission on Maps and the Internet of the International Cartographic Association (ICA) and with a collaborative interest from the ICA Commission on Ubiquitous Cartography, the Working Group 4.1.2 of the International Association of Geodesy and the Working Group V TC 2 of the International Society for Photogrammetry and Remote Sensing. As the world authoritative body on cartography, the International Cartographic Association attempts to advance the use of maps in society and science. The ICA Commission on Maps and the Internet, formed in 1999, brings together international specialists in the field of Internet mapping and disseminates information to a broader audience on new developments and major areas of research.

As the activities in the field of LBS and TeleCartography can be understood as an expansion of methods for Internet mapping and techniques for the mobile Internet, the ICA Commission on Maps and the Internet has a common interest in supporting the meetings in Vienna, Austria, dedicated to the issues of LBS and TeleCartography.

The 5th Symposium on Location Based Services and TeleCartography is held on November 26–28, 2008 with the purpose of offering a forum for research-driven activities related to the context of location and map-based services. Such activities emerged in the last years especially around issues of positioning, spatial modelling, and cartographic communication as well as in the fields of ubiquitous cartography, geo-pervasive services, user-centred modelling or geo-wiki activities. The innovative and contemporary character of the conference leads to a great variety of contributions in terms of interdisciplinarity. Presenters representing more than 15 countries with backgrounds varying from academia to business, from computer science to geodesy, cover an enormous number of topics with a heterogeneous relation to the conference main topic.

---

[1] Research Group Cartography and Geo-Media techniques within the Department of Geoinformation and Cartography since 1. 1. 2004

While contemporary cartography examines new and efficient ways on communicating spatial information, the development and availability of technologies like mobile networking, mobile devices or short-range sensors lead to interesting new possibilities of achieving this aim. Cartographers and researchers from a variety of disciplines try to make use of the available technologies by looking specifically at user-centred and context-aware system developments as well as new forms of supporting wayfinding and navigation systems.

Selected results of the interesting contributions to the 5th Symposium on LBS and TeleCartography, mirroring the main elements and trends within this field, are brought together in this book.

1.2 Progression of Research

1.2.1 Terms

The key terms being used in this book reflect the iterative way of applying terms and names to developments in the context of new technologies. In general a flexible approach in terms of defining the frame of the area of interest has been used, thus the main terms are applicable to most of the contributions.

- *Cartographic Model:* Presentation model of geospatial data, derived from the geo data model by means of cartographic generalization and symbolization.
- *Location Based Service:* Services that exploit knowledge about where an information device user is located.
- *TeleCartography:* Distribution of cartographic presentation forms via wireless data transfer interfaces and mobile devices.
- *Ubiquitous Cartography:* Ability for users to create and use maps in any place and at any time to resolve geospatial problems.

In modern cartography, the main focus is on understanding the processes and methods of "how to communicate spatial information efficiently". In this concern, the "responsibility" of cartography exceeds the creation of cartographic presentation forms, but is rather focused on understanding the relations within the "whole system" of communicating spatial information, including the user, the models and the transmission processes. The engagement of modern cartography in fields like LBS and TeleCartography and the various multidisciplinary approaches including cartography have to be seen in this context.

Telecommunication infrastructure (mobile network), positioning methods, mobile in- and output devices and multimedia cartographic information systems are prerequisites for developing applications that incorporate the user's position as a variable of an information system. Normally, cartographic presentation forms are involved in integrating geospatial information into such a system.Thus, the resulting system can be called a *map-based location based service* (LBS). This chapter discusses the elements of a map-based LBS, outlines the main research topics and describes some experiences in the context of conceptual design and the development of map-based LBS.

1.2.2 Elements

A system can be called a Location Based Service (LBS), when the position of a mobile device – and therefore the position of the user – is somehow part of an information system. The derivable types of applications in this context can be stated as heterogenous and include simple and text-based applications, which use the *cell ID* (unique identification of the cell of a telecommunication network) for a rough positioning (e.g., Which petrol stations are there around me?) to map-based multimedia applications including routing functionality. In this context, different names for the context of telecommunication infrastructure, location-based applications and cartography are used. In addition to *mobile cartography, ubiquitous cartography* the author proposes the term *TeleCartography,* to be understood as issues involved by the distribution of cartographic presentation forms via wireless data transfer interfaces and mobile devices.

Independent from the level of complexity of the system architecture, every map-based LBS needs some basic elements to handle the main tasks of positioning, data modelling and information presentation.

Positioning

The determination of the position of a mobile in/output device is a direct requirement for every system to be called LBS. Positioning has to be adequate to the service in terms of a dependent relationship and adapted to the tasks. For various applications the necessary level of accuracy needed can be served by the cell-ID of a telecommunication network and the thus derivable position, that gives an accuracy of positioning between 50 and 100 meters in urban areas. For navigation purposes – in particular in the context of pedestrian navigation – the accuracy demands increases to values of at least 25 meters and less. For indoor navigation, the requirements for the position determination are even greater.

Various methods of positioning are available for different levels of accuracy:

- Satellite-based positioning
- Positioning by radio network
- Alternative methods
- Combinations

Modelling and Presentation of Information

The possibilities of transmitting and visualising geospatial information in context of a determined position is primarily restricted by the limitations of the particular mobile device. The basic conditions of the cartographic communication process have to be fulfilled when using map-based LBS: The cartographic model has to be clearly perceivable while it is permanently scale-dependent and has to present the task-dependent appropriate geometric and semantic information.

This fact in combination with restrictions in size and format of current mobile devices leads to different levels of solutions for presenting information within map-based LBS:

- Cartographic presentation forms without specific adaptions
- Cartographic presentation forms adapted to specific requirements of screen display
- New and adapted cartographic presentation forms
- Multimedia add-ons, replacements and alternative presentation forms

Users and Adaptation

Experiences with LBS developments have led to various suggestions for a more user-centered system conception. Modelling parameters in the context of the *user* and the *usage situation* are seen as fundaments of such user-adequate attempts, that can be summarized as *concepts of adaptation.*

The adaptation of cartographic visualisations in this context can be understood as, for example, the automatic selection of adequate scales, algorithms for adequate symbolization, or even the change to text-only output of information in case of shortcomings in size or resolution of the output device. Adapting to the needs of the user results from the user profile, selected in advance from a list or entered manually by the user to influence the graphical presentation (size of lettering, used colours) or to provide pre-defined map elements. Adapting the visualisation to the situation would include the current day and time and the speed of travel.

Towards Ubiquitous Environments

The development of technologies like telecommunication infrastructures, wireless networks, radio frequency identification or innovative displays like electronic paper can all be seen as ways of developing a ubiquitous environment, where location-based services and ubiquitous cartographic applications can be applied.

1.3 Structure of the Book

The book is structured in four main sections, reflecting the results of the symposia and the main progression of research:

- Modelling
- TeleCartography
- Services and Applications
- Positioning and Sensor Fusion

1.3.1 The Section Modelling

In the first section *Modelling* aspects of modelling geo-data, determining routes and other elements for LBS and navigation systems and the consideration of user-dependent variables for developing adaptive or context-aware systems are discussed.

In *Chapter 2* **Martin Raubal** et al. propose a formal conceptual model for automatic mobile map adaptation that can be employed for different applications, such as pedestrian navigation. Through specified adaptation operations it aims at reduction of both the user interactions with the service and the cognitive load for the user. **Huang Haosheng** et al. present in *Chapter 3* an activity theory motivated approach to model context parameters by analyzing activity for context-awareness. and by identifying relevant context parameters. In *Chapter 4* **Masria Mustafa** et al. attempt to develop an analytical congestion visualization system using GIS with the help of probe vehicles as the source of data. **Stefano De Sabbata** et al. concentrate in *Chapter 5* on trajectory prediction by space rank and physics models. A method is proposed, where data related to position of people is mined and the importance of single positions is derived. The importance is then used to estimate users trajectories and future destinations. In *Chapter 6* **Alexandra Millonig** et al. propose an advanced method of developing pedestrian typology for personalised mobile information systems by applying a methodological set-up including qualitative and quantitative methods. In *Chapter 7* **Stefan van der Spek** looks on the concept of mapping pedestrian movement by user tracking technologies, applied in a project in the City of Koblenz.

1.3.2 The Section TeleCartography

In the second section *TeleCartography* various aspects of communication, map models and multimedia presentation forms are discussed in the context of LBS and TeleCartography. The range of presentation forms include maps, 3D presentations as well as augmented reality presentations.

In *Chapter 8* **Birgit Elias** et al. discuss the evaluation of user variables in topographic feature recall for the informed selection of personalized landmarks.

Eduardo Dias et al. propose a methodology to determine the impact on adoption of location-aware information in *Chapter 9.* When applying the Technology Acceptance Model the authors demonstrate, that a significant positive impact on the acceptance of mobile information systems can be achieved by location-awareness. In *Chapter 10* **Georg Gartner** et al. report on the impact of restricted display size on spatial knowledge acquisition in the context of pedestrian navigation. **Annu-Maaria Nivala** et al. report on user requirements for location-based services to support hiking activities in *Chapter 11.* By applying three usability engineering methods two main research questions are tackled, the question on changes during an activity and the question on community and content needs of hikers. In *Chapter 12* **Ioannis Delikostidis** et al. outline a study on how users proceed about geo-identification and pedestrian navigation with geo-mobile applications. **Markus Jobst** et al. discuss the influence of neo-cartography on map communication in LBS in *Chapter 13,* focusing on consequences on spanning ubiquitous cartography, user participation and considerations for geo-media techniques.

1.3.3 The Section Services and Applications

In the third section *Services and Applications* a broad variety of developed prototypes, attempts and services are described as well as the development of ideas of using LBS and TeleCartography services as data acquisition tools.

In *Chapter 14* **Filippo Dal Fiore** et al. describe results of the evaluation of a ilot initiative at the Dutch Police of adopting a mobile location-aware system for field police work. **Arend Ligtenberg** et al. describe in *Chapter 15* some aspects of enhancing the experience of the landscape by applying the "digital dowsing rod". With this the exploration of the cultural heritage of a region is possible within a framework of a Location Based Service. In *Chapter 16* **Vassilis Papataxiahis** et al. describe a framework for an indoor Location Based Service for all, called MNISIKLIS. The overall architecture and selected conception details are described. **Michael Lippautz** et al. describe in *Chapter 17* the components of a real-time monitoring system for action forces during disaster operations. In *Chapter 18* **Rein Ahas** et al. present a service for modelling home and work locations of population using passive mobile positioning data. **Stylianides** et al. describe in *Chapter 19* GEOVAL, a geographical information system for real estate, which can be applied on mobile GIS solutions as well.

1.3.4 The Section Positioning and Sensor Fusion

In the fourth section *Positioning and Sensor Fusion* various aspects of determining the position of a mobile user are investigated.

In *Chapter 20* **Shahram Moafipoor** et al. focus on quality assurance and quality control analysis of dead reckoning paramters in a personal navigator. **Amine Houyou** et al.'s *Chapter 21* briefly investigates methods to efficiently overlay mediation for mobile location based services. A design methodology is proposed to ensure the localized impact of mobility-lead communication overhead. In *Chapter 22* **Kefei Zhang** et al. propose a RFID/INS integration method, that can provide more reliable estimation of a position of a mobile user in a multi-storey building. **Frank Kleijer** et al. present a model to predict the GNSS availability and accuracy in urban environments in *Chapter 23*. The case study Schiphol Airport is used to find the effect of different parameters and variability of accuracy. In *Chapter 24* **Kefei Zhang** et al. present an investigation of the signal performance of the current and the future GNSS in typical urban canyons in Australia using a high fidelity 3D urban model. In *Chapter 25* **Qing Fu** et al. describe concepts and test results of using and combining RFID and INS for indoor positioning. In *Chapter 26* **Cheong Joon Wayn** et al. aim for combining GPS and WiFI in a real time positioning device and describe the conceptualization of such an aim and initial outcomes.

1.4 Summary

The concept of the book has been outlined in this chapter. It is discussed that the results of the international symposia on LBS and TeleCartography mirror the heterogenous variety of disciplines and backgrounds – which have contributed aspects and issues within the "umbrella terms" of Location-based Services and TeleCartography. The main areas of research, that can be derived from these results, are identified and the structure of the book is explained.

References

Gartner G (2004) Location Based Services and TeleCartography. Proceedings of the 2nd Symposium on LBS and TeleCartography. Geowissenschaftliche Mitteilungen, 66, Wien 2004.

Gartner G (2005) Location Based Services and TeleCartography. Proceedings of the 3rd Symposium on LBS and TeleCartography. Geowissenschaftliche Mitteilungen, 74, Wien 2005.

Gartner G, Cartwright W, Peterson M (2007) Location Based Services and TeleCartography. Lecture Notes in Geoinformation and Cartography. Springer, Berlin-Heidelberg, 2007.

Gartner G, Mok E (2008) 4. Conference on LBS and TeleCartography. Journal on LBS, Special Edition, 2, 1, 2008.

Kelnhofer F, Lechthaler M, Brunner K (2002) TeleKartographie und Location Based Services. Beiträge zum 1.Symposium TeleKartographie und LBS. Geowissenschaftliche Mitteilungen, 58, Wien 2002.

Section I: Modelling

2 A Formal Model for Mobile Map Adaptation

Martin Raubal[1] and Ilija Panov[2]

[1] Department of Geography, University of California, Santa Barbara, CA, USA
[2] Institute for Geoinformatics, University of Münster, Germany

Abstract

Computing has become increasingly mobile and pervasive, which implies that services must be aware of and adapt to changing contexts in highly dynamic environments. Services that require a lot of user interaction have less potential of being used, because they tend to be obstructive. Thus, context-awareness and adaptation are important research issues in the area of mobile computing. A major goal is to minimize user interaction through service adaptation, and to provide context-sensitive and personalized information to the user. Adaptations for mobile map applications must consider a wide range of factors – from technical requirements to cognitive abilities and goals of the user. However, specifying contextual facts in an accurate and traceable manner is challenging. Initial approaches have focused on information visualization for mobile map applications through context information. These typically focus on simplifying and generalizing route segments rather than adapting to personal information. In this paper we propose a formal conceptual model for automatic mobile map adaptation that can be employed for different applications, such as pedestrian navigation. This model is composed of three components – a *context* model, a *user* model, and a *task* model. Through specified adaptation operations it aims at a reduction of both the user interaction with the service and the cognitive load for the user.

Keywords: mobile computing, location-based services, mobile maps, adaptation, context-awareness, formal specifications

2.1 Introduction

Computing has become increasingly mobile and pervasive, and the emerging technologies provide 'anytime/anywhere' information. These changes imply that applications and services must be aware of and adapt to changing contexts in highly dynamic environments. A mobile user is potentially more distracted, and different constraints and limitations exist, such as small display, and limited energy and bandwidth. Users often need to make decisions on the spot and therefore require current personalized and context-sensitive information on their mobile devices, i.e., 'the presentation [...] must be conditioned by the users' activities and by the state of the world around them.' (Lake 2001, p.1) A prime example is pedestrian navigation. Finding ways for services to adapt appropriately within a wide range of possible user situations in order to best support human–computer interaction has been identified as an important research problem (Dey & Abowd 2000). It has been pointed out though that research on accurately discovering and efficiently disseminating contextual information is still at an early stage (Strang & Linnhoff-Popien 2004). Thus, context-awareness and adaptation comprise key research topics in the area of mobile computing and location-based services (LBS) (Raper, Gartner, Karimi, & Rizos 2007).

The user and her activities in a particular context define the amount and detail of necessary information, the degree of generalization, and the way such information is visualized on a mobile map. First attempts of adapting visualization for mobile services have been described in (Zipf 2002). Maps are of great value for people as they have the potential to represent large amounts of information about an area of interest within a single frame in a comprehensible form. Examples of where maps are useful for pedestrians range from searching for points-of-interest (POIs) to navigating in unfamiliar environments. These different tasks and circumstances require a large amount of user interaction, such as changing program settings and receiving personalized information. Therefore services have less potential of being used, because they tend to be obstructive. Consequently, a major goal in the field of mobile computing is to minimize user interaction through service adaptation, and to provide context-sensitive and personalized information to the user in a changing environment. In this paper, we propose an abstract formal model for mobile map adaptation, which takes these issues into account. Based on a number of adaptation operations this model aims at reducing both the user interaction with a service and the cognitive load for the user.

Section 2.2 presents a use case and derives current problems with respect to user interaction with mobile LBS. In *Section 2.3* we discuss previous work on context and adaptation. *Section 2.4* develops an abstract conceptual model for mobile map adaptation based on three components – a context model, a user model, and a task model. In *Section 2.5,* this model is applied to a LBS for pedestrian navigation by formally specifying the model components and their operations. *Section 2.6* compares the formal model to the use-case service. The final section presents conclusions and directions for future work.

2.2 Use Case and Problem Statement

This section introduces a use case for mobile map adaptation and identifies current problems regarding user interaction with mobile LBS.

2.2.1 Scenario

The presented scenario is based on previous work regarding prototypical implementations of a user-oriented pedestrian navigation service[1] *(utopian)* and a *HotelFinder* service (Rinner & Raubal 2004). *Utopian* is a LBS for recreation facilities and gastronomy offers combined with a navigation service for pedestrians. LBS assist users in the performance of spatio-temporal tasks and provide location-dependent information.

Alice visits the city of Münster for the first time and wants to stay for a few days. She has arrived at the train station and starts the *HotelFinder* software on her mobile device. This service supports Alice in finding a suitable hotel. After the decision-making process with the *HotelFinder, utopian* starts to navigate Alice from her current position to the chosen hotel. This navigation service provides a series of pictures with landmarks of every decision point along the way. Brief written instructions provide Alice with directions of turns. On her way through the city Alice passes several historical buildings. She is very interested in historical monuments, buildings and places, and therefore wants to get some information about them. The navigation tool does not support any kind of information retrieval beyond the pure navigation task. Therefore, Alice has to start a standard search engine to get more information. Continuing her way to the hotel, Alice receives the instruction to turn right at the next landmark. She reaches a place with a large building. The large and scattered environment confuses Alice and she is unsure what 'right at the building' exactly means. To get an overview of the current situation Alice has to switch to the digital map to verify her current position and look for the direction of the next decision point. This map always provides Alice an overview of the complete route from her starting point to the destination. The map scale does not deliver a realistic impression of her current location on the map. She has to manually zoom in to get a detailed view of the location and then zoom out to get back to the route overview.

[1] http://utopian-online.de

Alice feels uncomfortable when following the navigation instructions, because she must orient herself several times. This is due to the fact that the map is always oriented to the North direction and not aligned with her current walking direction.

2.2.2 Problems

Based on the described use case, we identify several problems that occur during user interaction with mobile LBS regarding map extent and alignment, zooming, personalized information, and time. These issues have also been recognized by others (Radoczky 2003, Wealands, Miller, Benda & Cartwright 2007).

Problem 1: Map Alignment with North Direction

Most maps are aligned with the North direction. When using an analog map, users often turn the map around every few minutes or at every direction change, because this facilitates orientation (Nivala & Sarjakoski 2003, Wealands et al. 2007). This is also true when using digital maps on mobile devices, which are aligned with the North direction *(Figure 2.1)*. Therefore, one of the most useful pieces of context information is the user's walking direction, which could be measured by the Global Positioning System (GPS) or different types of sensors (Baus & Kray 2002). 86% of the participants in the work by Radoczky (2003) stated that a track-up oriented map is indispensable.

Fig. 2.1. Initial map extent after starting *utopian* with North alignment.

Problem 2: Manual Zooming and Static Map Extent

When using *utopian* the user has to constantly change the map scale to get an overview of a larger area or a detailed view of an area of interest. In order to do this, the user must interact with the device by spanning a rectangle with the stylus on the screen. Some experience and skills are needed to get the correct map scale with this technique. In general, it is not known how to fit maps on small screens and which technique is best (Gutwin & Fedak 2004). Therefore, zooming at decision points is one important aspect of the model presented in our work. When the user reaches a decision point an automatic zooming function delivers a detailed view. In the survey done by Radoczky (2003), 45% of the participants stated this function as indispensable.

After the route calculation, an overview map for a general presentation of the whole route has been found vital by 64% of the participants in the survey by Radoczky (2003). But this overview map of the calculated route should only serve as a starting point at the beginning of the guiding process. In our use case the service constantly delivers the overview map with the entire route. The user must therefore manually zoom in and out to get more detailed views and to get back to the entire route. This results in a large amount of user interaction with the device. Additionally, a small-scale map results in a high cognitive load for the user. A smaller viewable map extent would reduce the cognitive load, because it reduces the amount of visualized information. The following map extent should depend on the velocity of the user, so that the actual map extent shows the area that the user can reach within a certain amount of time.

Problem 3: Visualization of Personalized Information

In *utopian* the user gets information about the POIs the service calculates for the tour. This means that the user only gets information about locations she is interested in for the current tour. Such information about short-term interests is not visualized automatically, but through direct interaction with the context menu of the service. The user has no facility to determine and set personal long-term interests, e.g., through preference settings via a user profile.

Problem 4: Daytime-Independent Landmark Visualization

People make use of salient objects in the environment to orient themselves and navigate through space (Denis, Michon & Tom 2006, Lovelace, Hegarty & Montello 1999). *Utopian* provides landmark-based navigation instructions using point landmarks at decision points. 73% of the survey participants in (Radoczky 2003) voted for multi-encoded navigation instructions, in particular the integration of landmark

Fig. 2.2. Same facade by day and night. Illumination at night increases visual attraction.

photographs in case of decision points. Raubal & Winter (2002) provided a formal measure to specify landmark saliency of buildings and mention visibility as one of the components. The visibility of buildings and other kinds of salient objects is different for day and night (Winter, Raubal & Nothegger 2005). Whereas some facades have low visual attraction during daytime, their visual attraction and saliency increases when illuminated at night *(Figure 2.2). Utopian* does not support this functionality of switching landmarks depending on the time of day.

2.3 Related Work

This section reviews *context* and *adaptation* from the perspective of geospatial mobile applications. The ability of services to use context information allows for the adaptation of the available information in order to generate a benefit for the user.

2.3.1 Context

The dynamic changes of service and user states cause a change in context, and therefore context-awareness is an important factor in mobile computing. Several definitions for mobile computing regard context as the changing execution environment, which is divided into *computing* (e.g., computation resources), *user* (e.g., social situation), and *physical environment* (e.g., weather). Dey & Abowd (2000, pp.3–4) presented a more generic context definition for ubiquitous and mobile computing: 'Context is any information that can be used to characterize the situation of an entity. An entity is a person, place, or object that is considered relevant to the interaction between a user and an application, including the user and applications themselves.' It offers two important advantages for designing a context model for the pedestrian navigation task described in *Section 2.2.1.* First, if a piece of information can be used to characterize the situation of the user or task, then this information is context. Location and time are examples for useful information within the model. Second, the definition allows for context to be either explicitly or implicitly indicated by the user. For example, the location of nearby objects can be detected implicitly by the service or explicitly through user input. The main aspects in the above definition are identity (user), activity (interaction with service and environment), location, and time (as temporal constraints). These aspects build the basis for *context-aware computing* – 'the ability of a mobile user's applications to discover and react to changes in the environment they are situated in.' (Schilit & Theimer 1994, p.3)

2.3.2 Adaptation

One of the goals of LBS should be to take up as little of the user's attention as possible. There are mechanisms available to fit the services to the current situation and make them flexible, called *adaptation.* Adaptation is therefore the answer to a changing context (Reichenbacher 2003). There are two ways of achieving knowledge sharing between a service and its user – making the service *adaptive* or *adaptable* (Fischer 1993, Oppermann 1994) *(Table 2.1).* One of the dominant factors for adaptation is the user's task. All relevant factors need to be formally represented within the service.

Table 2.1. Characteristics of adaptive and adaptable services, modified from (Fischer 1993).

	Adaptive	Adaptable
Definition	dynamic adaptation by the service to current task and user	user changes functionality of the service
Strengths	little (or no) effort by the user	user is in control
Weaknesses	loss of control	user must do substantial work

The visualization of geoinformation and its interactive use on mobile devices is adapted to either one or all components of the actual context (user, location, information, etc.) (Gartner 2004). However, the visualization does not need to adapt to all factors at once. Zipf (2002) argues that when adapting maps for mobile services it is insufficient to focus only on technical parameters such as device characteristics, but maps have to be dynamically generated according to a wider range of variables, including user preferences, task, and location. To achieve this goal the service needs to exploit a user model and context knowledge. Different tasks have different requirements regarding map design. While it is important for an overview map to show many features (but not necessarily in great detail), a route map must display important decision points or landmarks (Agrawala & Stolte 2001). Incorporating adaptation within the visualization process solves several usability problems encountered in the mobile environment.

2.4 A Conceptual Model for Mobile Map Adaptation

This section presents a conceptual model for mobile map adaptation, which will be applied to the use case of navigation and formally specified in *Section 2.5*. The adaptation model is designed in an abstract way to be used for different tasks, constraints, and requirements in the domain of mobile LBS.

2.4.1 Design Considerations

Our main hypothesis in this research is that *a formal model for mobile map adaptation predicts a reduction of user interaction and cognitive load during location-based tasks*. Different design decisions have to be made with respect to the following questions:

- What is the user's task?
- What are the user's requirements, needs, and preferences?
- Which context features are needed for the task and satisfy the user's preferences?
- Which kinds of operations are needed, so that the user can successfully accomplish the task?

Reduction of Cognitive Load

Cognitive load can be understood as the amount of work needed to acquire and use information. In the case of mobile pedestrian navigation services this corresponds to the visualized information and navigation instructions displayed on the screen. The *cognitive load theory* (CLT) offers designers of different services a way of assessing and affecting some critical components during the design process of digital maps (Bunch & Lloyd 2006). Several aspects of CLT were implemented in the area of mobile digital maps (Mayer & Moreno 2002). These techniques concentrate on the reduction of the visualized information. The pitfall in this information reduction is that it stands in contrast to adaptation methods, because the key to successful adaptation is to collect as much information about the user and her environment as possible (Hampe & Paelke 2005). Therefore, the challenge in the reduction of cognitive load is to find the appropriate amount of information, i.e., determining the right *level of detail* (LoD). Here, this LoD will be achieved through a task-driven adaptation model. The model and its components will be filled with features required for the specific task.

Reduction of User Interaction with the Device

The usability and usefulness of mobile map services is highly dependent on the appropriate *graphical user interface* (GUI) design including the visualization of spatial and non-spatial information. The visualization of the different elements is constrained by the limited resolution and small display size, therefore the GUI design on a small display must balance space requirements of both a map and a set of tools (Rinner, Raubal & Spigel 2005). Further constraints are imposed through the limited processing power and low resolution of pointing devices. Methods and techniques for GUI design for mobile devices have been proposed by Cartwright et al. (2001). Here, we focus on minimizing the use of pointing devices to achieve a reduction of user interaction.

2.4.2 Adaptation Model

The *AdaptationModel* (*A*) is composed of three submodels – the *ContextModel (C),* the *UserModel (U),* and the *TaskModel (T).* These models are classified into dynamic and static elements *(Figure 2.3).* The *ContextModel* represents the dynamic elements of the model, because in most cases the *Situation (S)* of using a mobile device implies that the surrounding context changes (e.g., the user's position), whereas the user and the task remain the same (Zipf & Jöst 2006).

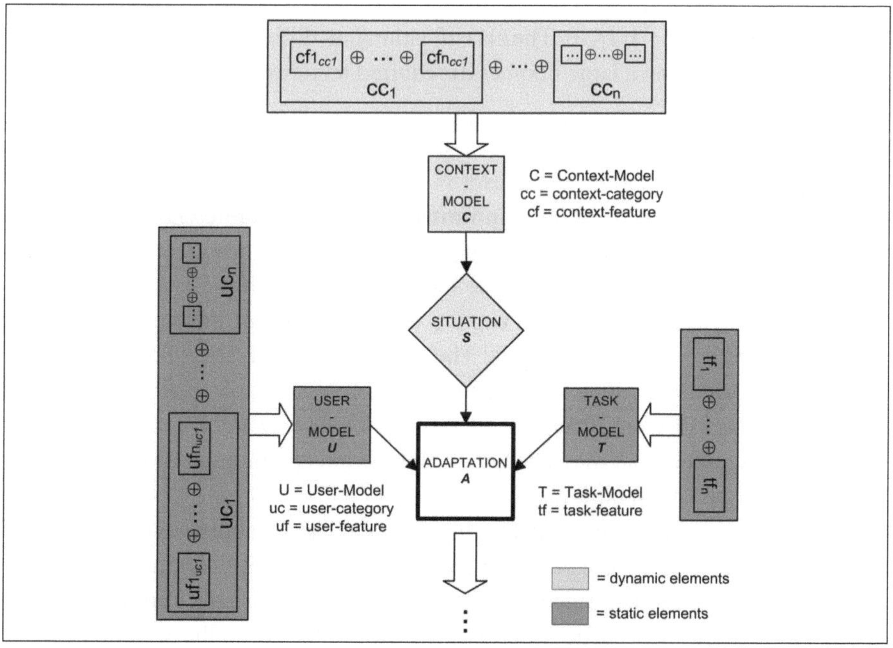

Fig. 2.3. Abstract *AdaptationModel* for modeling context-sensitive and user-centered LBS.

The *ContextFeatures (cf)* underlie dynamic actions and are always available in an explicit form, because they can be sensed by the device (e.g., position through GPS) or derived through other *ContextFeatures.* Their values change each time a *Situation* changes. In contrast to the *TaskModel,* both the *ContextModel* and the *UserModel* consist of different categories, and each category consists of an arbitrary number of features (e.g., the *ContextModel* consists of a *context-category cc1* with features *cf1,...,cfn)* (Schmidt, Beigl & Gellersen 1999). The categories serve as structural units to classify the features of the model. The values of these features serve as input parameters for the different types of operations, which can be either adaptive or adaptable, and strongly depend on the specific task.

The ContextModel

The *ContextFeatures* in the *ContextModel* describe the user's current situation with its various characteristics corresponding to the real world. With regard to the context definition given in *Section 2.3.1* this model should cover the two most important features, i.e., position and time, supplemented by further useful features such as direction of movement. The *Situation (S)* in *Figure 2.3* can be conveyed as the user being situated in a dynamic environment described by the *ContextFeatures* $(S = (cf_1, \ldots, cf_n))$ (Reichenbacher 2003). The ⊕ symbol represents a linking between several *ContextFeatures* (see *Figure 2.5* for examples). More available features make a more detailed description of the user's *Situation* possible. To manage the amount of possible and useful *ContextFeatures,* the *ContextModel* is classified into categories $(C = (cc_k, cf_l)$ *where* $1 \leq k, l \leq n)$. The classification into different categories is particularly useful when using a large number of required features for the task.

The UserModel

The service's representation of the user is incorporated through a *UserModel* that describes the user with predefined information about her preferences. This information is represented by the *UserFeatures* $(uf_1, \ldots, uf_n)$. These features represent all characteristics, which fall under the identity–category of the context definition *(Section 2.3.1).* The model should capture different types of information, such as user needs, preferences, and interests. The ⊕ symbol stands for a linking between several *UserFeatures,* which are organized into categories $(U = (uc_i, uf_j)$ *where* $1 \leq i, j \leq n)$. The different features representing the *UserModel* are static elements, because it is unlikely that they change during the navigation process (e.g., a user's preference for historical buildings).

Another approach to user modeling involves detecting patterns in their behavior (Zipf & Jöst 2006). This is a complex approach based on artificial intelligence and ubiquitous computing. The outcome of the current work focuses on a complementary approach where the service designer decides which changes in the *Situation* should lead to service adaptation based on available *ContextFeatures* and the predefined user information (Göker & Myrhaug 2002). It is a personalization approach where the service lets the user specify her own settings for how it should behave (Barkhuus & Dey 2003); e.g., the service designer specifies different recreation types or categories and the user chooses her favorite ones[2].

The TaskModel

One of the most challenging parts of the *AdaptationModel* is accounting for the user's purpose of using the map. The task mainly affects the determination of adaptive and adaptable operations based on the *ContextFeatures* and *UserFeatures.* To clarify the relationship between task and operations an appropriate approach is needed. The structure of the model in *Figure 2.4* is a simplification of the activity theory for cartography (Dransch 2002). It is a hierarchical framework where the activity builds the root element supplemented by goals, subgoals, and actions to accomplish the different activities. To reach the goal, several activities must be performed, which comprise the interactions of the user with the environment and the service. Hence, the operations depend on the specific task with its activities. Take, for example, a simple wayfinding task from a starting point A to a destination B. The goal is to reach B. An action could be that the user has to orient at a decision point and find the correct walking direction as a subgoal. The actions are represented by the different operations, which are either adaptive or adaptable.

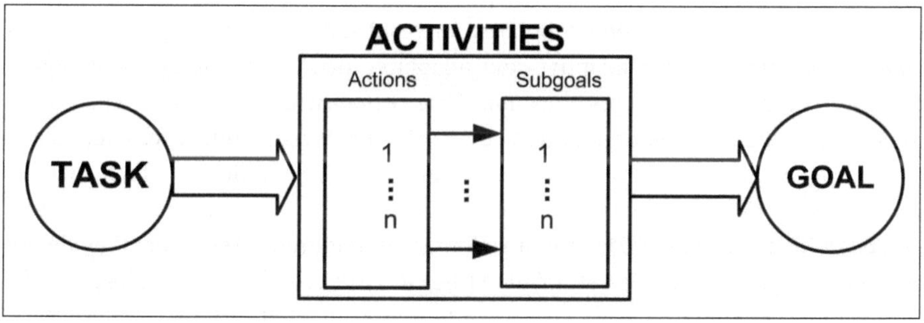

Fig. 2.4. Task-dependent activities to reach the goal. Simplified *activity theory* for cartography (Dransch 2002).

The representation of the *TaskModel* within the *AdaptationModel* is similar to the other submodels. The *TaskModel* is a static element of the *AdaptationModel,* but in contrast to the former two, a combination or selection of several *TaskFeatures* (*tf*) is not allowed ($T = (tf_n)$ *where* $n = 1$). This means that the ⊕ symbol should be read as a XOR-operator (exclusive disjunction: either tf_n or tf_{n+1}). This makes the definition of operations – and the required *ContextFeatures* – less complex than for an entire application such as *utopian.*

[2] See http://www.heidelberg-mobil.de for examples.

2.5 Application and Formal Model

This section demonstrates the applicability of the conceptual model for mobile map adaptation using the scenario of a pedestrian navigation service. The main focus lies on the formal specification of the adaptation operations. Our method of formalization uses algebraic specifications, which have proven useful for specifying data abstractions in spatio-temporal domains (Frank 2000, Raubal & Kuhn 2004). Entities are described in terms of their operations, depicting how they behave. The tool chosen here is Hugs, a dialect of the purely functional language Haskell (Hudak 2000). The result is a formal model that can be used as a basis for implementing mobile map adaptation for pedestrian navigation.

2.5.1 Adaptation Model

The specification of the adaptation model follows the conceptual model introduced in *Section 2.4.2.* We assign explicit features to each of its components regarding context, user, and task *(Figure 2.5).*

The pedestrian passes through different *Situations* during the wayfinding process, such as different decision points. These are described exclusively by the *ContextModel.* Therefore, the operations depend mainly on the explicit *ContextFeatures.* The *ContextFeatures* are subsumed in the spatio-temporal *context-category.* For the adaptive and adaptable operations the spatio-temporal category contains the features *Position, Time, Velocity,* and *Direction.* The ability to determine the *Position* of a mobile device is a direct requirement for every LBS. Nowadays, many mobile phones have integrated GPS modules, which provide accuracies from a few meters in stand-alone mode to sub-meter in differential mode (DGPS) (Gartner 2004). A major limitation of current LBS is that they do not consider temporal properties (Raubal, Miller & Bridwell 2004). The closest Café, for example, may not be open. *Time* can be represented at different scales. Here, we focus on time of day because it is the main parameter for deciding whether to visualize day- or nighttime pictures of landmarks. This context information can be obtained automatically by the built-in clock of the mobile device. *Velocity* is defined as the rate of positional change and can be calculated based on GPS positions in time. It is an important feature for calculating the current map extent.

The majority of people who use a mobile map rotate the device while walking (Schmidt et al. 1999). This is due to the fact that most mobile map services do not support an automatic track-up orientation of the digital map. Constantly orienting the current viewable map to the direction of travel may be confusing to the user and requires permanent attention. Therefore, we argue that the track-up orientation of the mobile map should depend on the projected walking *Direction* from one deci-

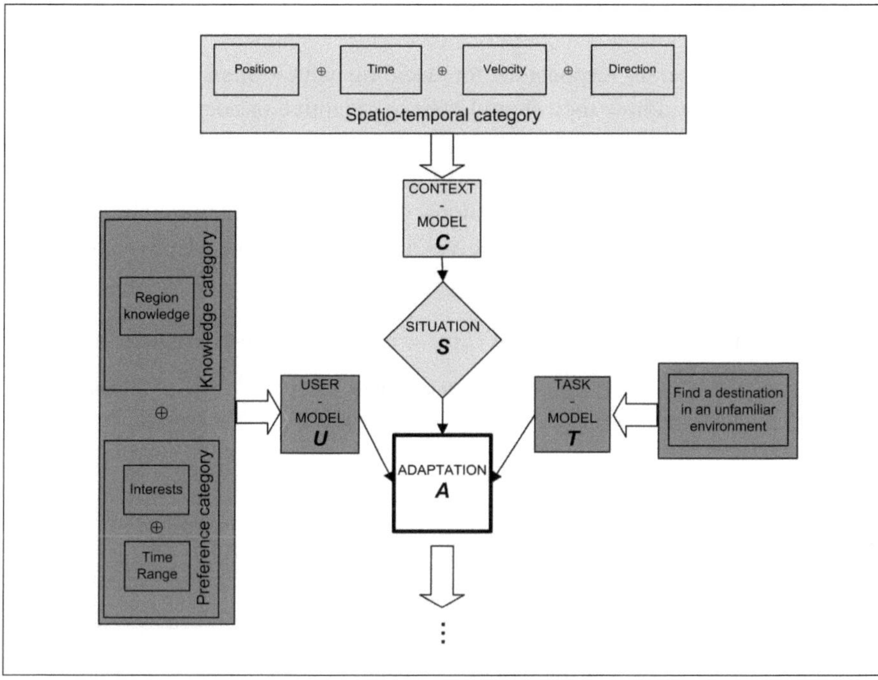

Fig. 2.5. The *AdaptationModel* for pedestrian navigation services.

sion point to the next. Such direction can be measured by determining the user's current position and the next decision point to be reached.

One of the most difficult characteristics to be interpreted and modeled is the user. Users vary with regard to physical abilities, cognitive and perceptual abilities, as well as in terms of personality differences (Shneiderman 1992). The service could automatically employ various pieces of user information to improve navigation instructions. Here, we consider information about the user's region knowledge and her interests, both classified in separate categories within the *UserModel (Figure 2.5)*. Even if people are unfamiliar with a particular environment, they may have conceptual representations about the location type (e.g., general structure of urban areas). The *Preference* category contains the features *Interests* and *TimeRange. Interests* determine the kinds of nearby objects to be visualized on the map, such as historical buildings or other attractions. This work focuses on the long-term interests (Zipf & Jöst, 2006), which do not change during the guiding process. *TimeRange* is responsible for the determination of the LoD *(Section 2.4.1)*. It enables the user to manipulate a parameter for both the *MapExtent* and the *Zooming* operations *(Section 2.5.2)*. The user can determine the preferred *TimeRange* (e.g., 10 minutes) and thereby the viewable map extent of the map. Similarly, the user can set the *TimeRange* for the *Zooming* operation (e.g., 1 minute) to predefine the local detail at decision points.

2.5.2 Adaptation Operations

The system designer is responsible for deciding which operations are required (task–dependent) and how they should behave (adaptive or adaptable). The determination of the *ContextFeatures* and *UserFeatures* for each operation is shown in *Table 2.2.* The *TaskFeatures* are not listed as input parameters for the operations; although they affect the kinds of operations, they do not deliver explicit features as input parameters for them. *Figure 2.6* gives an overview of the adaptive and adaptable operations.

Table 2.2. Required *ContextFeatures* and *UserFeatures* to formalize the suggested operations.

	ContextFeature				UserFeature
Operation	Position	Time	Velocity	Direction	TimeRange
Zooming	X				X
MapExtent	X	X	X		X
POIVis	X				
TrackUp	X			X	
LandmarkSwitching		X			

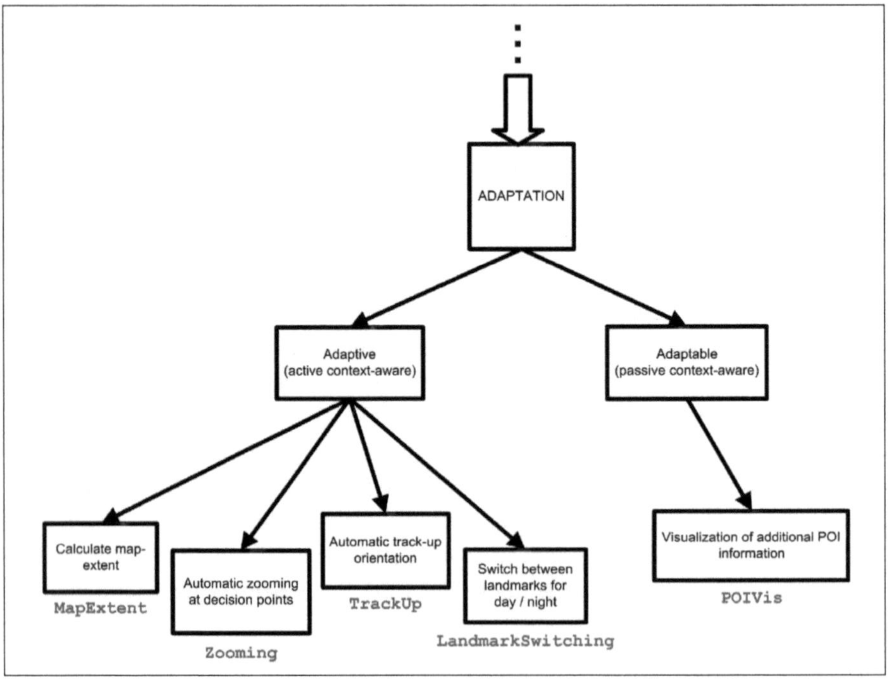

Fig. 2.6. Operations for mobile map adaptation in pedestrian navigation services. Automatic zooming at decision points.

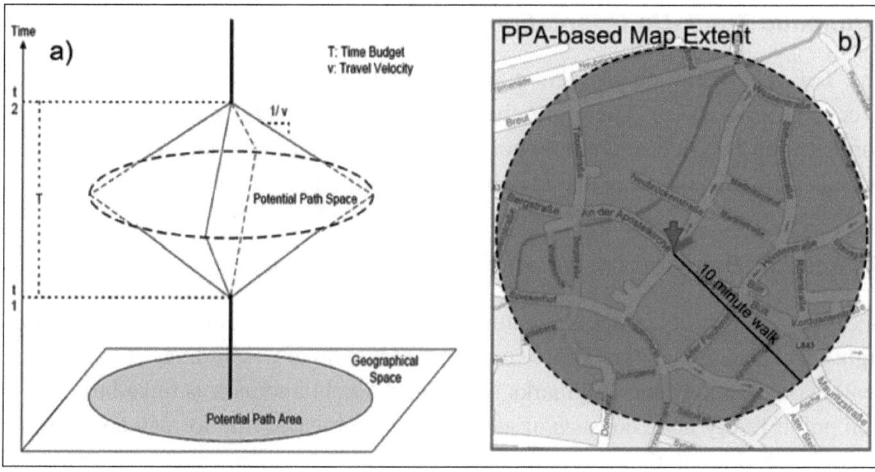

Fig. 2.7. a) Potential Path Space and Area. b) Example for a viewable map extent on a mobile device derived from the PPA.

Automatic Zooming at Decision Points

The *Zooming* operation provides the user with smooth multiple scaling at decision points. When reaching a decision point the service automatically zooms to local detail, predefined through the *TimeRange.* The small displays of mobile devices necessitate the use of different map scales ranging from overview to local detail. Additionally, a landmark will be displayed in a separate window. The correct direction can be represented through an arrow on the map or by written instructions.

Calculate Map Extent

Previous investigations have considered parameters for the selection and generalization of visualized information, but not for map scale and viewable extent (Hampe & Elias 2004). In this work the calculation of an appropriate map extent is essential, because other operations such as *POIVis* are directly related to it. The calculation of an appropriate map extent is necessary after leaving the start point of the route, while walking between decision points, and before arriving at the destination. It depends on the traveler's current *Velocity* (e.g., 5 km/h), her current *Position,* and a *TimeRange* (e.g., 10 minutes). Time geography (Hägerstrand, 1970) offers a way to calculate the map extent: the Potential Path Space (PPS) delimits all locations in space and time that an individual can possibly occupy, assuming some travel velocity. The Potential Path Area (PPA) results by projecting the PPS to the two-dimensional geographic plane (Miller 1991) *(Figure 2.7).*

Automatic Track-Up Orientation

The *TrackUp* operation delivers a map visualization depending on the walking direction *(Section 2.2)*. Therefore, the viewable map extent is aligned to the direction between two decision points but not to the viewing direction. This allows for orienting the map in a way that may be easier to interpret by the user.

Distinction Between Landmarks for Day and Night

Because the visibility of salient features varies according to time of day (e.g., day/night), and because landmarks are so important in human wayfinding, a time-dependent distinction between landmarks is required. Each landmark is linked to a decision point. Every node consists of at least two landmarks, one for day and one for night. Depending on the current time, the appropriate landmark will be displayed.

Visualization of POI Information

The *POIVis* operation is the only adaptable operation. It informs the user about nearby POIs, but does not automatically visualize additional information about them. The difficulty in developing this operation results from the meaning of 'nearby'. The location of nearby objects falls in the category of proximate selection (Schilit, Adams & Want 1994). It involves entering two variables, the 'locus' (user's current *Position)* and the 'selection' *(Interests* as nearby objects). To overcome and simplify the issue of what 'nearby' means to the user, the visualization of *Interests* as nearby objects depends on the calculated map extent. Using this approach, we follow the definition by Schilit et al. (1994): in context-aware services the most usefully located objects are close at hand, either co-located or requiring a short time to get to. The visualization of POIs that are close at hand depends on the area the user can reach during the predefined *TimeRange.* POIs are represented in the current map extent through icons. The user can click on the icon and receive the requested information.

2.5.3 Formalization

The *AdaptationModel* is formally represented[3] as a data type, which is constructed from different types. *Edge* represents the *Direction* of the *ContextModel,* because of the track–up orientation of the digital mobile map along the *Edge.* We consider the *UserFeature* region knowledge an implicit feature that does not directly affect

[3] The complete Hugs code is available at http://www.geog.ucsb.edu/~raubal/Downloads/MobileMapAdaptation_Hugs.rar. Hugs interpreters can be downloaded freely from http://www.haskell.org.

the formalization process. Because the perceived distances shrink when the user knows the area better (Zipf 2002), it is left to the user to define the two *TimeRanges* to receive appropriate map extents. The *MapState* is specified through three components. *ViewableExtent* changes at every *Node* (zooming) and for the calculation of the map extent. *Angle* represents the orientation of an *Edge.* Also, the appropriate *Landmark* for each *Node* is represented. The constructor of the *UserModel* gets the type *TimeRange* twice, representing the parameters for the *MapExtent* and the *Zooming* operations.

```
data AdaptationModel =
AdaptationModel ContextModel UserModel TaskModel MapState
  data ContextModel = ContextModel Position Edge Time Velocity
  data UserModel = UserModel Interests TimeRange TimeRange
  data TaskModel = TaskModel Task
  data MapState = MapState ViewableExtent Angle Landmark
```

The wayfinding environment is formally specified as a graph with *Nodes* and *Edges,* denoting decision points and transitions between them. The data type *Environment* is constructed from a list of *Edges.* Every *Edge* is constructed from two *Nodes* and an *Angle,* which provides the *Direction* of the *Edge.* Every *Node* has a *Position* represented by geographic coordinates and a list of *Landmarks.* These are constructed from a *Name* as an identifier and a specified *TimeSpan* to allow for a distinction between day and night.

```
data Environment = Environment [Edge]
  data Edge = Edge Node Node Angle
  data Node = Node Position [Landmark]
  data Landmark = Landmark Name TimeSpan
  data TimeSpan = Day | Night
```

The abstract type signatures for the operations are implementation independent and can be implemented for different types of pedestrian navigation services. In the following, the operations will be implemented for the data types *AdaptationModel* and *Environment* as presented above.

The *Zooming* operation represents the changing of viewable map extent to local detail at decision points. For this implementation it is the same as the *MapExtent* operation. Applying the function has the following effects:

1. The operation checks whether the destination has been reached (this is initially done by all operations and because the code is the same, we only represent it here). If yes, then the wayfinding task is completed.

2. The changes in the *ViewableExtent* depend on the current *Position*, *Velocity*, and *TimeRange*. The function *getTimeRangeAtNode* determines the extent of local detail defined by the user.

```
instance AdaptationModels AdaptationModel where
  zooming environment (AdaptationModel cm um tm (MapState ve a lm))
     = if isDestination (Node (getPosition cm) (Landmark “default” Day:
     [Landmark “default” Night])) environment
     then error (“The destination is reached.”)
     else (AdaptationModel cm um tm (MapState veChange a lm)) where
        veChange = ViewableExtent (getPosition cm) (getVelocity cm)
        (getTimeRangeAtNode um)
```

The *TrackUp* operation delivers a map visualization depending on the calculated walking *Direction* between decision points. The viewable map extent is aligned to the *Direction* between two decision points along an *Edge*. The function *getCurrentEdge* provides the *Edge* related to the current *Position* of the user. The *Angle* of the map will be replaced by the orientation of this *Edge*.

```
instance AdaptationModels AdaptationModel where
  trackUp environment (AdaptationModel cm um tm (MapState ve a lm))
     = if isDestination …
     else (AdaptationModel cm um tm (MapState ve aChange lm)) where
        aChange = getOrientation (getCurrentEdge cm)
```

The *LandmarkSwitching* operation enables a time-dependent extraction of landmarks for a decision point. The operation checks which *Node* is related to the current *Position* of the user. The *getTime* operation delivers the current *Time*, so that the required *Landmark* can be extracted.

```
instance AdaptationModels AdaptationModel where
  landmarkSwitching environment (AdaptationModel cm um tm
     (MapState ve a lm))
  = if isDestination …
  else (AdaptationModel cm um tm (MapState ve a lmChange)) where
     lmChange = getRecentLandmark (getNodeAtPosition environment
        (getPosition cm)) (dayOrNight (getTime cm))
```

The *POIVis* operation gives the user information about nearby POIs. The *Interests* of the user are visualized using the preselected *Interests,* which are retrieved by the *getInterests* function. The *getVisablePOIs* function uses an auxiliary function *(liesWithin)* to check whether *Nodes* are within the current map extent. The operation retrieves those *Nodes* in the list that are POIs and returns them.

```
instance AdaptationModels AdaptationModel where
  poiVis environment (AdaptationModel cm um tm (MapState ve a lm))
  = if isDestination …
  else (AdaptationModel cm um tm (MapState ve a lm)) where
     poiChange = getVisablePOIs (getInterests um)(MapState ve a lm)
```

2.6 Discussion

The reduction of user interaction with the device is mainly achieved through the *Zooming, MapExtent,* and *POIVis* operations. In order to quantify the effects of the model, we compare and measure the user interactions for *utopian* and the applied *AdaptationModel.* Interaction is represented through a pointing device such as a stylus. User interaction for both is measured by calculating/counting the individual clicks. If a survey were performed, one could also measure the interaction time by using methods such as the one based on the keystroke-level model (Haunold & Kuhn 1994). Our comparison consists of three actions, which are performed with *utopian.* These actions correspond to the formal operations of the *AdaptationModel. Table 2.3* gives an overview of the results.

Table 2.3. Comparison of *utopian* and the applied *AdaptationModel* regarding user interaction.

Operation	AdaptationModel	Utopian
Zooming	no interaction needed	minimum 4 clicks
MapExtent	no interaction needed	minimum 4 clicks
POIVis	1 click	3 clicks

Section 2.2 has shown that in *utopian* basic functionalities are only reachable via the context menu. This means that the user is forced to perform 2 clicks for each action – one to open the context menu and another to select the desired operation. In order to change the current map extent the user must span a rectangular area on the map. This rectangle determines the desired map scale and extent, and requires 2 more clicks. The applied *AdaptationModel* does not require a context menu for these frequently used operations. The *MapExtent* and *Zooming* operations are adaptive and do not necessitate any user-device interaction.

The second comparison deals with the visualization of POIs. As with the other two operations in *utopian* an interaction with the context menu is needed. The third click on the desired POI delivers the corresponding information. The *AdaptationModel* provides an adaptable functionality for the visualization of POIs. If a POI is located within the current map extent, an icon will be visualized on the map. A click on the icon delivers the requested information.

Evaluating the effects of changes in the cognitive load for the user also requires a quantification. Such reduction can be measured by using different subjective and objective measurement techniques (Bunch & Lloyd 2006). Both are based on surveys with participants. The *AdaptationModel* presented here does not support surveys, because it is not an implemented service that is executable on a mobile device. The reduction of cognitive load will therefore be evaluated through existing survey results (Radoczky 2003, Wealands et al. 2007). As mentioned in *Section 2.2.2,* many participants voted for multi-encoded navigation instructions *(SwitchingLandmark* operation) and a track-up oriented map *(TrackUp* operation). Both operations are not provided by *utopian.* In combination with the *MapExtent* operation, the number of perceived elements in the map is lower. Hence, a reduction of cognitive load is achieved. The *Zooming* operation provides the user with a detailed cartographic image. This reduces the amount of perceived elements for the user at decision points. Therefore the discussion so far indicates a verification of the hypothesis. An implemented pedestrian navigation service, which provides navigation instructions based on the *AdaptationModel* adapts to the user's situation, instead of forcing the user to adapt to the service.

2.7 Conclusions and Future Work

We presented an abstract formal model for mobile map adaptation, which can be used as a basis for implementing context-aware LBS. The useful combination of adaptive and adaptable functionality achieves a user-centered design. The *AdaptationModel* can act as a guideline for both simple (e.g., basic routing functionalities) and complex (e.g., manifold functionalities such as with *utopian)* LBS. The classification into submodels and components regarding context, user, and task makes the *AdaptationModel* manageable and flexible. To demonstrate the latter two characteristics the concrete *AdaptationModel* for a pedestrian navigation service was developed. Concrete operations were specified to represent user interaction with the service. The identified features of the submodels served as input parameters with the aim of achieving context-sensitive map adaptation. The functional programming language Haskell was used to express the algebraic specifications in a formal manner. The comparison of *utopian* to the operations of the developed model confirmed the reduction of both the user interaction with the device and the cognitive load for the user. We come to the conclusion that taking a formal view of context-aware computing enables reasoning about the foundational relationships that process context. In particular, the formal model in this work serves as a rigorous basis for further development of a formal framework to design and evaluate context-aware services.

We made various simplifying assumptions and only those features necessary to confirm the hypothesis were taken into consideration. One of the most important aspects for future work is the implementation and evaluation of the *AdaptationModel.* After implementing the operations in an executable pedestrian navigation service, a survey could be done to evaluate the proposed adaptation features. Further investigations are necessary to decide whether adaptive user interfaces may be confusing for the user or not, especially when the user is not aware of the underlying mechanisms and algorithms. With a GIS behind the map representation for a mobile device, the potential types and amount of information for display are unlimited. Additionally, digital mobile maps have the ability to display different levels of detail. More research is needed on how people interact with digital maps on mobile devices.

Another direction for future work is making the *AdaptationModel* more dynamic, i.e., realizing the *UserModel* as a dynamic part similar to the *ContextModel.* Design environments offer possibilities for integrating adaptive and adaptable components to increase the shared knowledge between users and computers. This could be realized through a user profile. Behavioral user data could be acquired automatically through sensors. To provide personalization and adaptation capabilities, systems need to be able to reason about their users (e.g., applying methods such as neural networks or machine learning techniques). There are also important user privacy and ethical issues that need to be addressed. 'Perfect' privacy guarantees are in general hard to achieve, therefore a balance between service enhancement and privacy concerns has to be found.

In the current version the *AdaptationModel* only supports one task to determine the required features. A next step could be the consideration of an entire application such as *utopian.* A possible incorporation of several tasks will lead to an increased number of features and operations. Complex applications will also require a deeper investigation of the relationships between the three submodels. Controlling the information visualization on the small mobile device may require weighting the available features. A research goal in this respect is finding appropriate weights for the different input parameters (i.e., adaptation targets).

Acknowledgements

The comments from two anonymous reviewers provided useful suggestions to improve the content of the paper. We also thank Gwen Raubal for improving the grammar and style of this work.

References

Agrawala M, Stolte C (2001) Rendering effective route maps: improving usability through generalization. In Pocock L (Ed.), *Proceedings of the 28th Annual Conference on Computer Graphics and Interactive Techniques (SIGGRAPH 2001)* (pp. 241–249). New York: ACM Press.

Barkhuus L, Dey A (2003) Is context-aware computing taking control away from the user? Three levels of interactivity examined. In Dey A, Schmidt A,. McCarthy J (Eds.), *Proceedings of 5th International Conference on Ubiquitous Computing 2003 (Ubicomp), Seattle, Washington* (Vol. 2864, pp. 159–166). Berlin: Springer.

Baus J, Kray C (2002) *Frames of Reference, Positional Information and Navigational Assistance.* Paper presented at the Fifteenth International Florida Artificial Intelligence Research Society Conference (Flairs 2002), Pensacola Beach, Florida, USA.

Bunch R, Lloyd R (2006) The Cognitive Load of Geographic Information. *The Professional Geographer, 58*(2), 209–220.

Cartwright W, Crampton J, Gartner G, Miller S, Mitchell K, Siekierska E, Wood J (2001) Geospatial Information Visualization User Interface Issues. *Cartography and Geographic Information Science, 28*(1), 45–60.

Denis M, Michon PE, Tom A (2006) Assisting Pedestrian Wayfinding in Urban Settings: Why References to Landmarks are Crucial in Direction-Giving. In Allen G (Ed.), *Applied Spatial Cognition – From Research to Cognitive Technology* (pp. 25–51). Mahwah, NJ: Lawrence Erlbaum.

Dey A, Abowd G (2000) Towards a Better Understanding of Context and Context-Awareness, *CHI 2000 Workshop on the What, Who, Where, When, Why and How of Context-Awareness.* The Hague, The Netherlands.

Dransch D (2002) *Handlungsorientierte Mensch-Computer-Interaktion für die kartographische Informationsverarbeitung in Geo-informationssystemen*: Fachbereich Geowissenschaften, Freie Universität Berlin.

Fischer G (1993) Shared knowledge in cooperative problem-solving systems – integrating adaptive and adaptable components. In Schneider-Hufschmidt M, Kühme T, Malinowski U (Eds.), *Adaptive user interfaces: Principles and practice* (pp. 49–68). Amsterdam: North-Holland.

Frank A (2000) Spatial Communication with Maps: Defining the Correctness of Maps Using a Multi-Agent Simulation. In Freksa C, Brauer W, Habel C, Wender K (Eds.), *Spatial Cognition II – Integrating Abstract Theories, Empirical Studies, Formal Methods, and Practical Applications* (Vol. 1849, pp. 80–99). Berlin, Heidelberg, Germany: Springer.

Gartner G (2004) *Location-based mobile pedestrian navigation services – the role of multimedia cartography.* Paper presented at the International Joint Workshop on Ubiquitous, Pervasive and Internet Mapping 2004 (UPIMap), Tokyo, Japan.

Göker A, Myrhaug H (2002) User context and personalization. In *6th European Conference on Case Based Reasoning and Personalization (ECCBR 2002), Aberdeen, Scotland.* Berlin: Springer.

Gutwin C, Fedak C (2004) Interacting with big interfaces on small screens: a comparison of fisheye, zoom, and panning techniques. In Heidrich W, Balakrishan R (Eds.), *Proceedings of the 2004 Conference on Graphics interface, 17.–19. May 2004, London, ON, Canada* (pp. 145–152). Ontario, Canada: Canadian Human-Computer Communications Society.

Hägerstrand T (1970) What about people in regional science? *Papers of the Regional Science Association, 24*, 7–21.

Hampe M, Elias B (2004) *Integrating Topographic Information and Landmarks for Mobile Applications.* Paper presented at the Symposium on Location Based Services and Telecartography, Vienna, Austria.

Hampe M, Paelke V (2005) *Adaptive methods for mobile applications.* Paper presented at the 7th International Conference on Human Computer Interaction with Mobile Devices and Services (MobileHCI 2005), Salzburg, Austria.

Haunold P, Kuhn W (1994) A Keystroke Level Analysis of a Graphics Application: Manual Map Digitizing. In Adelson B, Dumais S, Olson J (Eds.), *CHI'94 Conference on Human Factors in Computing Systems in Boston, MA* (pp. 337–343): ACM.

Hudak P (2000) *The Haskell School of Expression: Learning Functional Programming through Multimedia.* New York: Cambridge University Press.

Lake R (2001) *Location-Based Services & GML* (Technical report). Vancouver: Galdos Systems.

Lovelace K, Hegarty M, Montello D (1999) Elements of Good Route Directions in Familiar and Unfamiliar Environments. In Freksa C, Mark D (Eds.), *Spatial Information Theory – Cognitive and Computational Foundations of Geographic Information Science, International Conference COSIT '99, Stade, Germany* (Vol. 1661, pp. 65–82). Berlin, Heidelberg, New York: Springer.

Mayer R, Moreno R (2002) Aids to computer-based multimedia learning. *Learning and Instruction, 12*(1), 107–119.

Miller H (1991) Modeling accessibility using space-time prism concepts within geographical information systems. *International Journal of Geographical Information Systems, 5*(3), 287–301.

Nivala A, Sarjakoski L (2003) Need for Context-Aware Topographic Maps in Mobile Devices. In Virrantaus K, Tveite H (Eds.), *ScanGIS, The 9th Scandinavian Research Conference on Geographical Information Science, 4.–6. June 2003, Espoo Kivenlahti, Finland,* (pp. 15–29): Springer.

Oppermann R (1994) *Adaptive User Support Pod: Ergonomic Design of Manually and Automatically Adaptable Software*: Lawrence Erlbaum Associates.

Radoczky V (2003) *Kartographische Unterstützungsmöglichkeiten zur Routenbeschreibung von Fußgängernavigationssystemen im In- und Outdoorbereich.* Unpublished Master Thesis (German), Vienna University of Technology, Vienna, Austria.

Raper J, Gartner G, Karimi H, Rizos C (2007) A critical evaluation of location based services and their potential. *Journal of Location Based Services, 1*(1), 5–45.

Raubal M, Kuhn W (2004) Ontology-Based Task Simulation. *Spatial Cognition and Computation, 4*(1), 15–37.

Raubal M, Miller H, Bridwell S (2004) User-Centred Time Geography For Location-Based Services. *Geografiska Annaler B, 86*(4), 245–265.

Raubal M, Winter S (2002) Enriching Wayfinding Instructions with Local Landmarks. In Egenhofer M, Mark D (Eds.), *Geographic Information Science – Second International Conference, GIScience 2002, Boulder, CO, USA, September 2002* (Vol. 2478, pp. 243–259). Berlin: Springer.

Reichenbacher T (2003) *Adaptive Methods for Mobile Cartography.* Paper presented at the 21st International Cartographic Conference (ICC 2003), Durban, South Africa.

Rinner C, Raubal M (2004) Personalized Multi-Criteria Decision Strategies in Location-Based Decision Support. *Journal of Geographic Information Sciences, 10*(2), 149–156.

Rinner C, Raubal M, Spigel B (2005) *User Interface Design for Location-Based Decision Services.* Paper presented at the 13th International Conference on GeoInformatics, 17–19 August 2005, Proceedings (CD-ROM), Toronto, Canada.

Schilit B, Adams N, Want R (1994) *Context-aware computing applications.* Paper presented at the Workshop on Mobile Computing Systems and Applications.

Schilit B, Theimer M (1994) Disseminating Active Map Information to Mobile Hosts. *IEEE Network, 8*(5), 22–32.

Schmidt A, Beigl M, Gellersen H (1999) There is more to context than location. *Computers and Graphics, 23*(6), 893–901.

Shneiderman B (1992) *Designing the user interface: strategies for effective human-computer interaction.* Boston, MA: Addison-Wesley.

Strang T, Linnhoff-Popien C (2004) A Context Modeling Survey. In *Workshop on Advanced Context Modeling (UbiComp 2004), Nottingham, U.K.* (Vol. 3205): Springer.

Wealands K, Miller S, Benda P, Cartwright W (2007) User Assessment as Input for Useful Geospatial Representations within Mobile Location-Based Services. *Transactions in GIS, 11*(2), 283–309.

Winter S, Raubal M, Nothegger C (2005) Focalizing Measures of Salience for Wayfinding. In Meng L, Zipf A, Reichenbacher T (Eds.), *Map-based Mobile Services – Theories, Methods and Implementations* (pp. 127–142). Berlin: Springer.

Zipf A (2002) User-Adaptive Maps for Location-Based Services (LBS) for Tourism. In Wöber K, Frew A, Hitz M (Eds.), *Proceedings of the 9th International Conference for Information and Communication Technologies in Tourism, Innsbruck, Austria.* Heidelberg, Berlin: Springer.

Zipf A, Jöst M (2006) Implementing Adaptive Mobile GI Services based on Ontologies – Examples for pedestrian navigation support. *Computers, Environment and Urban Systems (CEUS), 30*(6), 784–798.

3 Using Activity Theory to Identify Relevant Context Parameters

Haosheng Huang and Georg Gartner

Institute of Geoinformation and Cartography
Vienna University of Technology, Austria

Abstract

One of the most important aspects of ubiquitous computing is context-awareness. In this paper, we adopt an interactional perspective on context: 1) Something is context because it is used for adapting the interaction between the human and the current system. 2) Activity is central to context. 3) Context differs in each occasion of the activity. Based on this understanding, this paper proposes an Activity Theory based method which attempts to answer the following questions: how to analyze activity for context-awareness, and how to identify relevant context parameters. This method includes two steps: 1) Decomposing activity into actions, which we take as units for identifying context parameters, by using Activity Theory's hierachical structure of activity. 2) Identifying relevant context parameters for each action by our extended Activity Theory's framework. Finally, this paper gives an outlook how this method can be used in designing context-aware pedestrian wayfinding services.

Keywords: context-aware computing, context modeling, Activity Theory, relevance, pedestrian wayfinding service

3.1 Introduction

Recent years have witnessed rapid advances in enabling technologies for ubiquitous computing. Ubiquitous computing demands applications that are capable of recognizing and adapting to highly dynamic environments while placing fewer demands on user's attention (Henricksen et al. 2002). It is widely acknowledged that context-awareness for ubiquitous computing can meet these requirements. As one type of ubiquitous computing, in order to improve usability, Location Based Services (LBS) should be context-aware and adapt to dynamic environments.

The term context-aware computing was first introduced by Schilit et al. (1994). Since then, lots of definitions of the term context have been proposed. Basically the majority of existing definitions of the term context can be categorized into definition by synonyms which suffers from self-referencing in loops, and definition by examples which suffers from incompleteness (Zimmermann et al. 2007). Dey and Abowd define context as "any information that can be used to characterize the situation of entities" (1999), which is used most widely in the current context-aware applications. However, in using "any information" the definition becomes too general (Zimmermann et al. 2007). The practical usefulness of such definitions is limited when developing some specific context-aware systems. An important view about context can be found in Dourish (2004), which identified two views for context: representational and interactional. He argues that the correct focus for research is on interaction between objects and activities and not solely on the representation of objects (Chen & Atwood 2007). We agree with this perspective and also with Winograd's point (Winograd 2001) that something is context because it is used for adapting the interaction between human and applications.

When developing context-aware systems the developer must pre-determine what aspects of the world can be used as context parameters. Often, researchers start with comprehensive definitions but operationalize much simpler concepts of context in their actual implementations. Such simplifications appear to be necessary when developing specific systems (Lueg 2002). These kinds of actual implementations always focus on context representation, such as using data structure (key-value, markup scheme, ontology-based models, etc.) to store the relevant context parameters or adding some metadata (reliability, spatial precision, etc.) to improve the quality of context (Henricksen et al. 2002). They choose some aspects of the world as their context parameters from their own views. What is missing, however, is a method about how to identify relevant context parameters. After all, this is a fundamental question for developers and has an important relationship to the usability of systems. At the same time, while many authors acknowledge that activity is central to context (Schmidt 2002, Reichenbacher 2004, Dransch 2005, Zipf & von Hunolstein 2003), little work has been done to develop methods to support modeling activities in context-aware systems.

This paper attempts to solve the two problems mentioned above. As activity is central to context, we therefore focus on analyzing activity. This will enable a better understanding of context and context-awareness (Kofod-Peterson & Cassens 2005), and help us to identify possible context parameters which are relevant to the activity at hand. Activity Theory (Engeström 1987) is one of the most popular theories for modeling activity. In this paper, we propose a method based on Activity Theory.

With this method, it is possible to decide which context parameters can be used in context-aware systems.

The paper is arranged as follows. *Section 3.2* describes our understanding of context. *Section 3.3* introduces some basic concepts of Activity Theory, and proposes our Activity Theory based method to identify relevant context parameters. *Section 3.4* shows how this method can be used in designing context-aware pedestrian wayfinding services. Finally, we draw the conclusions and present future work.

3.2 What's Context?

A fundamental question for any research on context-aware computing is "what is context". A better understanding of context will enable developers to identify what context to be used in their systems (Dey & Abowd 1999). We agree with Dourish's (Dourish 2004) and Winograd's (Winograd 2001) perspectives on context:

1. Something is context (parameter) because it is used for adapting the interaction between human and the current system, and features of the world become context (parameters) through their use (Winograd 2001). For example, the humidity of the room is a context parameter only if the adaptation of the interaction between human and the current system depends on it, but otherwise it is just a feature of the world.
2. Context arises from the activity (Dourish 2004). It interacts and constraints activities happened within it (Chen & Atwood 2007). An entity's activity determines its current needs. Context-aware systems are developed to meet human's needs and then assist human activities.
3. Context is dynamic, and it is particular to each occasion of activity (Dourish 2004), thus context parameters may be different in different actions in an activity. We also propose that the abstraction levels of context parameters' values may be different from occasions of activity, e.g. the values of the context *"location"* may be pair of longitude/latitude coordinates or "outdoor" in different actions.

Because the world is infinitely rich, we can not model every feature of context. In the following, we use the term *context parameters* to refer to the possible features of context.

3.3 A Method Based on Activity Theory (AT)

In this section, we concentrate on using AT to analyze human activity. This will help us to identify the possible environmental, social, and technological context parameters which are involved in performing the activity.

3.3.1 Activity Theory

Modern Activity Theory is based on the works of Leont'ev (Leont'ev 1987) and Engeström (Engeström 1987). It is a descriptive tool that is useful for analyzing and understanding activity in general, independently of any specific field of application (Barthelmess & Anderson 2002).

Activity, as defined by AT, provides enough contextual information to make an analysis meaningful. Activity has a hierarchical structure. It is composed of goal-directed actions, which are performed consciously. Correspondingly, actions consist of non-conscious operations (Kofod-Peterson & Cassens 2005). In AT *(Figure 3.1)*, activities or actions are performed by *subjects*, motivated by a goal, transforming an *object* into an *outcome*. An *object* may be shared by a *community* of actors, that work together to reach a desired *outcome*. *Tools*, *rules* and *divisions of labour* mediate the relationship between *subjects*, *community* and *object* (Barthelmess & Anderson 2002). *Object* and *outcome*, reflecting the activity's goal, are central to activity.

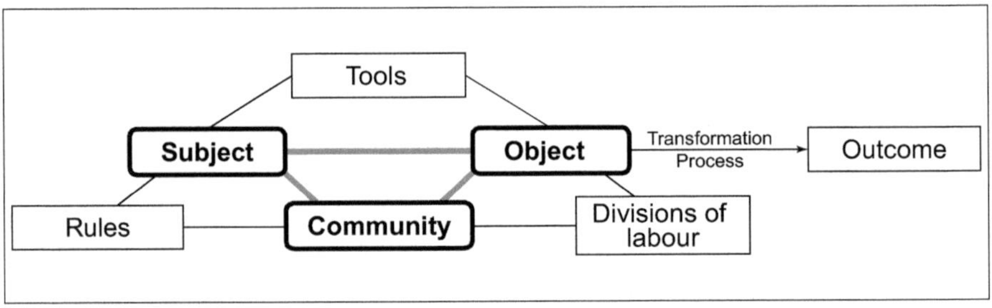

Fig. 3.1. Activity Theory's framework introduced by Engeström (Engeström 1987).

Although AT doesn't offer "ready-made techniques and procedures" (Engeström 1993), we can use AT's framework to identify key elements (*subjects, Tools, object, outcome, community, rules, and divisions of labour)* which influence the current human activity. Context-aware systems (tools) designed to facilitate and achieve human activity should also reflect these key elements. Therefore, AT can be used to identify possible context parameters.

Other works on the use of AT in identifying context are Kofod-Peterson & Cassens (2005) and Kaenampornpan & O'Neill (2004). However, both of them did't consider that context differs in each occasion of activity. In the meantime, as AT doesn't model the physical environment, such as location, light, sound, both of them also don't model the context parameters of the physical environment.

3.3.2 Using Activity Theory to Identify Relevant Context Parameters

In this section, we use AT's hierarchical structure of activity to decompose activity into actions. Then we extend the AT's framework and use it to identify relevant context parameters for each action.

3.3.2.1 Action as a Unit of Analysis

In AT, activities are longer-term formations; their objects are transformed into outcome not at once but through a process that typically consists of several steps or phases (Nardi 1996). There is also a need for shorter-term processes: activities are composed of goal-directed actions, which are performed consciously; actions, in turn, consist of non-conscious operations (Kofod-Peterson & Cassens 2005). The activity is similar to the task which users are trying to accomplish, actions can be considered as steps of achieving it, and operations are procedures under each step. Context differs in each step (Chen & Atwood 2007).

At the activity level we can identify the activities the context-aware system should support. This is always shaped by the *"software requirements specifications"*. By this, we can restrict the world view of the system and make the task of developing a context model manageable. We can use the notion of actions to identify the different situations the system can encounter, which will help us to identify context parameters for the current action (Kofod-Peterson & Cassens 2005). Every action has a subgoal, and all the subgoals form the activity's overall goal. Operations under the same action share the same subgoal, and thus share the same context parameters. We want to take *action* as a unit for identifying context parameters.

It is possible to identify actions which belong to the current activity (from AT's point of view). The decomposition of activity can use the methods used in *task analysis,* such as Hierarchical Task Analysis (HTA). It is always carried out by human-factors professionals in specific domains.

In order to facilitate and achieve human activity, context-aware systems (tools) should therefore know which action the user is engaged in at all times (Kofod-Peterson & Cassens 2005), and then adapt to the current action's relevant context parameters. The following section discusses about how to use the AT's framework to analyze each action, and identify relevant context parameters.

3.3.2.2 Using AT's Framework to Identify Relevant Context Parameters for each Action

AT doesn't model the physical environment – but features of physical environment, such as location, weather, noise, humidity, also have an influence on human activity.

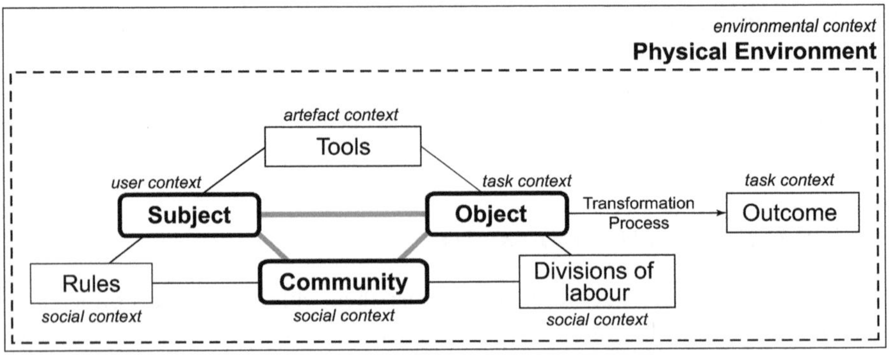

Fig. 3.2. The extended framework of Activity Theory (After Engeström (1987)).

Therefore we make an extension to the AT's framework: the physical environment in which the activity happens is added to the framework. *Figure 3.2* shows the extended framework of Activity Theory.

The above extended framework covers the key elements which influence human action. Features (attributes) of these elements can be viewed as different types of context parameters which are relevant to the action. Therefore, we propose a possible mapping from the elements of AT to context categories *(Table 3.1).*

As mentioned in Activity Theory, history is an important factor to the action. History will become *subject's* experience or knowledge. Tools can be viewed as transmission of human experiences for performing this activity or action, thus *tool* also reflects history. As a result, we don't want to model history explicitly. History will be represented in *user context* and *artifact context.*

Table 3.1. Mapping context categories from the elements of AT.

Elements of AT	Context category	Description of the context category
Subject	user context	Users' mental and physical information, such as preference, mood, skill, demographic information, and disabilities.
Object, Outcome	task context	This describes what action the user is engaged in, it includes the user's goals and tasks.
Community, Rules, Divisions of Labour	social context	This describes the social (especially organizational) aspects of the user, may include the information about people (in the community) who influence user's action or share the same object, norms, organizational rules and legislations the user should follow.
Tools	artefact context	Profiles of the tool and its availability, especially the information about systems and devices.
Physical environment	environmental context	This describes the information about the physical environment, such as location, time, weather, humidity, noise, etc. It also includes the common legal rules (behavior) or legislations in this environment, such as navigation rules, pedestrian rules which will change according to the different environments (countries).

The above table can be viewed as the *possible context dimensions* which may have impacts on the context-aware systems. Since the world is infinitely rich, when we try to choose some context parameters to be used in our context-aware system, we should carefully consider the notion of "relevance". If some aspects of the world have significant impacts on the current context-aware computing usage, these aspects are relevant to the current context-aware system.

"Relevance" is always shaped by the goal (reflected by object and outcome) of the current action, and sometimes is additionally defined by *"software requirements specifications"*. The requirements of the goal are always defined by some *empirical results (such as user expectations and preferences)* on this goal, thus by focusing on these, we can identify some relevant context parameters which are common for this type of goal. When concentrating on *"software requirements specifications"*, we can identify some other relevant context parameters which are specific to the current system. For example, some *empirical results* on the topic of pedestrian wayfinding show that, when communicating the route information with maps, wayfinders prefer "track up" oriented map to "north up" map. As a result, walking direction is commonly relevant to pedestrian wayfinding applications, and should be considered as a common relevant context parameter. With more empirical results available, the common relevant context parameters will be enriched.

In the design process, developers should:

1. Use the extended framework to identify the contents of each key element (such as *subject*, *community*, *tools*..., especially *object* and *outcome* which reflect the goal) for every action;
2. For every action, identify each key element's relevant features (attributes) by analyzing the current action's goal (featured by the empirical results on the corresponding topic) and *"software requirements specifications"*, and classify these relevant features (attributes) into related context category according to *Table 3.1*. Empirical results of the goal define the relevant context parameters which are common for this type of goal, while *"software requirements specifications"* define some other relevant context parameters which are specific to the current system.

At this moment it is impossible to define a context model which will empower the system to be universally context-aware, meaning that it will be able to build its own context model on the fly (Kofod-Peterson & Cassens 2005). Developers may choose some of the context parameters to be used in their system based on *"software requirements specifications"* versus implementation costs.

3.4 Context-Aware Pedestrian Wayfinding Services

Technological advance in mobile devices and mobile communication triggered a move towards Location Based Services (LBS). In order to improve the usability, LBS should also be context-aware, and only provide relevant information to the users.

A lot of research has been done on context-awareness in Location Based Services (Reichenbacher 2004, Dransch 2005, Zipf & von Hunolstein 2003, Sarjakoski & Nivala 2005). Reichenbacher (2004) provided a generic context model for mobile cartography which includes the dimensions of User, Activity, Situation, Information, System, and Meta-information, and he also discussed the process of adaptive visualization of geographic information. Nivala (Sarjakoski & Nivala 2005) suggested a categorisation of contexts for mobile maps which includes computing (System), User (Purpose of use, User, Social, Cultural), Physical (Physical surroundings, Location, Orientation), Time, History (Navigation history). All these Context models (categories) can be viewed as possible context dimensions for mobile cartography. But as mentioned in the first section, there is little research on how to analyse activity and how to identify relevant context parameters for context-aware LBS.

As one of the most important applications in LBS, mobile pedestrian wayfinding services aim at providing navigation guidance through a mobile device in an unfamiliar environment. Pedestrian wayfinding services should be context-aware, and adapt to the dynamic environment. They are designed to assist people's wayfinding activity. Since Activity Theory can help us to analyse human wayfinding and identify key elements which affect the accomplishment of human wayfinding, we can use our Activity Theory-based method to identify the relevant context parameters for mobile pedestrian wayfinding services. This section describes how this suggested method can be used in designing such services.

1. Hierarchical structure of wayfinding activity

From AT's point of view, the activity is wayfinding, or routing from origin to destination. According to Downs & Stea (1977), wayfinding activity includes four processes: *orientation* (determining one's position), *planning the route*, *keeping on the right track, discovering the destination*. The last two processes can be combined together as *moving from origin to destination*. They correspondingly relate to three modules in wayfinding services: positioning, route calculation, and route communication.

2. Using AT's framework to identify relevant context parameters for each action

In this section, we use the method described in *Section 3.3.2.2* to analyze the three above actions. First, we use the extended framework to identify the contents of each key element for every action. If the acting person is performing a wayfinding activity with some other people, the element "*community*" will include all the people who have an influence on this activity, and the element "*rules*" contains the explicit or implicit regulations, norms and conventions related to each action, and the element "*divisions of labour*" is the member's role in each action. To make it simple, we assume our example as individual wayfinding, where the acting person carries out the wayfinding activity mainly by the help of the wayfinding services, and he or she doesn't need the help of other people (such as pedestrians). Therefore, it is unnecessary to consider the elements *community*, *rules* and *divisions of labour*. *Table 3.2* shows the contents of key elements for each action.

Table 3.2. Contents of key elements for each action.

	Orientation	Planning the route	Moving from origin to destination
Object, Outcome	Position of the user for outdoor application	Route plan meeting the requirements	Keeping on the right track: take correct directions at every decision point
Subject	The acting person	The acting person	The acting person
Tool	Wayfinding services and positioning technologies	Wayfinding services	Wayfinding services, positioning technologies, environmental wayfinding signage, etc.
Physical Environment	At the start point	At the start point	Along the route

The next step is to identify each key element's relevant features by analyzing the goal of the current action and *"software requirements specifications"*, and classify them into related context category. In the following, we take the "*Moving from origin to destination*" action as an example. The goal of this action is "Keeping on the right track: take correct directions at every decision point". We can find some indispensable conceptions and functionalities from lots of empirical results on this topic (Gartner & Uhlirz 2005, Zipf & Joest 2003), such as automatic scrolling (automatic adaptation of the presented map section to the position of the user), egocentric mapview (the map is always adapted to the user's direction of move), velocity dependence (the level of details of the map is adapted to the user's moving velocity), device dependence (the map is adapted to the profile of the device, such as size of display, color, etc.), etc. Since position of the user *(environmental context),* moving direction (*environmental context*), velocity (*environmental context*), size of the display (*artefact context*), color (*artefact context*) are used to implement the above functionalities, they can be considered as common relevant context parameters for the "*Moving from origin to destination*" action. In the meantime, if some functionalities such as "the content of the route map should be changed according to daytime/night time, summer/winter" are specified in the *"software requirements specifications"*, "daytime/night time" and "summer/winter" (*envi-*

ronmental context) will be considered as specific relevant context parameters which are specific to the current system. Using the same method, the relevant context parameters (common and specific) for other actions can be identified.

3.5 Conclusions and Future Work

In this paper, we adopt an interactional perspective on context. First, something is context because it is used for adapting the interaction between human and applications. Second, activity is central to context. Third, context differs in each occasion of the activity. Based on this perspective, we propose a method using Activity Theory (AT) to identify relevant context parameters: 1) An activity is decomposed into actions, which we take as units for identifying context parameters by using AT's hierarchical structure of activity. 2) We make an extension to the AT's framework, and use this extended framework to identify relevant context parameters for each action. Finally, a simple application of this method in designing context-aware pedestrian wayfinding services is outlined. By using this method, developers can effectively identify relevant context parameters for context-aware systems.

Our next step is to develop some actual applications to evaluate this method. Furthermore, as relevance is very important in context modeling, we are also interested in using our extended Activity Theory framework to define different types of relevance.

Acknowledgements

This work has been supported by the SemWay project (funded by Austrian BMVIT) and the UCPNavi project (funded by Austrian FWF). We also gratefully acknowledge the contributions of our anonymous reviewers who helped to substantially improve this paper.

References

Barthelmess P, Anderson KM (2002) A view of software development environments based on activity theory, Computer Supported Cooperative Work (CSCW), vol. 11, no. 1-2, pp.13-37.

Chen Y, Atwood ME (2007) Context-centered design: bridging the gap between understanding and designing, In Jacko J (Eds.), HCII 2007, LNCS 4550, pp. 40–48.

Dey AK, Abowd GD (1999) Towards a Better Understanding of Context and Context-Awareness, College of Computing, Georgia Institute of Technology, Tech. Rep. GIT-GVU-99-22.

Dourish P (2004) What we talk about when we talk about context, Personal and Ubiquitous Computing, vol. 8, no. 1, pp. 19–30.

Downs RM, Stea D (1977) Maps in minds: Reflections on cognitive mapping, Harper & Row.

Dransch D (2005) Activity and Context – A Conceptual Framework for Mobile Geoservices, In Meng L, Zipf A, Reichenbacher T (Eds), Map-based Mobile Services – Theories, Methods, and Implementations, Springer, pp. 31-42.

Engeström Y (1987) Learning by expanding, Helsinki: Orienta-konsultit.

Engeström, Y.: Developmental studies of work as a testbench of activity theory. In Chaiklin S and Lave J (Eds.), Understanding practice: perspectives on activity and context, Cambridge: Cambridge University Press, pp. 64–103 (1993)

Gartner G, Uhlirz S (2005) Cartographic Location Based Services. In Meng L, Zipf A, Reichenbacher T (Eds), Map-based Mobile Services – Theories, Methods, and Implementations, Springer, pp. 159-169.

Henricksen K, Indulska J, Rakotonirainy A (2002) Modeling context information in pervasive computing systems, In Mattern F, Naghshineh M (Eds.), Pervasive 2002, LNCS 2414, pp. 167–180.

Kaenampornpan M, O'Neill E (2004) Modelling context: an activity theory approach, In Markopoulos P, Eggen B, Aarts E, Croeley JL (Eds), EUSAI 2004. LNCS 3295, pp. 367–374.

Kofod-Peterson A, Cassens J (2005) Using activity theory to model context awareness, In Roth-Berghofer TR, Schulz S, Leake DB (Eds.), MRC 2005, LNAI 3946, pp. 1–17.

Leont'ev AN (1987) Activity, consciousness, and personality, Prentice Hall.

Lueg C (2002) Operationalizing context in context-aware artifacts: benefits and pitfalls, Informing Science, vol. 5, no. 2, pp. 43-47.

Nardi AB (1996) Activity theory and human-computer interaction, In Nardi AB (Eds.), Context and consciousness: activity theory and human-computer interaction, MIT Press, pp. 4-8.

Reichenbacher T (2004) Mobile cartography - adaptive visualisation of geographic information on mobile devices, PhD Thesis, Technical University of Munich.

Sarjakoski LT, Nivala AM (2005) Adaptation to Context – A Way to Improve the Usability of Mobile Maps, In Meng L, Zipf A, Reichenbacher T (Eds), Map-based Mobile Services – Theories, Methods, and Implementations, Springer, pp. 107-123.

Schilit BN, Adams NI, Want R (1994) Context-aware computing applications, IEEE WMCSA, pp. 85-90.

Schmidt A (2002) Ubiquitous Computing – Computing in Context, PhD Thesis, Lancaster University.

Winograd T (2001) Architectures for context, Human-Computer Interaction, vol. 16, no. 2/4, pp. 401-419.

Zimmermann A, Lorenz A, Oppermann R (2007) An operational definition of context, In Kokinov B, Richardson DC, Roth-Berghofer T, Vieu L (Eds.), CONTEXT 2007, LNAI 4635, pp. 558–571.

Zipf A, von Hunolstein S (2003) Task oriented map-based mobile tourist guides, In Mobile Guides Workshop at Mobile HCI 2003, Undine, Itlay.

Zipf A, Joest M (2003) User expectations and preferences regarding location bases services – results of a survey, In Gartner G (Eds), LBS & TeleCartography, Geowissenschaftliche Mitteilungen, vol. 66, pp 63-68.

4 Congestion Visualization Based on Probe Vehicle Data

Masria Mustafa, Meng Zhang and Liqiu Meng

Department of Cartography, Technical University of Munich, Germany

Abstract

The purposes of this study was to develop an analytical congestion visualization system using GIS with the help of probe vehicles as the source of data. Vissim 5.0 was used to generate travel time data on a hypothetical network. Average travel time on links were analyzed for various percentages of probe vehicles and compared to the 'true' average travel time using 'bootstrapping' technique. We employed ArcGIS designed for use by transportation professionals to display the results of travel time provided by probe in a more understandable visual fashion.

Keywords: probe vehicle, Vissim 5.0, bootstrapping, ArcGIS

4.1 Introduction

Traffic congestion is one of the major problems in transportation system and the problem may be alleviated by providing the drivers with useful and effective information on the travel routes, predicted travel times and alternate routes. Visualization plays an important role in the process of understanding large and complex data sets. In transportation, visualization supports the road user to receive and interpret the information in a more effective way. Several tools are available for visualizing information. The use of Geographic Information System (GIS) in transportation field has become more and more popular. Researches associated with the use of GIS are widely spread both in transportation planning and traffic engineering. Claramunt and Jiang et al. (2000) came up with the new framework for the integration, analysis and visualization of urban traffic data within geographic information systems. Their research explored some experimental methods for the real-time integration (i.e., preprocessing), manipulation, visualisation and animation of dynamic phenomena within VDGIS. Rao and Sunitha et al. (2005) addressed the exploration of Intelligent

Traffic Guidance and Monitoring System under GIS environment. They used the Dynamic ITGS to promote online monitoring of traffic mobility, equilibrium conditions of dynamic (vehicular) characteristics and static (network) features.

In addition, Byon et al. (2006) developed GISTT (GPS-GIS Integrated System for Travel Time Surveys), a system for the estimation of link travel time, including static estimation for planning purposes and dynamic estimation for ITS applications. Their results revealed that the Dynamic GISTT provides information of probe vehicles in real time that matched to certain links. Bertini and El-Geneidy (2004) integrated GIS and intelligent transport system (ITS) to improve incident management and life safety where they highlighted the power of an existing GIS-T data model and integrate new highway incident (e.g., accidents and breakdowns) data model. Taylor et al. (2004) developed an integrated Global Positioning System (GPS) and GIS for collecting on-road traffic data such as travel time,delay, and congestion data from a probe vehicle.

From literatures, researchers proved the capability of GIS and its application for transportation analysis. However, which information should be visualized? The basic idea of this study was derived from the real time interactive map that has been used world wide. Certain output of roadway level of services is visually displayed, for example using "Traffic Congestion Map," which displays the traffic congestion in real time. The source data of this output can be observed by roadside sensors or manual counts. However, probe vehicle is expected to be an effective way to collect traffic information in the area where detectors have never been fully activated.

Over decades, many researchers focused on the utilization of probe vehicle in collecting information. Kolbi et al. (2002) compared results of motorway performance when using probe vehicle and other motorway flow detection techniques. Travel time was measured using induction MIDAS system and number plate matching techniques together with probe vehicle on United Kingdom motorway. Their results showed the influence of the measurement system in terms of counting and headway, travel speed as well as travel time. Linnartz and Westerman (1994) discussed a method of monitoring a metropolitan freeway system using probe vehicles and random access radio channel at San Francisco Bay area. Their results concluded that random access (ALOHA) transmission of traffic reports is an inexpensive and a flexible method of data collection that can provide accurate real-time link travel times and perform the Automatic Incident Detection. Additionally, Cohen et al. (2002) presented the dimensioning of a fleet of probe vehicles on motorway equipped with fixed measuring station which focused on the probe vehicle sample sizes for travel time estimation on motorways. Despite of that, Li and McDonald (2002) proposed link travel time estimation using a single probe vehicle equipped with GPS. The estimated travel times were calculated by means of a mathematical model which combines travel time of probe vehicles and movement characteristics based on the analysis of speed profile. The increasing number of road users requires accurate, timely traffic information especially using probe vehicle system. Taking this situation into account, Ferman, Blumenfeld, and Dai (2003) tested the feasibility of such a system by developing a simple analytical or statistical model. Their results proved the feasibility of a real-time traffic information system based on probe vehicles and should work for highways at penetrations of over 3%, while surface roads require more than 5%.

Previous studies show the credibility of probe vehicle as source of traffic data. In general, many researchers focus on providing travel time. Since visualization road travel time has been one of the important elements for traffic information systems, we believed that it is important to make sure the information is successfully received and fully utilized. Therefore this study was conducted to develop an analytical congestion visualization system using GIS with the help of probe vehicles as the data source. Travel time generated from probe vehicle was used as a performance indicator of the level of congestion in the road network due to the reason that it is the most common way that users measure the quality of their trip. Besides, it is a simple measure to use for traffic monitoring.

4.2 Methodology

The following sections describe the framework and method used in this study. This includes the general experiment procedures, equipment and design of traffic environment, design of experiments, and analysis methods. *Figure 4.1* shows the general methodology employed in this study.

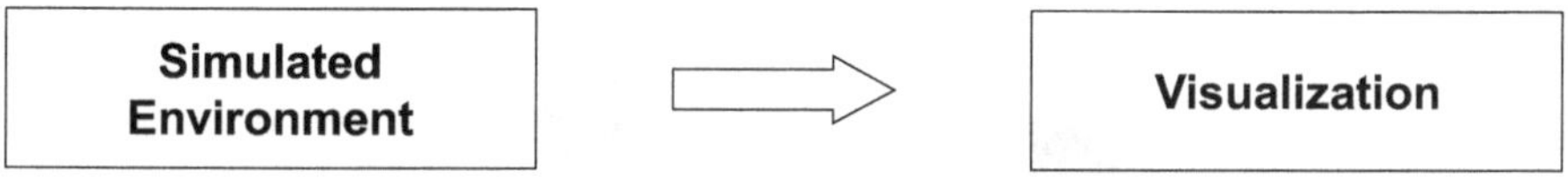

Fig. 4.1. General methodology employed in this study.

In this study, we combined results generated from traffic simulation software with ArcGIS. In order to illustrate the simulation of the potential of a probe vehicle in providing traffic information, an example network is required. To permit the analysis, a simple network was defined. In order to provide meaningful results, the network characteristics were developed to be as realistic and objective as possible. The main benefit of using traffic simulation software is to get data under various traffic conditions without disturbing the current transport environment. In terms of travel time study, it provides a potential source to generate traffic data in both good quality and sufficient quantity. A wide variety of models are available, both microscopic and macroscopic. However, a microscopic model is more suitable for our purpose because of its capability of modeling with great detail each individual vehicle in a network. Even though simulation has its own problems (high level of detail and input data requirements in addition to the need for proper calibration and validation), it offers a viable alternative to field data collection. The major advantage of traffic simulation is that once a model has been developed and validated, it is possible to generate data for a wide range of situations. A well-known microscopic traffic simulation model, Vissim 5.0 was selected for application in this study.

A hypothetical network was created to be modeled in Vissim V5 as a test bed for the generation of research data (link travel time) as shown in *Figure 4.2.* Three different scenarios (different levels of service) were created as a source of information need in this study. *Table 4.1* shows that by loading extra vehicles in the network, the level of service of every intersection becomes low. The road network has two lanes in each direction with six intersections. Each lane is 3.5m wide with 80km/h speed specification. For every network with different levels of congestion setting, only one run with the same seed number for each traffic level was set. The simulation was run for 5400 seconds from which 600 seconds were the warming-up time. Synchro 4, Traffic Signal Coordination Software was used to optimize the traffic signal settings for every intersection. The advantage of the signal optimization was to reduce overall intersection delay and to provide the optimal signal setting for the entire network.

The optimum signal setting from Synchro 4, Traffic Signal Coordination Software was determined for each scenario. It should be noted that the optimization was done in order to create traffic scenarios with different congestion levels. The recommended signal timing was the result of an optimization strategy. It involves several

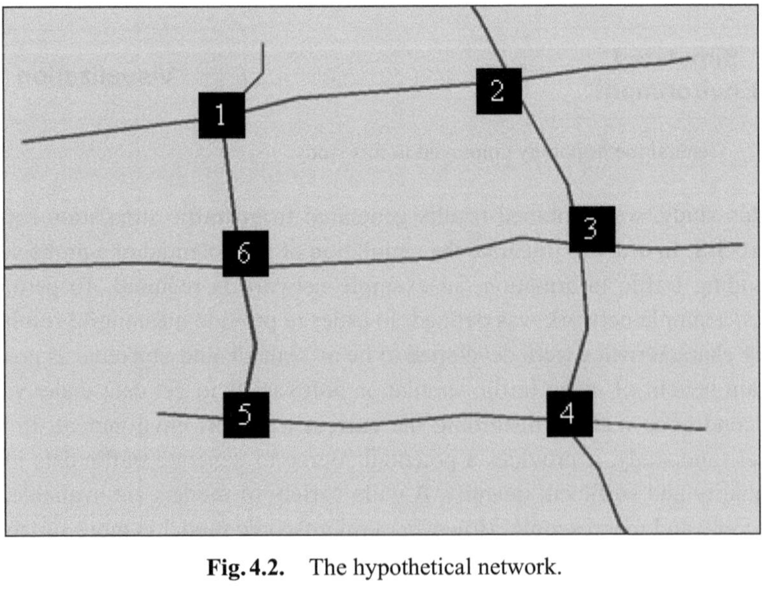

Fig. 4.2. The hypothetical network.

performance measurements including the length of queue and delay. *Actuated traffic controller* was specified in the network. Once the level of service for every network was obtained, the optimum setting for the cycle length together with the offset setting were transferred into Vissim for further simulation.

Table 4.1. The levels of service (LOS) for each intersection.

Intersection 1	Intersection 2	Intersection 3	Intersection 4	Intersection 5	Intersection 6	Average LOS
A	A	A	A	A	A	A
D	D	C	E	A	C	D
F	E	F	F	A	F	F

The design of the theoretical network was determined so that it represents typical road for a traffic situation which might give the various effects on the probe vehicle travel times. The theoretical network was investigated to cover a broader range of traffic volume. Our study has the purpose to detect the difference between the probe vehicle link travel time and the 'true' link travel time. The 'true' average link travel time was measured by summing up the travel time from all vehicles passing the link divided by the number of vehicles passing the link. On the other hand, since all vehicles in the network came with a unique ID (identity number), the probe vehicle travel time were measured by matching the ID of the vehicle in the link with the ID of the vehicle specified as probes, and the average travel time of probe vehicles was calculated. The study expects that the results from travel time determination from probe vehicles on a hypothetical network could reveal travel time accuracy obtained by probe method. The accuracy can be examined under various traffic (congestion) conditions. Although not attempting to replicate the real world, the study could reveals the actual gain of accuracy (of travel time from probes) if the data collection method is implemented in the real world.

As the first step, we input "travel time" stations in Vissim 5.0 to perform exactly the same task of producing travel time .By placing such stations in Vissim 5.0, the whole road network has been divided into many short segments. We collected simulated travel time of the vehicles in 1 second interval. However, the probe vehicle was not specified at the beginning of the simulation, but the iteration of producing and assuming some vehicles to be probe was set after the simulation stopped and data were generated. For every dataset (different levels of congestion), 'bootstrapping' was made for 500 times. This meant that a set of vehicle ID was selected for 500 times randomly and the travel time of each ID was identified in order to get the variation of travel time. The 'bootstrapping' was conducted for 10%, 20%, 30% and 40% sample of probe vehicle in every network. Application of bootstrapping to estimate probe vehicle travel time might allow us to use smaller samples in validation studies, thereby reducing the costs. Further discussion on how bootstrapped sampling was conducted for this study was discussed in Narupiti and Mustafa (2007).

Finally, it is important to make sure that information received by the user is fully utilized and successfully received. Taking into account the power of GIS as a visualization tool, in this study, we employ ArcGIS designed for use by transportation professionals to display the results of travel time provided by probe vehicle in a more understandable visual fashion. As described in previous paragraph, the travel time provided by probe vehicle is stored in database. Therefore, we can visualize the road network using color coded map for the entire test network. The flow diagram is shown in *Figure 4.3.*

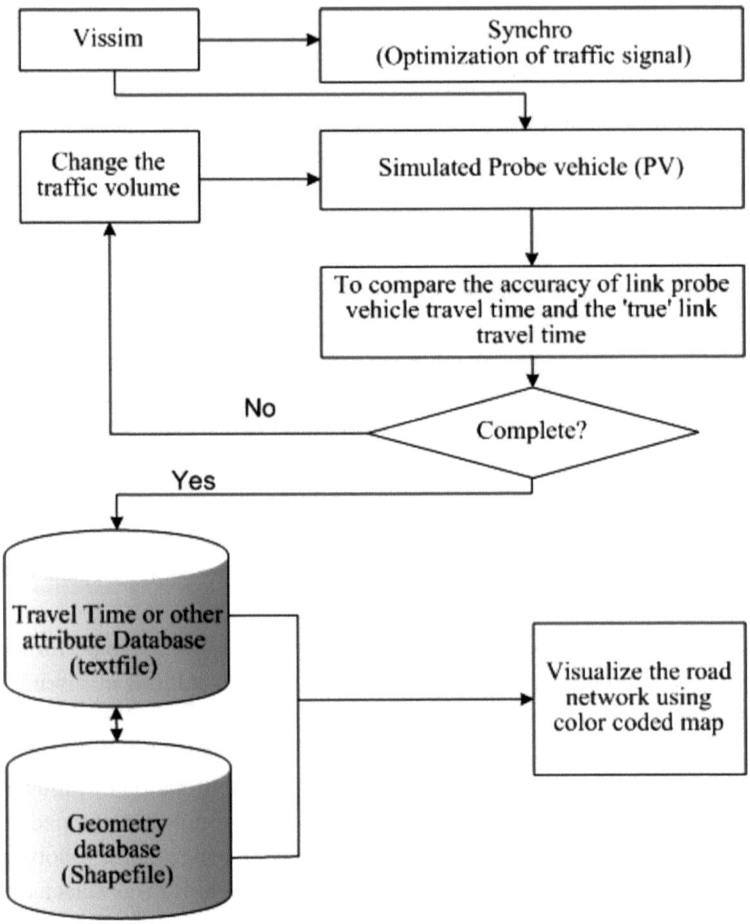

Fig. 4.3. Flow diagram of our study.

4.3 Results and Discussions

4.3.1 Statistical Analysis of Link Travel Time

Our study has the tasks to assess the effectiveness of probe vehicle in providing traffic information (travel time) under various traffic conditions and to develop an analytical congestion visualization system using GIS as a platform. They serve the purpose of assuring the accuracy of probe vehicle travel time as compared to the 'true' travel time. Travel time on link were analyzed and compared to the 'true' travel time. From the data generated by Vissim or microscopic simulation, we obtained datasets consisting of individual vehicle travel time. Different sample sizes of probe vehicle data were considered. The benchmark is, of course, the 100% data set. However, we only considered 10%, 20%, 30% and 40%, which in terms of number of vehicles are quite significant. Various traffic scenarios were simulated in Vissim simulation model. The simulation yielded datasets including vehicle identification numbers, travel times and other traffic data for validation. Several statistical analyses for the corresponding data were carried out. The key examination was the quality of the travel time received from probe vehicles under various traffic and probe (market penetration) conditions.

Note that the "true" link travel time is defined as the travel time calculated from *all* vehicles traversing on the link. This resembles the travel time data obtained from vehicles when they complete the entire route. It is analogous to the 100% probe vehicle market penetration case. At each market penetration case (% probe vehicle), some vehicles are assumed to be probe vehicle. Then travel time determination is basically the average of travel times from all probes. It is noted that the number of probe vehicle is random, based on percentage of probe. It is natural that travel times from probes vary due to their traffic and driving situations, thus variation in travel times can be seen (as indicated by range and standard deviation of travel times). Since the probe vehicles are re-sampled, the bootstrap of probe vehicles resulted in various numbers of probes in the same percentage of probe vehicle case. In this study, the probe vehicles are randomly selected for 500 times, producing various travel times and their variations.

As shown in *Figure 4.4a* to *c,* the plots of probe vehicle travel time versus 'true' link travel time agree well for different levels of service. This indicates the correlations between accuracy of the travel times and different congestion setting and also implies that the travel time from probe vehicles can be used to represent the 'true' travel time. However, the average travel time given by probe vehicle is not equal to the 'true' travel time when the level of service is poor where the network starts to load with vehicles. During this time, congestion level increases and the capacity of the roads will be exceeded.

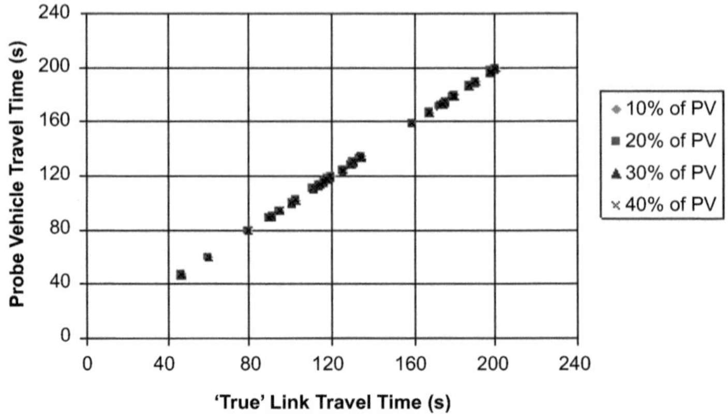

a) Level of service A

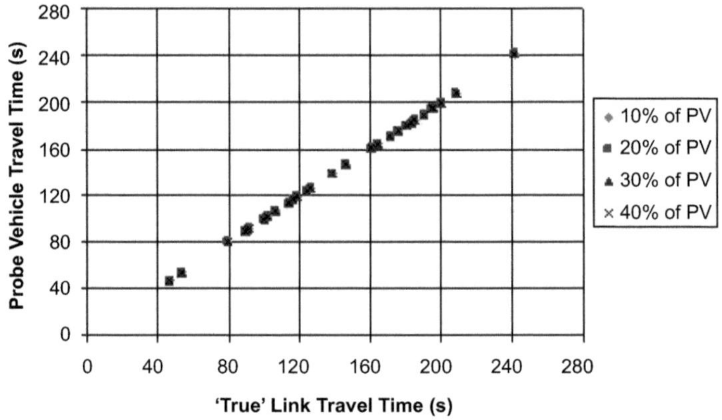

b) Level of service D

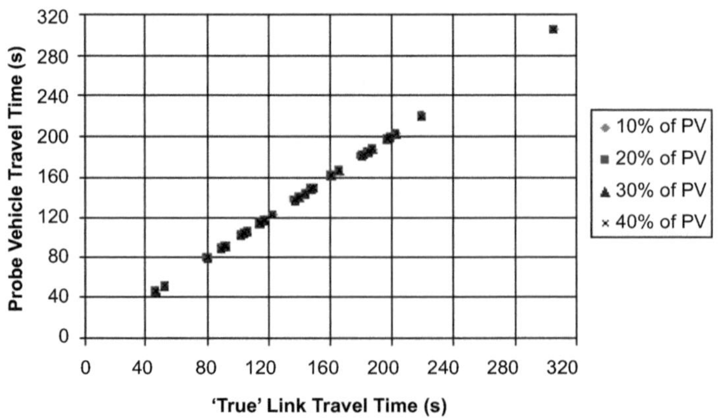

c) Level of service F

Fig. 4.4. Average probe vehicle travel time (bootstrapped) versus 'true' link travel time.

The accuracy of the link travel time obtained from the probe data is described by the magnitude of its standard deviation. Supposed that n_p probe vehicles traveling along a link and that each of these probe vehicles reports the time required to traverse the link. Then, the standard deviation of the average reported link travel times from the probe vehicles is as follows:

$$\sigma = \frac{\sigma_T}{\sqrt{n_p}} \tag{4.1}$$

where:

σ_T = standard deviation of the mean link travel time reported by the probe vehicles

n_p = number of probe vehicles.

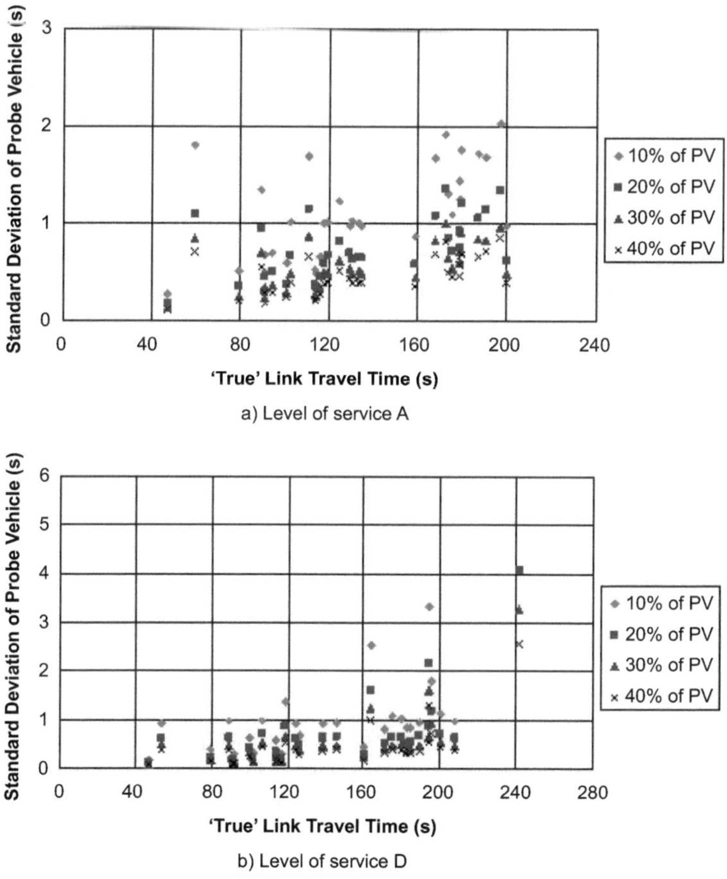

Fig. 4.5. *Continued on the next page.*

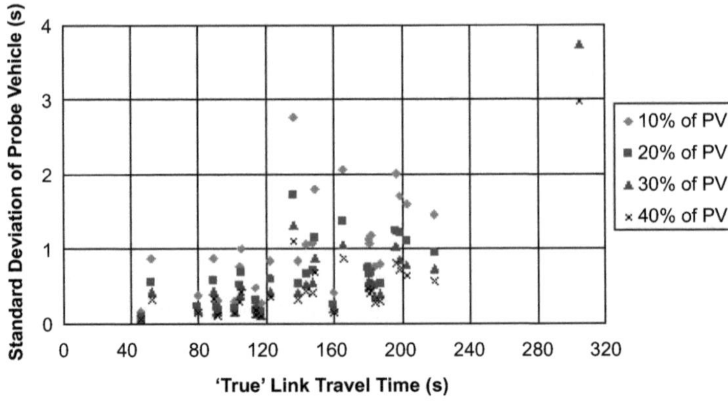

Fig. 4.5. Standard deviation of travel times from probe vehicles compared to 'true' link travel times under various traffic conditions.

The standard deviation and the average value of the travel times of probe vehicles are shown in *Figure 4.5.* Travel times of probe vehicles are less accurate with the increasing of traffic volume in the network. Statistically, the increase in standard deviation indicates that the 'true' travel times are more dispersed. Larger variation of standard deviation can be observed with the increasing of 'true' travel time while having a large number of probe vehicles resulted in less deviation of travel time. Other measures of dispersion are the range and variance. Knowing the standard deviation for a set of travel times is important because it is a good indicator for judging the mean as a representation of the "average" response. To further examine the variation (standard deviation) of travel time from the probe vehicles at various traffic conditions, the coefficient of variation is considered which can be expressed as follows:

$$\mathrm{COV} = \mathrm{SD} / \mathrm{average} \tag{4.2}$$

Where,

COV = coefficient of variation
SD = standard deviation
average = average value

The values of COV are illustrated in *Figure 4.6 a* through *c* where no patterns can be detected. This implies that the standard deviation (variation) of the travel times reported by probe vehicles is not correlated with the amount of link travel time.

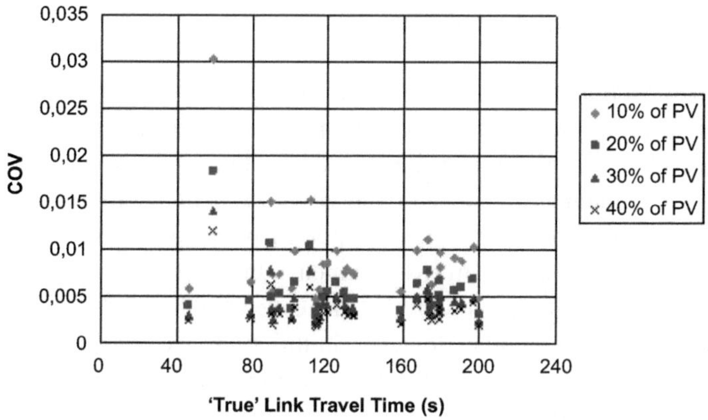

a) Level of service A

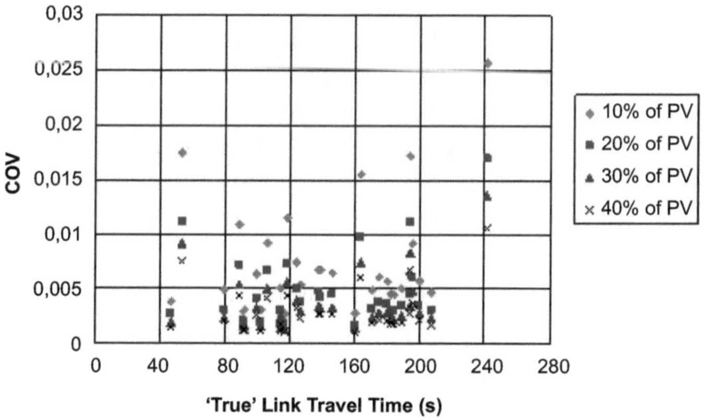

a) Level of service D

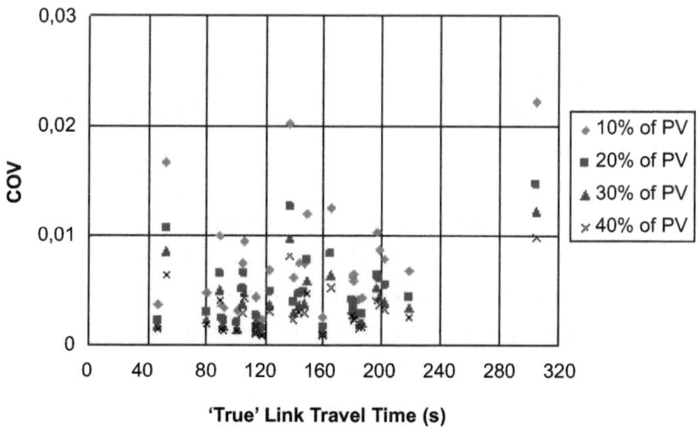

a) Level of service F

Fig. 4.6. Coefficient of variation of the travel time estimates.

4.3.2 Visualization of Travel Time

The results in *Section 4.3.1* have made it clear that computing the travel time from probe vehicle under various traffic conditions is a reasonable method. Therefore, if only link travel times from probe vehicle could be obtained in the entire network, we could still utilize them as an indicator and visualize them in a more understandable way. Our next step is to develop efficient reporting procedures. ArcGIS package was used to display travel time on all road segments. In general, we can build a variety of relational database queries to derive data needed for the production of graphical and tabular travel time reports. However, in this study we considered color coded maps as the visualization platform for the travel time. These color coded maps show the variation of travel time along the road segments.

Vissim has one log file which save all information about the created network. The file consists of coordinates of the roads and also other attributes. It is rather straightforward to export the road network from Vissim to ArcGIS . In this study, an interface was developed to realize the export of road network from Vissim to ArcGIS. Since ArcGIS was used for visualization of travel times, three indication classes were set in ArcGIS as seen in *Table 4.2.* If travel time is below 50 seconds, then the network will be grey, travel time below 150 seconds will appear to be dark grey and black will appear if the travel time is more than 150 seconds.

Table 4.2. Classes of travel time indication in ArcGIS.

Travel time (s)	Color
<50	Grey
<150	Dark grey
>150	Black

Figure 4.7a through *c* shows the color-coded road network for three different levels of service. The visualizations are based on 40% of probe vehicle penetration rate for every level of service. The black color depicts the high congestion road which the travel time is more than 150 seconds, dark grey for road with less congested (travel time less than 150 seconds) and grey for free flow condition (travel time less than 50seconds). Travel times of probe vehicle are less reliable with the increasing of traffic volume in the network. The visualization of the road network with different levels of congestion can make this statement more apparent (see *Figure 4.7).* During high traffic volume, the vehicles start to queue to pass the intersection. Some of the vehicles need to wait until a few cycle times in order to pass through that intersection. Beside, there are also vehicles starts to queue back into the release zones. On the other hand, in case of low traffic flow, all vehicles can easily pass through the intersections and the vehicle does not have any obstruction at the end of the network which can delay them to reach to their destination.

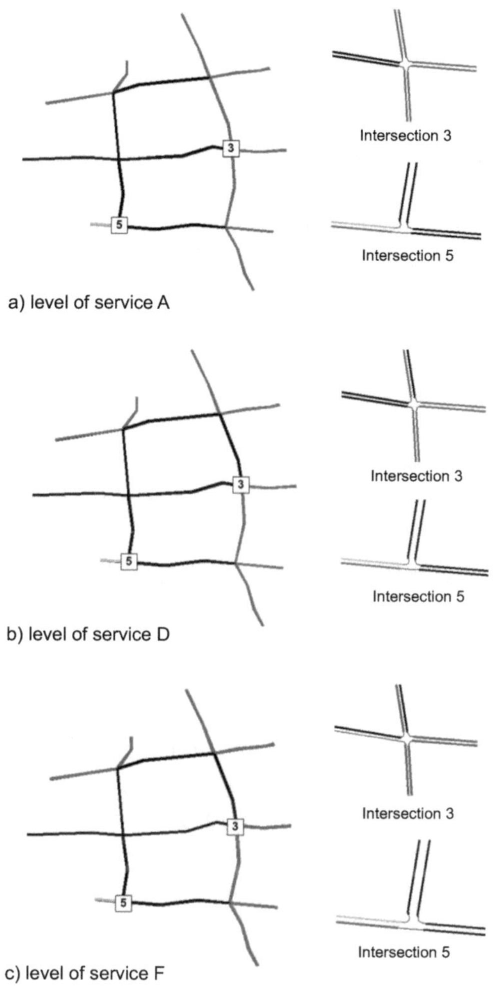

Fig. 4.7. The visualization in ArcGIS for different level of service for 40% of probe vehicle.

One of the interesting issue about probe vehicle as a data source of congestion visualization is the coverage area. It can be seen, in this study, having 40% of probe vehicle from the total vehicles traveling on the streets cover all links for LOS A and D. However, no probe can be detected in high congested network where most of the probe vehicles could not complete their journey until the end of simulation as in *Figure 4.7c*. This condition implies the truth in real world where there is a possibility of having no probe to collect such information in certain links. Therefore, it is significant to consider how to collect traffic information of these links. One of the solutions to be implemented is to replace the missing data by placing another type of traffic data detection.

4.4 Conclusions

This study tested the assumption that the probe vehicle could provide reliable and sufficient amount of data that could represent travel time information which then can be visualized in ArcGIS. Although this study is based on simulation, once proven successful by further testing and evaluation, it can be a help for the long-term savings in system operation and maintenance. The simulation data in our study has given us a clear view how travel times are temporally correlated. The visualization technique described in this study provides an idea of typical visualization available to gain understanding and insights of probe vehicle data sets. The technique allows users to get a visual overview of travel time variation. The approach which combined traffic simulation for travel time analysis and ArcGIS to provide visualization can be easily implemented. The travel time accuracy can be assured, given that we have only probe vehicle as a travel time measurement. The visualization scheme showed an overview of an alternative solution which totally relies on probe vehicle as the data source.

Further theoretical developments are possible with the model. For example, statistical techniques can be applied to smooth the probe vehicle data. Traffic flow theory can be applied to find the parameters that may affect the accuracy of the data by probe vehicle. While the simulation as a test bed provides an unlimited environment to collect, test and validate the travel time, the model cannot be put into practice without a validation with 'real' traffic environment. This study has used a small amount of travel time data. In the validation stage, a large amount of data is needed. In our future work, a 'real' environment with a sufficiently large network will be tested and validated at various time periods, under various traffic conditions with various types of road characteristics. We will then focus on the development and application of advanced visualization techniques. Instead of simply color-coding classified data, analytical visualization tools based on techniques such as 3D color surface with z-axis representing the speed, dynamic linking and brushing etc. will be developed and tested with user participation.

References

Bertini RL, El-Geneidy AM (2004) Toward validation of freeway loop detector speed measurements using transit probe data, IEEE 7th International Conference on Intelligent Transportation Systems, pp.779-784

Byon YJ, Shalaby A, Abdulhai B (2006) GISTT: GPS-GIS Integrated System for Travel Time Surveys, TRB annual meeting CD-ROM.

Claramunt C, Jiang B, Bargiela A (2000) A new framework for the integration, analysis and visualisation of urban traffic data within GIS, Transportation Research Part C: Emerging Technologies.

Cohen S, Bosseboeuf JL, Schwab N (2002) Probe vehicle sample sizes for travel time estimation on equipped motorways, 11th International Conference on Road Transport Information and Control, pp. 172- 176

Ferman MA, Blumenfeld DE, Dai X (2003) A simple analytical model of a probe-based traffic information system, Intelligent Transportation Systems, IEEE, Vol. 1, pp.263- 268

Kolbi R, McDonald M, Fisher G, Brackstone M (2002) Probe vehicle: a comparison of motorway performance measure with other motorway flow detection analysis, 11th International Conference on Road Transport Information and Control, pp. 182- 186

Li Y, McDonald M (2002) Link travel time estimation using single GPS equipped probe vehicle, IEEE 5th International Conference on Intelligent Transportation Systems, pp. 932- 937

Linnartz JPMG, Westerman M (1994) Monitoring a metropolitan freeway system using probe vehicles and random access radio channel, IEEE 44th Vehicular Technology Conference,Vol.1, pp. 410-414

Narupiti S, Mustafa M (2007) Quality of Travel Time Estimation from Probe Vehicles: A Simulation Study, Journal of the Eastern Asia Society for Transportation Studies, Vol. 7, pp.2517-2532

Rao KML, Sunitha K, Jayasree K (2005) Dynamic geographic information system for intelligent traffic guidance systems, Geoscience and Remote Sensing Symposium. IEEE International, Vol.2, pp. 866-869

Taylor MAP, Woolley JE, Zito R (2004) Integration of the global positioning system and geographical information systems for traffic congestion studies, Transportation Research Part C, pp. 257-285

5 Where do you Roll Today? Trajectory Prediction by SpaceRank and Physics Models

Stefano De Sabbata, Stefano Mizzaro and Luca Vassena

Department of Mathematics and Computer Science, University of Udine, Italy

Abstract

Pre-destination, the prediction of a user's future destination, is recently gaining interest and importance in location-aware, ubiquitous, and mobile computing. An increasing amount of data related to position of people is becoming available because people usually take their mobile devices (phones, smart-phones, PDAs, etc.) with them. We propose to mine these data to derive the importance of the single locations in an area of interest, given by either a single user or a community. Then we use the importance of locations as basis for our approach to pre-destination, where well-known physics models (namely gravitation and electrical force) are exploited to estimate users trajectories and future destinations.

Keywords: location-awareness, location importance, physics models, trajectory, destination prevision

5.1 Introduction

Pre-destination can be defined as the attempt to predict the future location of the user at a given time point. In the location-aware, ubiquitous, and mobile computing communities, pre-destination is recently gaining interest and importance due to actual pervasiveness of mobile devices in conjunction with growing people's need of location-based information systems and services, for example in traffic control, transportation planning, computational advertising, proximity marketing, etc. The location of a user is an important feature for many systems, and it is studied in several respects (Eagle & Pentland 2006). In principle, pre-destination allows to implement more efficient systems, since it allows to set up the destination information environment before the destination is reached. At the same time, pre-destination can be useful in reducing uncertainty when determining users or items locations.

An increasing amount of data related to position of people (for example the usual route taken when driving to work, the usual path followed by people walking in the mountains, etc.) is becoming available because people usually take their mobile devices (phones, smartphones, PDAs, etc.) with them, and these mobile devices can (and indeed they do) leave a record of their past positions. Several technologies allow the generation of this huge amount of data: GSM, UMTS, HSDPA, etc. phones frequently exchange data with the network antennas; Bluetooth and Wi-Fi devices can see and be seen by other devices in close proximity; GPS and Galileo can determine the device position with good accuracy; triangulation and trilateration can be used with all these technologies to provide a more accurate estimate; and so on.

Recently, a large amount of this kind of data has been collected and analyzed, deriving that it is generally more likely that a user will go in a place more related to his/her habits (Gonzalez et al. 2008). Starting from these considerations, we propose to mine the data related to position of people to derive the importance of the single locations in an area of interest, given by either a single user or a community. Then we use the importance of locations as the basis for our approach to pre-destination, where well-known physics models (namely gravitation and electrical force) are exploited to estimate users trajectories and future destinations.

The paper is structured as follows. We first briefly survey related work in *Section 5.2*, describing our approach, named *SpaceRank*, in which we show the encoding of users behavior with graphs and the PageRank algorithm, in order to associate importance values to locations (De Sabbata et al. 2008). In *Section 5.3* we present our physics based approach to pre-destination, while in *Section 5.4* we use our approach to analyze experimental data. At the end we present some conclusions and the future work.

5.2 Related Work

New mobile devices and new technologies allow to record the movements of single users. All these records are an interesting data source that can be used to determine past people positions; in turn, mining people positions can allow to derive the importance, or popularity, of places in the real world; and, finally, importance/popularity of locations in the real world can be very useful in order to predict future users locations.

5.2.1 Give Importance to Locations

Past user (or users) behavior can be exploited in this respect: generally speaking, it is more likely that a user's position in the future will be a location where he or she has already been (or is used to go to) than a position where he or she has never been (or usually does not go to). Also, if one has to predict the user next position, it is generally more likely that a user will go in a place more related with his/her habits and interests (Gonzalez et al. 2008). So a first step consists in determine the importance of the locations to the user(s).

5.2.1.1 Classical Indexes

In order to define a location importance to a user, it is straightforward to think of three indexes:

- *#visits*: Number of visits by the user in the location.
- *avgTime*: Average time spent by the user in the location.
- *totTime*: Total time spent by the user in the location.

These indexes can be computed on the basis of the data collected during a certain period of time (one day, one week, one month, etc.). This procedure can be tailored either to a single user or to a community: thus, the results will be related to the single user habits or to the community behavior. This approach is followed in several studies (Eagle & Pentland 2006, Ashbrook & Starner 2003, Chan et al. 1998).

If these indexes are considered separately, they just give a partial vision of the user's behavior, and for some applications this could not be enough. For example, if we consider *#visits* only, a location where the user passes by without stopping and a location where the user stops for a long time would be indistinguishable. Similarly, using *avgTime* could put at the same level locations with rare visits and a location with frequent visits. The last index, *totTime*, could be a good importance index since a location should be visited frequently or for long periods to have a high *totTime* value. However, for example, we would like to give different importance to locations having the same *totTime* value, on the basis of one or both of the other two indexes.

One approach could be to use a combination (e.g., a linear combination) of the three indexes, or of some of them. However, this would rise the problem of the choice of both the indexes and the weights to assign to each of them.

5.2.1.2 SpaceRank

Starting from the limitations of the previous indexes, we have proposed a novel approach to define a location importance to a user. Our approach, named SpaceRank (De Sabbata et al. 2008), is based on the PageRank algorithm.

PageRank (Page et al. 1998) is a well known algorithm for the analysis of links on a hyper-linked set of documents, whose goal is to measure the importance of each document within the set. PageRank assigns a numerical weighting to each element in the set and so represents the likelihood that a person will arrive at that page, randomly clicking. PageRank is used, together with thousands of other features, by search engines to rank the pages retrieved after a user query.

Our aim is to determine each location importance on the basis of geographical properties and users past movements among the locations; past data can concern either a single user (private importance) or a group of users (social importance). We start by concentrating on an area of interest, that we divide into a finite n contiguous sub-areas, defined as *locations*. We suppose that the locations are contiguous and there are no disconnected locations. The number of neighbors for each location is dependent on how the area is divided into locations; in our experiments we use square locations of the same size.

Our approach is based on the linear combination of two $n \times n$ matrix matrixes, where n is the number of locations in the area, according with the *equation 5.1.*

$$SpaceRank = (1 - d) \times HabitsMatrix + d \times TransitionMatrix \qquad (5.1)$$

The *Transition Matrix* encodes the relations of contiguity among locations, represented as a graph *(Figure 5.1a),* while the second *(Habits Matrix)* encodes the habits of the user(s) *(Figure 5.1b)*. The d parameter is used to make the results more related to user habits or to geographical properties.

To create the *Habits Matrix* we record users' movements, sampling at regular intervals position and speed. For example if we have recorded temporally subse-

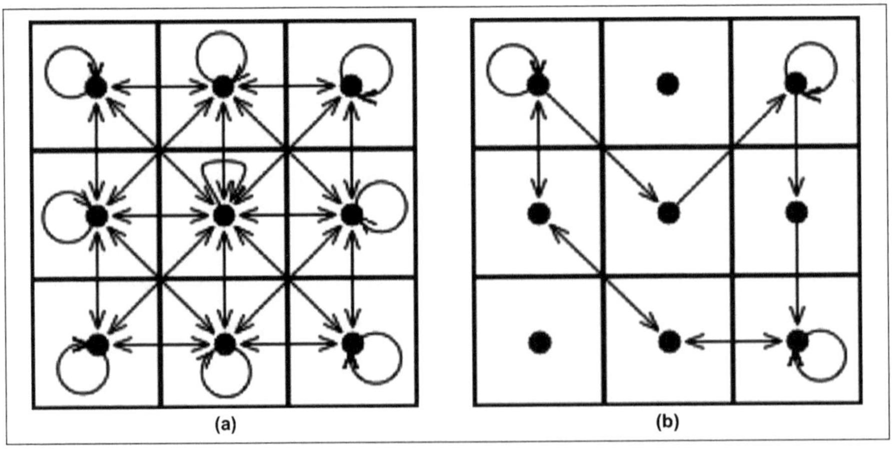

Fig. 5.1. An example of a base transitions graph (without edges weights) with nine square locations (a). An example of an historical data registry based graph (without edges weights) with nine square locations (b).

quent user's position related to the same location, the probability that the user will remain in that location will increase. On the contrary if we have recorded temporally subsequent user's positions related to different locations, the probability of transition between the locations will increase.

Speed is also used to discriminate if the user is moving or if he is standing still in a location. For example, assuming that we have registered two temporally subsequent elements with the same location, $speed > 0$ suggests that user is moving, while $speed = 0$ suggests that the user is standing still in the location. If we consider as permanences (i.e., loops on the same location) only the records with $speed = 0$, the importance of locations where the user has stopped is increased, while the importance of locations where the user has just passed through is decreased. In the same way if we consider only the records with $speed > 0$, the results, obtained with the computation of the importance, will highlight the areas that require a high amount of time to be crossed, reflecting locations with high traffic density because of traffic jams.

5.2.2 Prevision of the Destination

The knowledge of future users locations in the real world can be very useful for several applications. For example, in traffic control (Taylor et al. 2000), in itineraries planning (Bierlairea & Frejinger 2008), in network handover (Zaidi & Mark 2005), and, in general, in location-aware applications (Hazas et al. 2004), knowing which location is the more likely future destination would allow resources optimization, more efficient services, and even more features. Indeed, in the field of context-aware applications, although several factors are taken into account in order to define user's current context, location is usually the main one.

The predestination algorithm is presented in (Krumm & Horvitz 2006) as a procedure that aims at predicting a driver destination point using the history of driver's past destinations and driver's behavior data. More generally, we can define pre-destination as the problem of calculating, given an initial point, direction, and speed, a set of possible destination points with their own probability value.

Mountain, in his Ph.D. thesis (Mountain 2005), presents an extensive study of location-based filters for information retrieval and introduces a simple system to predict the user's next position, speed and heading in order to improve spatial proximity, temporal proximity and speed-heading criteria.

In (Krumm & Horvitz 2006) is proposed a pre-destination approach for a car driver, based on Bayesian inference driven by data about driver's (i.e., the moving object in the space taken into account) previous destinations and behavior, to produce a probabilistic map of possible destinations.

All these proposals are based on a large set of heuristic data. Indeed, so far, pre-destination is mainly an experimental discipline lacking a complete formalization.

5.3 Trajectory Prevision

5.3.1 Problem Formalization and Working Hypotheses

We can provide a first formalization of the problem as following. We take into account a generic moving object, be it a human walking, cycling, or driving; a car; a bus; an airplane; a train; etc. We refer to the generic moving object as *Pre-Destined Body* (PDB). We assume that it moves on a two dimensional surface, but the extension to the three-dimensional case is straightforward.

Our goal is to determine the probability that the PDB will be at a specific (x, y) location of the surface at some time t in the future, given its position (x_0, y_0) and its movement vector $\vec{v}_0$ at time t_0. At first glance, we can say that we are searching for a function like:

$$p(x_0, y_0, t_0, \vec{v}_0, x, y, t) \tag{5.2}$$

So, that function should compute the probability that (x, y) will be PDB position at time t.

However, people normally do not move around without being attracted by some places. So it seems obvious that the placement of interesting objects will affect in some way the movement of the PDB: on the surface are placed *n Points Of Interest* (POIs), that could be houses, offices, shops, cities, etc.

So, the function we are looking for could be something like:

$$p(x_0, y_0, t_0, \vec{v}_0, x, y, t, [k_1, \dots, k_n]), \tag{5.3}$$

where $[k_1, \dots, k_n]$ is a list of k_i elements that represent all the information about singles POIs, such as their position (x_i, y_i). Or, in an equivalent way, we can take $[k_1, \dots, k_n]$ as fixed and define $p(x, y, t)$ in the space shaped by $[k_1, \dots, k_n]$:

$$p_{[k_1, \dots, k_n]}(x_0, y_0, t_0, \vec{v}_0, x, y, t) \tag{5.4}$$

5.3.2 Physics Metaphors

The aim of an abstract model is to simplify the analysis of the problems and provide a common basis for the interpretation of different phenomena. In our opinion, a suitable way to do that is to rely on a metaphor derived from real world or widely known theories. In this paper we try to show how concepts and ideas drawn from physics can provide both intuitive metaphors and well established theories, that can be useful to understand and formalize in a more precise way the above defined pre-destination problem. We first propose our basic model, and then present several improvements to make it less abstract and more related to the real world.

5.3.2.1 Classic Gravitation

If you are walking around in a museum, some items will attract you, and some items will not capture your attention. So you are not so different from a moving object in a gravitational system, or an electric charge in an electrostatic field directed towards a specific point. The basic intuition behind our approach is to use the well-known geometrical theory of gravitation, published by Albert Einstein as the Theory of General Relativity, as a metaphor for general pre-destination scenarios.

In this metaphor, the PDB is like a ball rolling with no friction and negligible mass and diameter (as a point particle) on the surface. POIs are static bodies, with a mass and a diameter that encode some properties of the object as, for example, the number of previous visits, the number of people that are currently visiting it, or the average of time spent in that place. So, POIs act like a planet in a gravitation system and, because of their mass and diameter, create a distortion of the surface, that can change the trajectory of the PDB as it is sketched in *Figure 5.2.* More important objects have a distortion that is both wider and deeper.

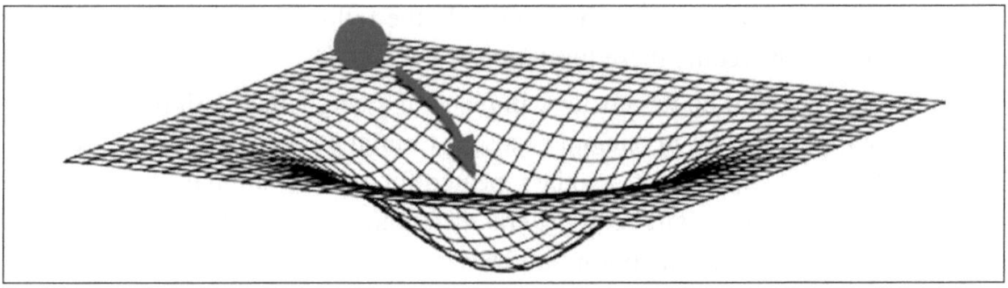

Fig. 5.2. A POI changing the shape of the surface and attracting the PDB.

5.3.2.2 Attraction and Repulsion: from Gravitation to Electrical Force

While the gravitational force is an attractive force, the electrical or Coulomb one can be both attractive and repulsive, depending on the electric charges. From this point of view, a pre-destination model based on Coulomb's law is more general and complete than the classic gravitational model. In the gravitational model, a POI can just attract a PDB as it is important for it, while in the electrical model a POI can also repulse PDBs, as in *Figure 5.3*. To illustrate this situation, we can think about a supermarket: a user could change his path if it leads to a too crowded lane.

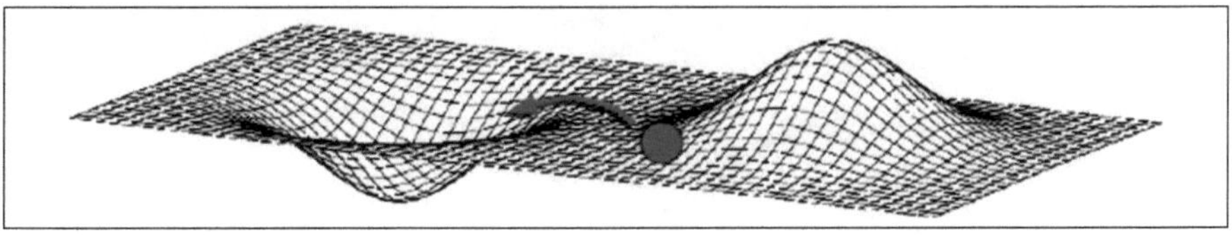

Fig. 5.3. A PDB in a Coulomb model.

Thus, the surface presents depressions and peaks, of different height and depth. Sometimes repulsion is certain (a man cannot walk through a wall; a metro car must move inside a tunnel network; etc.). Attraction can be certain too in some – indeed more specific – cases (e.g., a train moving at high speed and thus having a high inertia; etc.). Thus peaks are sometimes walls of infinite height and depressions are sometimes holes of infinite depth.

5.3.2.3 More Objects in Movement

The previously described model presents a limitation: it is a deterministic model. Given the initial (time t_0) parameters of position, direction and speed, there will be only one point where the PDB can be at a time $t > t_0$. Also, the (x, y) position is determined with no uncertainty, which is usually not true in real world applications and situations (e.g., as in GPS systems), and is unrealistic, since usually the future cannot be predicted with certainty. In practice, the probability distribution degenerates to a Dirac's Delta function, having a 1 value in the (x, y) point where the ball is at time t, and a zero value in all other points. In the following we will improve our model in order to overcome these limitations.

A first attempt could be to consider, in place of a single ball with precise initial position, a ball that can be in one out of a set of initial points. Then the ball can be let to roll, and this "experiment" can be repeated for every point in the initial points set. The final destinations will be in general different, and the different experiments results can be averaged in an appropriate way.

This could work, but is rather inconvenient. A simple improvement is to consider a set of balls, rolling contemporarily, with no friction, and no interferences among the balls. In general, the differences in the initial points could lead, after some time, to locations that could be different, even to a great extent (see *Figure 5.4*). Using this simple model we can assume that the probability for PDB is proportional to the number of balls at (x, y) at time t.

To represent this situation we:

- Associate to a user more PDBs arranged uniformly on the region representing the hypothetic user location.

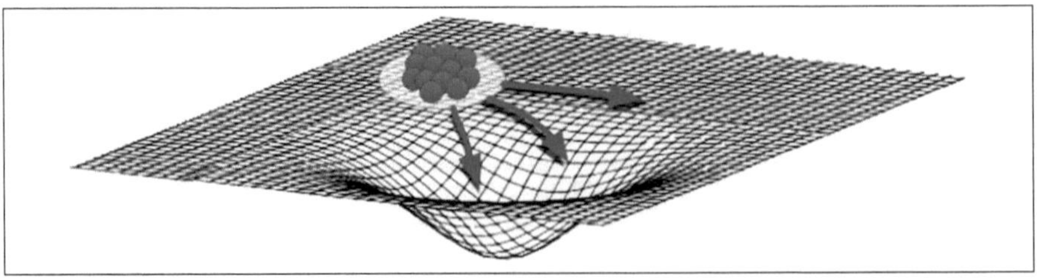

Fig. 5.4. More objects in movement.

- Assign to each PDB a probability value, representing the probability that the PDB reflects the real user position.
- Compute the prevision for each PDB.

In this way, at each time, the user position is represented by a probability distribution associated to different locations, as sketched in *Figure 5.4*.

5.3.3 Pre-Destination Algorithm

5.3.3.1 From Importance to Acceleration

We now propose our approach to pre-destination, named ARDA (A Rolling to Destination Algorithm). We start from the results obtained analyzing user's habits with our SpaceRank algorithm, although this is not a strict choice as our algorithm is flexible and independent from how the importance values have been determined: the only requirement is that the importance values are stored in a matrix representing the locations grid. Let's call this matrix M_{imp}.

Having adopted the electrical physics model, we need a matrix M_{pot}, representing the electrical potential energy in each location. In this model, a PDB will be attracted by lower potential areas, so, in order to have the PDB attracted by more important locations, given that more important locations have higher values in M_{imp} matrix, we need to use opposite importance values in our physics model. So, starting from M_{imp} we generate a matrix with opposite values M_{pot}, representing the locations potentials. We use the function $f(x) = -x$ for simplicity, but any monotonically decreasing function could be used. Then we compute the matrix M_{acc} of accelerations depending on the potentials in the area of interest. Each cell of M_{acc} contains the acceleration vector related to the location represented by the matrix cell. M_{acc} is computed using the gradient:

$$M_{acc} = -\vec{\nabla}(M_{pot}) \tag{5.5}$$

5.3.3.2 Trajectory Computation

We can now compute the future trajectory on the basis of simple physics laws. We start from:

1. The user position (x_0, y_0) at time t_0.
2. The movement vector $\vec{v}_0$ at time t_0.
3. The acceleration matrix M_{acc}.
4. A function $acc(x, y)$ that returns the acceleration value referred to the location (x, y). In this function an attenuation factor γ_{acc} is used to adapt vectors magnitude to the magnitudes in the physic simulation. In fact physics simulation's data on user position and moving vector are related to real world values fitted to loca-

tion's size in the grid, whereas acceleration vectors' magnitude is related to locations' importance estimation values, which are not related to real world sizes.

5. A period T_C to use in user state's next sampling simulation.

We use the uniformly accelerated motion law to simulate the itinerary of the user and therefore to predict the future user position (x_1, y_1) and movement $\vec{v}_1$, at time $t > t_0$:

$$(x_1, y_1) = (x_0, y_0) + \vec{v}_0 T_c + acc(x_0, y_0) T_c^2 \tag{5.6}$$

$$\vec{v}_1 = \vec{v}_0 + acc(x_0, y_0) T_c \tag{5.7}$$

If we consider a probability distribution as initial position (as described in *Section 5.3.2.3*), we have to apply the described procedure to each PDB and, for each location in the area of interest, the probability to be the user position is equal to the sum of the probabilities assigned by the computation of the procedure on PDBs.

5.3.3.3 Holding Factor

We then introduce a "holding factor" as a procedure to slow down user's speed on the basis of the importance of the location the user is in.

The holding factor can be interpreted as the probability that, given the importance of the location the user is in, the user will slow down and stop in that location. The holding factor is simulated computing the movement vector as follow:

$$\vec{v}_{j+1} = \frac{\varphi}{\gamma_\varphi} \vec{v}_j \tag{5.8}$$

where, given the importance value l_{imp} of the location where the PDB is in, $\varphi = 1 - l_{imp}$ is the value associated to the holding factor and γ_φ is the value used to attenuate the same factor (to adapt the importance value to the magnitudes used to encode the information related to the real world).

5.3.3.4 Considerations

The described algorithm uses as a main information source the matrix M_{imp} with the importance values computed by SpaceRank. However several other different approaches can be used to define the matrix of importance. For example, we can integrate SpaceRank with information in the user calendar, or we can use the user history to improve the computation of the importance. We can also take into account social information together with the user profile: for example if a location is important for most students and I am a student, maybe that location is important for me too.

From this point of view the model we propose is flexible and we can compute the importance matrix on the basis of different sources:

$$M_{imp} = \alpha_1 M_{source1} + \alpha_2 M_{source2} + \ldots \tag{5.9}$$

Moreover our ARDA algorithm can be used both for the trajectory prediction in a short time (next minutes) and for the final route destination prediction. The main problem of our algorithm is represented by the attraction of the starting location, as in most cases it is an important location. This location applies an attractive force to the PDB that will affect all the previsions related to the first part of the route.

The solution we have adopted is to smooth the starting location attraction by means of a monotonically decreasing function related to the distance from the starting point. As there is a probability $< 10\,\%$ that a route will last less then 4 minutes (Krumm & Horvitz 2006), we apply the smooth function to all the locations reachable in 4 minutes.

5.4 Preliminary Evaluation

In this section, some preliminary experiments are presented; the aim of these experiments is to test our ARDA algorithm and verify its correctness and effectiveness.

5.4.1 Experimental Settings

In these experiments, the log data concern a typical week of one of the authors of this paper, collected manually. The relevant area, Udine town and his surroundings, is divided into 9×16 squared locations, each of them having 700 meters as side length. The ARDA algorithm has been developed using Python and the following modules: PyXML for XML processing, RPy for statistics computation, SciPy and NumPy for matrixes computation.

The aim of these experiments is to test our ARDA algorithm and verify its correctness and effectiveness; future experiments will have more users, more itineraries, and a more detailed location subdivision.

We started by computing the locations importance using SpaceRank, with different parameter values; the result is shown in *Figure 5.5*.

5.4.2 Short Time Prediction

In this first experiment we have applied our ARDA algorithm to predict the user location in the next minute. This computation has been applied to 23 samples of user position and moving vector taken every 20–40 seconds (in order to avoid subsequent equal previsions) during the route from university to author's home.

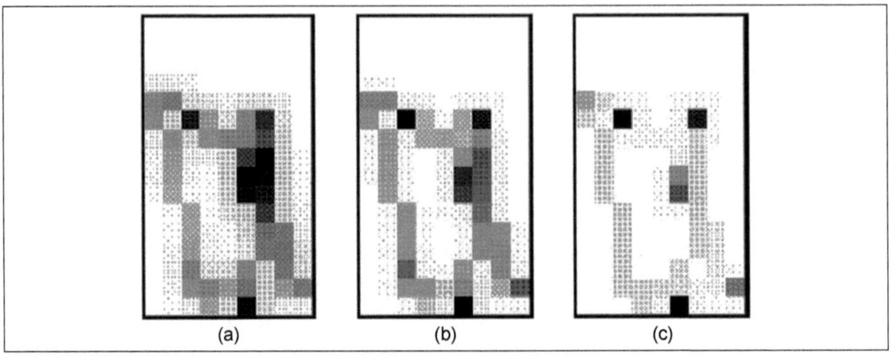

Fig. 5.5. Importance computed by SpaceRank with d=0.15 (a), with d=0.05 (b) and giving more importance to locations where the user stops (c).

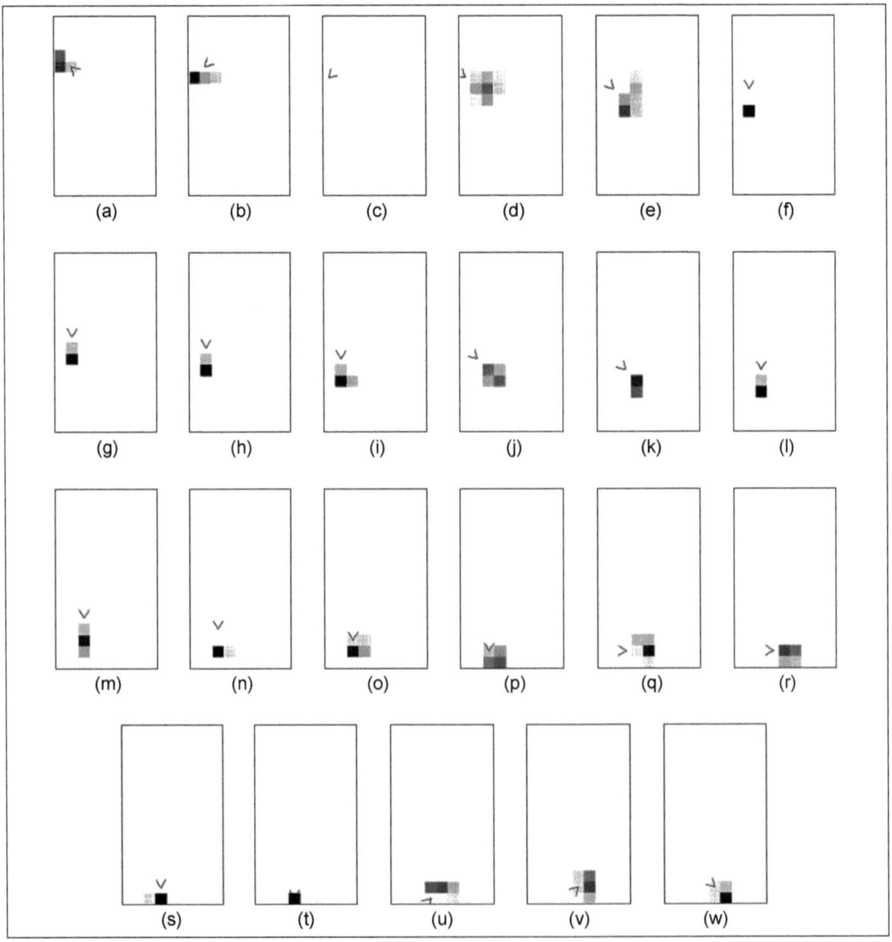

Fig. 5.6. Results obtained predicting the user position in the next minute.

In the SpaceRank equation we have used d=0.15 (*Figure 5.5a*): this means that we consider not only the locations where the user has stopped, but also the locations the user has passed through. A high importance location (darker squares in the location, in the bottom part of the figure) is the home of the user; another high importance location is the University (in the top left hand side), where the user stays about 9 hours per day 5 days per week. The other high importance locations have been visited for hours by the user. The lighter gray level locations are the roads where the user drives usually. White squares locations have never been visited.

On the basis of position and speed values obtained from the real data, we have experimentally seen that the most suitable values for the parameter in the ARDA algorithm are $\gamma_{acc} = 0.01$ and $\gamma_{\varphi} = 0.3$.

In *Figure 5.6* the results of this experiment are shown. Each image shows one of the 23 prevision computations: each square represents a location, the location color is related to the assigned probability (a darker color means a higher probability that it will represent the user future position) and the little arrow represents the actual position and movement direction of the user.

With a "to 60 seconds" prevision it is possible to obtain a satisfactory prediction of the user's position in the next 40–80 seconds. However because of the granularity chosen in space subdivision, we have observed that it is difficult to have a prediction with accurate timings. A more detailed space subdivision would have signified a more detailed knowledge of the road shape and therefore a more detailed prediction.

5.4.3 Destination Prediction

In this experiment we have applied our ARDA algorithm to a set of positions and movement vectors recorded along the way from university (location (5, 0)) to home (location (2, 10)). The aim of this experiment is to predict the final destination of the user, so the computation stops only when the speed of the user in the simulation is 0.

We have computed the locations importance using SpaceRank with d=0.05, therefore considering all location where user has been (*Figure 5.5b*). On the basis of position and speed values obtained from the real data, we have experimentally seen that the most suitable values for the parameter in the ARDA algorithm are γ_{acc}=0.05 and γ_{φ}=0.3. As the starting location is important, we introduced a factor to smooth the starting location attraction, as described in *Section 5.3.3.4*.

In *Figure 5.7* the results of this second experiment are shown. The images have to be interpreted as in *Figure 5.6;* however while in *Figure 5.6* the user position in the next minute is predicted, in *Figure 5.7* the user final destination is predicted.

In general, especially during the first part of the itinerary, we have noticed a strong dependence between the user direction in the instant of the prediction and the obtained prediction result. This is due to the SpaceRank algorithm that assigns to the location where the user moves through importance values very near to the values assigned to the locations where the users stops. In this case, with a low γ_{acc} value there is not enough force to attract the user towards the final destination, and with a high γ_{acc} value the prediction is disturbed by the accelerations related to the locations in the itinerary.

To overcome this limitation, we have performed a further experiment, with a different importance index. In this case, we assign more importance to locations where the user usually stops *(speed = 0),* decreasing the values related to the locations where the user just passes through *(Figure 5.5c).* This version of the algorithm improves the prediction of the final destination. In particular we can notice that even in the first part of the itinerary, the algorithm assigns high probability values to the

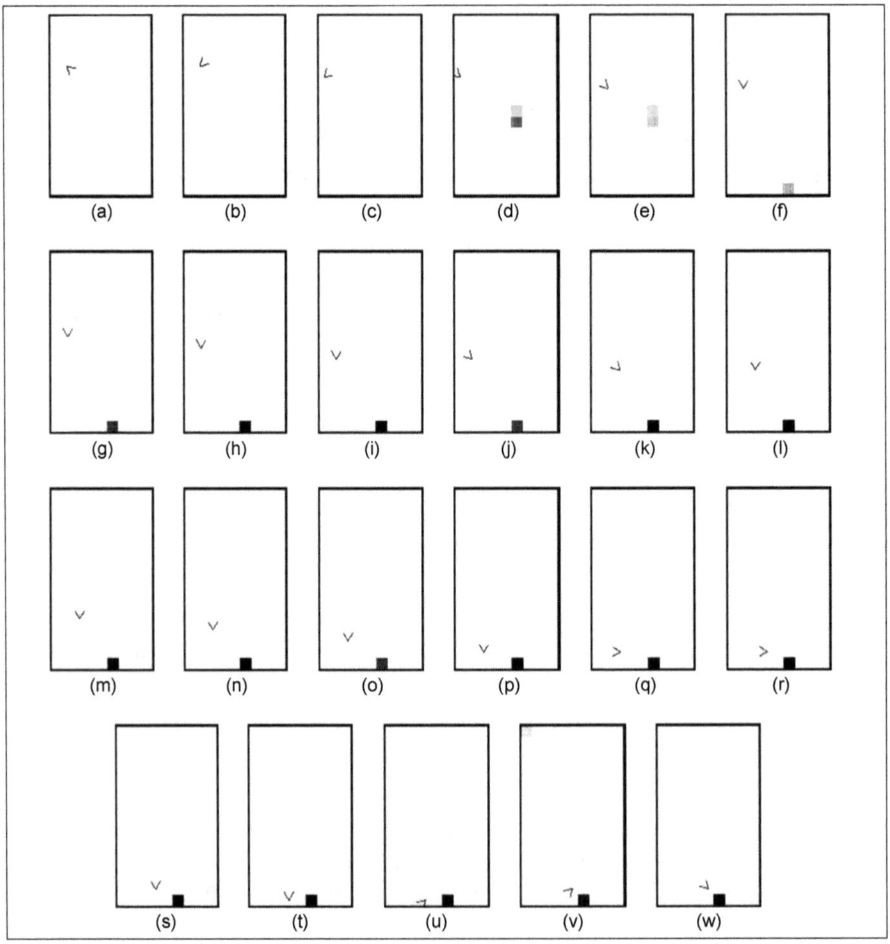

Fig. 5.7. Results obtained predicting the user final destination.

location of effective destination. In particular it returns good results, comparable to the other algorithms in literature. Moreover, if the final destination is an important location far from other important locations, our algorithm assigns it a high probability; on the contrary if the final destination is an important location near other important locations, our algorithm tends to equally distribute the probability among these locations.

5.5 Conclusions and Future Work

We have emphasized how the prediction of a user's future location and destination is gaining importance due to actual pervasiveness of mobile devices in conjunction with growing people's need of location-based information systems and services. Moving from this point, we have proposed an algorithm to predict future user locations. The algorithm has two steps: (i) computation of the importance of the locations, based on geographic properties and user habits; and (ii) prediction of future locations, based on physics models. Our approach is different from the existing solutions, it is intuitively reasonable and, accordingly to preliminary experimental results, it seems effective.

In the future we plan to compare our algorithms with other existing approaches both for importance estimation and for trajectories previsions. Moreover, we intend to study the problem of understanding human behaviors and habits at broader and deeper levels. We plan to mine location-based logs and discover knowledge in these data, in order to better estimate places importance and trajectories not only in real world but also in virtual worlds like the web graph, search engines networks, social networks, and online role play games. For instance, interesting problems would be destination prediction in SecondLife, in Web browsing, in search engine querying, etc.

Our future studies will also focus on what is "importance", how "global importance" is different from "personal importance", how we can estimate places "importance" and how we can use these information in the development of novel context-aware applications both for mobile devices and networking tools.

Finally, we intend to improve our trajectories prediction algorithm both by including alternative importance information sources, such as personal calendar or friends position, and by considering others physics' fields, such as weather forecast, in search of better approaches and models.

References

Ashbrook D, Starner T (2003) Using GPS to learn significant locations and predict movement across multiple users, *Personal and Ubiquitous Computing* 7(5), 275–286.

Bierlairea M, Frejinger E (2008) Route choice modeling with network-free data, *Transportation Research Part C: Emerging Technologies* 16, 187–198.

Chan J, Zhou S, Seneviratne A (1998) A QoS adaptive mobility prediction scheme for wireless networks, *in* Global Telecommunications Conference, 1998. GLOBECOM 98. The Bridge to Global Integration. IEEE, Vol. 3, pp. 1414–1419.

De Sabbata S, Mizzaro S, Vassena L (2008) SpaceRank: Using PageRank to estimate location importance, *in* Proceedings of ECAI '08 Workshop on Mining Social Data (MSoDa '08), pp. 1–5, University of Patras, Greece.

Eagle N, Pentland AS (2006) Reality mining: sensing complex social systems, *Personal and Ubiquitous Computing* 10(4), 255–268.

Gonzalez MC, Hidalgo CA, Barabasi AL (2008) Understanding individual human mobility patterns, *Nature* 435(7196), 799–782.

Hazas M, Scott J, Krumm J (2004) Location-aware computing comes of age, *Computer* 37(2), 95–97.

Krumm J, Horvitz E (2006) Predestination: Inferring destinations from partial trajectories, *in* Lecture Notes in Computer Science, Vol. 4206, Springer, pp. 243–260.

Mountain DM (2005) Exploring mobile trajectories: An investigation of individual spatial behaviour and geographic filters for information retrieval, PhD thesis, City University.

Page L, Brin S, Motwani R, Winograd T (1998) The PageRank citation ranking: Bringing order to the web, Technical report, Stanford Digital Library Technologies Project.

Taylor MAP, Woolley JE, Zito R (2000) Integration of the global positioning system and geographical information systems for traffic congestion studies, *Transportation Research Part C: Emerging Technologies* 8, 257–285.

Zaidi ZR, Mark BL (2005) Real-time mobility tracking algorithms for cellular networks based on kalman filtering, *IEEE Transactions on Mobile Computing* 4(2), 195–208.

6 Ways of Walking – Developing a Pedestrian Typology for Personalised Mobile Information Systems

Alexandra Millonig[1,2] and Georg Gartner[1]

[1] Department of Geoinformation and Cartography
Vienna University of Technology, Austria
[2] Human Centered Mobility Technologies, arsenal research, Vienna, Austria

Abstract

In recent years, technological progress and an increasing amount of ubiquitously available information set the stage for the development of mobile navigation tools for pedestrians. However, the vast quantity of accessible navigational and environmental information aggravates effective information extraction and necessitates tailoring wayfinding instructions and additional location based information to individual needs. In order to facilitate the provision of customised information and to avoid redundant information, we currently determine a pedestrian typology using a multi-method approach considering motion behaviour as well as underlying preferences and individual attitudes. We developed a methodological set-up including qualitative-interpretative and quantitative-statistical data, which will lead to the determination of a typology of lifestyle-based pedestrian mobility styles. In this contribution we present results from the first of two consecutive empirical phases based on datasets of over 100 trajectories observed by shadowing methods and 130 interviews; furthermore we highlight differences in the outcomes resulting from data collected by different empirical methods and in different investigation areas (indoor and outdoor).

Keywords: pedestrian navigation, spatio-temporal behaviour, methodological triangulation, typology

6.1 Introduction

Recent years have seen raising interest in mobile navigational services for pedestrians. Especially the parallel increase of ubiquitously available environmental information offers the possibility to provide walking users with useful information concerning optimal routes and interesting facilities in vicinity. However, the consideration of "optimal routes" and "useful information" varies with a number of factors like context, time constraints, walking preferences, or individual interests and attitudes. Findings in pedestrian behaviour research have proven that pedestrians – especially when having plenty of time – favour different route qualities than simple shortness, e.g. attractiveness, convenience, safety, simplicity (fewest turns), or novelty (difference from previous routes) (Golledge 1995, Helbing et al. 2001, Thomas 2003, Millonig and Schechtner 2007a).

Successful wayfinding and information services have to take these factors into account in order to efficiently support pedestrian wayfinding and to avoid potential information overload. Hence, it is necessary to comprehensively understand what influences human spatio-temporal behaviour. To provide a basis for developing customised information provision in mobile information services for pedestrians, a current research project aims at the determination of a typology of lifestyle-based pedestrian mobility styles, exploring the subject from different perspectives by the use of various complementary empirical methods. The outcomes are of great interest for many different research fields: Results will disclose coherences between walking patterns and related self-perceptions of walking behaviour, form the basis for personalising navigational and environmental information provision, and can be used for determining type-related parameters in order to develop realistic agent-based pedestrian simulation models.

The study comprises a multi-stage approach employing a combination of qualitative-interpretative and quantitative-statistical methods in order to gain comprehensive insight to human spatio-temporal behaviour and decisive influence factors. In this contribution the results of the completed first empirical phase of the project are presented. We describe initial typologies we have compiled from statistical analysis of observation and interview data, compare results based upon discriminative empirical data collection techniques and point out similarities and differences in observed indoor and outdoor behaviour.

6.2 Pedestrian Behaviour Monitoring

The investigation of human spatio-temporal behaviour has already been focused on in several different scientific approaches, following diverse purposes and applying a variety of methodological techniques. Relevant fields of interest include for example tourism research, tracking people for security purposes, monitoring evacuation behaviour, or the development of design elements of pedestrian facilities, planning guidelines or navigation and guiding systems (Helbing et al. 2001, Millonig and Schechtner 2007b, O'Connor et al. 2005). Many approaches are aiming at the analysis of pedestrian behaviour in specific context situations; a lot of effort has been made for example to explore the behaviour of visitors of museums and exhibitions in order to unveil behaviour patterns and to design wearable guides and "augment" the visitor's experience in a personalised way (Kanda et al. 2007, Sparacino 2002). Other research groups concentrate on the investigation of tourist behaviour (Hartmann 1988, Keul and Kühberger 1997, O'Connor et al. 2005) or walking patterns of shoppers (Koike et al. 2003, Kitazawa et al. 2003).

First attempts of analysing pedestrian behaviour have mainly been carried out by using direct observation and questionnaire methods (e.g. Hill 1984). Still interview techniques belong to the most commonly used methods in pedestrian behaviour research. With emerging positioning technologies like the Global Positioning System (GPS) or land-based technologies (e.g. cell identification of mobile phones, RFID, Bluetooth) new approaches have been developed in order to overcome methodological drawbacks of conventional techniques, as direct observations are very time-consuming and only allow for small samples, and questionnaire data reflect subjective impressions which may be incorrect or constructed ex post – especially as human behaviour is never fully determined by verbalised structures (Nisbett and Wilson 1977) and people tend to adapt their answers (consciously or subconsciously) to what they expect to be socially desired behaviour (Esser 1985). For a general overview of applicable methods and related studies see Millonig and Gartner (2007).

Nevertheless, each empirical method possesses its specific strengths and limitations. Hence, a combination of at least two complementary methods is recommended in order to maximise the benefits from different techniques (Janssens et al. 2007, Fielding and Schreier 2001). In our project, we use a combination of several qualitative and quantitative methods following the concept of "across-method" triangulation (Jakob 2001).

6.3 Current Triangulation Approach

The study starts with two consecutive stages of empirical data collection; *Figure 6.1* shows the study design including the specific methods we use in each empirical phase of the project.

In the heuristic phase unobtrusive observations and standardised interviews are used to identify basic types of pedestrian spatio-temporal behaviour, whereas the deductive phase of the study is to verify the previously defined initial typology by combining localisation and detailed semi-standardised interview techniques. Results

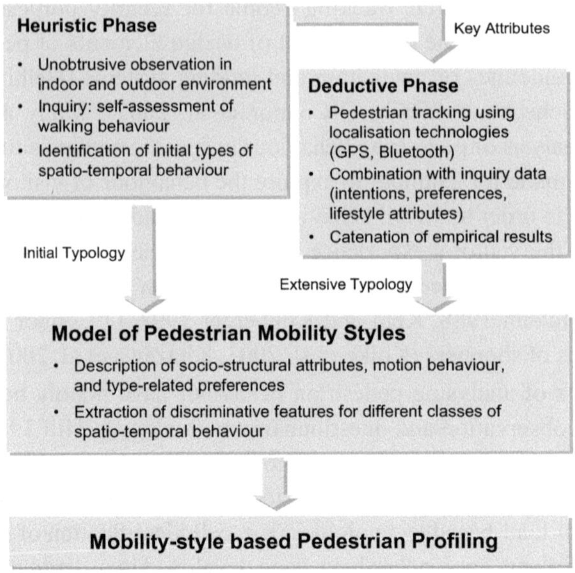

Fig. 6.1. Methodology.

of both empirical phases are subsequently related to each other, which leads to the development of a typology of pedestrian walking styles and underlying interests and preferences. Discriminative features will be determined and extracted in order to create pedestrian spatio-temporal profiles which can be used for customising spatial information in a mobile wayfinding system.

Following the assumption that the individual behaviour of a person is to a certain extent influenced by the context a person is acting within, all investigations are conducted in shopping environments (a shopping mall and a shopping street). This is to minimise the risk of observing differences in behaviour that are largely caused by the influence of various context situations. The outcomes of the investigations will later be tested with regard to their validity in other context situations.

The analysis of data collected during the heuristic phase has been completed. Identified key factors have been used to specify the methods applied in the currently ongoing heuristic phase of data collection. We now particularly explain the data collection techniques we used and exemplify results from the analysis of collected data.

6.4 Heuristic Phase

Tracking (also known as "unobtrusive observation" or "shadowing") has been one of the two essential empirical methods used during the first phase of the project. The other applied method consisted of brief standardised interviews which have not been taken simultaneously, as a combination of both techniques could not be realised at this stage of the study (due to the fact that subjects were to be followed as long as possible and it was not feasible to predefine the optimal point in time for terminating the observation and approaching the individual to ask for an interview).

6.4.1 Data Collection: Tracking

The unobtrusive observation method applied in the first phase of the study consisted of random selection of an unaccompanied walking person and following the individual as long as possible while mapping her path on a digital map. For a detailed description of the observation procedure see Millonig and Gartner (2007). The use of technology in this phase (digital map on a tablet PC, tracking software) offers mainly two major advantages: Firstly, a large investigation area can be covered without having to handle a large paper map, and secondly all points drawn in the map are recorded with time-stamps and map coordinates, which allows the calculation of average speeds and the detection of stops for each trajectory. In total trajectories of 111 individuals with a balanced gender and age ratio have been collected (57 observations on the shopping street, 54 in the shopping mall). Outdoor observations had an average length of 12 minutes (maximum: 62 minutes), indoor observations lasted approximately 16.5 minutes on average (maximum: 57 minutes).

6.4.2 Data Collection: Interviews

Interviews offer the only chance to gain insight to intentions, motives, and other determinants influencing human spatial behaviour. During the heuristic phase, brief standardised face-to-face interviews have been held. The questionnaire consisted of questions concerning sociodemographic attributes (age, gender, education, profession, etc.), goals and time budget for the current activities in the study site, frequency of visits, and questions referring to individual walking habits. Participants were asked to give a self-assessment of their preferences concerning walking behaviour and walking environments (e.g. rather "slow" or "fast", rather "exploring" or "goal-oriented", rather "bustling environments" or "calm environments").

In total, 130 individuals have been interviewed in the heuristic phase of the study; 100 interviews have been conducted in the indoor environment, 30 interviews have been given by visitors of the outdoor shopping area, as pedestrians walking the shopping street were conspicuously less willing to give an interview. The majority of the participants were female (61.5%). The sample possessed a balanced age distribution with an average age of around 36 years.

6.5 Results

The following subsections describe the results we obtained from performing multivariate analysis algorithms. Motion data were analysed with respect to velocities (histograms) and stops (frequency, duration, and position of stops), additionally visual attributes like appearance (dress style) have been regarded. Interview data referring to individual walking preferences have been analysed in order to derive classes of self-reported subjective behaviour.

6.5.1 Tracking Results

The collected datasets have been analysed according to the velocity computed between each marked point in the observed path, additionally locations and durations of stops within the trajectory have been detected (cf. *Figure 6.2)*. Subsequently, speed histograms of each trajectory have been compiled, showing the proportional amount of time an individual walked at a velocity within a specific time interval.

In order to gather initial classes of behaviour, the histograms of each investigation area have subsequently been classified using clustering algorithms (hierarchical clustering and k-means algorithm). In a second step, lifestyle-related factors (visual appearance, categories of visited facilities) have been included besides the speed

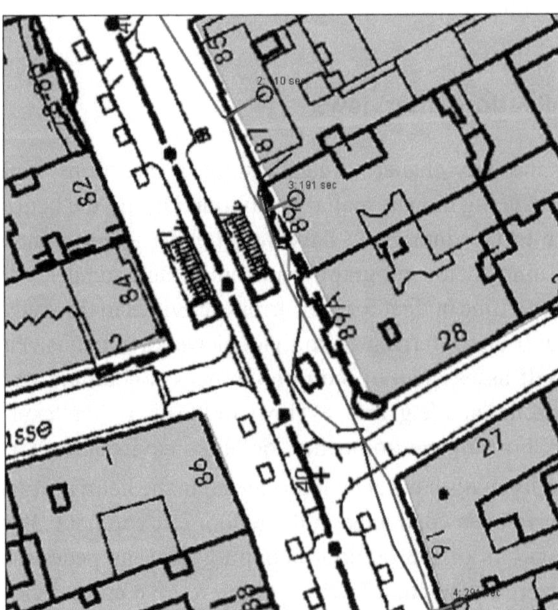

Fig. 6.2. Velocities and stops detected from a typical trajectory (outdoor area).

histograms. Interestingly, the lifestyle-related factors showed little impact on the results; the dendrograms illustrating the arrangement of the clusters produced by hierarchical analysis showed only marginal differences, and the clusters resulting from k-means clustering did not differ from the results based on speed histograms only. Hence, lifestyle-related factors that are determined solely by observation show no significant influence on the identification of specific classes. This may be based on two factors: Firstly, as observations took place in summertime during a fairly hot period, people tended to dress lightly and differences in style and taste were not so obvious. Secondly, the interpretation of individual interests by analysing visited shops and facilities is deficient, as the observed behaviour does not necessarily represent actual preferences, and current constraints influencing individual behaviour are completely unknown. To overcome this issue, the second phase (deductive phase) of the study will combine tracking data with interviews including lifestyle-related factors (see *Section 6.7).*

We identified three discriminative clusters for the indoor observations, and eight clusters for the outdoor observations. However, half of the clusters derived from the outdoor analysis consist of only a very small number of people; hence these classes have only limited significance. In *Table 6.1* the clusters compiled from the indoor analysis and the four main clusters derived from the outdoor analysis are juxtaposed. The following subsections present detailed descriptions of the clusters.

Table 6.1. Tracking-cluster descriptions.

Cluster		TI-1	TI-2	TI-3	TO-1	TO-2	TO-3	TO-4
No. of subjects		10	14	30	7	15	13	14
Gender:	female	40%	35.7%	66.7%	71.4%	53.3%	69.2%	14.3%
	male	60%	64.3%	33.3%	28.6%	46.7%	30.8%	85.7%
Age:	< 30	50%	35.7%	20%	57.1%	33.3%	30.8%	28.6%
	30–60	50%	64.3%	63.4%	42.9%	60%	61.6%	71.4%
	> 60	0%	0%	16.6%	0%	6.7%	7.7%	0%
Average age		25–30	30–35	35–40	25–30	35–40	30–35	30–35
Average speed		1.19 m/s	0.61 m/s	0.24 m/s	0.25 m/s	0.63 m/s	1.04 m/s	1.35 m/s
Av. no. of stops (max.)		0.3 (3)	1.36 (2)	3.57 (13)	4.14 (11)	2.93 (7)	2.23 (5)	0.71 (4)
Av. duration of stops		6.97 s	2.58 min	4.66 min	4.43 min	1.83 min	32.97 s	11.34 s
Fashion style		casual or convenient	casual or conservative	casual/ trendy or elegant	convenient or casual	conservat./ elegant (f); convenient /casual (m)	casual/ convenient or elegant/ proper	casual, trendy
Visited shops/facilities		food store	no main focus	fashion, specialities, drugstore, bookshop	fashion and accessories	no main focus	tendency to specialised /exclusive shops	no main focus

6.5.1.1 Indoor Clusters

The classification process resulted in three homogeneous clusters for the indoor datasets (containing 10, 14, and 30 individual observations per cluster). The cluster containing the largest number of cases (55.6% of all observed individuals in the shopping centre) shows a high percentage of time spent in front of or inside a shop (77.8%). Individuals falling in this category on average also walk at a lower speed than persons related to the other classes. The behaviour patterns in each of the three clusters can be interpreted as "swift shoppers", "convenient shoppers", and "passionate shoppers":

- *Tracking-indoor cluster TI-1* ("swift shoppers", 10 cases) contains 60% male participants and represents the youngest group with 50% of all subjects younger than 30 years. They walk at comparatively high speed (1.19 m/s on average; people most frequently walked at a speed of about 1.3 to 1.4 m/s) and stop rarely, for a very short time and mainly at supermarkets. Women belonging to this group generally dress fashionably casual; men appear to prefer a convenient and comfortable style. Apparently this group represents people following their daily routine who have little time or dislike extensive shopping.
- *Tracking-indoor cluster TI-2* ("convenient shoppers", 14 cases) includes almost two thirds of male subjects (64.3%); the participants are aged around 30 to 35 years with 64% younger than 40. They stop rather shortly and moderately often (on average 1.36 times for about 2.6 minutes); the person spending most time in or in front of different facilities spent about 8.2 minutes at each stop. They walk slower than the individuals belonging to the first cluster: Most frequently the individuals walked at a velocity between 1.2 and 1.3 m/s. The behaviour patterns observed within this cluster appear to illustrate mainly male shopping behaviour: Men belonging to this cluster show interest in a greater variety of different shops and facilities than women.
- *Tracking-indoor cluster TI-3* ("passionate shoppers", 30 cases) contains individuals who stop fairly often in front of or inside a shop or other facility (approximately 3.57 times, up to 11 times). One stop lasts on average 4.66 minutes (maximum: 17.37 minutes). The great number of stops results in a very low average speed (about 0.24 m/s); most frequently the subjects walked at a speed of about 1 to 1.1 m/s. Two thirds of the subjects belonging to this cluster are female. The estimated age lies beneath 35 years for most of them (56.7%), but the cluster also contains a greater number of elder individuals (26.6% aged between 55 and 70 years). Male subjects appear to prefer a comfortable or casual fashion style, younger female individuals mainly dress trendy or casually, and elder females prefer a traditional, classic or elegant style. Apparently this cluster comprises a large amount of people who are quite keen on shopping or simply have enough time to spend, like to stroll around and to visit a large number of shops, and pause

from time to time to have a refreshment. Most interest is attracted by fashion shops (clothing, shoes, and accessories) as well as shops offering culinary specialities.

6.5.1.2 Outdoor Clusters

The analyses of outdoor dataset produced eight clusters, with a vast majority (86%) of observations belonging to four classes. The other four clusters solely comprise a number of one, two or three subjects; basically people who could not be observed for a very long time. Individuals related to these clusters (TO-5 to TO-8) predominantly walk at a rather high average speed (up to 1.35 m/s), stop either not at all or rarely and for a relatively short time. The difference to the four larger clusters lies mainly in either unusual values of lifestyle-related factors (e.g. explicitly unconventional appearance) or uncommon motion behaviour (e.g. alternating very different speed levels). Clusters TO-1 to TO-4 partly show similarities with clusters compiled from the indoor analysis. To a certain extend all three indoor clusters can be identified when analysing the outdoor behaviour; additionally a cluster of specific behaviour patterns has been observed.

- *Tracking-outdoor cluster TO-1* ("passionate shoppers", 7 cases) seems highly reminiscent of cluster TI-3. It mainly consists of female participants (71.4%) who walk at comparatively low speed on average (0.25 m/s), stop quite frequently (about 4.14 times per observation, up to 11 times), and stay for relatively long time (on average 4.43 minutes). Subjects are however significantly younger than those in cluster TI-3 (on average 25-30 years), seem to prefer a casual or convenient style, and stop most frequently at fashion shops.
- *Tracking-outdoor cluster TO-2* ("convenient shoppers", 15 cases) shows most similarities with indoor cluster TI-2. The gender ratio is quite balanced, just as well as the age distribution. Their average speed is moderate (0.63 m/s), and subjects are either predominantly dressed casual (males) or conservative (females). Still, their stopping behaviour differs from behaviour observed in TI-2: Individuals related to TO-2 stop more frequently (2.93 times, up to 7 times), but for a shorter average duration (1.58 minutes).
- *Tracking-outdoor cluster TO-3* ("discerning shoppers", 13 cases) again consists of mainly female subjects and shows a rather balanced age distribution. Individuals walk at comparatively high speed (approximately 1.04 m/s), but still stop almost as frequently as people belonging to cluster TO-2 (on average 2.23 times, up to 5 times), although the stops are very short on average (32.97 seconds). To a large extend subjects within this cluster prefer either casual and convenient style or dress elegant and proper. Regarding the visited shops and facilities, a slight tendency towards specialised and exclusive shops is observed.

- *Tracking-outdoor cluster TO-4* ("swift shoppers", 14 cases) reminds of indoor cluster TI-1; in this case the amount of male individuals is very dominant (85.7%). The average age lies by 30-35 years, no subject appears to be older than 60. Participants walk very fast (1.35 m/s) and stop rarely and shortly (approximately 0.71 times per observation, up to 4 times; average duration: 11.34 seconds). Their fashion style is mainly casual or trendy, and no main focus could be identified regarding the categories of shops they visited.
- *Tracking-outdoor clusters TO-5 to TO-8* (number of cases: 1, 3, 2, 2) constitute outliers in measured walking behaviour. TO-5 includes only one subject walking at very high speed without stopping; the observation apparently did not take long enough to unveil significant patterns. TO-6 consists of three individuals showing comparable behaviour to cluster TO-2, but stop significantly shorter (on average 23.89 seconds). The two individuals in cluster TO-7 walk at very high speed (1.22 m/s), but do not stop at all. Finally, the two females belonging to cluster TO-8 appear to be quite similar to subjects observed in TO-3, but stop less frequently and shorter.

6.5.2 Interview Results

The self-assessment profile interrogated in the interview contained 17 pairs of opposed attributes. To simplify classification processes, common data reduction methods have been applied in order to obtain a smaller number of variables. However, responses turned out to appear quite arbitrary: Although some of the items had very similar meanings, no significant correlation could be observed. This confirms the assumption that pedestrians have only little knowledge about their own behaviour and seem to guess what they prefer. Data reduction processes resulted in downsizing the total number of variables to 9 factors; 5 of them had to be excluded from the Principal Component Analysis (PCA) as they showed no correlations at all, the remaining 12 variables could be reduced to 4 factors including factors related to motion behaviour, personality-related factors, preferences concerning environmental qualities, and shopping behaviour patterns. Subsequently, the profiles were classified using hierarchical and k-means clustering algorithms. For each of the observation areas the analysis resulted in 4 clusters. Some of the original datasets had to be excluded from classification during hierarchical clustering, as they showed too little similarities with other datasets. In general, the classes derived from the clustering process do not show very distinct patterns when regarding the reported self-assessment of walking preferences. This is presumably caused by the previously discussed limitations of interview data analysis. *Table 6.2* juxtaposes the main results for each class compiled from classifying indoor and outdoor questionnaire datasets.

Table 6.2. Interview-cluster descriptions.

Cluster		QI-1	QI-2	QI-3	QI-4	QO-1	QO-2	QO-3	QO-4
No. of subjects		22	18	28	30	5	6	5	13
Gender:	female	45.5%	77.8%	63.4%	63.3%	100%	50%	40%	53.8%
	male	54.5%	22.2%	35.7%	36.7%	0%	50%	60%	46.2%
Age:	< 30	45.5%	44.4%	35.7%	60%	60%	83.3%	20%	46.2%
	30–60	50%	38.9%	39.3%	20%	40%	16.7%	60%	38.5%
	> 60	4.5%	16.7%	25%	20%	0%	0%	20%	15.4%
Average age		33.7	38.1	40	33.5	30.4	27	42	37.2
Level of education:									
compulsory school		18.2%	33.3%	28.6%	36.7%	40%	0%	0%	23.1%
grammar school		72.7%	38.9%	46.4%	50%	20%	83.3%	100%	46.2%
higher education		9.1%	27.8%	21.4%	13.3%	40%	16.7%	0%	30.8%
other		0%	0%	3.6%	0%	0%	0%	0%	0%
Self-assessment of walking behaviour (preferences)		following own way, hardly anxious, little curiosity	curious, hardly anxious, weak orienta-tion skills	curious, self-deter-mined, fast, own way, hardly anxious, strong orienta-tion skills	fast, curious, modern, self-deter-mined, strong orienta-tion skills	weak orienta-tion skills, own way, self-deter-mined, modern, fun-oriented	not anxious, short distances, swift, curious, modern, own way, planning	not curious, own way, urban and busy envi-ronments, long distances, not anxious, flexible, quality-conscious, complexity, strong orientation skills	strong orientation skills, not anxious, curious, long distances, slow, flexible, price-conscious, simplicity

Although the use of questionnaires is very popular in investigating spatio-temporal behaviour and walking preferences, these results do not confirm the opinion that interviews alone belong to the most appropriate methods in pedestrian behaviour research. Some of the clusters show similarities to classes identified by analysing trajectories; others seem to have little in common with patterns described in tracking clusters. The following section will highlight the differences between indoor and outdoor results as well as tracking and interview outcomes.

6.6 Discussion and Comparison of Results

As shown in *Section 6.5,* especially the analysis of tracking data results in several distinctive classes of behaviour. Outdoor results bear resemblance to indoor outcomes, although several significant differences have been identified.

A comparison of all outdoor and indoor histograms illustrates that individuals observed in the indoor environment spend significantly more time stopping (in front

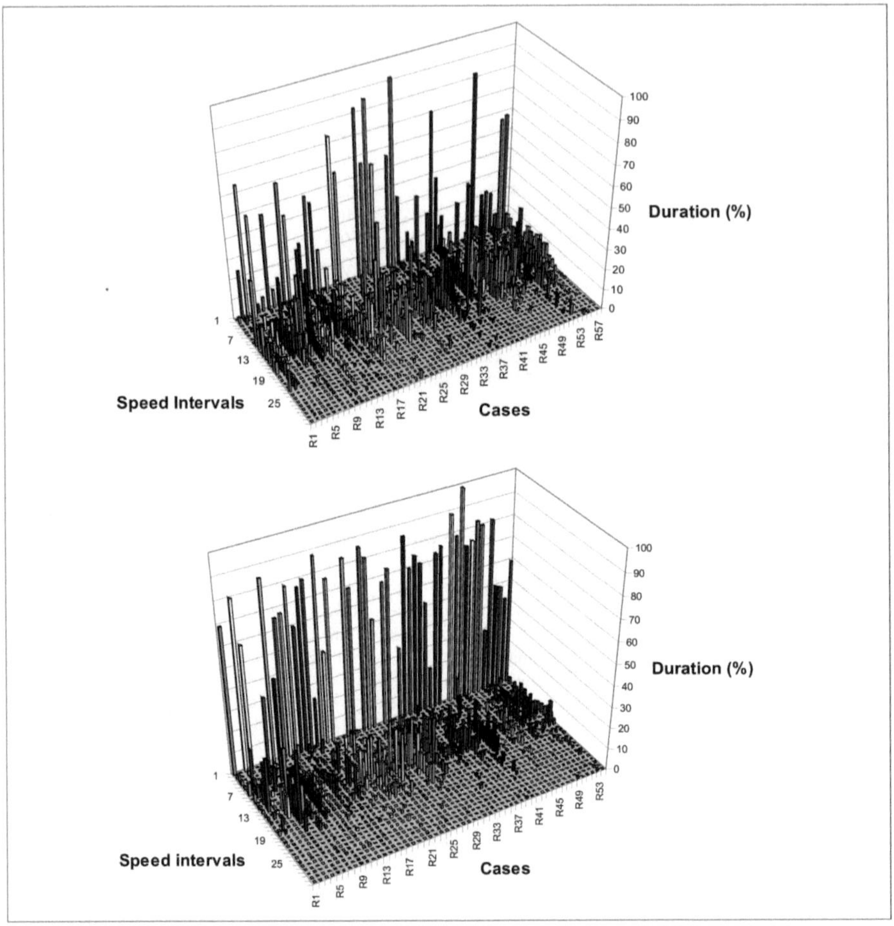

Fig. 6.3. Histograms of outdoor (above) and indoor (below) observations.

of or inside a shop or other facility) and walk in general at lower speed than subjects observed in the outside area. This is a first indicator for different behaviour in a shopping mall compared to a shopping street, which may be caused by the fact that a shopping mall is usually visited by shoppers only, while a street (even if a great number of shops is located there) is used by people following a greater variety of intentions, hence a greater number of context situations can be assumed. *Figure 6.3* shows diagrams consisting of all histograms compiled from indoor and outdoor observations (speed intervals: 0.1 m/s steps, 30 intervals; duration: proportional amount of time spent walking within a speed interval).

During the classification process, outside observations produce a greater number of different classes of behaviour than indoor observations. Although still many individuals are spending a considerable amount of time inside a shop or standing in front of it, more time is spent walking at a greater number of different speed levels than in the indoor observation field. These differences in behaviour also confirm consequences caused by a greater variety of context influences: As the outside investigation area consists of an urban street, observed individuals might have aimed for other objectives than shopping. A person who enters a shopping mall, however, seldom pursues other goals than shopping, which leads to the identification of a smaller number of discriminative behaviour categories.

A comparison of indoor and main outdoor clusters shows strong similarities; the analysis of outdoor observations additionally produced one behaviour cluster which could not be identified in the indoor environment. Without knowledge concerning underlying motivations and intentions it is not possible to give a sound interpretation of these differences. The combination of observations and interviews in the following deductive phase is to overcome these limitations.

The analysis and classification of self-assessment profiles in interviews could not reproduce the clusters derived from tracking datasets. The reported behaviour preferences for each cluster differ only marginally from each other. A comparison of indoor and outdoor results unveils slight resemblances between individual clusters: Cluster QI-1 reminds of cluster QO-3, cluster QI-4 shows certain similarities with cluster QO-2, cluster QI-3 is somehow similar to cluster QO-4, and finally cluster QI-2 has most in common with cluster QO-1. Still, both differences and similarities are quite small, and hence conclusions are limited. This applies accordingly to comparisons between tracking and interview results; although some of the interview clusters seem to be comparable to tracking clusters, similarities do not possess satisfactory significance.

6.7 Outlook: Deductive Phase

The analysis of data collected in the currently ongoing deductive phase of the project aims at the expansion of insights gained from the heuristic phase. The interpretation of results will include current findings in mobility behaviour research. Although recent research focusing on time geography and travel behaviour mainly examine the overall mobility behaviour regarding all transport modes (e.g. Flamm and Kaufmann 2006, Axhausen et al. 2002), aspects of data interpretation of related studies will be used to broaden the knowledge gained in this project.

During this phase, pedestrians are tracked with the help of localisation technologies: Bluetooth for the indoor environment, GPS in the outdoor area. Participating individuals are also asked to give a semi-standardised interview; hence observable behaviour can be interpreted in consideration of survey data. Results of the previous heuristic phase are used to specify observation methods and questionnaires. The method comprises the collection of data by equipping participants with tracking devices and continuously recording their position, velocity, and moving direction. After returning the device to the observer, participants have to complete a semi-standardised face-to-face-interview which explores current intentions, attitudes, and lifestyle and socio-structural attributes. The questionnaire has largely been inspired by a concept developed by German sociologists researching on the conduct of everyday life. Based on qualitative methods, the subjectively perceived daily life of people is classified according to their beliefs, feelings, and habits in order to position groups of people with regard to social status and basic values (tradition, modernisation, or re-orientation) (Krason et al. 2003). In combination with collected motion data, relations between lifestyle characteristics and motion behaviour are investigated. This combination of observation techniques with interviews offers two major advantages: Firstly, inaccuracies in observations can be validated with the help of interview responses, and secondly, distortions in the reported self-assessment of motion behaviour can be identified by analysing motion data.

6.8 Conclusions

The results of the first empirical phase indicate that a number of homogenous behaviour patterns can be identified, especially in consistent context situations. Data collected by observation methods produce more distinct results than interview data. Although multivariate analysis of tracking data results in different quantities of discriminative clusters for indoor and outdoor environments, significant similarities between particular clusters could be identified. Currently ongoing investigations using a non-disguised form of observation combined with detailed interviews include and test basic findings of the first analyses. The combination of several complementary empirical techniques is a very promising approach to gain comprehensive insight to human spatio-temporal behaviour patterns, even though some limitations have to be accepted. Although technological progress has led to the development of new tools for monitoring and analysing pedestrian behaviour, data collection is still quite laborious, as pedestrians have to be persuaded to participate in the study. Other limitations are caused by the costs of tracking equipments: To cover a large investigation area, which is necessary to consistently track pedestrians over a sufficient amount of time, a considerable amount of material has to be supplied.

Further empirical analyses of more data during the currently ongoing second empirical phase as well as a careful examination of the results in different context situations during the final stage of the study are expected to lead to a comprehensive interpretation of pedestrian spatio-temporal behaviour. This can on the one hand be used in future mobile navigation services to provide customised route suggestions and location based information, and on the other hand serve as a basis for determining parameters for pedestrian simulation models.

Acknowledgements

This work is part of the "UCPNavi" project, a cooperation project between the Vienna University of Technology and arsenal research, Vienna. The project is supported by the Austrian Funds for Scientific Research (FWF). The authors would like to thank M. Ray (arsenal research) for developing the shadowing tool; furthermore the authors want to thank the team at salzburg research for developing and providing the Bluetooth tracking tool. Special thanks go to N. Brändle (arsenal research) for his help and advice concerning data analysis. The digital map used in *Figure 6.2* has been provided by Stadt Wien – ViennaGIS (www.wien.gv.at/viennagis/).

References

Axhausen KW, Zimmermann A, Schönfelder S, Rindsfüser G, Haupt T (2002) Observing the rhythms of daily life: a six-week travel diary. *Transportation.* 29 (2), 95–124.

Esser H (1985) Befragtenverhalten als „rationales Handeln" – Zur Erklärung von Antwortverzerrungen in Interviews. *ZUMA-Arbeitsbericht.* 85/01, Zentrum für Umfragen, Methoden und Analysen, Gesellschaft Sozialwissenschaftlicher Infrastruktureinrichtungen (GESIS).

Fielding N, Schreier M (2001) *Introduction: On the Compatibility between Qualitative and Quantitative Research Methods.* Available: http://www.qualitative-research.net/fqs-texte/1-01/1-01hrsg-e.htm. Last accessed 03 July 2008.

Flamm M, Kaufmann V (2006) Operationalising the Concept of Motility: A Qualitative Study. *Mobilities.* 1 (2), 167-189.

Golledge RG (1995) Defining the criteria used in path selection, *Technical Report UCTC No. 78*, University of California Transportation Center.

Hartmann R (1988) Combining Field Methods in Tourism Research. *Annals of Tourism Research.* 15, 88-105.

Helbing D, Molnár P, Farkas IJ, Bolay K (2001) Self-organizing pedestrian movement. *Environment and Planning B: Planning and Design.* 28, 361-383.

Jakob, A. (2001) *Möglichkeiten und Grenzen der Triangulation quantitativer und qualitativer Daten am Beispiel der (Re-) Konstruktion einer Typologie erwerbsbiographischer Sicherheitskonzepte.* Available: http://www.qualitative-research.net/fqs-texte/1-01/1-01hrsg-e.htm. Last accessed 03 July 2008.

Janssens D, Hannes E, Wets G (2007) Planning interventions in the interactions between individual activity patterns, patterns of functions and infrastructure. Position paper, *Urbanism On Track – Expert meeting on the application in urban design and planning of GPS-based and other tracking-based research.* Delft, NL.

Kanda T, Shiomi M, Perrin L, Nomura T, Ishiguro H, Hagita N (2007) Analysis of People Trajectories with Ubiquitous Sensors in a Science Museum. *Proceedings IEEE Internation Conference on Robotics and Automation (ICRA 2007)* Rome, Italy, 2007.

Keul A, Kühberger A (1997) Tracking the Salzburg Tourist. *Annals of Tourism Research.* 24, 1008-1012.

Kitazawa K, Zhao H, Shibasaki R (2003) A Study for Agent-based Modeling of Migration Behavior of Shoppers. *Proceedings of the 8th International Conference on Computers in Urban Planning and Urban Management.*

Koike H, Morimoto A, Inoue T, Kawano T (2003) Studies on the Characteristic Differences of Pedestrian Behaviors among Cities and Between City Districts such as Downtown and Suburban Shopping Centers. *Proceedings Walk21 IV.* Portland, Oregon, United States, 2003.

Krason A, Mayr M, Schipperges M (2003) The modern upper middle classes in Western and Central Eastern EuropeCross-cultural targeting in the case of Bank Austria Creditanstalt. ESOMAR-Congress "Excellence in Consumer Insights", Madrid, Spain.

Millonig A, Schechtner K (2007a) Decision Loads and Route Qualities for Pedestrians – Key Requirements for the Design of Pedestrian Navigation Services. In: Waldau, N., Gattermann, P., Knoflacher H, Schreckenberg M (eds.) *Pedestrian and Evacuation Dynamics 2005.* Springer Berlin Heidelberg, 109-118.

Millonig A, Schechtner K (2007b) Developing Landmark-based Pedestrian Navigation Systems. *IEEE Transactions on Intelligent Transport Systems (IST)* 8(1), Special Selection on ITSC'05, 43–49.

Millonig A, Gartner G (2007) Monitoring Pedestrian Spatio-Temporal Behaviour. *Technical Report.* 42, TZI Technologie-Zentrum Informatik, Universität Bremen.

Nisbett RE, Wilson TD (1977) Telling more than We can Know: Verbal Reports on Mental Processes. *Psychological Review.* 84, 231-259.

O'Connor A, Zerger A, Itami R (2005) Geo-Temporal Tracking and Analysis of Tourist Movement. *Mathematics and Computers in Simulation.* 69, 135-150.

Sparacino F (2002) The Museum Wearable: real-time sensor driven understanding of visitors' interests for personalized visually-augmented museum experiences. *Proceedings of Museums and the Web (MW 2002)* Boston, 2002.

Thomas C (2003) *Zu Fuss Einkaufen.* Available: http://www.fussverkehr.ch/presse/zufuss_schlussbericht.pdf. Last accessed 03 July 2008.

7 Mapping Pedestrian Movement: Using Tracking Technologies in Koblenz

Stefan van der Spek

Faculty of Architecture, Urbanism and Building Sciences
Delft University of Technology, The Netherlands

Abstract

The enhancement of GPS technology enables the use of GPS devices not only as navigation tools, but also instruments used to capture a travelled route. In the Spatial Metro project, this ability has been used to develop a method to track pedestrians in city centres. Where have you been, and what have you been doing?

In combination with traditional methods of analysis, these tracks result in a valuable instrument that shows the actual movement of people within an urban context. Individual tracks produce quantitative data in terms of location in time and space, resulting in the duration and the distance of visits. The track logs can be projected onto geographical information in GIS, generating maps showing the collective use of space, destinations and intensities.

This chapter focuses on the results of tracking pedestrians in Koblenz. The research question was how actual patterns of use are related to the physical conditions of the context and proposed investments. The outcomes of tracking were cross-matched with four analysis drawings: access points, commercial activities, points of interest and investments. The cross-match delivered useful conclusions with regard to missing links, barriers, lacking programme and the focus for future investments in public space.

Keywords: GPS, tracking, pedestrians, mapping, movement

7.1 Introduction

This article describes the results of a pedestrian observation study carried out in connection with the Spatial Metro project in Koblenz from 8 till 14 October 2007.

The article is part of a series describing the results of the tracking research carried out in three cities from different points of view: a general introduction and spatial outcomes (Van der Spek 2008a), a comparison between cities (Van der Spek 2008b) and a technical evaluation (Van der Spek 2008c). This article particularly focuses on the outcomes in Koblenz regarding the physical environment.

7.1.1 Context

The Spatial Metro project brings together a transnational group of partners, enabling them to co-operate with a view to improving city centres for pedestrians. The theme of Spatial Metro is 'Discovering the City on Foot'. The project aims to render city visits more enjoyable for pedestrians by making the cities easier to navigate, easier to walk around and easier to understand and appreciate.

The project has been allocated European Regional Development Funding through the INTERREG III/B Community Initiative. Spatial Metro has a group of ten participants. These are the lead city of Norwich (UK) and the cities of Rouen (F), Koblenz (D), Bristol (UK), Biel/Bienne (CH), as well as academics from the University of East Anglia (UK), the Delft University of Technology (NL), the University of Koblenz (D) and the Swiss Pedestrian Association (Spatial Metro 2008).

The main role of the chair of Urban Design was to develop instruments to evaluate visitor experience and to observe the use of public space. The purpose of the observation studies was to evaluate the use of space in relation to investments, opportunities and threads in the city; the outcome focuses on a comparison between the actual situation and real use based on monitoring and the experiences of the visitors.

For the observation of public space, a specific method using Global Positioning System (GPS) devices capturing the movement of pedestrians was developed, starting in 2006 and put into practice for the first time in Norwich in June 2007. The recording of pedestrian behaviour was accompanied by a questionnaire adding background information on the participants' particular trip and on the participants themselves. (Van der Spek 2008a)

GPS is a globally available system, but devices are required to record the trips people make. The deployment of the devices entails their distribution among the participants. In order to work efficiently and be a minimum burden on participants, the most logical locations for distribution and collection are the so-called arrival or access points to the city centre such as train stations, bus stations, parking facilities or bike storage sheds. Similar research was carried out in the Old City of Akko, Israel (Shoval & Isaacson 2007) and in a park study in Aalborg, Denmark by Hovgesen and Nielsen in August 2007 (Hovgesen et al. 2008). To ensure that people would return the expensive devices containing valuable information at the end of their visit, they were distributed in parking facilities. Potential participants were asked to take part if the purpose of their visit was either shopping (retail) or leisure. The devices were handed out to participants when they were leaving the car park and collected on return (Van der Spek 2008a). This also made it possible to add qualitative research in terms of a questionnaire in addition to quantitative data.

Two access points were selected for distribution. In Koblenz these were a shopping centre (Löhr-Center, 1400 cars) and an underground car park (Görresplatz, 350 cars). The facilities are situated on opposite sides of the city centre.

7.1.2 Objective and Research Question

The main objective of the study carried out by Delft University with regard to the Spatial Metro project was, if feasible, the development of an instrument that could be used to collect, observe and analyse processes of movement in historic city centres. Key issues of the study were the collection of data, the processing of the acquired data and the visualisation of the results. The goal of the results was to produce 'an image of actual use of the city'. Important aspects were the walked route in time and space, the destinations in time and space, the overall duration and the total trip length.

The main research question was how the actual patterns of use are related to the physical conditions of the context and the proposed investments. The physical environment was defined by the network of public spaces, access points, the activities and the Points of Interest (POIs).

Sub-questions of the study were whether particular groups use the city in different ways and whether this leads to specific, deviating spatial patterns. One primary distinction between particular groups was based on travel aspects: *origin* (local, regional or (inter) national), *familiarity* (fist, occasional or regular visit), *purpose* (shopping or leisure) and *duration* (<2 hrs, 2–4 hrs or 4+ hrs) (Van der Spek 2008a, 2008b). A secondary division was based on personal aspects: *age* (<20, 20–40, 40–60 or 60+), *gender* (male or female) and *group composition* (alone, partner, family or other). The processing of these three sets of data is in progress. All categories are defined based on the offered choices in the questionnaire, except for the duration.

7.1.3 The Focus of this Article

In order to research the results of the tracking study in relation to the physical context, illustrations have been made based on five major themes: a Google Earth map illustrating the spatial characteristics; access points illustrating the type and location of arrival points in the city; activities illustrating the type and location of services in the city; attractions illustrating the location of points of interest and investments illustrating future developments. Each theme has four illustrations: the current situation showing the city without a pattern of use, all movement showing the accumulated use generated by both locations, the pattern of use of Löhr-Center showing the exposure of the first location and the pattern of use of Görresplatz showing the exposure of the second location. Opposite to earlier articles, the tracks are visualised by dots instead of density analysis drawings. This delivers the best visual representation in greyscale of the data for this purpose.

The result is a matrix of twenty drawings representing the actual movement and the physical conditions of the context. The investment illustrations show the future conditions of this context and these illustrations will therefore be used to evaluate the results of the analysis.

7.2 Research

7.2.1 Outcomes

An approximate total of three hundred people participated in the Koblenz study, 180 of whom started in Löhr-Center and 120 from Görresplatz. After the assessment of the data, only one hundred valid tracks remained per location due to technical problems with batteries, the fixation of the GPS, blurring or other inconsistencies (Van der Spek 2008a, 2008c). Especially Time to First Fix (TtFF) and the clouding of points due to poor reception inside buildings were major issues in Koblenz. These problems are related to the distribution location in the underground parking facilities or in the shopping centre, where the GPS devices had no reception at all.

All tracks were cleaned manually and corrected where necessary. Generally, impossible coordinates in time and space were deleted. Incomplete or unreadable tracks were marked as invalid and were left out of the analysis. No data were added. This explains the high amount of invalid data.

7.2.2 Profiling the Visitor

The profile of the participating visitors clearly differs per location within each city and from one observed city to the next (Van der Spek 2008a, 2008b). In Koblenz, people parking in Löhr-Center (LC) mainly came from the region (70%), although in Görresplatz (GP) the figure was nearly 55%, mainly due to the fact that here, about 35% arrived from outside the region compared to 15%. Löhr-Center also attracts more local visitors (17%) compared to Görresplatz (11%). This also resulted in an

Table 7.1. Visitor profile.

ASPECT	LÖHR-CENTER	GÖRRESPLATZ	total	percent
ORIGIN				
local	30	13	43	14%
region	122	64	186	62%
nation/global	26	41	67	22%
unknown*	2	1	3	1%
FAMILIARITY				
no/first visit	19	42	61	20%
occasional	99	39	138	46%
regular	59	37	96	32%
unknown*	3	1	4	1%
PURPOSE				
shopping	147	53	200	67%
leisure	33	55	88	29%
other*	–	11	11	4%
DURATION				
<2 hrs	45	46	91	30%
2-4 hrs	101	57	158	53%
>4 hrs	34	16	50	17%
GENDER				
male	42	61	103	34%
female	133	58	191	64%
unknown*	5	–	5	2%
AGE				
<20	12	2	14	5%
20-40	68	42	110	37%
40-60	87	55	142	47%
60+	11	20	31	10%
unknown*	2	–	2	1%
GROUP				
alone	24	22	46	15%
partner	111	78	189	63%
children	39	17	56	19%
other	3	2	5	2%
unknown*	3	–	3	1%
TOURIST				
no	154	79	233	78%
one day	22	30	52	17%
multiple days	4	10	14	5%

* Unfortunately, not all participants did fill in all questions. Lacking data were not adjusted or corrected afterwards.

approximately three times higher ranking of first-visits in Görresplatz (LC 10%, GP 35%), although the percentage of regular visitors was similar.

Not unexpectedly, the percentage of visitors coming to shop in Löhr-Center (80%) was much higher than the percentage of shoppers starting from Görresplatz (45%). This is also reflected in the number of female participants (75% in LC, 50% in GP) and the group type (families: 25% in LC, 15% in GP), in my opinion explaining the higher number of shoppers.

There was also a very clear difference in the duration of visits; on average, participants starting in the shopping centre stayed longer. Shorter stays occurred more frequently in Görresplatz in particular (LC: 25%, GP: 40%), and this factor was compensated by a lesser percentage of longer stays (LC: 20%, GP: 10%).

Overall, visitors to Koblenz participating in the study can be profiled as people from the region who visit the city occasionally, mainly for a shopping trip with a duration of two to four hours. There was one significant distinction – Görresplatz attracted far more tourists, both in terms of percentage and in actual number.

7.3 Results

7.3.1 Context: Situation

Aerial Image

Koblenz is a historic city of around 106,000 inhabitants. The city centre is bordered by the River Mosel in the north and the River Rhine in the east. The spot where the rivers merge is known as the 'Deutsche Ecke – the German Corner' (Wikipedia 2008).

The river front and palace form soft green edges on one side of the city centre, while arteries of railway and highway infrastructure cut through the city on the other side. The city centre itself has an orthogonal grid structure with a mix of functions. As a result of several wars many buildings were lost. The historic core can be recognised in the grid by its fine, curved pattern of streets. Exceptions in the grid are the mega-block of the Löhr-Center Shopping Centre in the west and the palace in the east. Both objects are connected by one of the two long lines, the east-west oriented 'Schlossstrasse'. The other long line runs north-south. Both lines cross at 'Löhrrondell', the square near the southern side of Löhr-Center *(see Figure 7.1)*.

Access

The N9 encloses the city centre on the western side, giving access from the north, south and west. On the southern side, the N49 gives access from the east. A major

Fig. 7.1. Google Earth map of Koblenz (© Google Earth 2008).

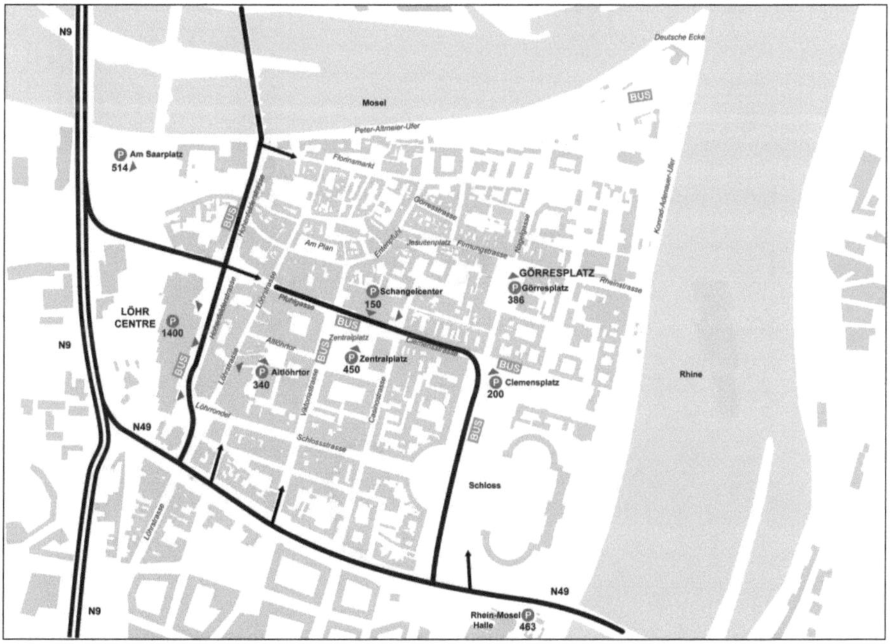

Fig. 7.2. Access map of Koblenz (black).

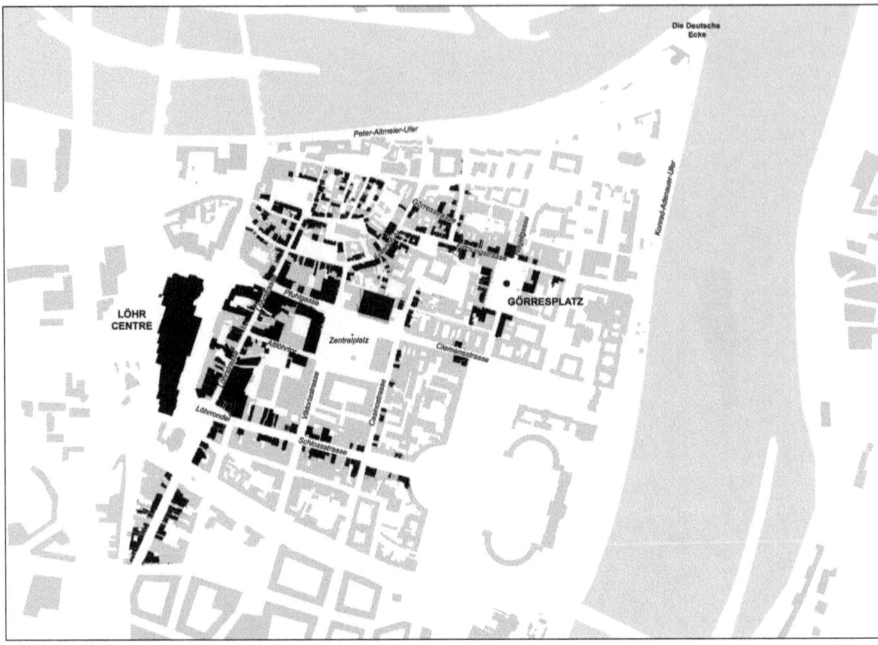

Fig. 7.3. Activities in Koblenz (black).

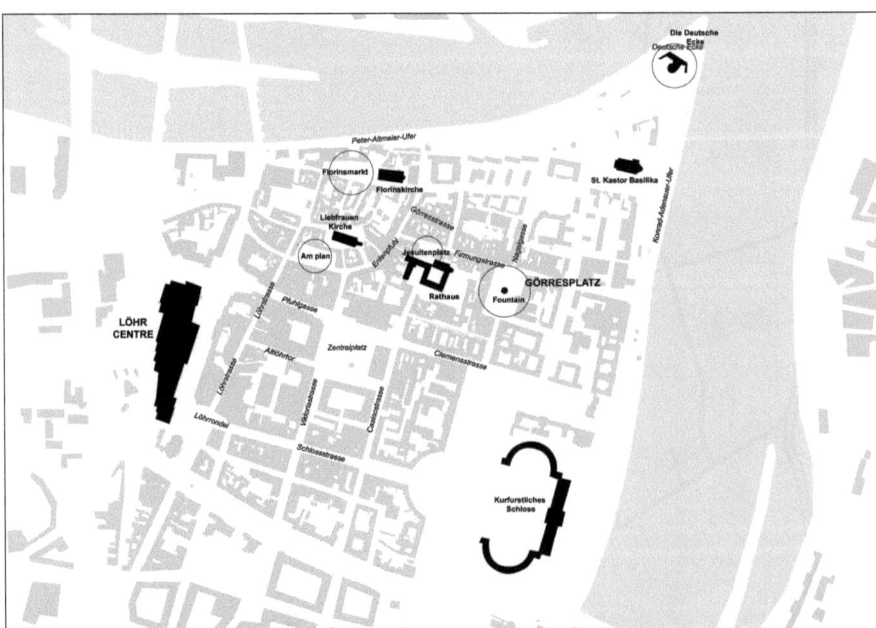

Fig. 7.4. Points of Interest in Koblenz (black).

road cuts through the city centre giving access to the parking facilities (which are mostly located on the southern and western side of the city centre) and providing access by public transport. The railway station is located to the south outside the reach of the city centre (see *Figure 7.2).*

Commercial Activities

The commercial activities are situated on a limited number of streets. The main commercial streets are 'Löhrstrasse' – the main shopping street running north-south, 'Schlossstrasse' – running east-west, 'Entenpfuhl', on the edge of the historic core and 'Firmungsstrasse', running between Görresplatz and the historic core. Further, commercial activities can be found in the historic core itself, where a great deal of bars and restaurants are also located.

Remarkably, there are hardly any commercial activities in the zones between the core, river front and palace area (see *Figure 7.3).*

Attractions

Interesting spaces and objects in Koblenz are the 'Florinsmarkt' and 'Florinskirche', 'Liebfrauenkirche' and 'Am Plan' in the historic core. Outside the historic core a cluster of 'Rathaus' (City Hall), 'Jesuitenplatz' and Görresplatz represent the most famous spaces (Wikipedia 2008). However, the best known is the German Corner with its prominent statue and St. Kastor Basilik. Other landmarks are the palace, which is closed to the public at the present, and the Löhr-Center Shopping Centre (see *Figure 7.4).*

7.3.2 Google Earth Aerial Map

All Tracks

Figure 7.5 shows the results of all accumulated valid tracks projected on top of the Google Earth aerial image. The main visited areas are the shopping streets Löhrstrasse, Am Plan, Entenpfuhl and Firmungsgasse. The streets to and from Zentralplatz, the historic core, Schlossstrasse and the river front are of secondary importance. Particularly the access to the river front has many alternatives. Apart from the Rheinstrasse, the most obvious route from Görresplatz, there are no other clear alternatives.

A number of people visited the surrounding area, e.g. the south, moving in the direction of the train station, the east moving in the direction of the palace and the Conference Centre (Rhein-Mosel hall), and the west continuing along the quay and returning via alternative routes.

Fig. 7.5. Google Earth (© Google Earth 2008) map with all tracks projected.

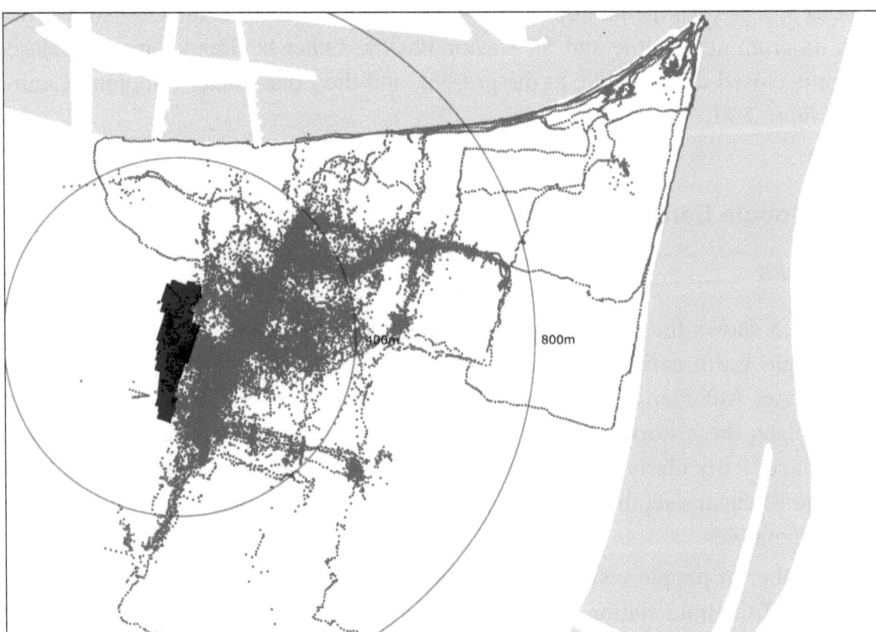

Fig. 7.6. Map based on tracks only, Löhr-Center.

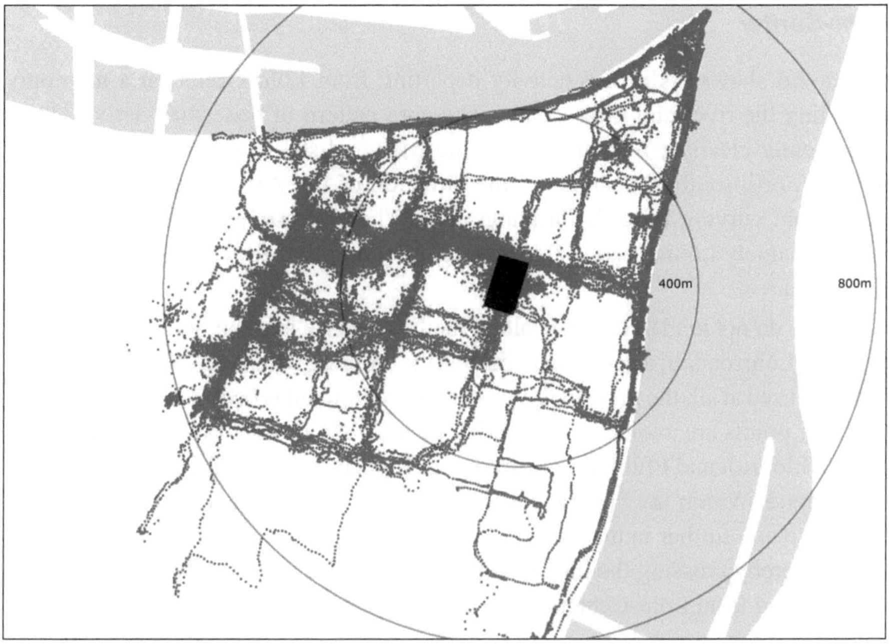

Fig. 7.7. Map based on tracks only, Görresplatz.

Fig. 7.8. Access map with all tracks projected.

Löhr-Center

Figure 7.6 shows pedestrian activity departing from Löhr-Center on a map only indicating the river and Löhr-Center. The core pattern of pedestrian activity from Löhr-Center clearly covers the main pedestrian and shopping streets: Löhrstrasse running north-south, the streets in the direction of the Zentralplatz and Am Plan Entenpfuhl curving towards Firmungsgasse. There are some major breaks in the pattern, namely Löhrstrasse, Schlossstrasse, Zentralplatz, the historic core, and the river front.

People do not tend to continue along the Löhrstrasse in a southern direction after reaching Löhrrondell; a particular barrier is the crossing with the N49. Schlossstrasse is hardly used at all, the first block away from Löhrrondell is occasionally visited, but the other points are avoided in particular. People tend to walk towards Zentralplatz along Altlöhrtor and Pfuhlgasse or from the 'Rathaus' (City Hall), but then stop and turn around. Within the historic core, people visit the Liebfrauenkirche but do not tend to go any further in that part of the city. Some people accidently visit the river fronts, thereby crossing the historic city core or palace area. Generally, these areas area avoided from Löhr-Center.

Görresplatz

Figure 7.7 shows the pedestrian activity from Görresplatz on a map only indicating the river and Görresplatz. The core of pedestrian activity starts at Görresplatz and continues in the direction of the historic core, along the Firmungsgasse and Entenpfuhl to Am Plan and from there along the Löhrstrasse until Löhrrondell. Here, twenty percent of the participants from Görresplatz also entered Löhr-Center. From Löhrstrasse, people use different routes to return to Görresplatz, making the orthogonal grid structure visible. Zentralplatz was also crossed diagonally several times.

Another major flow is along the Rheinstrasse directly to the Rhine quays and from here to the German Corner and then along the Mosel quays.

Comparison of the Patterns

The comparison between the pattern of activity in Löhr-Center and Görresplatz shows some major dissimilarities. Firstly, the river fronts are more frequently visited from Görresplatz. This is probably due to the background of visitors (tourists, (inter) national origin) and the location of the access point. Secondly, participants from Görresplatz used the city more broadly, taking several smaller streets, particularly in the area between Löhrstrasse and Görresplatz. Finally, this broader use of the grid also resulted in a better integration of Schlossstrasse and Zentralplatz in the pattern of use from Görresplatz.

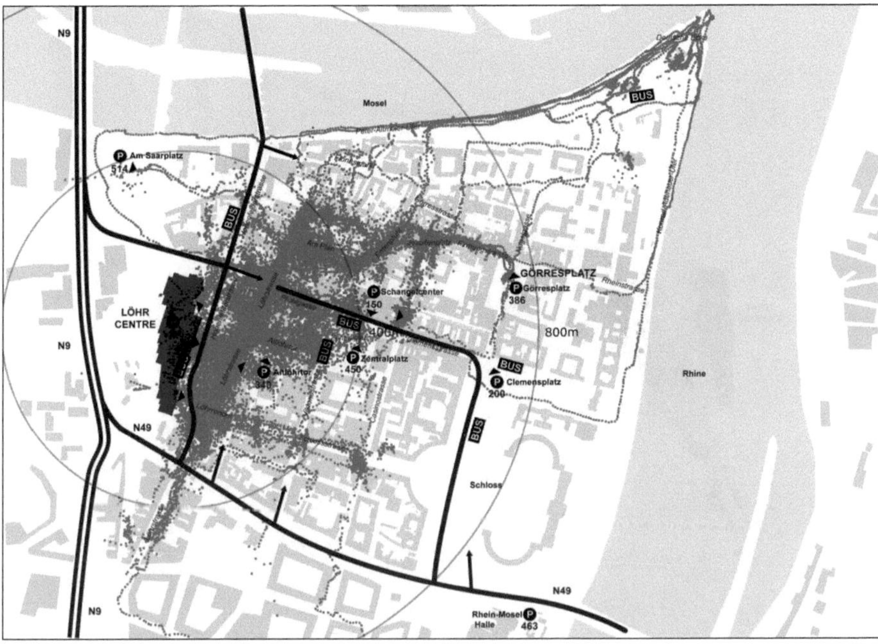

Fig. 7.9. Access map with projection of tracks from Löhr-Center.

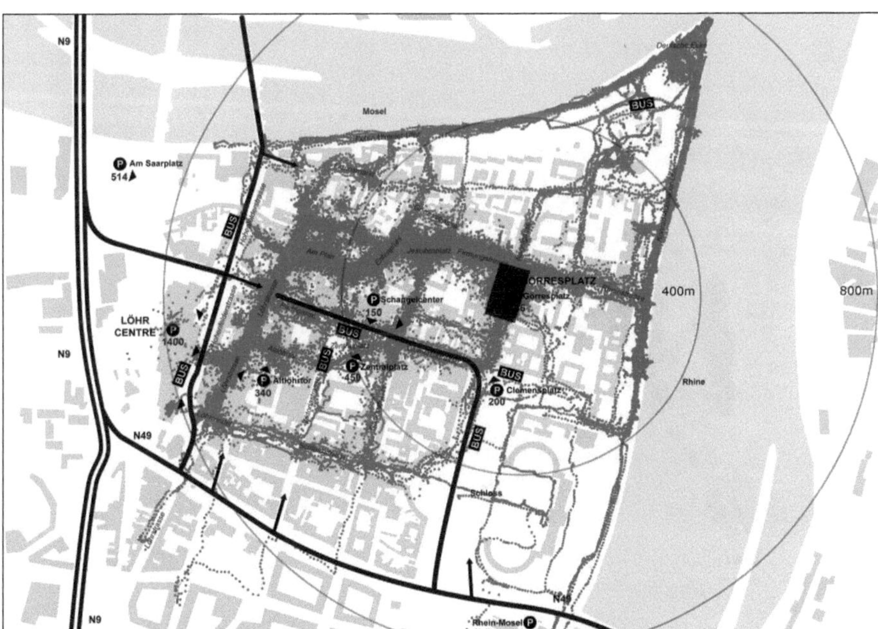

Fig. 7.10. Access map with projection of tracks from Görresplatz.

Fig. 7.11. Activity map with all tracks projected.

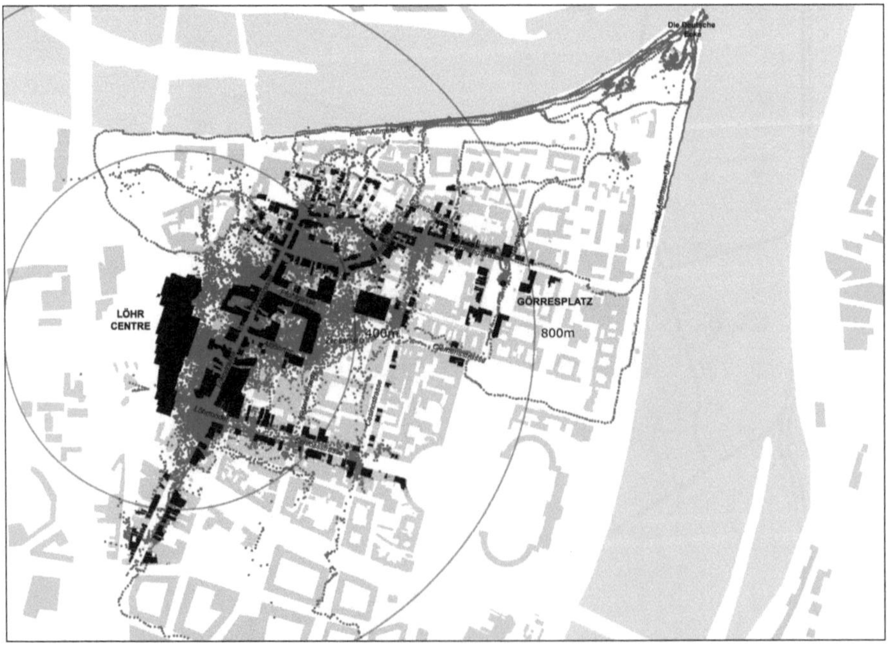

Fig. 7.12. Activity map with projection of tracks from Löhr-Center.

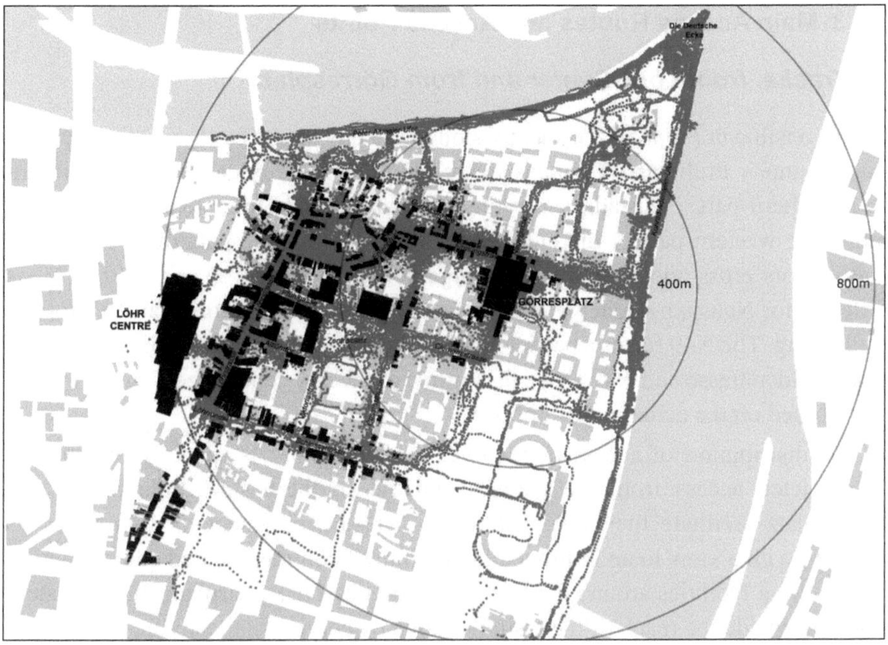

Fig. 7.13. Activity map with projection of tracks from Görresplatz.

Fig. 7.14. Points of Interest map with all tracks projected.

7.3.3 Main Access Routes and Access Points

All Tracks, from Löhr-Center and from Görresplatz

Clearly visible in *Figure 7.8* is the division of the city centre by three intensely used roads, namely Pfuhlgasse/Clemensstrasse, dividing the city centre into a northern and southern part, Neustadt, separating the palace, and Hohenfelderstrasse separating the western part of the city centre. Nevertheless, the pattern of use shows that visitors cross and walk along Pfuhlgasse (see *Figures 7.9* and *7.10)*. This is different for Neustadt, Hohenfelderstrasse and N49 which visitors scarcely cross or walk along. The N49 forms a barrier in the Löhrstrasse (see *Figure 7.8)*. Particularly Hohenfelderstrasse and N49 shape the outline of the city centre (see *Figures 7.9* and *7.10)* based on the actual pattern of use.

The distribution of access points over the city centre is spatially limited due to restricted access from the western and the southern side. Löhr-Center and Görresplatz facilitate their own public. The (re-)development of Zentralplatz is required with a view to an improved facilitation of commercial and cultural activities. Parking facilities around the German Corner and Am Saarplatz on the western side are outside the reach of the visitors drawn by commercial activities.

7.3.4 Commercial Activities

All Tracks, from Löhr-Center and from Görresplatz

The pattern of use (see *Figure 7.11)* covers the location of commercial activities well. Nevertheless, some spots, the southern part of the Löhrstrasse (south of the N49, see *Figures 7.12* and *7.13)* and the historic city core (see *Figure 7.12)* clearly have fewer visitors. The pattern from Löhr-Center also lacks activity in the Schlossstrasse and Casinostrasse running in a northern direction from the Schlossstrasse. Further, Görresplatz is not a destination from Löhr-Center, although visitors do journey from Görresplatz to Löhr-Center (see *Figure 7.12)*.

In *Figure 7.13,* it can be seen that Löhrrondell outlines the use from Görresplatz. From here, people stop walking south, instead walking in an eastern direction into Schlossstrasse (see *Figure 7.13)*. This probably explains the more intense use of Schlossstrasse, Casinostrasse and Zentralplatz from Görresplatz, resulting in a broader use of the city centre. The pattern in *Figure 7.13* shows a lack of activities (destinations) on and around Zentralplatz, around the palace and along the Rheinstrasse as a major access to the Rhine quay.

7.3.5 Points of Interest (POI)

All Tracks, from Löhr-Center and from Görresplatz

The overall pattern of use shows that most Points of Interest were visited, except for the palace (see *Figure 7.14).* However, the analysis per access location indicates that participants from Löhr-Center incidentally visited the destinations in the historic city core such as Florinsmarkt, Florinskirche and Görresplatz or the destinations along the quays such as St. Kastor Basilika and the German Corner (see *Figure 7.15).* These destinations, including the shopping centre, were well embedded in the pattern of use from Görresplatz.

It is remarkable that nearly 20% of the participants from Görresplatz also entered Löhr-Center. The other way round, nearly 60% of the participants from Löhr-Center visited both the city and the shopping centre, and 40% only visited the city if they parked in Löhr-Center. Although visitors not leaving the shopping centre have not been accounted for, Löhr-Center seems to contribute to the city as well.

7.4 Conclusions and Evaluation

7.4.1 Investments (ERDF)

Koblenz previously made investments in the quality of the public space during the Spatial Metro project and the city council intends to continue making such investments, driven by the Federal Garden Show 2011 (Bundes-Garten-Schau 2011, BUGA).

During the Spatial Metro project, investments enabled by the European Regional Development Fund (ERDF) were made in information systems (blue dots), street and object lighting (orange), pedestrian zones and shopping streets (yellow/blue) and a competition on the redevelopment of Löhrrondell (blue), including the surrounding area (Van der Spek 2008, Kallenbach 2008).

Firstly, a digital information system called City Guide was developed. This system consists of facilities offering Location Based Services for mobile systems using Bluetooth or Wifi. Several access points were installed on main locations in the urban fabric (Furbach et al. 2008).

Secondly, a lighting master plan was designed to implement a system for street lighting and to improve the lighting of Points of Interest (Kallenbach 2008). This included new street lighting for Löhrstrasse and Schlossstrasse, and lighting schemes for Am Plan, Jesuitenplatz (including Rathaus), Münzplatz and Görresplatz.

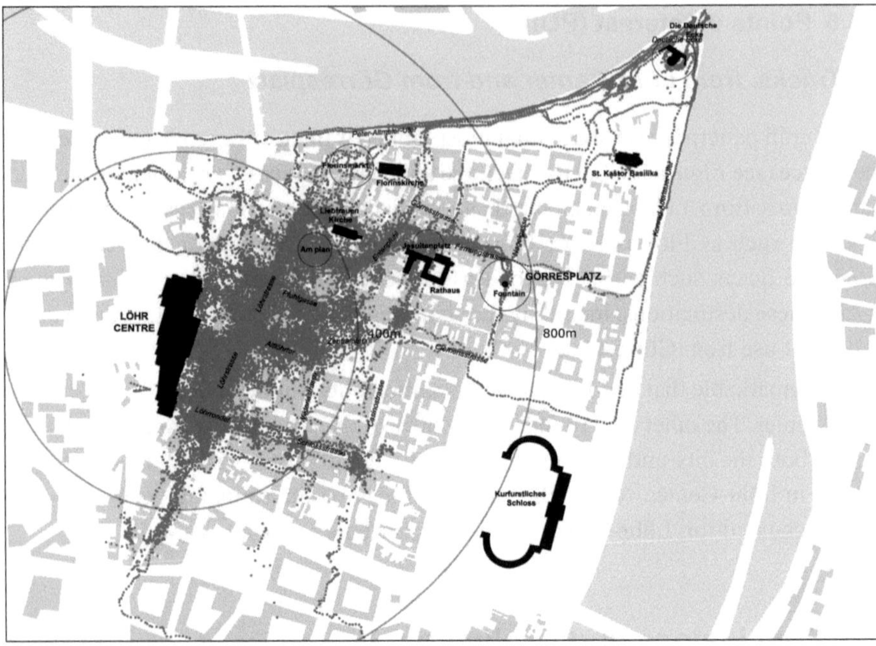

Fig. 7.15. Points of Interest map with projection of tracks from Löhr-Center.

Fig. 7.16. Points of Interest map with projection of tracks from Görresplatz.

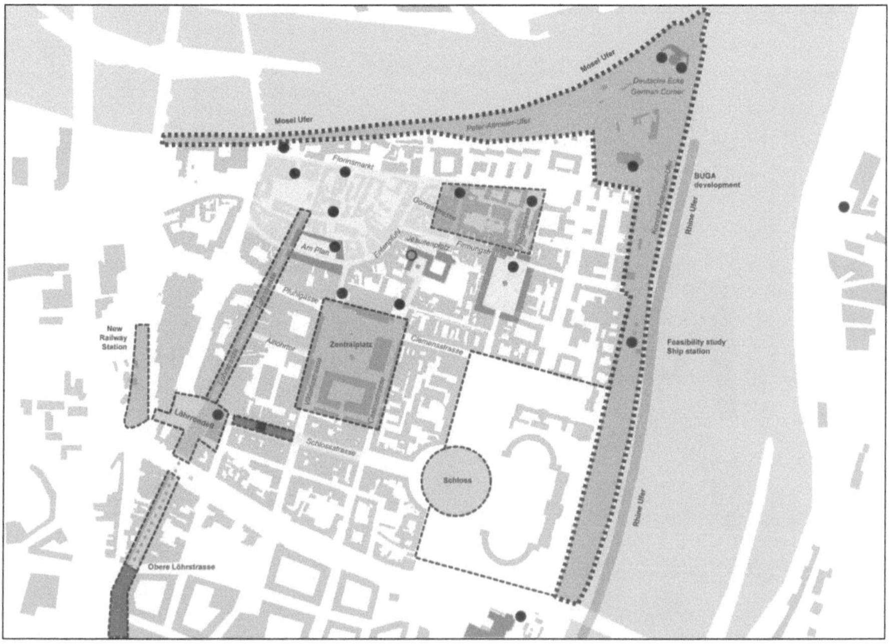

Fig. 7.17. Investment map of Koblenz.

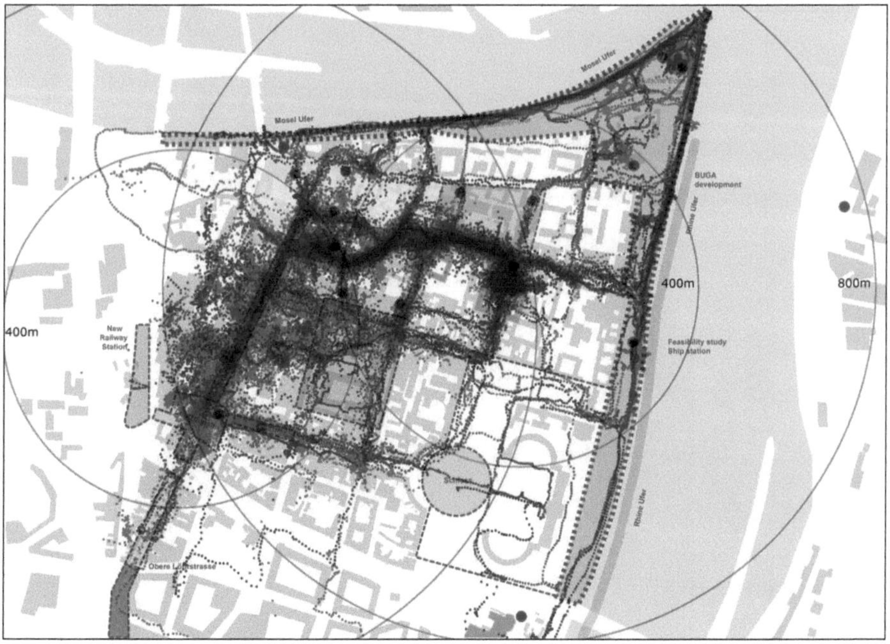

Fig. 7.18. Investment map with all tracks projected.

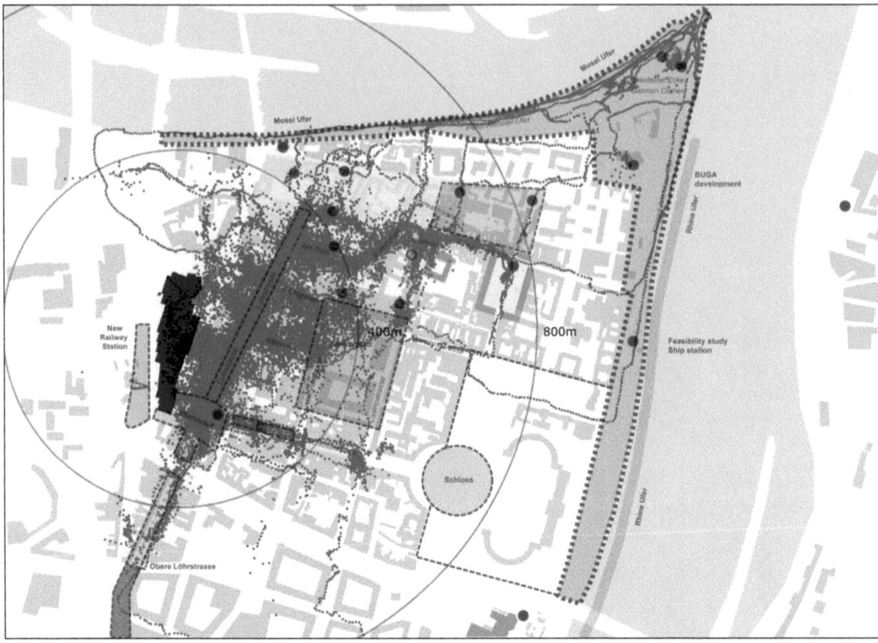

Fig. 7.19. Investment map with projection of tracks from, Löhr-Center.

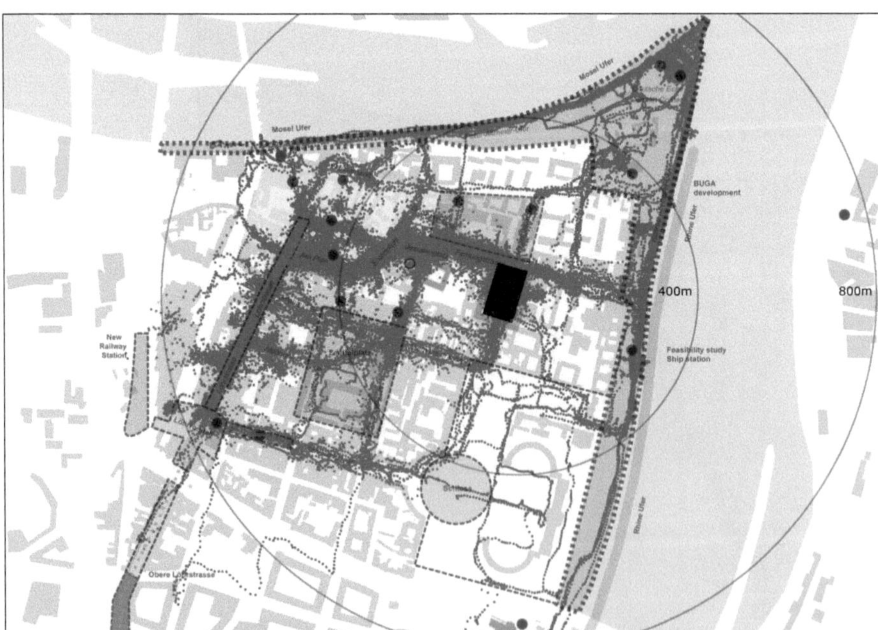

Fig. 7.20. Investment map with projection of tracks from Görresplatz.

Thirdly, public space was improved by new engineering in the Schlossstrasse and a new pavement in the Löhrstrasse. A new conflict-free crossing will guide pedestrians from Schlossstrasse to the palace without visual interruption.

Finally, a competition was organised on the redevelopment of Löhrrondell, the square were both Löhrstrasse and Schlossstrasse meet. This crucial square will be changed from a major car crossing to a pedestrian-friendly area.

Future investment projects (pink) in connection with the coming Federal Garden Show are a new train station, the redesigning of Zentralplatz and the redesigning of both the Rhine and Mosel quays. Firstly, a new train station will be projected behind Löhr-Center, offering access to Löhrrondell. Secondly, a competition for the redevelopment of Zentralplatz has been announced. Finally, the quays and access to the city from the quays will be improved.

7.4.2 Investments and Pattern of Use

All Tracks, from Löhr-Center and from Görresplatz

For Löhr-Center, particularly the investments in Löhrstrasse, Schlossstrasse and the historic core are crucial for the functioning of the area around Löhrrondell. The redevelopment of Löhrrondell, Zentralplatz, the palace and the development of a new train station will offer new opportunities. All types of investments will potentially expand and intensify its use, thus changing the pattern of use.

Görresplatz will benefit most from the investments in the historic core and the redevelopment of the Zentralplatz area, integrating it better into the urban tissue. The redevelopment of the quays will potentially attract more visitors to this side of the city.

7.5 Concluding Remarks

One of the remarkable discoveries made in Koblenz was the distance that people walk. Görresplatz is well positioned, covering the whole city centre within its 800 metre radius and with its main activities at both a 400 metre and a 800 metre range. Löhr-Center only covers the commercial activities at a 800 metre radius, and only Löhrstrasse and Schlossstrasse are within a 400 metre range.

A second discovery was the lack of activities in the middle of the city centre, Zentralplatz. Here, a new development will take place that will potentially fill the gap. Zentralplatz is within a 400 metre range of both Görresplatz and Löhr-Center and well connected to both access points. Zentralplatz has been found by visitors coming from both locations, but making it an attractive area and destination may

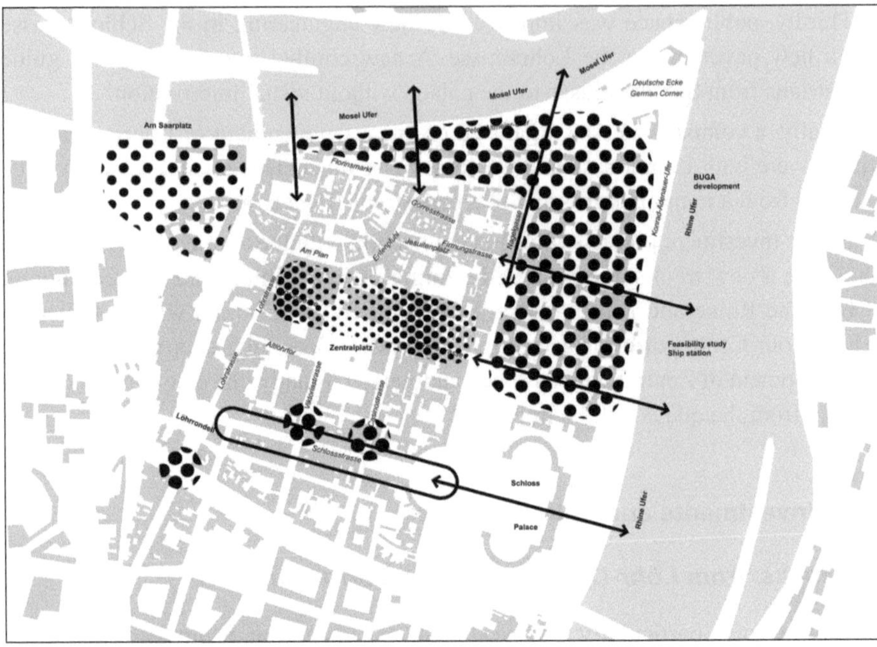

Fig. 7.21. Conclusion: Future development?

change the pattern of use the most for Löhr-Center. For the optimal integration of Zentralplatz, a number of conditions still need to be changed, namely the amount of and the continuation of commercial activities in the Schlossstrasse and commercial activities in Viktoriastrasse, Casinostrasse and Clemensstrasse. Further, the traffic function and spatial design of the Pfuhlgasse should be considered *(Figure 7.21)*.

A third discovery was the segregation of functions. Commercial activities can be found in the heart of the city, with one previously mentioned exception, namely Zentralplatz. Points of interest can be found on the edge of and outside this area. The river fronts are separated in space and by their programming from the commercial and historic city core by means of an in-between zone of housing, the palace and a hotel. There is virtually no visual or spatial relation with the city centre. A suggestion has been put forward to connect the water front by the prolongation of long, visual lines, facilitated by a clear continuation in space, a spatial character and the programming of commercial and cultural activities for example. Such lines could be Schlossstrasse, Clemensstrasse, Rheinstrasse and Nagelgasse (see *Figure 7.21)*.

A fourth discovery was the black spot in the Löhrstrasse. Investments in place and space only make sense if the network is taken into account. The definition of investments in the Löhrstrasse, making it a connection between the city centre and the main train station, requires the consideration of the barrier with the N49 (see *Figure 7.21)*.

A fifth discovery concerned the palace and Schlossstrasse. The palace is a huge object in the city and is not accessible at the present. The traffic infrastructure surrounding the palace makes it an area not easily accessible for pedestrians. The Schlossstrasse is separated into three five parts, namely Löhrrondell, three spatially different build parts and the crossing with Neustadt (see *Figure 7.21).*

Finally, the Mosel quay ends at the Am Saarplatz area. This area lacks continuation in its connections with the city centre and forms a separated unity, making it a backside area (see *Figure 7.21).*

The current and future investments make it desirable to repeat the same study at a later date. Future research should include tracking pedestrians from the same locations to enable a comparison of the outcomes with former results. However, the new situation also makes it possible to conduct research from new or upgraded locations such as Zentralplatz with a view to collecting larger amounts of data.

Acknowledgements

The collection of data was initiated by the Spatial Metro project and partly financed by European Regional Development Funding (ERDF) through the Interreg IIIb community initiative. The project is called City on Foot in Germany.

All data were required anonymously. All outcomes should be considered as results of the working method and the limited participating population. The outcomes only represent the activities of the participants from two car parks and do therefore NOT represent the overall behaviour of visitors or pedestrians in Koblenz.

References

Furbach U et al (2008) Information Systems for Spatial Metro, in: *Street Level Desires*. Delft: Urbanism.

Harder H et al (2008) Experiences from GPS tracking of visitors in Public Parks in Denmark based on GPS technologies, in: *Urbanism on Track*. Delft: IOS Press.

Kallenbach R (2008) Driven by the federal garden show, in: *Street Level Desires*. Delft: Urbanism.

Shoval N, Isaacson M (2007) Sensing Human Society,, in: *Environment and Planning B: Planning and Design* 34 (2007: 191-195)

Spatial Metro (2008) *Spatial Metro project*. Available: http://www.spatialmetro.org. Accessed 24 April 2008

Van der Spek SC (2008a) Tracking Pedestrians in Historic City Centres using GPS, in: *Street Level Desires*. Delft: Urbanism.

Van der Spek SC (2008b) Spatial Metro: Tracking Pedestrians in Historic City Centres, in: *Urbanism on Track*. Delft: IOS Press.

Van der Spek SC (2008c) Tracking Technologies: Truth or Dare,. Paper, to be published: *SPM2008*. Estonia.

Wikipedia (2008) *Koblenz*. Available: http://nl.wikipedia.org/wiki/Koblenz_%28Duitsland%29. Accessed 24 April 2008

Section II: TeleCartography

8 Evaluation of User Variables in Topographic Feature Recall for the Informed Selection of Personalized Landmarks

Birgit Elias, Volker Paelke and Marcel Chaouali

Institute of Cartography and Geoinformatics (ikg)
Leibniz University of Hannover, Germany

Abstract

Landmarks are a very important part of effective wayfinding instructions. Therefore much research is conducted analyzing the nature of landmarks and studying how they can be integrated appropriately into navigation services. The question whether the use of objects as landmarks differs between varying user groups becomes especially relevant if wayfinding instructions are to be adapted to context and users. In this paper the planning, execution and interpretation of a user-questionnaire regarding the mental representation of salient topographic features in the city centre of Hanover is presented. The user test focuses on recalling the important landmark features in the test area from memory and investigates if there are significant differences according to sex, age or familiarity with the environment.

Keywords: navigation, landmarks, user adaptation, user test, questionnaire

8.1 Introduction

One of the central features of location based services is to provide a routing component to navigate a user from his current location to a specific place. This functionality is provided by pedestrian navigation systems and much recent research in pedestrian guidance and mobile cartography has addressed the appropriate communication of routes and the associated wayfinding instructions (Meng et al. 2007). Research in the field of spatial cognition investigates the structure and elements of wayfinding instructions and provides another important foundation for the design of effective pedestrian navigation systems. Daniel and Denis (1998) have identified route actions (instructions about the next movement), orientations and landmarks as the basic components of (verbal) route directions. Further experiments have shown that the integration of landmarks into routing instructions enhances the perceived quality of a description (Denis et al. 1999).

Currently, landmarks are not part of commercial navigation data sets. In fact, all available route planning and guidance applications use data sets that are tailored to the requirements of car navigation. With the increasing amount of pedestrian navigation applications on mobile devices, the need to integrate relevant information for pedestrians rises. If information about landmarks were available, they could be integrated into the databases and used for generating more intuitive wayfinding descriptions.

However, there is a common understanding that the quality of an object to be a landmark varies according to context and time. It is strongly related to the specific course of the route, its wayfinding context and the personal characteristics of the user for whom a route description with landmarks is generated. Therefore, different research approaches try to identify the relevant influence factors of context and personalization on landmark choice. In this paper, we focus on the aspect of personalization. In particular we analyse if there are significant differences in remembering salient topographic objects in the inner city environment of Hanover depending on age, sex and familiarity with the environment.

The remain of the paper is structured as follows: In the next section related work regarding landmarks and adaptation is reviewed. We also shortly introduce previous studies conducted in the same test area, because they inspired the direction of this study and are used as a reference. After that the user questionnaire is introduced and the underlying hypotheses about the differences according to sex, age or familiarity are presented. Following that we describe the implementation and analysis of the study. The results of the statistic analysis are presented and compared to previous studies. The paper concludes with a discussion and an outlook.

8.2 Related Work

8.2.1 Landmarks for Navigation

Landmark-based navigation is the most natural concept to navigate for humans. Landmarks are significant physical, built, or culturally defined objects that stand out from their surroundings and therefore help locating the geographic position (Golledge 1999). In this context, a landmark may be defined by its particular visual characteristics, by its unique purpose or meaning, or its central or prominent location (Sorrows & Hirtle 1999).

The integration of landmarks into wayfinding descriptions requires a detailed analysis of the elements and structure of verbal wayfinding instructions. Research in this direction proposes an ontology for the wayfinding task (Winter 2002). Different research approaches try to develop formal models or extract landmarks automatically from databases and focus on local landmarks at (potential) decision points (Raubal & Winter 2001, Elias & Brenner 2004, Elias 2006). In that context, all indications are that the chosen landmark objects depend on the specific wayfinding context (e.g. way of movement, day or night) and user group (age, gender, level of education) (Zipf 2002, Elias et al. 2003). For example, the type of movement (going by car or walking) dictates which type of objects are chosen as landmarks for the routing instructions. The study of Burnett et al. (2001) reveals, that in applications for car navigation 'road furniture', such as traffic lights, pedestrian crossings and petrol stations plays a vital role as landmarks. In contrast, according to the research of Michon and Denis (2001) wayfinding instructions for pedestrians should include objects like roads, squares, buildings, shops and parks. According to this, it is necessary to adapt the selection of landmarks to the mode of movement. Also an initial study found a significant impact of daylight or night time onto people's choice of objects being an appropriate landmark (Winter et al. 2004).

8.2.2 Adaption – Personalizing and Context

The effective design of a system that aims to support users in navigation must take the specific circumstances and constraints of the intended use into account. Navigation systems, like many other mobile applications, are used for short episodes, possibly as one task among many. The intentions of users may vary from effective A to B navigation to more entertainment oriented uses with an integrated wayfinding component, as in mobile tourist guides or mobile games. In all use cases the system should be functional and usable. Depending on the usage situation this can lead to a variety of possibly contradicting requirements, e.g. with regards to physical device characteristics (e.g. waterproofing, readable displays under varying lighting conditions, size, weight) and user interfaces design (e.g. interaction and presentation techniques used). Since every design decision involves a tradeoff, it is useful to defer design decisions to run-time – where possible – in order to create systems that can adapt to the specific conditions of a use situation. To personalize a system adaptation is possible with regards to the following variables:

- The technological constraints of the user's device
- The user's physical context
- The user's current activities
- The user's personal characteristics

In recent years much research has been devoted to the development of adaptation techniques for mobile devices. Techniques were developed to address typical constraints of mobile devices like small displays and limited interaction modalities and to adapt a system to the available hardware resources (Schmidt et al. 1999, Reichenbacher 2003, Paelke et al. 2005, Wu & Wu 2006).

Adaptation to the physical context in which a user is operating is addressed in a number of location-based systems (e.g. Rhodes 2003, Reimann & Paelke 2005) and in work on ubiquitous computing (e.g. Want 1996, Niemelä & Latvakoski 2004). Many systems also allow to change functionality and representation depending on the task at hand, either manually (e.g. switching from road-based routing to simple positioning for off-road use in GPS devices (Magellan 2008)) or automatically (based on the user's current activities) in mixed-initiative user interfaces and sensing systems (Gajos et al. 2006, Oviatt et al. 2008). Another interesting domain is the adaptation to the personal characteristics of the user. In a user interface the content, it's presentation and also the offered interaction possibilities can be varied according to a user profile. A common use of this is seen in personalized web services (e.g. Ho & Kwok 2002, Espinoza 2003). In the work presented in this paper we studied the impact of several user characteristics on landmark selection and recall. In future navigation systems the results can help to adapt the selection of landmarks on the content level to user characteristics. For the adaptation of landmarks at the presentation level we refer to our previous studies (Elias et al. 2006, Elias & Paelke 2008).

8.2.3 Previous Work / User Tests in Hanover

In addition to studies regarding the presentation of landmarks, we also conducted user questionnaires examining landmark choice and alternative ways of descriptions in wayfinding instructions. In the first study by Lübke (2004) 20 students were asked to describe two different routes, one of them leading through the inner city centre of Hanover. All mentioned landmarks were recorded and assigned to the decision points of the route to which they belong. This data was used to evaluate automatic extraction procedures (Elias 2006).

A second study focused on the personalization aspects of landmark choice, especially the question if sex, age and familiarity with the environment makes a difference (Köpke 2007). For that purpose a two-fold approach was pursued carrying out a web-based questionnaire and a field test. In the web-based questionnaire subjects were asked to answer if they are able to recognize landmarks given by their name or a description in photo panoramic views of intersections. The photo panoramas were provided in a software environment which enables the subjects to see a cut-out of the scene that represents the human field of view and allows them to navigate themselves in the scene (make a 360° degree turn around the viewpoint, zoom in and out to some extent). About 220 subjects rated whether objects situated at 8 different intersections are recognizable or not. The results show that there are no differences between men and women, but a few statistic significant differences in landmark objects between different age groups. But the most significant differences showed up in the analysis with regards to different levels of familiarity with the test environment. In order to back up the (artificial) online test results a field test was also conducted, questioning 50 people about their memorized objects of one of the intersections. This time the only significant divergence showed up in the analysis of the relevance of gender. Of course the number of subjects was too small to be evaluated statistically, but it clearly indicated the need for further study to reveal whether sex, age or familiarity have a relevant impact on landmark choice. Therefore, we designed the study reported here to back up and refine the results to be able to establish guidelines for landmark adaptation in future.

8.3 User-Questionnaire – Planning and Performing

8.3.1 Hypothesis: Differences Depending on Sex or Age

Many studies have addressed the impact of personal user characteristics like gender and age on spatial knowledge acquisition, navigation and orientation (Bia et al. 2007, Holding & Holding 1989, Sharps & Gollin 1988). Unfortunately, there is no coherent model available that would allow accurate predictions of user preferences for a specific use case like pedestrian navigation. We have therefore conducted experiments in the specific use context of an envisioned pedestrian navigation system to determine the influence and preferences with regards to three user characteristics (sex, age and familiarity with the environment). The following sections describe the specific questions examined in the study:

8.3.1.1 Sex

The first question that we address in our experiment is whether sex has an impact on the selection and recall of topographic features as landmarks. Some studies suggest that women differ in their selection of navigation techniques from men, which might impact landmark objects selection and recall. In our experiment we aimed to test if this leads to the selection and use of different topographic objects as landmarks in navigation instructions. Based on the results of a previous pilot study our test hypothesis was that there is no significant impact of gender on the suitability of objects as landmarks.

8.3.1.2 Age

In the second part of our experiment we examine the impact that the age group of a person has on the selection and recall of topographic features as landmarks. As with sex there are some research results that seem to indicate that age has a significant influence on the use of different navigation strategies. For practical reasons we limit our current study to the comparison of two age groups, namely teenagers and the age group over 20. This choice was made to compare age group differences in populations that are typical users of modern technology products like navigation systems (as the intended application domain) and to avoid possible distortions due to ongoing learning of spatial skills (in children) and effects of physical deterioration (for elderly people). Due to limitations of the subject group we didn't divide the over 20 category into more detailed age groups. Results from a previous pilot study indicated a significant impact of age on landmark selection and use, that we aimed to study in more detail. Our hypothesis was that the types of topographic objects used as landmarks differ between teenagers and over 20s.

8.3.1.3 Familiarity

The third part of our experiment aimed to clarify the impact that the degree of familiarity with a spatial environment has on the selection, recall and description of topographic objects as landmarks. While a general impact is obvious in the extreme recall case (a user who is entirely unfamiliar with a location can't pick landmarks at all) we aim to clarify the differences in use and descriptions between different categories of users according to familiarity with the environment. For practical purposes we were especially interested in the differences between "locals" (people who are very familiar with a location and visit it frequently) and "knowledgeable visitors" (who are just familiar enough with an area to provide meaningful landmarks). As a previous pilot study found interesting differences, we wanted to examine this issue further and formulated the hypothesis that different groups use different objects as well as different descriptions.

8.3.2 Planning the Questionnaire

Because the results of the survey from Köpke (2007) point at the strong connection between landmark memory and familiarity with the environment, this study was designed to focus on subjects who do not live in the test city, but have enough city experience so that they can remember landmark objects at all. This is not an easy task to accomplish, because people with just this kind of rough knowledge about the city are difficult to find and approach. Two different approaches are pursued: one was conducting interviews at a large fair which is held outside of the inner city at the Hanover fair ground and is visited mostly by non-residents, the other was questioning pupils in a town about 30 km away from Hanover. Because Hanover is the regional capital and the biggest city in the area, it can be assumed that even young people have visited the city once in a while.

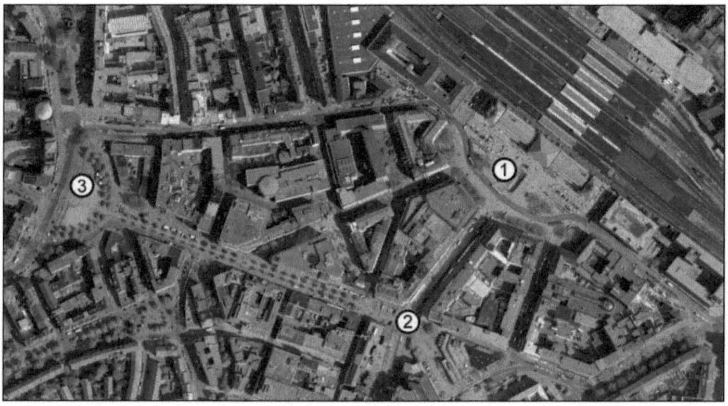

Fig. 8.1. Test area: Three locations in Hanover (image taken from Google Maps): 1 – Ernst-August-Platz (main train station), 2 – Kröpcke (main underground interchange and central place in city), 3 – Steintor.

The questionnaire includes an introduction to the topic and asks for general information about the subjects (sex, group of age, own assessment of familiarity with Hanover). In the actual question part the participants should recall their memorized landmark objects about three different intersections / plazas which are located in the inner city centre of Hanover and are believed to be among the most well-known places in town (see *Figure 8.1).*

The subjects were asked to name all objects which they think are important at that place. For each location five free lines were given in the form to fill in the names or descriptions.

8.3.3 Execution

The survey was conducted in two steps. In the first step the interviews of visitors of the industry fair at the fair ground were carried out. Unfortunately, it was difficult to get hold of people with the kind of rough knowledge of the inner city centre of Hanover that we aimed for. Either they were inhabitants of Hanover or complete strangers with no Hanover experience at all. Therefore, we moved the survey to a nearby university campus (approximately 20 kms from the city centre) attended by many students who do not live in Hanover, but commute each day to the campus from other towns nearby.

The second part of the survey was executed as a questionnaire. At a school located in a small town about 30 km from Hanover we were allowed to hand out questionnaires to pupils of the 10th up to 12th grade (age 16-18 years) while they were in class. Three complete classes participated in the test.

8.4 Data Analysis

8.4.1 General Statistics

Altogether 350 subjects participated in the test (see *Table 8.1)*. The school questionnaire was filled in by 252 pupils and their teachers (representing the older than 20 years aged subjects) and 98 people were interviewed on the Hanover fairground during a major industrial fair or on the Campus of the University of Applied Sciences and Arts Hanover, located 20km south of the city centre. Because of the unbalanced distribution of age, the age groups were only separated into two age groups 'age < 20' and 'age > 20' for the following analysis.

The interdependency between the basic subject properties (given in *Table 8.2)* shows that there are too few participants in the subgroups 'age > 20' and 'high familiarity'. So the statistical analysis of this data sets is not as well grounded as the other ones. This is especially relevant if the data sets are split up according to more than one property at a time. In this paper we confine our further analysis to the evaluation of the influence of one single property alone.

Table 8.1. General statistics about test subjects.

	total	sex		age						familiarity		
		male	female	<20	21-30	31-40	41-50	51-60	>60	high	average	low
school	252	122	130	244	0	2	3	3	0	44	95	113
interview	98	48	50	51	27	10	8	2	0	19	34	45
total	**350**	**170**	**180**	**295**	**27**	**12**	**11**	**5**	**0**	**63**	**129**	**158**

Table 8.2. Distribution of subject characteristics (in grey: substantially smaller samples compared to other subgroups).

sex	age		sex	familiarity			age	familiarity		
	< 20	> 20		high	average	low		high	average	low
M	141	29	m	34	57	79	< 20	48	104	143
F	154	26	f	29	72	79	> 20	12	22	21

8.4.2 Data Analysis

The subjects were queried about three different locations in Hanover. Each place was analysed separately to compare the landmark objects which were named by the participants. There are several alternative ways to look at the data: The original answers were recorded (either written by the subjects themselves on the questionnaire or written down by the interviewer), so the landmark descriptions can be evaluated (examining what kind of name or description is given). But we avoid the difficulties of a qualitative analysis altogether and lay the focus of this paper onto the quantity evaluation of references for each object at each place. On the one hand it is possible to rank the objects according to their number of references and compare these rankings for different subject properties. On the other hand it can be assumed, that the participants would name those landmarks first, that are the most important ones for them and therefore first come to mind while memorizing the place. For that reason it is possible to compare the order in which the objects are given and compute average rank positions for each landmark. In this paper we focus mainly on the evaluation of the quantity of references, but show one example of the rank placing analysis. For the complete analysis of the data see Chaouali (2008).

First of all, we introduce the data more generally (see *Table 8.3)*. From the 350 participants 314 have given landmarks for the Ernst-August-Platz (90 %), 260 for the Kröpcke (74%) and 183 for the Steintor (52%). This indicates that the first two locations are more well-known places in Hanover city than the third one. And in fact, place (1) is the plaza in front of the main train station and (2) – not far away from it – is the main underground exchange and most central place in the shopping area of the city. The data also provides indication that the number of landmarks that are given by the participants depends on the global degree of popularity (see second column of *Table 8.3)*. There is no clear indication that men or women give more or less landmarks (except at Kröpcke, where on average female subjects give nearly one landmark object more than male ones do), or that the age affects the number. But the familiarity of the subjects with the environment plays a vital and clear role: the better the subjects knew the place, the more landmark objects they referred to in the test.

Table 8.3. General statistics: average number of landmarks given.

intersection	how many subjects answered	number of landmarks	sex		age		familiarity		
	%	(average)	male	female	<20	>20	high	average	low
(1) Ernst-August-Platz	**90%**	3,1	2,9	3,3	3,2	2,9	3,5	3,3	2,8
(2) Kröpcke	**74%**	3,0	2,6	3,5	3,0	3,4	3,8	3,4	2,5
(3) Steintor	**52%**	2,0	2,2	2,0	2,1	2,2	2,9	2,1	1,7

A more detailed analysis of the referenced landmarks at each location shows that between 12 (Steintor) and 18 (Ernst-August-Platz) different objects were used. We summarize these in ranking lists, that assign the highest rank to the most often named object. These lists were compiled splitting the data according to subject properties, e.g. all answers from males against females. The visual inspection showed that the different lists are altogether surprisingly homogenous. All subgroups and their counterparts (male / female, age < 20 and age > 20, etc.) use the same objects as landmarks. There also tends to be a common understanding what the really important landmarks are, indicated by the observation that the top 5 landmarks are identical for all subgroups most of the time. Only in the lower ranks do the landmark choice really differ. In order to provide evidence that these lists are very similar to each other, we present the Spearman-rank-correlation coefficients in *Table 8.4,* which allows to compute the correlation between two ranked lists (Bortz 2005). (For the computation the lists were cut off to the top 10 objects.) A correlation coefficient of value 1 reflects the total identity of two rankings, while a coefficient with value 0.5 expresses, that the ranks in the lists are not correlated at all.

Table 8.4. Spearman-rank-correlation coefficients.

Comparison between:	(1) Ernst-August-Platz	(2) Kröpcke	(3) Steintor
male / female	0.85	0.82	0.58
age <20 / age >20	0.90	0.88	0.89
familiarity high / average	0.98	0.94	0.89
familiarity average / low	0.94	0.92	0.94
familiarity high / low	0.90	0.84	0.93

As it is clearly visible in *Table 8.4,* most of the rankings are in fact very similar to each other, indicated by correlation coefficients higher than 0.90. The ones that are below that threshold are marked in grey (lower than 0.60 in dark grey), because these values can be seen as an indicator that the ranks of the compared lists have some discrepancies between each other and therefore seem promising for our analysis.

The data analysis presented in this paper focuses on the Steintor plaza, because obviously there seems to be something notable in the landmark choice between the subgroups. To explain this we compared the rank placement of landmark objects between Ernst-August-Platz and Steintor (see *Figure 8.2).* At the Ernst-August-Platz there is one dominating landmark object, namely the 'Ernst-August-Denkmal' (a large historic bronze statue of prince Ernst-August of Hanover on a horse). Nearly

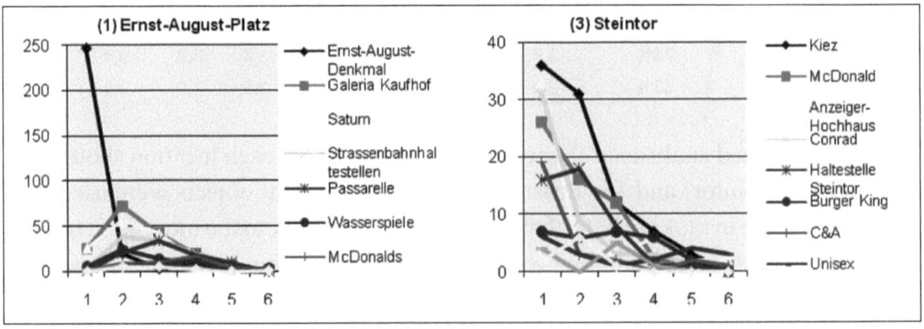

Fig. 8.2. Varying importance of landmarks (x-axis: placing in rank, y-axis: quantity of subject references). Left: (1) Ernst-August-Platz: one landmark object dominates all others, (3) Steintor: several objects may occur in first place (rank 1).

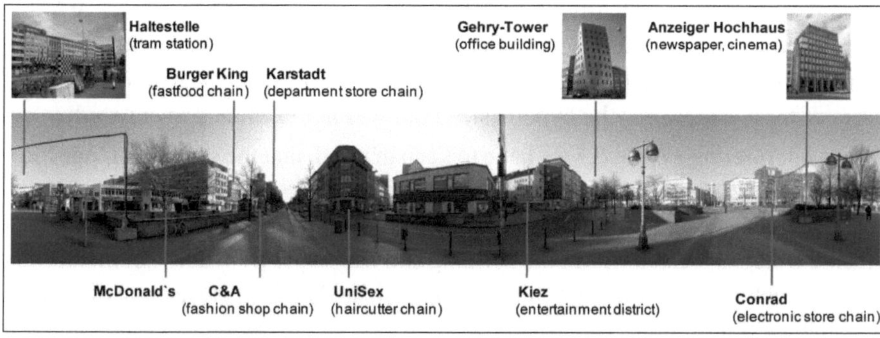

Fig. 8.3. Referenced landmarks at Steintor.

all subjects reference to this landmark and most of the time they give it in the first place, so this object is a rather strong landmark that dominates that place. Quite the opposite happens at the Steintor. There is no common understanding about the single best landmark for that place, indicated by the wide variety of answers (especially on first rank, see *Figure 8.2)*. These are now evaluated, whether there is a significant dependency between landmark choice and sex, age or familiarity or not.

In *Figure 8.3* a photo panoramic view of the Steintor plaza and its referenced landmarks are shown. The place is at the border of the shopping area of Hanover, at the beginning of the pedestrian zone leading directly to Kröpcke. There is a wide variety of buildings and connected activities around the place: on one side there are many shops, on the other is an important street, there is an underground and a tram station as well as an entertainment district nearby.

In the following figures *(Figure 8.4-6)* the quantity of references to the different landmarks split up by the characteristics sex, age or familiarity are given. The small black triangles above specific bars of the diagrams indicate when statistically significant differences occur (checked with a chi-square test and confidence interval of 95%, see Bortz (2005)). So in fact there exist differences depending on the three different subject properties.

In *Figure 8.4* the divergence happens at the landmarks Conrad, C&A and Karstadt, which are all different types of stores. The first incident is easy to explain, because

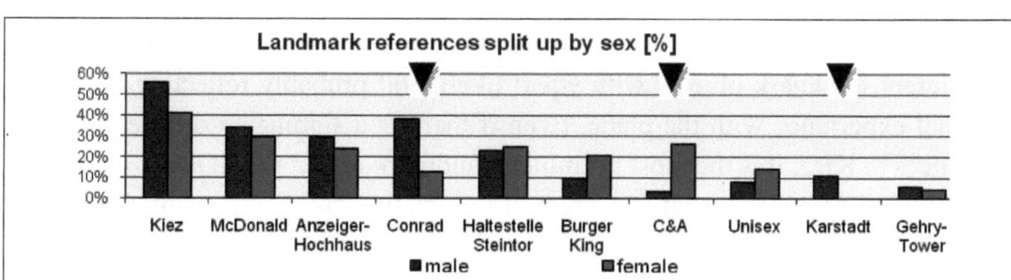

Fig. 8.4. Differences in referencing landmarks depending on sex.

Conrad is an electronic hardware store quite frequently visited by men. C&A is a fashion shop chain and Karstadt a general department store (also including fashion). Whether the groups that found these more important as a landmark are more frequent users of these stores, may be hypothesized but was not examined in our study.

For interpreting *Figure 8.5* we have to keep in mind, that the sample for 'age>20' is not very large and therefore the results may be not representative. There are significant differences in using the Anzeiger Hochaus as a landmark, probably caused by a cinema which is located within the building and whose program does not include the typical blockbusters, but targets a more sophisticated audience. Also the difference in using the tram station 'Haltestelle Steintor' as a landmark indicates the influence of different personal experiences, interests or backround knowledge of the groups. This tram and bus station is a visually very distinctive and eye-catching construction, designed by an award winning architect 15 years ago and the most striking in a series of especially designed bus stops in Hanover. On the other side Unisex is a new haircutter chain in Hanover, that seems to have an especially young clientele as their customers.

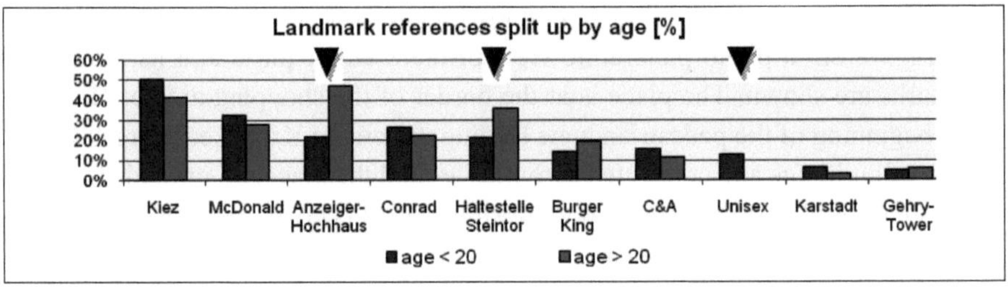

Fig. 8.5. Differences in referencing landmarks depending on age.

Because the degree of familiarity with the environment influences the number of landmarks given (check *Table 8.3* again), the percentage of references to landmarks is most of the time higher than in the other groups, as each subject simply names more objects. But it seems, that with increasing familiarity with the place the subjects build up a common understanding which landmarks define the Steintor plaza appropriately and represent its unique function in the city – namely evening entertainment (Kiez (night clubs), McDonald (fast food even late night), Anzeiger Hochhaus (cinema)). On the other side the more unfamiliar subjects name a variety of different landmark objects with equal likelihood probably reflecting their own personal experience with the place. Even if there is a significant difference in the references to Kiez, it is the most prominent landmark object for all groups (in fact it is more of an area, because it is a street of houses that build up a small amusement district). The Anzeiger Hochhaus, on the other hand, is clearly more important for people familiar with Hanover than to others. That is due to the fact that it is one of the architectural sights of the city and quite popular with locals, but this quality is

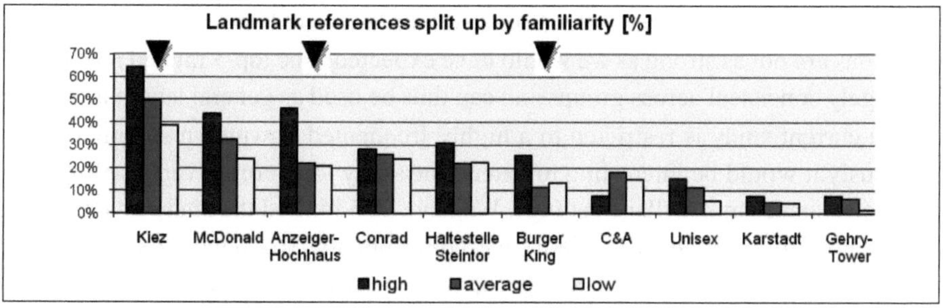

Fig. 8.6. Differences in referencing landmarks depending on familiarity.

probably not observable at first glance (because it is embedded in a row of buildings) and rises with increasing knowledge of the place.

For the last divergence, there seems to be no reasonable explanation why Burger King is more salient when you are familiar with the location. One possible explanation is that the location was previously occupied by a well-known shop, so that the recent change in the well-known environment might causes the difference here.

8.4.3 Results

Altogether the data reflects a clear homogeneity of the landmarks used in all different groups. There seem to be a common understanding what the important landmarks at a specific place are. All subgroups refer to the same landmark objects, although there is a variety of up to 18 different objects for one place. Additionally, subjects assign them more or less the same relevance (reflected in the rank), that means round about the top 5 landmarks are consistent throughout all subgroups. Only when analysing the data in more detail, some user specific differences show up in the ranks (about 2 or 3 objects per location have statistically significant differences in rank). Beyond the discussion in *Section 8.4.2* about the Steintor also at Kröpcke and Ernst-August-Platz differences were established, which back up the hypotheses about user preferences, we have assumed so far: H&M (a fashion shop chain) is more important for women than for men, while vice versa Saturn (a media and electro-equipment store) is a more important landmark for men. McDonald's is ranked higher by younger people, while the opera building is more important for people older than 20 years.

8.5 Conclusions and Outlook

In this paper we have analysed if there are significant differences in remembering salient topographic objects in the inner city environment of Hanover depending on age, sex and familiarity with the environment as the basis for better adapted landmark selection for different user groups. While some significant differences were identified for the different variables (age, sex, familiarity with the environment) these variations are not as strong as we would have expected. The top-5 ranked landmarks are largely consistent across groups and can thus be used as general landmarks.

The current study is restricted to a highly frequented downtown shopping area. Obviously it would be interesting to extend the study to a more diverse set of locations, e.g. suburban or village settings. We have also limited the study to a selected age group. Given the data available we could not make a detailed study across different age groups in the 'over 20' group. The current data indicates that the top-5 landmarks would be suitable for the different sub-categories of this group but interesting variation could exist. We have explicitly excluded very young subjects and the elderly. For these groups we would expect more significant differences and thus a much more pronounced role for adaptation to users.

Summarising, we can say that the top-5 landmarks for all locations will work for all groups studied (acceptable as landmarks) while the identified significant variations might be exploited for a more personalized experience (preferred as landmarks). For practicality we limited the current study to the most relevant target audience (early adopters) and application (inner city navigation) in pedestrian navigation. Future studies should extend the coverage to different user groups and spatial as well as usage contexts.

References

Bia K, Sewon L, Lee J (2007) Gender Differences in Spatial Navigation, in: Proc. WASET Vol. 25, Nov. 2007.

Bortz J (2005) Statistik für Human- und Sozialwissenschaftler, 6th Edition, Springer.

Burnett G, Smith D, May A (2001) Supporting the Navigation Task: Characteristics of 'Good' Landmarks, *in:* Hanson MA, Hrsg., Contempory Ergonomics 2001, Taylor and Francis, Seiten 441–446.

Chaouali M (2008) Personalisierte Landmarken, Diploma thesis, Leibniz University of Hanover.

Daniel MP, Denis M (1998) Spatial Descriptions as Navigational Aids: A Cognitive Analysis of Route Directions, *Kognitionswissenschaft* 7(1), 45–52.

Davidyuk O, Riekki J, Rautio V, Sun J (2004) Context-aware middleware for mobile multimedia applications. In Proceedings of the 3rd international Conference on Mobile and Ubiquitous Multimedia [College Park, Maryland, October 27 – 29, 2004]. MUM '04, vol. 83. ACM, New York, NY, 213-220.

Denis M, Pazzaglia F, Cornoldi C, Bertolo L (1999) Spatial Discourse and Navigation: An Analysis of Route Directions in the City of Venice, *Applied Cognitive Psychology* 13, 145–174.

Elias B (2006) Extraktion von Landmarken für die Navigation, Wissenschaftliche Arbeiten der Fachrichtung Geodäsie und Geoinformatik der Universität Hannover, Nr. 260, ISSN 0174 – 1454, gleichzeitig erschienen in der Deutschen Geodätischen Kommission, Reihe C, Nr. 596.

Elias B, Brenner C (2004) Automatic Generation and Application of Landmarks in Navigation Data Sets, in: Peter Fisher (Edt.): Developments in Spatial Data Handling – 11th International Symposium on Spatial Data Handling, Springer Verlag, pp. 469-480.

Elias B, Hampe M, Sester M (2004) Adaptive Visualisation of Landmarks using an MRDB. In: Meng L, Zipf A, Reichenbacher T (Eds.): Map-based Mobile Services Theories, Methods and Implementations, Springer, pp. 75-88

Elias B, Paelke V (2008) User-Centred Design of Landmark Visualizations, in: Meng L, Zipf A, Winter S (Eds.): Map-Based Mobile Services – Interactivity, Usability and Case Studies, Springer Berlin.

Elias B, Paelke V, Kuhnt S (2006) Kartographische Visualisierung von Landmarken, in: Aktuelle Entwicklungen in Geoinformation und Visualisierung, GEOVIS 2006, 5./6. April 2006, Potsdam, Kartographische Schriften Band 10, Kirschbaum Verlag.

Espinoza F (2003) Towards individual service provisioning. In Proceedings of the 8th international Conference on intelligent User interfaces [Miami, Florida, USA, January 12 – 15, 2003]. IUI '03. ACM, New York, NY, 239-241.

Gajos KZ, Czerwinski M, Tan DS, Weld DS (2006) Exploring the design space for adaptive graphical user interfaces. In Proceedings of the Working Conference on Advanced Visual interfaces [Venezia, Italy, May 23 – 26, 2006]. AVI '06. ACM, New York, NY, 201-208.

Golledge R (1999) Human Wayfinding and Cognitive Maps, In: Golledge, R., Edt., Wayfinding Behavior, The John Hopkins University Press, Baltimore and London, pp. 5-46.

Ho SY, Kwok SH (2002) The attraction of personalized service for users in mobile commerce: an empirical study. SIGecom Exch. 3, 4 [Dec. 2002] 10-18.

Holding CS, Holding DH (1989) Acquisition of route network knowledged by males and females. Journal of General Psychology, 116, 29-41.

Köpke J (2007) Personalisierte Landmarken, Diploma thesis, Leibniz University of Hannover.

Lübke C (2004) Extraktion von Landmarken aus ATKIS-Daten, Diploma thesis, University of Hannover.

Magellan (2008) http://pro.magellangps.com/en/news/releases/viewRelease.asp?id=584, Press release February 4, 2008, last access 30.5.2008.

Meng L, Zipf A, Winter S (Eds.) (2007) Map-Based Mobile Services – Interactivity, Usability and Case Studies, Springer Berlin.

Michon P, Denis M (2001) When and Why Are Visual Landmarks Used in Giving Directions?.- In: Montello D, Ed.: Spatial Information Theory, International Conference COSIT 2001, Proceedings, Springer Verlag, S. 292-305

Niemelä E, Latvakoski J (2004) Survey of requirements and solutions for ubiquitous software. In Proceedings of the 3rd international Conference on Mobile and Ubiquitous Multimedia [College Park, Maryland, October 27 – 29, 2004]. MUM '04, vol. 83. ACM, New York, NY, 71-78.

Oviatt S, Swindells C, Arthur A (2008) Implicit user-adaptive system engagement in speech and pen interfaces. In Proceeding of the Twenty-Sixth Annual SIGCHI Conference on Human Factors in Computing Systems [Florence, Italy, April 05 – 10, 2008]. CHI '08. ACM, New York, NY, 969-978.

Paelke V, Elias B, Hampe M (2005) The CityInfo Pedestrian Information System, in: Proceedings of 22nd International Cartographic Conference, 9–16 July 2005, La Coruña/ Spain.

Peternier A, Righetti X, Hopmann M, Thalmann D, Repettoy M, Papagiannakis G, Davy P, Lim M, Magnenat-Thalmann N, Barsocchi P, Fragopoulos T, Serpanos D, Gialelis Y,

Kirykou A (2007) Chloe@University: an indoor, mobile mixed reality guidance system. In Proceedings of the 2007 ACM Symposium on Virtual Reality Software and Technology [Newport Beach, California, November 05–07, 2007] Spencer SN, Ed. VRST '07. ACM, New York, NY, 227-228.

Raubal M, Winter S (2002) Enriching Wayfinding Instructions with Local Landmarks. In: Egenhofer MJ, Mark DM (Eds.), Geographic Information Science. Lecture Notes in Computer Science, Vol. 2478. Springer, Berlin, pp. 243-259.

Reichenbacher T (2003) Mobile Kartographie und ihre Nutzer: Szenarien adaptiver Visualisierungsmöglichkeiten für mobile Benutzeraktivitäten, in: Visualisierung und Erschließung von Geodaten. GEOVIS 2003, Vol. 7 of Kartographische Schriften, Kirschbaum Verlag.

Reimann C, Paelke V (2005) Adaptive Mixed Reality Games, in: Proc. ACM SIGCHI International Conference on Advances in Computer Entertainment Technology [ACE 2005) 15th–17th June 2005, Valencia, Spain.

Rhodes B (2003) Using Physical Context for Just-in-Time Information Retrieval. IEEE Trans. Comput. 52, 8 [Aug. 2003) 1011-1014.

R S, Sugumaran V (2007) Adaptive middleware architecture for information sharing on mobile phones. In Proceedings of the 2007 ACM Symposium on Applied Computing [Seoul, Korea, March 11 – 15, 2007]. SAC '07. ACM, New York, NY, 800-804.

Schmidt A, Aidoo KA, Takaluoma A, Tuomela U, Laerhoven KV, Velde WV (1999) Advanced Interaction in Context. In Proc. 1st int. Symposium on Handheld and Ubiquitous Computing, Lecture Notes In Computer Science, vol. 1707, Springer-Verlag.

Sharps MJ, Gollin ES (1988) Aging and free recall for objects located in space. Journal of Gerontology: Psychological Science, 43,8-11.

Sorrows M, Hirtle S (1999) The Nature of Landmarks for Real and Electronic Spaces.- In: Freksa C, Mark D, Eds.: Spatial Information Theory, International Conference COSIT '99, Proceedings, Springer Verlag, S. 37-50

Wahlster W (2007) Smartweb: multimodal web services on the road. In Proceedings of the 15th international Conference on Multimedia [Augsburg, Germany, September 25 – 29, 2007]. MULTIMEDIA '07. ACM, New York, NY, 16-16.

Want R, Schilit WN, Adams NI, Gold R, Petersen K, Goldberg D, Ellis JR, Weiser M (1996) The ParcTab Ubiquitous Computing Experiment, Mobile Computing, Imielinski T, Korth HF, Editors, Kluwer Academic Publishers, Boston, MA (1996) pp. 45-102.

Winter S (2002) Ontologisches Modellieren von Routen für mobile Navigationsdienste. In: Kelnhofer F, Lechthaler M (Eds.): TeleKartographie und Location-Based Services. Schriftenreihe der Studienrichtung Vermessungswesen und Geoinformation. Technische Universität Wien, Vienna, pp. 111-124

Winter S, Raubal M, Nothegger C (2005) Focalizing Measures of Salience for Route Directions. In: Meng L, Zipf A, Reichenbacher T (Eds.): Map-Based Mobile Services – Theories, Methods and Design Implementations, Springer Geosciences, Berlin, pp. 127-142.

Wu S, Wu K (2006) Effective location based services with dynamic data management in mobile environments. Wirel. Netw. 12, 3, May. 2006.

Zipf A (2002) User-Adaptive Maps for Location-Based Services (LBS) for Tourism, *in:* Woeber K, Frew A, Hitz M (Hrsg.): Proceedings of the 9th International Conference for Information and Communication Technologies in Tourism, ENTER 2002. Innsbruck, Austria, Springer Verlag.

9 Measuring the Impact of Location-Awareness in the Acceptance of Mobile Systems

Eduardo Dias[1] and Euro Beinat[2]

[1] Department of Spatial Economics, Vrije Universiteit Amsterdam, The Netherlands
[2] Centre for Geoinformatics, Salzburg University, Austria

Abstract

In this paper we propose a methodology to determine the impact on adoption of location-aware information. In a National Park setting digital information is already being served to its visitors. We evaluated if enhancing the information with location awareness has an impact on its acceptance. We tested two identical applications that aim to inform visitors to Nature areas: being one a simple informative digital tool, while the other is enhanced by location interactivity (it uses the position of the visitor to automatically deliver information). We modelled the future usage of the systems using the Technology Acceptance Model, an established model from the Information Systems literature. The results show that making a system location-aware produces a significant positive impact in the acceptance of the system and validates novel approaches to the measurement and comparison of mobile systems by quantifying the relative impact of location-awareness. This work also proposes a model to explain the usage of information in Natural Areas.

Keywords: LBS, Technology Acceptance Model (TAM), GPS, digital guide, protected areas

9.1 Introduction

While the main goal for most Natural Parks is the conservation of Nature, many areas have Education and Recreation as additional goals. The development of informative websites and multimedia CD-ROMs is a recognition from Park managers about the importance to inform the visitors, but do not work for the field visit when most questions arise. Mobile technology can be used to satisfy the visitors' information needs anytime anywhere and when connected to a GPS receiver it can intelligently filter or push the right information at the right place/time. But, is the added-effort of making a location-aware application compensated by its added value? In this paper we propose a methodology to determine the added-value of delivering location-aware information.

9.1.1 Motivation

In recent decades, recreational use of natural areas has grown rapidly from low intensive and relatively passive use to a situation where tourism is the dominant force driving change in many rural areas and their associated communities (Butler, Hall & Jenkins 1998). Nevertheless, excessive use of natural areas can have significant direct and indirect negative impacts, which include, but are not limited to environmental degradation (Farrell & Marion 2001) and diminishing of the quality of visitors' recreational experience (Lynn & Brown 2003). Mobile Information Services *(Figure 9.1* shows an example of a Mobile information system implemented in a Personal Digital Assistant) are seen as a mean of supplying park managers with both the possibility to monitor and manage the visitor distribution within parks and, additionally, to help visitors become more aware of the richness of the natural and cultural resources they are visiting. However, mobile information systems cannot help managers and visitors if they are not used.

Fig. 9.1. Visitors accessing information in the field visit.

In this paper we analyze the usage potential of such an information tool with and without location awareness. A methodology is proposed that allows the quantification of the added value that the visitors ascribe to the location awareness. This research tries to prove the following hypothesis:

Enhancing information with location awareness positively influences the digital information adoption in the park.

9.1.2 Research Context

Location-Based Services (LBS) allow access to information for which the content is filtered and tailored based on the location of the user. We tend to spend the majority of our time in known or familiar environments, where we either do not require information or know where to obtain it. Consequently, LBS can be particularly useful in the tourism and leisure domains where visitors are eager for information and also unfamiliar with the local environment.

9.1.3 Experimental Protocol

A controlled experiment was designed to measure the influence that the location-awareness component had on the acceptance of digital information in visitors to natural areas. In the experiment, subjects were issued with the same palm-sized digital devices (PDAs) and divided into a control and a test group. The control group was provided with digital information that described specific locations of the park in terms of particular interesting features about flora, fauna or particular interesting landscape features. This system had no location sensitivity (the users had to actively request the information). The test group had access to the same information in the same device, but its information was enhanced with location-sensitivity (the device was connected to a GPS receiver). This system has two main location components:

1. A cross indicating the exact location of the visitor while walking in the dune park on a map.
2. Information content is pushed to the visitor when they are at specific locations. A soft cuckoo-song-sound is emitted by the device at these locations and the relevant content page is automatically shown.

The "Texel Dunes National Park", located on an island in the north of the Netherlands, served as the testing ground for this work. Part of the dune park is only accessible via the EcoMare museum and visitor centre, which is visited by a large number of tourists during the summer period. Random visitors to the EcoMare museum were approached and asked if they would be interested in participating in

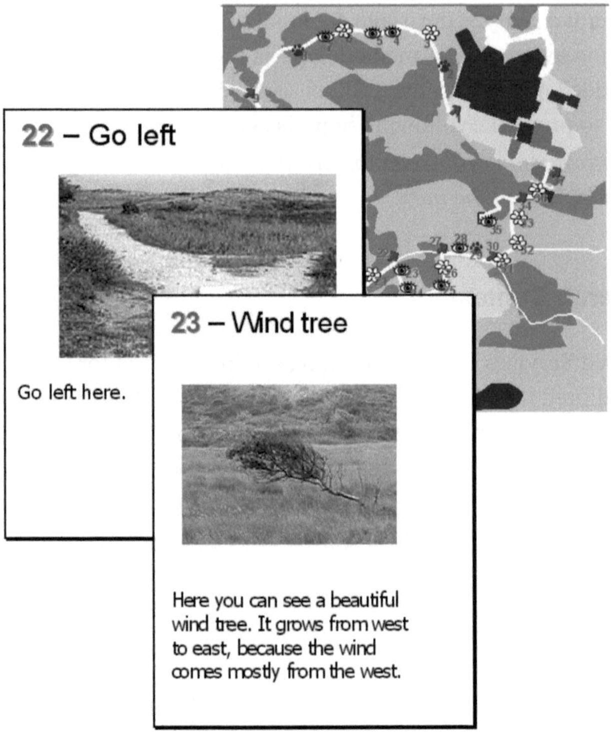

Fig. 9.2. Map and PoI content available to the visitors.

this research. The composition of the groups was controlled to ensure their profiles were as similar as possible. In addition, all subjects set out to follow the same route, in similar weather conditions. Therefore, we assume that if any differences in the results are found it is due to the presence of location sensitivity.

9.1.4 Information Content

The information provided to the subjects comprised of a map of the route with the locations of a number of Points-of-Interest (PoI) displayed, see *Figure 9.2*.

Detailed information about each of these was supplied in the subsequent information. This content consisted of a prominent title, a photo of the feature and a text description. The POIs were classified into four categories: "Directions" (indicating the path the subject should follow); "Plants" (information about a particular plant visible from the path); "Animals" (information about animals relevant at a particular point of the path) and "Landscape" (information about landscape features visible from a certain location).

9.2 Technology Acceptance Model

9.2.1 The Theory of Reasoned Action

The TAM was first described and applied by Davis in 1985 (Davis 1985). This model is based on the Theory of Reasoned Action (TRA) (Ajzen & Fishbein 1980, Fishbein & Ajzen 1975, Fishbein & Ajzen 1981), which states that a person forms an attitude about a situation, object or action on the basis of his/hers beliefs. The two hypotheses that form the TRA are: (1) intention positively affects usage; and (2) attitude positively affects intention. Additionally, on the basis of attitude, an intention is formed to handle the situation and this intention completely determines the actual behaviour. The TRA has been extensively researched in the past (Madden et al. 1992, Sheppard et al. 1988), and the model has been proven successful in predicting and explaining user behaviour in a varied range of domains and fields, such as National Parks management (Bright et al. 1993), environmentally-conscious behaviour (Goldenhar & Connell 1993) or wireless Internet adoption (Lu et al. 2003). According to the TRA, a person's behaviour is determined by her or his behavioural intention, which is determined by both the person's attitude and subjective norm concerning the behaviour. The intention can be considered a measure of the strength of a person's intention to adopt a specific type of behaviour. The attitude is defined as a person's positive or negative feelings about performing that behaviour, and the subjective norm refers to the person's perception that most people who are important to her or him think she or he should (or should not) perform that behaviour (Fishbein & Ajzen 1975). It is also important to note that the TRA is a general model and not specific to a given domain or technology type. Therefore, it is not particularly suited for a specific domain and requires adjustments to fit the particularities of a study. Thus, the beliefs that motivate behaviour in a particular case have to be explicitly specified (Davis et al. 1989).

9.2.2 TAM's Constructs and Assumptions

Davis developed the TAM as an adaptation of the TRA, specifically designed to test the acceptance of information systems (IS). The TAM uses the TRA as the theoretical underpinning for defining the links between the two basic constructs and attitude, intention and the adoption behaviour. The TAM defines that the acceptance of technology is dependent on two independent constructs: the perceived usefulness (PU), and the perceived ease-of-use (PEU), and on the causal chain from the TRA: attitude, intention, and, finally, usage behaviour. The perceived usefulness is defined as "the degree to which a person believes that using a particular system would enhance his or her job performance" (Davis 1989). It is a quantification of the users' perception of how the technology can help them perform their job better.

The perceived ease of use is defined as the "degree to which a person believes that using a particular system would be free of effort" (Davis 1989). This construct is extremely important because, even when a person considers a technology to be useful, this person can still reject it if she believes that the effort to use it is greater than its performance benefits.

Previous research has explained perceived ease of use to be based on a model composed of three anchors that determine early perceptions about the ease of use of a new system. These anchors are: control (internal and external – conceptualized as computer self-efficacy and facilitating conditions, respectively); intrinsic motivation (conceptualized as computer playfulness); and emotion (conceptualized as computer anxiety) (Venkatesh 2000).

The cost-benefit paradigm is an important concept to understand the relation between perceived usefulness and perceived ease of use. TAM is based on a rational evaluation, where the behavioural intentions are the outcome of the rational assessment of the presented software (balancing the PU and PEU), and the outcome determines the behavioural intention to use it (Davis 1989). According to Davis (Davis 1989), the perceived usefulness is a major determinant of people's intention to use the tool, whereas perceived ease of use is a (significant) secondary determinant of intention. *Figure 9.3* shows the TRA combined with the technology acceptance model. The arrows represent the relations that underlie the model. The first two relations are based on the TRA, while the others are TAM-specific:

- T1: Intention determines usage.
- T2: Attitude determines intention.
- T3: Perceived usefulness affects intention.
- T4: Perceived usefulness influences attitude.
- T5: Perceived ease of use affects attitude.
- T6: Perceived ease of use affects perceived usefulness.
- T7 and T8: External variables (that depend on the field of study) relate to perceived usefulness and ease of use.

The model was originally designed to measure the factors that explain the acceptance and usage of classic information technology in the desktop and office work. Nevertheless, it has proved to be robust enough, and has been extensively adapted and applied, including for wireless services (Lu et al. 2003, Wu & Wang 2005) and mobile devices (Sarker & Wells 2003). Junglas (2005) tested the adoption of LBS also using a protocol that included a test group and a control group (users with PDAs with and without location awareness), but different from the present study, we use a scenario with real users (N=204) of an existing system and carrying out their own task (exploring the park), in the Junglas study, students (N=58) were used and the tasks/scenarios were elaborated specifically for the study. It is interesting to observe that the results from this study are similar and therefore validate the previous study.

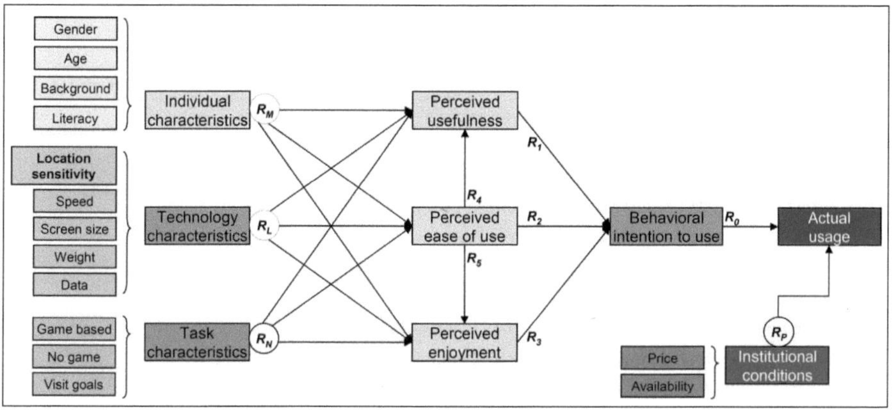

Fig. 9.3. The Technology Acceptance Model (Davis 1989).

9.2.3 Scaling Perceptions of Information Systems

The most common way to collect data on the constructs is by administering questionnaires. In a nutshell, the study subjects become acquainted with the information system (through a presentation and/or direct usage experience), and afterwards they fill in questionnaires by answering a series of questions that measure their perceptions of the technology. These questions are called 'scale items' and are typically Likert scales with five alternative multiple-choices, ranging from completely disagree to completely agree. As in most social research, the formulation and definition of the questions is very important as it is necessary to distinguish between the factors and not bias the data collection. Using the Likert scale makes it possible to quantify the relations between the different model constructs. Therefore, the questions have to be designed in such a way that they translate these relations. Davis's research was aimed at measuring acceptance of a new software tool by users and introduced a standard set of questions that can be used in standard TAM studies, called the Davis's scale (see *Table 9.1).*

Table 9.1. Classic Davis set of questions that compose the PU and PEU constructs (Davis 1989).

Scale item	Construct
Using <*TestTool*> in my job would enable me to accomplish tasks more quickly	
Using <*TestTool*> would improve my job performance	
Using <*TestTool*> in my job would increase my productivity	PU
Using <*TestTool*> would enhance my effectiveness on the job	
Using <*TestTool*> would make it easier to do my job	
I would find <*TestTool*> useful in my job	
Learning to operate <*TestTool*> would be easy to me	
I would find it easy to get <*TestTool*> to do what I want to do	
My interaction with <*TestTool*> would be clear and understandable	PEU
I would find <*TestTool*> to be flexible to interact with	
It would be easy for me to become skilful at using <*TestTool*>	
I would find <*TestTool*> easy to use	

9.2.4 The Hedonic Extension

Depending on the application, modified and/or extended versions of the model have been designed and applied by a number of authors (Gefen & Straub 1997, Venkatesh & Davis 2000). One particular extension is extremely relevant to the mobile information system discussed in this thesis: the perceived enjoyment construct. In 1992, Davis et al. (1992) introduced perceived enjoyment has an addendum and defined it as "the extent to which the activity of using the computer is perceived to be enjoyable in its own right, apart from any performance consequences that may be anticipated" (Davis et al. 1992, p. 1113). Most studies present perceived usefulness as the most significant predictor of user acceptance followed by perceived ease of use and perceived enjoyment (Igbaria et al. 1996, Shang et al. 2005). Nevertheless, the opposite has also been reported in the literature (Atkinson & Kydd 1997, Moon & Kim 2001, Van der Heijden 2004, Venkatesh 2000) where the subject systems are accepted mainly because of their perceived enjoyment and ease of use and less because of their perceived usefulness. A common denominator to these studies is the fact that the systems under study are pleasure-oriented rather than the classic productivity-oriented systems. The pleasure-oriented are also called Hedonic information systems. They are connected to leisure activities that focus on the fun aspect of information and aim to provide self-fulfilling value to the user, while the productivity-oriented systems focus on the utilitarian value of the information. These are mainly connected to work-related, productive usage and aim to provide instrumental value to the user (Van der Heijden 2004). Here, the instrumentality concept indicates that the system usage objective is external to the user, while the self-fulfilling concept indicates that using the system per se is an objective in itself.

Concluding, user acceptance depends on both the extrinsic motivation and the intrinsic motivation. Extrinsic motivation manifests itself when the user expects a reward or external benefit by interacting with the system. Intrinsic motivation occurs when a user derives benefit just from using the system (Brief & Aldag 1977, p. 497). Referring to the context of this study, even though receiving information during and about the walk can be perceived as useful, the visitors to the National Park were on holiday and, therefore, the usage of the system is based on voluntariness and has a strong leisure component: the visitors will only use the system if they perceive it to be enjoyable. It is a Hedonic information system. Consequently, the technology acceptance model should be extended with the perceived enjoyment construct.

9.3 Protected Area's Information System Acceptance Model

We propose a model for the acceptance of the Mobile Information System based on the technology acceptance model and its extensions. *Figure 9.4* shows the proposed model for this research. Note that the full Theory of Reasoned Action (TRA) was substituted by a simplified version. In order to simplify the TRA, the construct behavioural intention to use replaced the original two constructs attitude towards use and intention to use, as proposed in the theoretical extension of the technology acceptance model (Venkatesh & Davis 2000).

This model tries to explain Actual usage based on factors. The factors and variables are represented by coloured rectangles and the relationships between them are represented by arrows. When a certain factor is composed of several items, examples of these items are presented on its right, and a bracket aggregates these into the factor.

9.3.1 Model Relationships

As explained previously, the actual usage of an information system is determined by the behavioural intention to use (R_0 – BIU influences the actual usage). This relationship is established by the TRA and has been extensively studied in the literature; therefore it will not be empirically tested in this study. The behavioural intention to use is influenced by the perceived usefulness (R_1 – PU influences BIU), by the perceived ease-of-use (R_2 – PEU influences BIU), and by the perceived enjoyment (R_3 – PE influences BIU). It is also expected that the perceived ease of use influences both the perceived usefulness (R_4 – PEU influences PU) and the perceived enjoyment (R_5 – PEU influences PE).

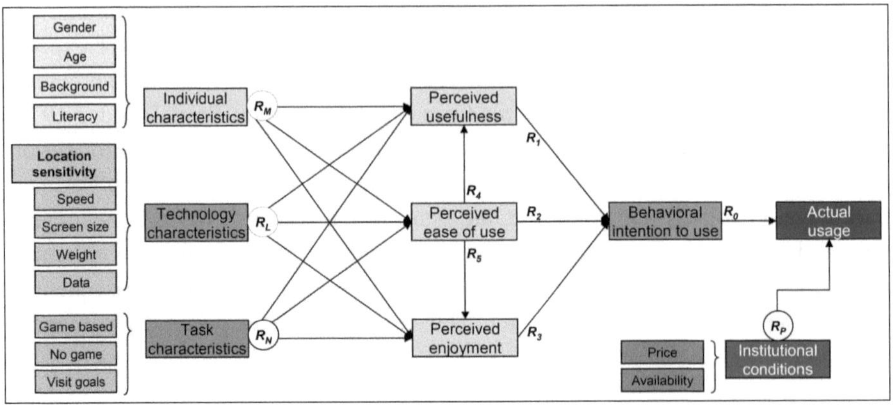

Fig. 9.4. Theoretical model to explain the actual usage of the mobile information system in Protected Areas.

Different people have different perceptions of the experiences and surrounding environments: certain individual characteristics have an impact on perceptions. Therefore, it is expected that all the TAM constructs, perceived usefulness, perceived ease of use, and perceived enjoyment would be influenced by some individual characteristics. Examples of characteristics expected to have great impact are Gender, Age, Professional background, and Literacy level, amongst others (R_M – Individual characteristics influence TAM perceptions).

Naturally, not all systems are alike and therefore the intrinsic technology characteristics are also expected to influence the subject's perception (R_L – Technology characteristics influence TAM perceptions). Some examples of technical characteristics that are expected to have an influence on the perception of users are the software response speed, the screen size, weight (because it is carried by the user), the data (one of the most important factors because it is what the user actually sees), the look-and-feel of the application (design and layout), and the location component for mobile applications. This last factor is what this research is focusing-on and will be explained in detail in the following section *(Section 9.3.2)*. Previous research (Dishaw & Strong 1999, Goodhue & Thompson 1995, Junglas 2003) have highlighted the importance of task-technology fit. This construct states that if a technology is to have a positive impact, it must be a good fit with the task it supports. Since, in the case of the information system for Protected Areas, we have a voluntary usage and not a clear task to fulfil, the technology fit will depend on the goals of the visitors when visiting the Park. For example, students will be looking forward to learning about the ecosystems they are visiting, while the leisure-oriented visitors are hoping not to get lost in the Park, and probably a game-based information system with questions/quizzes would have an impact on their perception of the technology (R_N – task characteristics variables influence TAM constructs).

The last construct is an external construct from the TAM model itself and the overall individual acceptance of the technology. It refers to the institutional conditions imposed by the Protected Area managers. Only if the system is made available, can the visitors actually use it in the future (R_P – Actual usage depends on institutional conditions).

9.3.2 Location as a Determinant of Acceptance

This paper focuses on the adoption of location-awareness to improve the information flows inside nature areas. It is hypothesized that location can increase the adoption of the information technology, and increase the motivation to learn and use technology. Therefore the location sensitivity was used as an independent variable and manipulated in the research, and the effects of this manipulation are measured acceptance levels (the TAM constructs). This was the only empirically tested external variable so far. As explained in the experimental protocol *(Section 9.1.3)* there were two different technological implementations of the system, one called Digital info and the other Location-based service. In both implementations the information was available in a portable handheld computer (PDA). In the Digital info implementation, the user had to pull the information by simply selecting the icon on the map corresponding to the place where he was at the time. For the LBS implementation, the system is aware of the location of the user (via GPS positioning) and it pushes the information to the user at the right place. This is the only difference between the systems, therefore, if differences in the acceptance of the technology are found they can be ascribed to the location component, more precisely to the location based push.

9.4 Results

As explained before, from the model presented on the previous section, not all the relationships were empirically tested. Because the TAM constructs need to be adapted to the technologies being tested, the inter-construct relationships of the technology acceptance model were empirically tested and its reliability, measured by means of scales, validated. Once confidence was established that the instruments are able to measure technology adoption, the location sensitivity was manipulated and its effects measured on the TAM constructs. This manipulation of the independent variable enabled the measurement of the impact of location on adoption. Therefore, the empirical analyses carried out in the framework of this study were the quantification of the relationships in the model presented in *Figure 9.4* with the labels: R_1, R_2, R_3, R_4, the R_5 and part of the R_L.

The values for PU, PEU and Intention were computed from the responses to the ex post questionnaires. The scores for these constructs are based on the average of the corresponding scale items responses.

9.4.1 Reliability

In social sciences, and especially when measuring attitudes and human perceptions, it is important to check for the reliability of the tests administered. Reliability tests have two main objectives: (1) to check if the group of questions or items used measure the same construct; and (2) to check if there are two or more items that are too similar, so that they duplicate the measurement and therefore can be removed from the scale. Because it is difficult or impossible to establish absolute standards for the meaning of human responses to a survey, the reliability analysis can at least measure the consistency of the scale.

The reliability analysis determines the properties of the measurement scale and its items by measuring the relationships between the scale's individual items. The Cronbach's alpha (Cronbach 1951) was the model chosen to determine the reliability level. This is a model of internal consistency, based on the average inter-item correlation. The alpha is a lower bound for the true reliability of the survey. In mathematics, reliability is the size of the variability in the responses that results from differences in the respondents. In other words, the answers to a reliable survey can have differences because the subjects have different opinions and not because of misinterpretations of the questions. The Cronbach's alpha is calculated by using the number of items in the survey (k) and the ratio of the average inter-item covariance to the average item variance, according to Equation:

$$\alpha = \frac{k(\text{cov/var})}{1+(k-1)(\text{cov/var})} \tag{9.1}$$

According to Nunnally and Bernstein (Nunnally & Bernstein 1994, p. 264), when interpreting the reliability coefficient, a value of 0.70 is sufficient for the early stages of research, but basic research should require test scores to have a reliability coefficient of 0.80 or higher. *Table 9.2* displays the reliability indicators for the constructs used in the model. The reliability for the five measured constructs is higher than 0.8, which indicates that the scales are reliable and the questions were well understood by the respondents. For the constructs Perceived Usefulness and Behavioural Intention to Use, the reliability was even higher than 0.9 and 0.95, respectively. This was an important result since the original scale was in the English language and the items were translated to the Dutch language.

Table 9.2. Reliability indicators for the constructs used in the model.

Construct	# Items	N	Alpha
Perceived Enjoyment	5	197	0.832
Perceived Usefulness	5	198	0.928
Perceived Ease of Use	4	203	0.877
Behavioural Intention to Use	2	200	0.964

In order to see if there were redundant items (items measuring exactly the same perception) or items that measured contradictory results, the correlations between the items for each scale were also measured. The results are shown in *Table 9.3, Table 9.4, Table 9.5* and *Table 9.6* for the constructs Perceived Enjoyment, Perceived Usefulness, Perceived Ease-of-Use, and Behavioural Intention to Use, respectively.

For all the items that comprise the four used constructs, there is positive correlation and no duplicating results were found. This means that the scale used was appropriate and yields reliable results that can be used for deeper analysis and to discover the external variables that influence these constructs.

Table 9.3. Inter-item correlation matrix for the Perceived Enjoyment items.

	Exciting	Enjoyable	Interesting	Amusing	Delightful
Exciting	1.00				
Enjoyable	0.49	1.00			
Interesting	0.48	0.32	1.00		
Amusing	0.57	0.51	0.48	1.00	
Delightful	0.51	0.42	0.45	0.78	1.00

Table 9.4. Inter-Item correlations matrix for the Perceived Usefulness items.

	Useful	Practical	Functional	Handy	Efficient
Useful	1.00				
Practical	0.67	1.00			
Functional	0.69	0.80	1.00		
Handy	0.62	0.80	0.82	1.00	
Efficient	0.62	0.72	0.75	0.74	1.00

Table 9.5. Inter-Item correlations matrix for the Perceived Ease of Use items.

	Clear	Effortless	Under-standable	Easy
Clear	1.00			
Effortless	0.71	1.00		
Understandable	0.71	0.70	1.00	
Easy	0.56	0.56	0.68	1.00

Table 9.6. Inter-Item correlations matrix for the Behavioural Intention to Use items.

	Future	Next time
Future	1.00	
Next time	0.49	1.00

9.4.2 Inter-Construct Correlations

The correlation between two variables reflects the degree to which the variables are related and gives a quantitative indication of the strength of correlation. The most common measure of correlation is the Pearson's correlation. This statistic relies on the assumption that the variables being tested have normal distributions, and that the relationship between the variables is a linear one, but these assumptions are not met for the scale data. Examining the descriptive statistics of the constructs (*Table 9.7*), it is observable that the skewness value is different from zero which indicates that the data are not normally distributed.

Table 9.7. Descriptive statistics for the construct scales.

	Mean	Std. Dev	Skewness	N
Hedonic	0.82	0.99	-0.86	199
PU	1.52	1.19	-0.83	199
PEU	1.77	1.04	-1.08	204
BIU	1.36	1.53	-1.33	202

Therefore, the Spearman's rank correlation coefficient (designated by the Greek letter ρ) was used to reveal the relationships. The Spearman's test is a non-parametric measure of correlation that does not make assumptions on the frequency distribution of the variables, and, additionally, it does not require that the relationship between the variables should be linear. This methodology has been extensively used in the TAM literature (Cartwright & Shepperd 2000, Konradt et al. 2003, Konradt et al. 2006, Mavri & Ioannou 2006). The results are displayed in *Figure 9.5*. As expected, all the relationships studied were found to have a positive correlation and be significant at the 1% level (2-tailed).

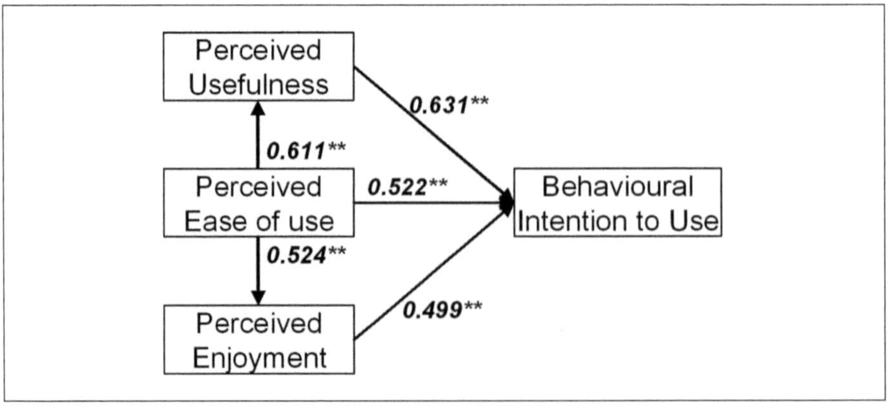

Fig. 9.5. Correlation's strength between the constructs (Spearman coefficients). Note: ** Correlation is significant at the 1% level (2-tailed).

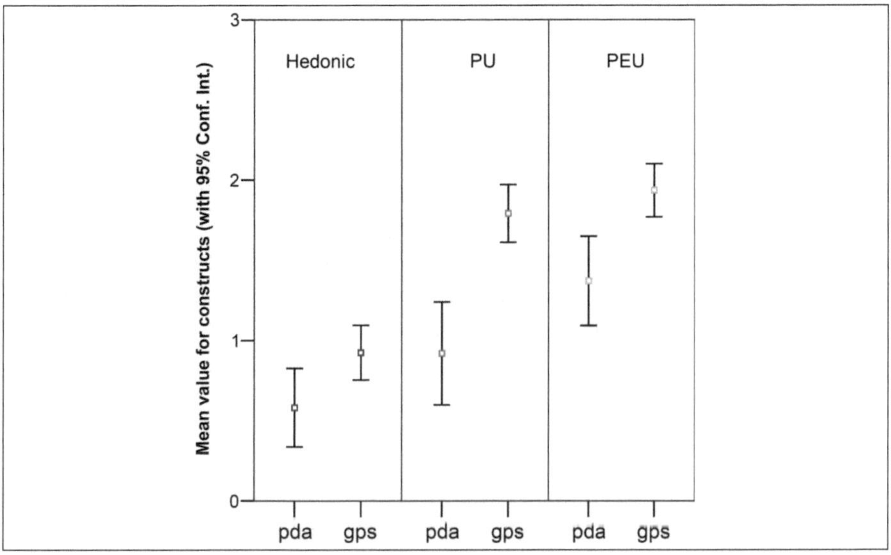

Fig. 9.6. Average scale results for the perception constructs for both LBS or location-based (gps) and Digital info or non-location based (pda) information services

In order to test if the location component of the application does influence the perception of the visitors, and therefore the adoption of the technology, the two groups (Digital info and LBS) were compared for significant differences. The average scale results for both groups were plotted in *Figure 9.6,* which indicates that the location-based group had higher values for the technology perception.

The two groups were also compared for significant differences using the non-parametric Mann-Whitney test (see results in *Table 9.8).* The Mann-Whitney test shows significance levels below 0.01, meaning that for all constructs there are significant differences at the 1% level.

Table 9.8. Significance of differences according to the Mann Whitney test

	Mann-Whitney U	Wilcoxon W	Sig. (2-tailed)
Hedonic	3347.5	5492.5	0.008
PU	2610.5	4755.5	0.000
PEU	3186.5	5532.5	0.000

It is observable, nevertheless, that the location component has different levels of influence on the constructs. The strongest influence is on the perceived usefulness: the users with location-enabled information perceive the application as much more useful. The smallest influence (but still significant) is on the perceived enjoyment: both systems were found to be 'fun to use', but the location-based system was 'enjoyable'.

9.5 Conclusions

An extended version of the Technology Acceptance Model was applied to the information service both for the Digital info and the Location-based service versions. The analyses of the results show that perceived usefulness, perceived ease of use and perceived enjoyment (or hedonic value) influence the intention to use the system in the future. On different strength levels, the location component was found to have a significant positive influence on all the constructs that influence adoption. This suggests that, if the information system is location-enabled, the visitors to Protected Areas are more likely to use it in the future.

9.5.1 Institutional Conditions as a Crucial Determinant of Usage

In our explanatory model *(Figure 9.4),* the institutional conditions play a crucial factor in restricting the actual usage. We assume usage will only take place if the system *is available* at reasonable conditions on the Protected Area site. Even if the individual motivations to use the system are very high, if the pricing is prohibitive, usage will not take place. Therefore, it is crucial that the Park Managers create an infrastructure to make the system available and affordable to its visitors. Park Managers recognize the need for such a tool in order to improve the information flows inside the Park (Dias et al. 2004), not only in order to inform tourists and to have more environmentally-aware visitors but also because this tool allows for the collection of valuable and highly detailed information on the visitors' behaviour which enables analysis on sustainable usage and environmental protection (Dias et al. 2007). Therefore, it is in the interest of the Park Management to create and maintain the infrastructure that facilitates the actual usage.

Acknowledgements

The authors gratefully acknowledges all who contributed to this research: mainly, the anonymous testers who were kind enough to fill-in the questionnaires during their visit to the park and namely to Ecomare (www.ecomare.nl), Camineo (www.camineo.fr) and Geodan (www.geodan.com) for developing the LBS system and allowing its usage in this research. The financial support from the Portuguese National Science Foundation (FCT/MCT) under the PhD grant SFRH/BD/12758/2003 is also gratefully acknowledged.

References

Ajzen I, Fishbein M (1980) A theory of reasoned action, *Understanding Attitudes and Predicting Social Behaviour*.

Atkinson MA, Kydd C (1997) Individual characteristics associated with World Wide Web use: an empirical study of playfulness and motivation, *ACM SIGMIS Database*, vol. 28, no. 2, pp. 53-62.

Brief AP, Aldag RJ (1977) The Intrinsic-Extrinsic Dichotomy: Toward Conceptual Clarity, *The Academy of Management Review*, vol. 2, no. 3, pp. 496-500.

Bright AD, Fishbein M, Manfredo MJ, Bath A (1993) Application of the Theory of Reasoned Action to the National Park Service's Controlled Burn Policy, *Journal of Leisure Research*, vol. 25, no. 3.

Butler R, Hall M, Jenkins J (1998) *Tourism and Recreation in Rural Areas* Wiley Europe.

Cartwright M, Shepperd M (2000) An empirical investigation of an object-oriented software system, *IEEE Transactions on Software Engineering*, vol. 26, no. 8, pp. 786-796.

Cronbach LJ (1951) Coefficient alpha and the internal structure of tests, *Psychometrika*, vol. 16, no. 3, pp. 297-334.

Davis FD (1985) *A Technology Acceptance Model for Empirically Testing New End-user Information Systems: Theory and Results,* Massachusetts Institute of Technology, Sloan School of Management.

Davis FD (1989) Perceived Usefulness, Perceived Ease of Use, and User Acceptance of Information Technology, *MIS Quarterly*, vol. 13, no. 3, pp. 319-340.

Davis FD, Bagozzi RP, Warshaw PR (1989) User Acceptance of Computer Technology: A Comparison of Two Theoretical Models, *Management Science*, vol. 35, no. 8, pp. 982-1003.

Davis FD, Bagozzi RP, Warshaw PR (1992) Extrinsic and intrinsic motivation to use computers in the workplace, *Journal of Applied Social Psychology*, vol. 22, no. 14, pp. 1111-1132.

Dias E, Beinat E, Scholten H Effects of Mobile Information Sharing in Natural Parks, in *EnviroInfo 2004: 18th International Conference Informatics for Environmental Protection*, Philippe Minier & Alberto Susini, eds., Tricorne Editions, Geneva, Switzerland, pp. 11-25.

Dias E, Edwardes A, Purves R (2007) Analysing and Aggregating Visitor Tracks in a Protected Area, in *5th International Symposium on Spatial Data Quality proceedings*, ISPRS, Enschede.

Dishaw MT, Strong DM (1999) Extending the technology acceptance model with task-technology fit constructs, *Information & Management*, vol. 36, no. 1, pp. 9-21.

Farrell TA, Marion JL (2001) Identifying and Assessing Ecotourism Visitor Impacts at Eight Protected Areas in Costa Rica and Belize, *Environmental Conservation*, vol. 28, no. 03, pp. 215-225.

Fishbein M, Ajzen I (1975) *Belief, Attitude, Intention, and Behavior: An Introduction to Theory and Research* Addison-Wesley.

Fishbein M, Ajzen I (1981) Attitudes and voting behavior: an application of the theory of reasoned action, *Progress in Applied Social Psychology*, vol. 1, pp. 253-313.

Gefen D, Straub DW (1997) Gender Differences in the Perception and Use of E-Mail: An Extension to the Technology Acceptance Model, *MIS Quarterly*, vol. 21, no. 4, pp. 389-400.

Goldenhar LM, Connell CM (1993) Understanding and predicting recycling behavior: An application of the theory of reasoned action, *Journal of Environmental Systems*, vol. 22, no. 1, pp. 9-103.

Goodhue DL, Thompson RL (1995) Task-Technology Fit and Individual Performance, *MIS Quarterly*, vol. 19, no. 2, pp. 213-236.

Igbaria M, Parasuraman S, Baroudi JJ (1996) A motivational model of microcomputer usage, *Journal of Management Information Systems*, vol. 13, no. 1, pp. 127-143.

Junglas IA (2003) *U-Commerce: An Experimental Investigation of Ubiquity and Uniqueness,* Ph.D., Graduate Faculty of The University of Georgia, Athens.

Junglas IA (2005) An Experimental Investigation of Location-Based Services, *System Sciences, 2005.HICSS'05.Proceedings of the 38th Annual Hawaii International Conference on* p. 85a.

Konradt U, Christophersen T, Schaeffer-Kuelz U (2006) Predicting user satisfaction, strain and system usage of employee self-services, *International Journal of Human-Computer Studies*, vol. 64, no. 11, pp. 1141-1153.

Konradt U, Wandke H, Alazs B, Christophersen T (2003) Usability in online shops: scale construction, validation and the influence on the buyers' intention and decision, *Behaviour & Information Technology*, vol. 22, no. 3, pp. 165-174.

Lu J, Yu CS, Liu C, Yao JE (2003) Technology acceptance model for wireless internet, *Internet Research: Electronic Networking Applications and Policy*, vol. 13, no. 3, pp. 206-222.

Lynn NA, Brown RD (2003) Effects of recreational use impacts on hiking experiences in natural areas, *Landscape and Urban Planning*, vol. 64, no. 1, pp. 77-87.

Madden TJ, Ellen PS, Ajzen I (1992) A Comparison of the Theory of Planned Behavior and the Theory of Reasoned Action, *Personality and Social Psychology Bulletin*, vol. 18, no. 1, p. 3.

Mavri M, Ioannou G (2006) Consumers' perspectives on online banking services, *International Journal of Consumer Studies*, vol. 30, no. 6, pp. 552-560.

Moon JW, Kim YG (2001) Extending the TAM for a World-Wide-Web context, *Information and Management*, vol. 38, no. 4, pp. 217-230.

Nunnally JC, Bernstein IH (1994) *Psychometric Theory*, 3rd edition edn, McGraw-Hill, New York, NY.

Sarker S, Wells JD (2003) Understanding mobile handheld device use and adoption, *Communications of the ACM*, vol. 46, no. 12, pp. 35-40.

Shang RA, Chen YC, Shen L (2005) Extrinsic versus intrinsic motivations for consumers to shop on-line, *Information and Management*, vol. 42, no. 3, pp. 401-413.

Sheppard BH, Hartwick J, Warshaw PR (1988) The Theory of Reasoned Action: A Meta-Analysis of Past Research with Recommendations for Modifications and Future Research, *The Journal of Consumer Research*, vol. 15, no. 3, pp. 325-343.

Van der Heijden H (2004) User acceptance of hedonic information systems, *MIS Quarterly*, vol. 28, no. 4, pp. 695-704.

Venkatesh V (2000) Determinants of Perceived Ease of Use: Integrating Control, Intrinsic Motivation, and Emotion into the Technology Acceptance Model, *Information Systems Research*, vol. 11, no. 4, pp. 342-365.

Venkatesh V, Davis FD (2000) A Theoretical Extension of the Technology Acceptance Model: Four Longitudinal Field Studies, *Management Science*, vol. 46, no. 2, pp. 186-204.

Wu JH, Wang SC (2005) What drives mobile commerce? An empirical evaluation of the revised technology acceptance model, *Information and Management*, vol. 42, no. 5, pp. 719-729.

10 Impact of Restricted Display Size on Spatial Knowledge Acquisition in the Context of Pedestrian Navigation

Georg Gartner and Wolfgang Hiller

Institute of Geoinformation and Cartography
Vienna University of Technology, Austria

Abstract

In this paper the influence of the display size on the acquisition of spatial knowledge is investigated. The acquisition of spatial knowledge based on presentation forms with different display sizes has been analysed in an empirical test for pedestrian navigation. This contribution describes and interprets the test and results and gives an outlook on the overall importance for future application developments in the context of LBS, as mobile systems typically use small displays and the overall concern exists, that these displays can not communicate any useful spatial information in typical map form to pedestrians.

Keywords: pedestrian navigation, mental maps, sketch maps, spatial knowlege acquisition, small display maps

10.1 Introduction

Navigation systems, including those for pedestrians, are becoming increasingly popular. Usually, mobile devices – such as mobile phones, pocket computers or similar compact equipment – are used as hardware platforms. These devices can be characterized by the fact, that the size of the display is usually small, as a result of keeping the device as compact as possible.

Maps are important presentation forms for routes and topographic features and, therefore, usually central parts of all kinds of navigation systems (Gartner et al. 2001). Characteristic attributes of maps are the scale and the graphical code used. It is of key importance that a balanced combination of scale, perceivable graphical coding and selected content is aimed at when attempting to represent the topography or a spatial scenario for a particular purpose. This balanced combination has to be applied on a media which transports the map, whether it be paper, the screen of a computer, or a mobile device. The size of the media plays an important role for this balance as well as the overall content and structure of maps.

Different sizes of maps have always been used. The size used is usually correlated with the purpose of the map. When using maps for navigation purposes and orientation in real world the conflict between representation of a relevant part of space (in terms of keeping the overview) and an "easy-to-use" format has led to various solutions in the past, including especially different forms of folding or the use of different scales.

When using electronic devices and screen displays the media not only allows the display of static representations but also of interactive applications. Scrolling, panning and zooming are usually used for changing the view on a map or even selecting a different map (with a different context of scale, graphical coding and content), when needed. Although these tools enable the user to access all parts of a map, the overall context of a map of a particular scale is not accessible at once, if this map is exceeding the display of a particular device in size.

In this paper we adress the influence of display size on spatial knowledge acquisition, especially in the context of orientation and navigation. Our hypothesis states that the acquisition of spatial knowledge and therefore the ability to orientate in real world is influenced by the size of the display on which maps are presented.

10.2 Spatial Cognition and Wayfinding

Questions related to pedestrian navigation systems are currently tackled by various disciplines, including computer science, cognitive science, geoinformation sciences, communication science as well as modern cartography. The analysis of relevant work on human acquisition of spatial knowledge and human orientation marked the starting point of our research.

Issues of spatial cognition can basically be classified in two categories, the egocentric and the exocentric strategy of spatial cognition. Theories on egocentric spatial cognition deal with aspects of how humans are able to reconstruct a three-dimensional model from the two-dimensional image recorded by the eye. It is stated that three core directions are applied from an egocentric perspective (the vertical body axis for distinguishing above and below, the horizontal body axis for distinguishing behind and in front of and another horizontal body axis in 90° distance for distinguishing left and right (Schweizer et al. 2006, Herrmann et al. 1998). Navigation instructions usually deal with the exocentric spatial cognition, where a point in space is used as origin from where the directions are given (Schweizer 1997). Instructions about directions refer to objects in space, which can be used to define and describe directions. Such objects are called landmarks and can be identified by outstanding attributes (Sorrows & Hirtle 1999). A distinction between local and global landmarks is useful in the context of orientation, as local landmarks are used for local decisions or confirmations (such as "turn right at the green building"), while global landmarks can be referred to from a bigger area. As Mallot (2000, 2004) has demonstrated, humans are able to adapt their orientation strategies if either only local or global landmarks are available. If a person is offered both landmarks types as reference objects, local landmarks play a more important role for orientation.

The strength of human wayfinding strategies lies in the ability to integrate different strategies, like referring to global and local landmarks or using integrative semantic descriptions or route based instructions (Mallot 2004).

The term cognitive or mental map, which was introduced by authors like Lynch (1960) or Downs and Stea (1982), is referring to the result of the process of spatial knowledge acquisition. Generally speaking, three distinct functions of mental maps can be defined as (a) spatial recognition and identification, (b) spatial localization and memory, and (c) planning of spatial actions (Poucet 1993). Helwich (2003) pointed out that mental representations are primarily topological descriptions of spatial relations, while Lynch (1960) distinguished five elements of mental maps as lines, borders, areas, nodes and landmarks. Based on these elements a mental map is structured as the result of individual combinations. Appleyard (1970) built empirical tests upon this theory, which confirm in general Lynch's theory but developed a further distinction by categorizing sequential (lines, nodes) from spatial elements (borders, areas, landmarks).

Mental maps of cognitive representations of space are referred to when humans have to act in space, e.g. find a way (Kuiper 1982). Tversky (1993) pointed out that humans are referring to various mental maps with different "scales" and "abstraction levels". For solving a spatial problem or acting in space, a human is dependent on the availability of information about "where", which includes the own location as well as the location of other relevant objects (target, decision points), as Downs & Stea (1982) have mentioned. They argue further, that beneath information about "where" information about "what" is essential as well. Focusing on way finding this means that the information needed includes such on general orientation, route decision, route confirmation and goal finding (Downs & Stea 1982). General orientation means the relevant information, which is needed to answer the question "where am I". Therefore elements of the stored mental map have to be connected to perceived elements of the real world. If successful, this will allow defining one's current location as the origin of the egocentric spatial reference system. The connection of cognitive representations with elements of the real world is done via reference points. Such reference points can be global or local landmarks or other significant objects. Gartner et al. (2005) have argued that relations of humans to particular objects can be used in this context as well. Route decision is the strategy that humans develop in order to find a connection between a start point and an end point. Therefore sequences beneath decision points are planned. When moving, this plan is monitored permanently by referring to objects of the real world and comparing them with the cognitive representation for route confirmation. Finally the relevance of mental maps for way finding can also be found in the identification of the target as such, the goal finding.

Spatial knowledge acquisition is needed to build mental representations which can be referred to for way finding. As Platzer (2005) and also Briggs (1973) pointed out, various methods of spatial knowledge acquisition can be used, including the sensual perception of the real world as well as the acquisition of spatial knowledge from models of the real world, such as maps. Another common way of building cognitive representations is the derivation of experiences from comparable situations. In this context Neisser (1976), Downs & Stea (1973), and Platzer (2005) pointed out that the active knowledge acquisition by human senses when moving through real space is resulting in more and better information. Whitaker & Whitaker (1972) and Platzer (2005) mentioned, that personal characteristics have a big influence on spatial knowledge acquisition.

When analysing mental maps, the method of sketching maps is often used. As mentioned by Golledge (1976) and Platzer (2005) the interpretation of sketch maps have to take into account many influences, most of them dealing with abilities and personal skills of test persons. Byrne (1979) demonstrated this influence in a test, where 90% of angles of crossings between 60–70° and 110–120° have been drawn as ninety degree angles in sketch maps. Tversky (1981) and Thorndyke et al. (1982) pointed out that the skills of drawing correct angles and finding a way are not significantly correlated. It is therefore useful to stick to topological interpretation of sketch maps only, as Lynch (1960) already mentioned, but enhance the possible results by additional methods of estimating distance or direction, such as pointing methods (Henry 1992, Popp 1998).

Research on the influence on various presentation forms on wayfinding (Radoczky 2003, Dillemuth 2007), on the usage of automated navigation systems (Parush et al. 2007), and on the influence of the size of displays (Dillemuth 2007) have demonstrated, that the focus is going beyond technical questions when developing navigation systems. These findings are are used as a fundament for this research.

10.3 Methodology

Based on the hypothesis that the display size is influencing the spatial knowledge acquisition an empirical test has been set up. Thereby possible differences in the ability of self-orientation in space – caused by the different sizes of the map – during navigation with the help of a map presented on media were detected. The basic concept was to test the hypothesis in a real world scenario.

A route in the city centre of Vienna was selected. It included a number of waypoints, which had to be passed by the tested persons. Major points of interests such as the St. Stephan's cathedral were close to the route but not passed on the route itself. The setting was chosen in a way that every test can be completed within one hour. The route was presented on a map, which displayed the route along with selected topographic features. The map was provided by the Institute for Geoinformation and Cartography of TU Vienna in the scale of 1 : 8500.

The test was performed with 30 test persons. The test setting included an instruction phase by the instructor, where the test conditions were explained. Then the test persons had to find the route by means of either using the map printed on paper (referred to in this paper as map1) in the size of 13.5 x 16.5 cm or using the map represented on a small display mobile device (referred to as map2) with a display size of 4.7 x 5.5 cm. In both cases identical maps were used, but test persons referring to the map on the mobile device were forced to scroll and pan in order to access other parts of the map.

The map for the mobile device was divided into rectangular tiles. The size of the rectangles was approximately the same as the size of the display (4.7 x 5.5 cm). The bottom part of the display was reserved for a text area, which was used to give information about the following target point of the route. The rectangles were over-

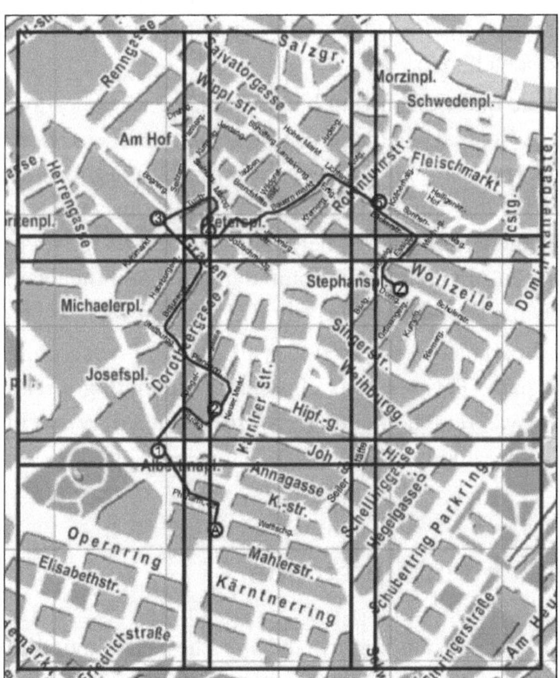

Fig. 10.1. Map of Vienna and marked tiles.

lapping, so that the orientation was facilitated when the user scrolled from one map tile to another. All map tiles together depicted the same area as the printed map.

To get the same conditions for all test persons, the test with the mobile device was done without any automatic positioning. The particular map tile displayed the route as well as the targets in an identical way as the printed map. The text information for the following waypoint was shown below the map on the display.

When the test persons reached the last waypoint, they were asked to solve some tasks and answer some questions. By thism means a general indication of the orientation skills was gained and the build mental model of the passed area was evaluated. Furthermore individual human influences were filtered out.

The first task that had to be solved by the test persons referred to the orientation skills. Subjects were asked to give an approximate direction to the starting point of the route, to the five main waypoints of the route and to the St. Stephan's cathedral by drawing arrows on a provided sheet of paper with the actual position being marked. None of these points were directly visible from the last waypoint. The cathedral was not passed directly, but is partly visible from some parts of the route.

Another task for the test persons was to draw a sketch map of the area they have passed. The focus was set primarily on the route but also on landmarks being noticed. The subjects were asked to draw the route as precisely as possible. Furthermore, they should mark the starting point of the route, the five waypoints, the final target and the St. Stephan's cathedral. Other individual memorised landmarks should also be marked in the sketch map.

Afterwards some questions had to be answered. These questions should give an indication about individual orientation skills and experiences. Besides some demographic questions (gender, age group) some specific questions about experiences in the use of maps and navigation systems have to be answered. In the later case a classification scheme was offered to help users ranking their experiences in reading maps, their ability to orientate, their computer skills and their familiarity with the test area. The scheme included grades from 1 (very good) to 5 (very bad).

Finally the test persons were asked to walk back to the second waypoint on the shortest possible path. This test was chosen for simulating a real world scenario. The chosen route was recorded and afterwards valuated by the instructor using a grade system (1...very good, 5...very bad).

10.4 Results and Interpretation

Generally speaking, the results confirm the hypothesis that the size of a map display has influence on spatial knowledge acquisition and therefore on the ability to orientate and navigate through space in an unfamiliar area.

Sense of Direction

The first task after reaching the target point was to draw the directions to the starting point, the main waypoints and the St. Stephan's cathedral, which was not passed directly but is located close to some parts of the route. The test persons had to draw the directions by using arrows on a provided piece of paper, where the actual position is marked. The analysis of the drawings was made by quantitative measures of angles.

The test analysis provides a clear result. The difference between the drawn directions and the correct directions was 21.6° on average for those having referred to the map presenting the whole area, while those referring to the tiled map show an average deviation of angle between the correct and estimated direction of 47.9°, which is more then twice as much.

Table 10.1. Average deviation of angle between correct and estimated directions.

	Deviation of angle
Map1	21.6
Map2	47.9

A statistical distribution of the classified deviations gives an identical result. The distribution also showed that there were no major mistakes (deviation of angle between correct and estimated direction of more than 80°) within the group of users referring to map1, while three users within the other group performed quite badly. These three subjects had a deviation between 80° and 100°, which is much higher than the average of 47.9°.

Another interesting detail was the result of the calculation of the average deviation for all points, where a direction had to be given. The user group referring to map1 achieved homogenous results in terms of quantitative measures for the different

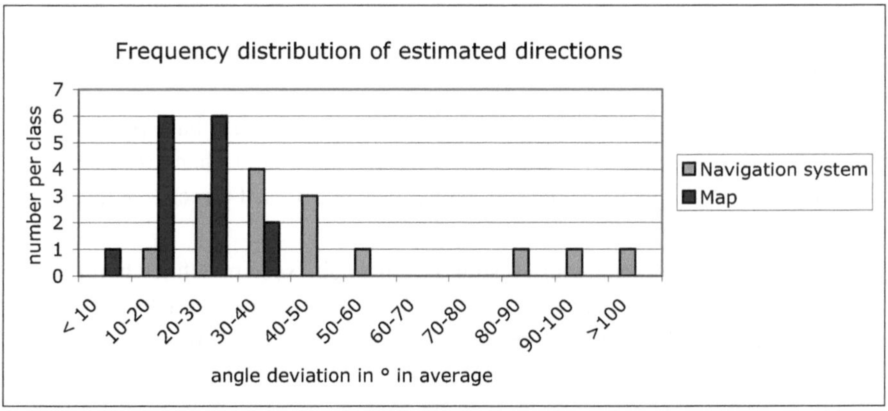

Fig. 10.2. Distribution of deviation of angle between estimated and correct direction.

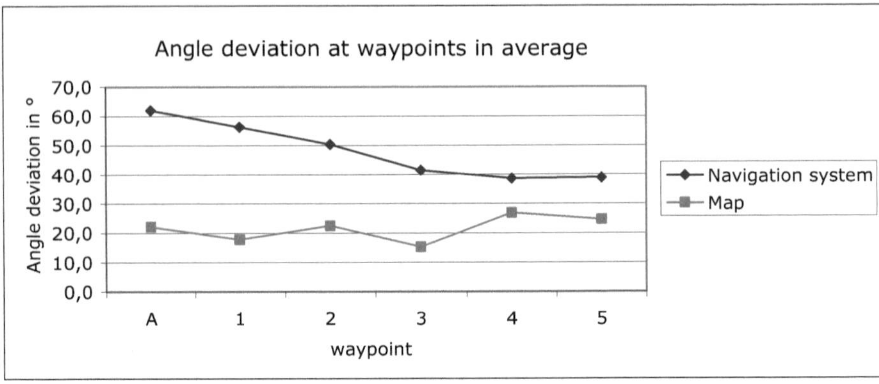

Fig. 10.3. Average deviation of angle for each waypoint.

targets, independent of their alignment on the route, while the group referring to map2 did not. In this group the deviation of angel is increasing with spatial distance to the waypoints. While the deviation is comparable with the average deviation for the last (most recent) passed waypoints, the starting point and the first waypoints to pass show a deviation much worse. This detail can be interpreted in a way that the more tiles are used, the worse the overall sense of direction gets.

The average deviation of the angle between the correct and the given direction from the last point of the route to the St. Stephan's cathedral resulted in 49.7° for the user group referring to map1 and 51.7° for the user group referring to map2. The aim of this task was to explore the ability to refer to a "global landmark", which was not passed directly. Surprisingly, the result is not showing a significant difference, although we expected clearer results. This could be explained by the fact, that the distance between the end point of the route and the St. Stephan's cathedral is rather close. A specific test of the ability to locate gobal landmarks after referring to maps of different sizes will be done in an ongoing project.

Sketch Maps

All test persons were asked to draw a sketch map of the area they have passed on the route. They were instructed to draw all waypoints and major changes of direction as well as the St. Stephan's cathedral, which was not passed directly on the route, as precisely as possible.

The analysis of the sketch maps resulted from a point system defined beforehand. Thus a quantification of differences in the sketch maps could be achieved, which can be used as an indication for interpretation. For every major change in direction, which is represented in the sketch maps, a point was given. If it is was not clearly possible to identify a specific change in the drawing, 0.5 points were given by the interpreter. Up to 18 points could be scored. In order to get a more objective result,

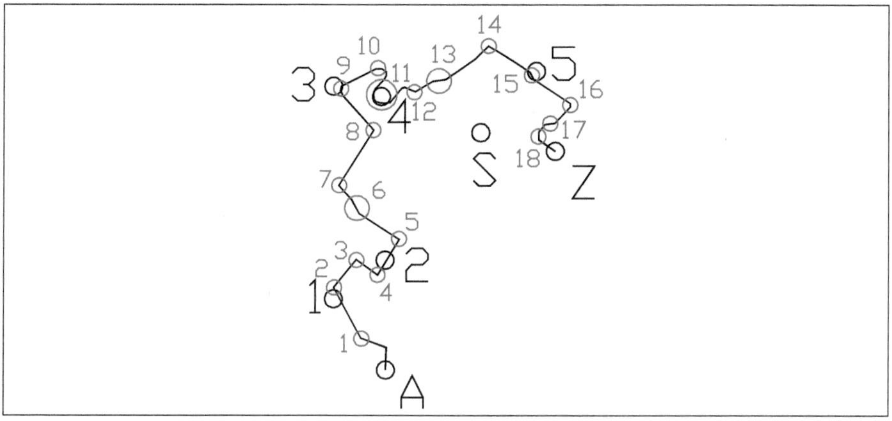

Fig. 10.4. Analysis of sketch maps.

the analysis was done by three independent persons. No major differences between the three interpreters have been identified.

The overall result is shown in *Table 10.2.* It demonstrates that test persons referring to map1 were able to represent more major changes of direction.

Table 10.2. Analysis of sketch maps by point system to quantify depicted major changes in direction.

	Points
Map1	8.5
Map2	6.8

Finally the location of St. Stephan's cathedral within the sketch maps was analysed by defining three geometrical figures with increasing granularity. Again points have been given by three independent interpreters depending on the location of the St. Stephan's cathedral either inside or outside the defined polygons.

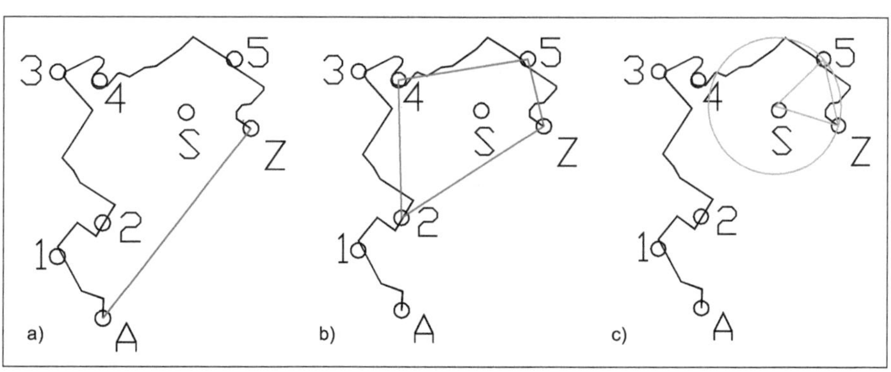

Fig. 10.5. Analysis of location of global landmark in sketch map.

The results of this analysis showed that test persons referring to the tiled maps (map2) receive 0.7 points on average, while the test persons referring to map1 get 1.7 points on average. This means, that most of the persons of group2 are only able to give a rough estimation of the location of the global landmark "St. Stephan's cathedral", which can be compared to the area given in *Figure 10.5a,* while group2 is able to estimate the location in a finer granularity, comparable with the area given in *Figure 10.5b.* As an interesting detail is has turned out, that two thirds of group2 were completely unable to locate St. Stephan's cathedral at all, while none of the test persons of group1 failed completely.

Spatial Action

In the final part of the test the test persons were asked to navigate back to waypoint 2 on a direct route without referring to any map or guiding instruction.

The results again show big differences between the two test groups. The test persons referring to map1 achieved an overall average of 2.1 points, while test group2 achieved an average of 2.9 points. Points have been given for the overall distance of the chosen route from 1 (direct route) to 5 (very long).

Two thirds of the test persons of group1 were rated with 1 or 2, indicating that they have been able to find a more or less direct route to waypoint 2 without referring to any map or guiding instruction, while only 20% of group2 get a rate of 1 or 2. 80% of those test persons were rated 3 (fair) or 4 (bad). Although they somehow were able to find the way back to waypoint 2, the route they chose is longer or much longer then the direct route.

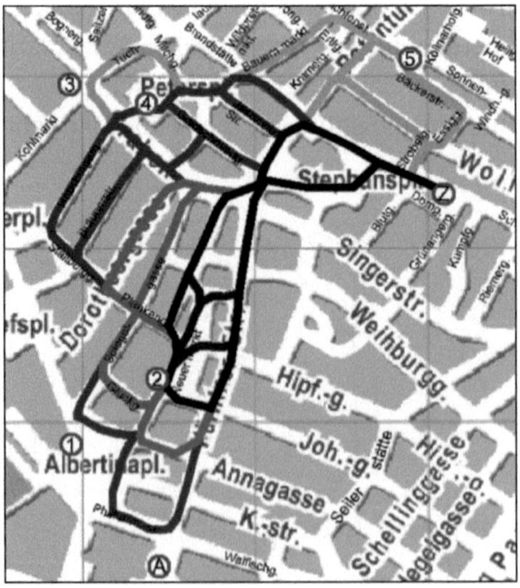

Fig. 10.6. Possible routes between waypoint 5 and waypoint 2.

In order to make sure, that there is no strong influence of individual knowledge or skills, a correlation of these results with the questions about the participants overall abilities and experiences was performed. In general no major differences have been detected. Detailed results show that those test persons of group2 which achieve good results in the tests consider themselves as having significantly higher abilities in map reading and more experiences in navigation. The results of test group1 are much less influenced by individual abilities. Test persons reported hardly any experiences in way finding and judge themselves as having none or very little map reading skills were also able to achieve quite good results.

10.5 Conclusions and Outlook

Navigation systems on mobile devices have become increasingly popular. Usually mobile devices can be characterized as having small displays, restricting the possible area and size of displayed maps. This has a strong influence on spatial data acquisition and thus on the ability to orientate and navigate in real world, as has been shown in this paper. More research is necessary on the influence on different presentation media in correlation with different display sizes.

References

Appleyard D (1970) Styles and methods of structuring a city. In: Environment and Behavior, 2, pp 100-117

Briggs R (1973) Urban Cognitive Distance. In: Image and Environment, Aldine Publishing Co., Chicago, pp. 361-388

Byrne RW (1979) Memory for urban geography, In: Quarterly Journal of Experimental Psychology, 31, pp 147-154

Dillemuth J et al. (2007) Choosing the scale and extent of maps for navigation with mobile computing systems. In: Journal of LBS, 1, 1, Taylor & Francis, pp 46-62

Downs R, Stea D (1973) Cognitive Maps and Spatial Behavior: Process and Products. In: Image and Environment, Aldine Publishing Co., Chicago, pp 8-26

Gartner G, Uhlirz S (2001) Cartograpic Concepts for Realizing a Location Based UMTS Service: Vienna City Guide Lol@. In: Mapping the 21st Century - The 20th International Cartographic Conference, (2001), ISBN: 7-5030-1018-5; pp 3229 - 3238

Gartner G, Oakley K (2005) Emotional Landmarks. In: Wiener Schriften zur Geographie und Kartographie, 17, pp 115-123

Golledge RG (1976) Methods and Methodological Issues in Environmental Cognition Research. In: Environmental Knowing, Dowden, Hutchinson and Ross, pp 300-313

Helwich J (2003) Graphenbasierte Navigation eines Geometrischen Agenten: Integration von Perzeption und Instruktion. Fachbereich Informatik der Universität Hamburg

Henry D (1992) Spatial Perception in Virtual Environments: Evaluating an Architectural Application. College of Engineering, University of Washington

Herrmann T, Schweizer K, Janzen G, Katz S (1998): Routen- und Überblickswissen – konzeptuelle Überlegungen. In: Kognitionswissenschaft, 7, pp 145-159

Kuipers BJ (1982) The »Map in the Head« Metaphor. In: Environment and Behavior, Nr. 14, pp.202-220

Lynch K (1960) Das Bild der Stadt. Birkhäuser Verlag

Mallot H (2000) Sehen und die Verarbeitung visueller Information. Vieweg, Braunschweig

Mallot H (2004) Kompass im Kopf, 5/2004, pp 18-27

Neisser U (1976) Cognition and Reality. WH Freeman, San Francisco

Parush A, Ahuvia S, Erev I (2007): Degradation in spatial knowledge acquisition when using automatic navigation systems. In: COSIT 2007, LNCS 4736, Springer, pp 238-254

Platzer E (2005) Spatial Cognition research: The human navigation process and its comparability in complex real and virtual environments. PhD, Universität der Bundeswehr München

Popp M (1998) Zeigeverhalten, Landmarken und mentale Raumpräsentationen. In: Visuelle Wahrnehmung, Beiträge zur 1. Tübinger Wahrnehmungskonferenz, Knirsch Verlag

Poucet B (1993) Spatial Cognitive Maps in Animals. In: Psychological Review, Vol. 100, No. 2, pp 163-182

Radoczky V (2003) Kartographische Unterstützungsmöglichkeiten zur Routenbeschreibung von Fußgängernavigationssystemen im In- und Outdoorbereich. MTh. TU Vienna.

Schweizer K (1997) Räumliche oder zeitliche Wissensorganisation? Zur Repräsentation der Blickpunktsequenzen bei räumlichen Anordnungen. Pabst Science Publishers Verlag

Schweizer K, Horn M (2006) Raumwahrnehmung und Raumvorstellung. In: Geographie und Schule, 164, pp 21-26

Sorrows M, Hirtle S (1999) The nature of landmarks for real and electronic spaces. In: Freksa C, Mark D (eds): Spatial Information Theory. COSIT Proceedings, Springer, pp 37-50

Thorndyke PW, Hayes-Roth B (1982) Differences in spatial knowledge acquired from maps and navigation. In: Cognitive Psychology, 14, pp 560-589

Tversky B (1981) Distortions in memory for maps. In: Cognitive Psychology, 13, pp. 407-433

Tversky B (1993) Cognitive maps, cognitive collages, and spatial mental models. In: Frank, A.U. and Campari, I. (Eds.) Spatial Information Theory: a Theoretical Basis for GIS. Springer, LNCS 716, Heidelberg, pp 14-24

Whittaker JO, Whittaker SJ (1972) A cross-cultured study of geocentrism". In: Journal of Cross-Culture Psychology, 3, pp 417-421

11 User Requirements for Location-Based Services to Support Hiking Activities

Annu-Maaria Nivala[1], Tiina Sarjakoski[1], Katri Laakso[2], Joonas Itäranta[3] and Pyry Kettunen[1]

[1] Geoinformatics and Cartography, Finnish Geodetic Institute, Masala, Finland
[2] Devices R&D, Nokia Corporation, Tampere, Finland
[3] Nokia Research Center, Nokia Corporation, Tampere, Finland

Abstract

Location-based services (LBS) are gaining popularity and being adapted to new domains. The aim of this study was to identify user requirements for location-based information and services to support outdoor hiking activities. Two issues were emphasized in the study: possible changes occurring during a hike, and community and content needs of the hikers. Three usability engineering methods (questionnaires, empathy probes and focus-group discussions) were used to identify the potential users and their tasks during a hike. Approximately 100 specific user requirements for development of future location-based services were recognized, from which nine major themes were identified: planning a hike, additional information on the hiking area, 'I am here' services, location of other hikers, changing conditions, emergency situations, saving experiences, sharing experiences, and integrated and adaptive services for hiking purposes. This paper describes the test set-up to identify user requirements and discusses the critical requirements from the hiking viewpoint. Fulfilling these requirements enables the development of improved applications, services and devices for outdoor activities.

Keywords: user requirement study, location-based information, mobile device, outdoor activity

11.1 Introduction

Today, location-based services (LBS) are an emerging research topic. The potential of location-based services has been recognized along with the rapid development of mobile Internet devices, next generation mobile networks and accurate positioning technologies. LBS can provide users of mobile devices with personalized services tailored to their current location or other location of interest. According to a market research study (Research and Markets 2005), the global LBS and Geographical Information System (GIS) market started to accelerate in 2005 and is considered to be the next 'killer' service after SMS. A review on topics and future trends presented at Mobile Location Services conferences in 2006 and 2007 shows that there are already many successful applications available today, especially in the domains of emergency and personal security, car and pedestrian navigation, tourist information, and fleet, asset and workforce management (MLS 2006, 2007).

However, there is great potential for many other location-based application domains as well, especially now with the growing number of GPS-equipped mobile phones. These mobile phones are becoming part of people's everyday lives, and their role is no longer that of a traditional phone but of a multimedia computer optimised for entertainment consumption. Similarly, the usefulness of LBS is no longer considered only in the traditional sense, for example in guiding the user in unfamiliar environment; instead, people want services that are also fun, entertaining and with aesthetically pleasing user interfaces. Information and communication technologies have also changed people's leisure-time behaviour, as socializing today often takes place through SMS messaging and online meetings. Another trend is the emerging social services like Jaiku and Facebook (Jaiku 2007, Facebook 2008), through which people create different communities and communicate, for instance, with friends or people with the same interests. People use their mobile devices for accessing these services, as well as for creating their own content and sharing their experiences with others (Lehikoinen et al. 2007).

In line with current trends, the potential applications for future LBS are related to friends, location-based gaming, commerce, communities and social networking (MLS 2006, 2007). The problem with these types of applications is that the perceived value for the consumer is often not so obvious. In order to make successful products, the key issue is how to understand the potential users of the service.

Hiking is a popular activity in many countries, and for example Finnish national parks have become increasingly popular during the past decades. The Finnish Forest Research Institute, Metla (2006), reports that nowadays there are over one million visits to national parks every year. Every fifth citizen uses the national parks and other state-owned areas for recreation, and the average person visits these areas seven days per year (Metla, 2006). Clearly, there is a large base of potential users who would benefit from location-based services designed especially for hikers.

11.1.1 The Goal of the Study

The research presented in this paper was part of the Nokia Research Center's (NRC) larger research project that aimed at studying different aspects of location-awareness in the context of outdoor navigation. The hypothesis of the present study was that hikers, and also people engaged in other rural outdoor activities, would benefit from location-based information that would – in addition to wayfinding instructions – support users' communication and social behaviour needs.

The goal of the study was to define the user requirements regarding the geospatial and other location-based information – focusing on the needs of people interested in hiking and related outdoor activities. The aim was to gather qualitative information about how hiking and other rural outdoor activities are currently planned and carried out and to observe the potential problems in carrying out these tasks today. The objective was to use the collected information to compile a set of user requirements to further develop LBS for hikers.

In addition to the general user requirements, the study focused on finding answers to the following research questions:

1. What type of unexpected changes may occur during the hike and what needs these situations cause for the hikers? What type of support actions would help hikers to recover from and adapt to these sudden changes?
2. What type of community and content needs the hikers may have while hiking? Are there some benefits that LBS could offer; for example; what type of needs do hikers have in relation to creating and sharing their own content while hiking? What kind of content should be provided for them, and how could it be used while carrying out an outdoor activity? How would hikers benefit from knowing the location and other information about the other hikers?

11.1.2 Related Work

Research on LBS has, so far, mainly focused on personal navigation in urban settings. The market potential of LBS supporting tourism and personal navigation has been recognized in many studies (Brown and Laurier 2005, Pospischil et al. 2002, Schmidt-Belz et al. 2003, Cheverst et al. 2000, van Setten et al. 2004). Furthermore, the user requirements have been identified for supporting personal navigation in an urban area (e.g. Baus et al. 2002, Bornträger et al. 2003), or for specific user groups (elderly people: Osman et al. 2003, blind people: Klante et al. 2004, Goodman et al. 2004). Studies on location-based information on mobile devices have also been carried out, such as the LoVEUS (2005). The research on identifying user requirements for future LBS, as carried out in this study, could provide preliminary input for the LBS research topics listed by Raper et al. (2007):

user profiling, orientation support, interaction design, geographical relevance or context, and geospatial content management.

Extending the research to the outdoor domain potentially reveals new ideas and designs that can be generalized for application and UI concepts. To date, only few studies have been published on hikers' needs for location-based information in non-urban areas, as presented in this paper. An LBS prototype for hikers was developed in a project named PARAMOUNT and validated in test areas in the Alps and Pyrenees (Sayda 2005). User requirements were gathered with a survey questionnaire posted on a website (PARAMOUNT 2002). The final prototype provided topographic maps, routing functions, tourist information on points of interest (PoIs), local weather forecasts, etc. Hikers' user requirements have also been studied in WebPark (Edwardes et al. 2003), and GiMoDig (Sarjakoski & Sarjakoski 2005) projects.

One of the major challenges for future LBS is to support user activities by providing contextually-adapted information through mobile devices. Discussion on how to utilize the information of a user's location in more intelligent ways has been ongoing for several years (e.g. Meng et al. 2005, Nivala & Sarjakoski 2003, 2007, Reichenbacher 2004, Zipf 2002). Nivala and Sarjakoski (2003) stated that besides the most important context for mobile map services today, the location of the user, there are also other important context to which the service should adapt to: the system used, purpose of use, time, physical surroundings, navigational history, orientation, user and cultural and social context. However, Beeharee and Steed (2007) observed that adaptive systems have, so far, been rather naive in relation to what information is relevant for the user. Gartner (2004) also stated that today adaptation to the 'user' means being limited to user profiles selected in advance from a list or entered manually by users themselves.

11.2 Materials and Methods

The potential future LBS users were contacted, and the user requirements were studied with three different usability engineering methods: questionnaires, focus-group discussions and empathy probes. The motivation to use multiple methods for answering the same research questions was to gather qualitative information on as many aspects as possible, as each of the method had a slightly different emphasis on the research topics. Each user participated in only one of the methods. The study was carried out together with the Nokia Research Center and the Finnish Geodetic Institute, Department of Geoinformatics and Cartography, during summer and fall 2007.

11.2.1 Questionnaires

Questionnaires are used to gather information about the users, such as, their skills, experience, work practices, preferences and opinions. The questionnaire in the present study was eight pages long and started by gathering quantitative information about users' backgrounds, such as age, gender, frequency of outdoor activities and previous experience with maps. The main part of the questionnaire consisted of open questions aiming to collect qualitative information on users. The topics of the questions were related to different aspects of hiking, such as route planning, guidance material taken on the hike, current use and opinions of real-time positioning, user experiences during the hike and attitudes to sharing their experiences with other people, surprising situations the users had encountered during their hikes and users' ideas of 'dream devices' to support hiking trips.

A total of 38 printed questionnaires were distributed to random hikers in two different locations: Nuuksio National Park in southern Finland, which mainly attracts daytime visitors, and Pallas-Yllästunturi National Park in northern Finland, where hikers usually stay overnight. The researchers recruited the participants at the camping sites in the national parks.

The participants were aged from 30 to 65 years in Nuuksio, and from 15 to 54 years in Pallas-Yllästunturi. In Nuuksio, nine females and nine males participated, and in Pallas-Yllästunturi nine females and eleven males. The length of their route varied from 1.5 to 20 km in Nuuksio, and from 10 to 60 km in Pallas-Yllästunturi. The duration of their hike was from 1 to 10 hours in Nuuksio and from 25 hours to 9 days in Pallas-Yllästunturi.

11.2.2 Focus-Group Discussions

Focus-group discussions are used to obtain user feedback and initial reactions to discussion topics (Caplan 1990). The interaction between the participants often prompts spontaneous reactions and ideas. In this study, the topics of the focus-group discussions were mainly the same as the topics in the questionnaires, but there were also additional questions about the community needs during a hike. Two researchers lead the two hour discussions, and audio data were recorded for analysis purposes.

The focus-group discussions were carried out in two separate sessions, with two different groups of people sharing similar interests in relation to hiking. Participants in the first group were interested in outdoor activities (e.g. hiking, cycling, canoeing, bird observing) and belonged to an association that aims to support outdoor activities among its members. The group comprised of five people aged 29 to 61 years.

The second group consisted of eight students from a sports institute, aged 18 to 20 years. Their outdoor activities were mainly related to their future work tasks, such as guiding customers at different outdoor events.

The lengths of typical outdoor activities varied between the two groups: the first group had experience of longer hiking trips to northern Finland, while the second group typically arranged shorter, one-day outdoor activities with different themes, such as activity days for students or recreation days for companies, but occasionally also longer hiking or canoeing trips. The hiking group sizes for both groups were typically between 10 and 20, depending on the event.

11.2.3 Empathy Probes

In order to gather information on and to understand the users' thoughts, feelings, dreams, experiences and lives, several participants were asked to keep diaries (i.e. empathy probes) during their hiking trip. The word 'probe' refers to a recording device that can be carried around by the user, i.e. to places where the researcher cannot go. The probe can be, for example, a recorder, diary or a disposable camera (Koskinen et al. 2003).

The empathy probe in this study was 24 pages long diary, which was designed to be attractive, fun to fill with lots of colourful pictures *(Figure 11.1),* and easy to carry during the hike (A5 size, waterproof covers). The structure of the diary followed the same topics as the questionnaires and group discussions, but there were more questions about, for example, the lifestyle and experiences of the users. The questions were also more open; for example, users were asked to draw a map of their hike and write anything that came to mind about different sets of pictures.

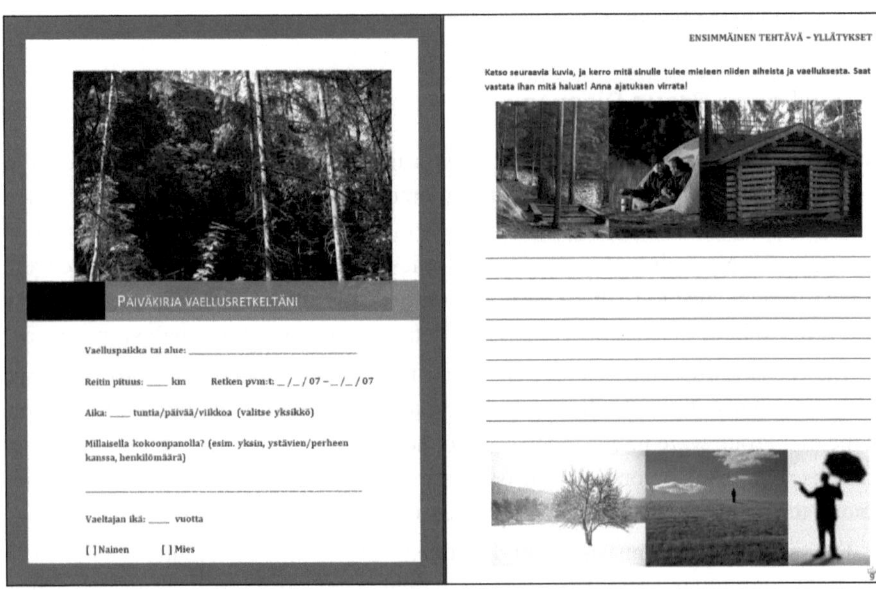

Fig. 11.1. Two sample pages of the probe.

Researchers scrutinized the hiking diaries together with the participants when they were returned.

The empathy probes were delivered to seven persons going for longer hiking trips. The participants were aged 24 to 33 years; four of them were females and three males. The length of their route varied between 8 and 150 km, and the duration of their hike was two to seven days. The hiking areas were located in northern and eastern Finland, and one in Iceland.

11.2.4 Analysis of the Data

The data gathered in the study were analysed in two phases. First, the user requirements were derived separately from the questionnaires, empathy probes, and focus-group discussions with affinity diagrams. In the second phase of the analysis, the identified user requirements were combined and regrouped into thematic entities. This generated a more general understanding of the user requirements for location-based information in the case of hiking and other outdoor activities.

11.2.4.1 Affinity Diagrams

Affinity diagrams are used to organize a large body of data into their natural relationships. According to Beyer and Holtzblatt (1998) the diagrams are suitable especially when analysing verbal data, such as, ideas, opinions or issues, and for dealing with many facts or ideas that seem to be in chaos. The tools used are sticky notes, marker pens and a large work surface. Ideas can be recorded on a separate sticky note or card with a marker pen. The notes are organised into groups of ideas that seem to be related.

In the present study, affinity diagram meetings were arranged with 5 to 6 researchers participating. The data were examined and the answers categorised according to their contents and relationships. As a result, a group of 'observations' were established (for example 'Sharing my location' and 'Sharing others' locations', as shown in *Figure 11.2),* which each had their own list of remarks or ideas from users (for example 'could share with people close to me', 'would give a feeling of safety', 'could share with limitations', 'could be useful if I get lost' and 'only in an emergency'). From these, individual user requirements were derived *(Figure 11.2).*

11.2.4.2 Synthesis of the User Requirements

A detailed list of altogether 101 individual user needs was gathered as a result of the data analysis. The user requirements were further elaborated into wider thematic categories *(Figure 11.2,* Synthesised user requirement themes). The results of this synthesis are presented and discussed in the following sections.

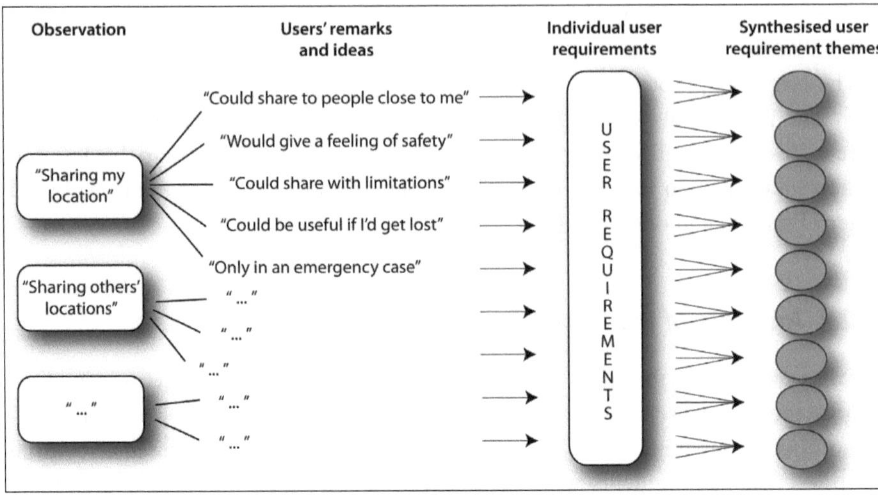

Fig. 11.2. The individual user requirements were established based on the observations from the data. The results were further categorised into synthesised user requirement themes.

11.3 Results

The user requirements resulted in nine thematic categories *(Figure 11.3):*

- Planning a hike
- Additional information on the area
- 'I am here' services
- Location of other hikers
- Changing conditions
- Emergency situations
- Saving experiences
- Sharing experiences
- Integrated and adaptive services.

The categories partly overlap, also in the sense that the same user needs are related to different phases of the hike (before, during or after). In the following, the main user requirements identified for using location-based information during outdoor activities are presented with clarifying examples from the users. In addition to these nine themes, users' ideas on the properties of their 'dream device' to support hiking are discussed at the end of this section.

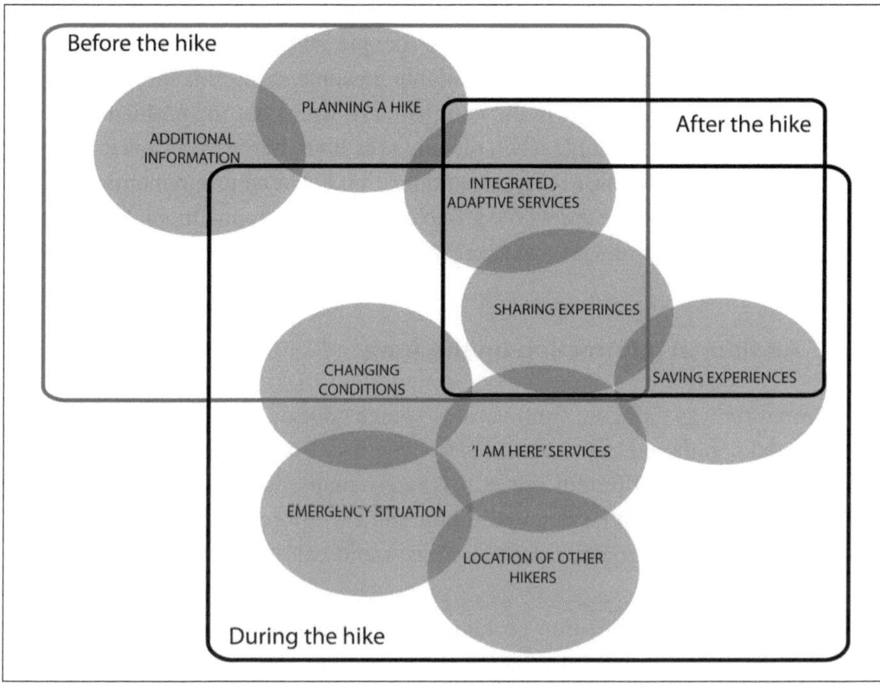

Fig. 11.3. Identified user requirements grouped into 9 thematic categories and their relationship to different phases of the hike (before, during and after).

11.3.1 Planning a Hike

Heterogeneous information was needed for planning a hike or other outdoor activity. No single media, book or other information source provided all the required information, and people gathered information from various services and sources, such as books, maps, Internet pages and magazines. Other people's experiences of the area also played an important role for some people.

When planning a hiking trip, people also needed information on public transport (routes, timetables), local weather forecasts and car parks (location, size, road maintenance during winter). For this, some people used Internet route-planning services, while others used traditional topographical and road maps. Information was also needed on taxi services in the area, in order to return to the car after the hike, or in case of an emergency.

Planning the actual route in the national park was also based on information gathered from different sources, for example, multiple websites, maps, books and phone calls to the national park services. Participants stated that automatic route suggestions based on different criteria would help in the planning as different participants had different preferences for planning the route, such as, 'the shortest', 'the nicest view', etc. There was criticism that information about the accessibility of the hiking routes for special user groups (people with disabilities, families with baby carriages) was often not readily available. In some cases, this information was available on information signs in the national park itself, but the problem was that this information is not available when planning the hike. Large overview maps were considered important to allow easy communication between group members in the route planning phase, since the whole group could gather around them. Maps on the small screens of mobile devices were considered insufficient for this purpose.

11.3.2 Additional Information on the Area

As information was gathered from various sources during the planning of a hiking trip, the data gathered were also heterogeneous. For instance, people would have needed maps at different scales: overview maps for general route planning and detailed maps for more detailed topographical information (e.g. locations of wetlands and swamps, rocks, river crossings, rapid categorisation and more accurate contour lines).

More descriptive information on hiking areas was also sought. For example, a map provides only limited information on a camping site, whereas people would often need a better idea of the surrounding views, the size and the facilities of the camping site, etc. Sometimes this type of descriptive information was gathered from others' experiences of the area. However, there was always the question of the validity of the information, as in the case of hiking reports published on the Internet.

People also needed various thematic information to support their outdoor activities, such as different maps for hiking, cycling and canoeing. These maps should have the information required for the activity in question and for the specific size and form of the hiking group. Furthermore, people are interested in different topics: some wanted information on geology, while others were interested in the history or vegetation of the area. Others were looking for more detailed information on swimming sites, cottages, reindeer fences in Lapland, locations for going ashore while canoeing or access to drinking water.

11.3.3 'I Am Here' Services

There was a recognised need for the 'I am here' types of service to support hikers during their trip. Such services were regarded as especially important in cases of emergency or when people are lost. In these situations, suggestions for alternative routes would be of use, and it was considered important for the rescue services to be able to locate lost or injured hikers easily. Accordingly, it appeared that current GPS devices for navigating were considered mainly as safety equipment for quickly reporting one's location in a possible emergency situation. Today, their value is not in the maps included in the device, as they are often too small and not usually tailored for hiking purposes.

'I am here' services would also be useful for helping people to follow their planned route and for guiding them back to the route, if they happened to veer away from it, either willingly or accidentally. The services would also need to log the hiking route and allow it to be shared with other people. Besides a route displayed on top of a map, an alternative way of showing hikers' routes to other people could be a verbal description of one's location, e.g. 'On top of mountain, -20 degrees Celsius'. For some people, location sharing was related to the enjoyment of sports and competitions, or meeting up with a friend.

11.3.4 Location of Other Hikers

Real-time information about other hikers' locations was occasionally considered necessary – mainly to avoid crowded cottages, tracks and parking sites in national parks. Many people stated that they were unwilling to meet too many other hikers, and the information about other people's locations could be used to avoid them. However, in emergency situations it was considered important to be aware of the location of others for safety reasons.

One of the user requirements for bigger hiking groups was the need to know other hikers' locations when the group was divided into smaller sub-groups. Sometimes members of the hiking group would be hiking in a long line, causing one end of the line to be unsure where the other end was, and the location information of others would be needed for communication.

11.3.5 Changing Conditions

Changes in conditions during an outdoor activity were often considered critical. It was observed that changing weather, light or temperature conditions emphasized the need for real-time location-based information both before and during the hike. Also information on the varying snow depth, rising and setting times of the sun, likelihood of seeing the Northern Lights and the phases of the moon were regarded as important information that could change during the hike. Other factors affecting the hike were movement of other people (e.g. crowded huts), dangerous animals and the conditions of the tracks and facilities in the hiking area. A new route plan might be required if the conditions changed or were not what was expected.

11.3.6 Emergency Situations

One of the most obvious needs for locating one's own or somebody else's position was related to emergency situations. The feeling of safety was considered important, and the knowledge that the emergency services would be able to locate the hiker easily was appreciated. The possibility of getting lost raised the need for supporting services, such as 'I am here' services. Sudden illness, accidents and encounters with wild animals were fears that raised the need for emergency support. One suggestion was a simple emergency button on a mobile phone that would transmit the necessary location information to the emergency services when pressed. Suggestions for alternative routes were also a topic often mentioned: quick and easy routes leading out of the wilderness should be provided on easy-to-read maps.

11.3.7 Saving Experiences

People take photos, write diaries and shoot videos while hiking. However, people would also like to save their experiences in various other ways. One of the wishes was to save personal PoI information, for example reminders of nice views, berry and mushroom picking places, or observations on animals, plants and natural phenomena. This type of information could be saved for the users' own purposes or also made available to other people. The automatic track logging feature of GPS was considered useful as a type of a 'diary'.

Along with technical restrictions, the need to manage the gathered data during and after the hike was considered problematic. The data should be synchronised with a home PC and organised on the fly. People wished to have easier and more automatic ways for saving the combined data, such as, a travel account together with photos and Internet links.

11.3.8 Sharing Experiences

Today, people share their experiences, such as nice views or sightings of birds, berries and geological formations, mostly by photos, videos, multimedia messages and text messages. Methods used more rarely included stories, drawings, poems and people's own music compositions. There was also interest in sharing experiences through mobile blogs, email, voice, web links and services like Google Earth.

The current hiking route, either shown on a map or given as a set of GPS waypoints, was among the most popular things to share. There was a need to share the progress of the hike on a daily basis, too. People also wanted to share their experiences in something like a blog, where photos or voice recordings could be combined with text. However, there is a need to manage the gathered information in a more efficient way to be able to share it easily with others.

The participants were especially willing to share their experiences with family members, friends and other close people. Sharing routes between friends was considered fun and an important way of obtaining tips on where to hike, whereas sharing the current route with family members was considered important mostly for safety reasons. Some people wanted to share their experiences with colleagues and with people sharing the same interests. There was also the need to restrict access to the shared information, as people wanted to share different data with different people.

The 'push' or subscription type of information sharing was also seen beneficial. People would like to be able to define the different information categories that interest them and then receive information on that subject, for example when approaching a certain location. People also want to be able to block information or apply some type of filtering to control the amount of information given to them. They also want to restrict the access to information that has been tagged for a specific location.

11.3.9 Integrated and Adaptive Services

A general user requirement related to many stages of the hike was the need to get all the necessary information for the hike from the same service. People also wanted an Internet service for downloading and printing outdoor maps at different scales. In order to satisfy the needs of individual people and their various outdoor activities, tailored maps would be needed for each user and use situation, e.g. detailed route planning, cycling, canoeing or just getting an overview of the area. Special needs for maps to take into account the requirements of special user groups, like families with baby carriage, or athletes, were also mentioned.

Not only should the users be provided with different types of maps for different activities, but the device itself and its user interface should also adapt to the use situation. For example, when skiing in cold weather, the map could be displayed on the surface of the user's sunglasses, and the user interface could be voice guided, in order to avoid using small device buttons with gloves on. Also, the usage situation of the service should be taken into account. For instance, audio guidance or a touch screen could be possible alternatives for interacting with a service during a hike.

At the same time, it was considered critical that the devices should not distract the user from the hiking experience. People also wanted adaptive services that could filter information based on the user's situation or preferences. One example was being able to get relevant phone numbers in unexpected situations (local taxi firm, etc.). Better interoperability between services was also considered important.

11.3.10 Dream Devices and Services

It is still a fact today that more reliable network connections, durable batteries and quicker processing capabilities would be needed in order to provide users with devices that they could trust in a hiking situation. The devices should also be resistant to moisture and temperature changes.

Easy-to-use devices were also required since many users suffer from failing eyesight due to their age. The devices should be small and light to carry, but at the same time the display should also be easy to read. A function where the device could project the map onto a suitable surface such as the roof of the tent was suggested. It was also noted that a digital paper map might be useful in the future.

An 'all-in-one' device, which would include functionalities such as a step-counter, compass, camera, GPS and mobile phone was on the wish list. A barometer, altitude meter, thermometer and radio were also regarded as being important.

Technological assistance was criticised by some participants who wanted to stay away from technical devices while visiting nature. Some respondents noted that it is part of the fun to use the paper maps during the hike-planning phase so that everyone can see the map all at the same time and point to different locations on a map and to get an overview of the area. Accordingly, collaborative map use should be supported with the maps on computer screens, too.

11.4 Discussion

The present research aimed to study how location-based information could be beneficial, especially to the hikers in rural areas. There seems to be a need for LBS that would be able to support hikers' various needs and tasks. Some of the user requirements identified in the study may seem already well know. The question is why do not the available services and devices already answer to these requirements of the users? Is it because of the lack of the technologies to provide these?

The next step after identifying the user requirements would usually be to design and implement LBS that would answer to the identified detailed user requirements. Since this was not the focus of this paper, a follow-up study would be needed, to be able to differentiate between different types of users, tasks and use situations in order to develop LBS that could answer to the identified adaptivity requirements etc.

In addition to general LBS user requirements, two research questions were specifically studied: the changes occurring during the hike, and the community and content needs of hikers. Changes in the hiking conditions and how to deal with them can be critical to the success of the hike. Sudden changes are often hard to predict, and in such situations people often need new information that they have not been able to obtain beforehand. To cope with the changes, people sometimes need to contact some external party for help or to make new arrangements, such as alternative transportation or accommodation. Sometimes it is a question of survival, but more frequently the changes just make things less easy or unpleasant. Accordingly, a good LBS should know what situations hikers might encounter and how they could be best supported to deal with the possible changes.

The most important user requirements in relation to the unexpected situations were assistance in planning a new route and general support for leaving the hiking area. In practice, this would mean new route descriptions, public transport timetables and contact details for the relevant services. Another important requirement was related to emergency situations, in which an easy and reliable system would be needed to deliver an emergency message to the rescue services. This was a more general problem of today – users require more reliable devices and services. Due to their technical shortcomings (e.g. short battery life, no resistance to water), the devices that are currently available are mainly carried as safety equipment.

By realising how critical the changing conditions are for the hike, an important user requirement was identified: the need for real-time and up-to-date information. First, in the planning phase of the hike, people wanted information on local weather forecasts, snow depth for skiing and river water levels for canoeing. Up-to-date information was also needed on the condition of huts and different tracks in the national parks, since the information currently available was not always to be trusted. People reported incidents where huts were uninhabitable, tracks had been changed or maps were not up-to-date. Webcams were proposed as a possible solution for the need for real-time information on whether the huts were currently occupied and how busy the tracks were.

The second central theme emphasised in this study was based on the observed trend that people are already saving and sharing their location information, photographs, etc. by using mobile devices. Some hikers found it important to share their own experiences and information related to their hike not only among close people but also among 'like-minded' people, mostly in the form of photos posted on the Internet. Besides having fun, reasons to share were also practical, such as sharing information on available facilities like access to drinking water or toilets. Also, GPS-based routes were saved and shared with others in order to report the progress of the hike.

Community needs may also become important in this domain, especially due to the different communities that form on the Internet nowadays and the social networking that they facilitate. The community needs of hikers were related to all the three phases of the hike. When planning a hike, information on other hiker's routes and experiences in the area was considered important. During a hike, the community needs were related to the communication between the sub-groups and knowing the locations of other hikers, either to avoid strangers or to keep in contact with friends. After the hike, the community needs were related to sharing the experiences.

11.5 Summary

Future LBS may have an important role in supporting leisure activities and providing services specific to the current location. The aim of this study was to define user requirements regarding the location-based information of hikers and people carrying out similar outdoor activities. Empirical usability-engineering methods, questionnaires, empathy probes and focus-group discussions, were used to gather qualitative information about potential users and their tasks.

As a result of the study, approximately 100 specific user requirements for future LBS were identified. These requirements were grouped into nine main categories that were identified as important from the users' viewpoint: planning a hike, additional information on the area, 'I am here' services, location of other hikers, changing conditions, emergency situations, saving experiences, sharing experiences, and integrated and adaptive services. Some issues were more important when planning the hike, some during the actual hike and navigation, and others after the hike when saving and sharing one's experiences with others in various ways.

People often choose the places they want to visit based on the available facilities and services. This research highlighted the current needs of hikers, and through that information, what type of services should be developed to attract more people to national parks. This paper discussed each of the user requirement themes to outline the critical requirements for the future LBS from the viewpoint of a hiking user. By designing devices and services that meet the identified user requirements, new 'killer' applications for outdoor leisure activities could become a reality with LBS.

References

Baus J, Krüger A, Wahlster W (2002) Resource-adaptive Mobile Navigation System, *Proceedings of the 7th International Conference on Intelligent User Interfaces,* San Francisco, California, USA, pp. 15–22.

Beeharee AK, Steed A (2007) Exploiting real world knowledge in ubiquitous applications, *Personal and Ubiquitous Computing*, vol 11, no. 6, pp. 429–437.

Beyer H, Holtzblatt K (1998) *Contextual Design. Defining Customer-Centred Systems,* NY: Morgan Kaufmann Publishers.

Bornträger C, Cheverst K, Davies N, Dix A, Friday A, Seitz J (2003) Experiments with Multi-Modal Interfaces in a Context-Aware City Guide, in Chittaro L (ed.), *Proceedings of MobileHCI03,* Udine, Italy, pp. 116–129.

Brown B, Laurier E (2004) Designing Electronic Maps: an Ethnographic Approach, in Meng L, Zipf A, Reichenbacher T (eds), *Map-based mobile services – Theories, Methods and Implementations,* Springer Verlag, pp. 247–265.

Caplan S (1990) Using Focus Groups Methodology for Ergonomic Design, *Ergonomics*, vol 33, no. 5, pp. 527–533.

Cheverst K, Davies N, Mitchell K, Friday A, Efstratiou C (2000) Developing a Context-aware Electronic Tourist Guide: Some Issues and Experiences, *Proceedings of the SIGCHI conference on Human factors in computing systems,* The Hague, The Netherlands, pp. 17–24.

Edwardes A, Burghardt D, Weibel R (2003) WebPark – Location Based Services for Species Search in Recreation Area, *Proceedings of the 21st International Cartographic Conference (ICC), Cartographic Renaissance,* Durban, South Africa, pp., 1012–1021, CD-ROM.

Facebook (2008) [Online], Available: http://www.facebook.com/ [31 January 2008].

Gartner G (2004) Location-Based Mobile Pedestrian Navigation Services – The Role of Multimedia Cartography, ICA UPIMap2004, Tokyo, [Online], Available: http://www.ubimap.net/upimap2004/ [31 January 2008].

Goodman J, Gray P, Khammampad K, Brewster S (2004) Using Landmarks to Support Older People in Navigation, *Proceedings of Mobile HCI 2004,* Springer-Verlag.

Jaiku (2007) [Online], Available: http://jaiku.com/ [Accessed 31 January 2008].

Klante P, Krösche J, Boll S (2004) Accessights – A Multimodal Location-aware Mobile Tourist Information System, *Proceedings of the 9th International Conference on Computers Helping People with Special Needs, ICCHP2004,* Paris, pp. 287–294.

Koskinen I, Battarbee K, Mattelmäki T (2003) *Empathic Design. User Experience in Product Design,* Helsinki: IT Press.

Lehikoinen J, Aaltonen A, Huuskonen P, Salminen I (2007) *Personal Content Experience. Managing Digital Life in the Mobile Age.* John Wiley & Sons.

LoVEUS (2002) User Requirements of the LoVEUS system, Deliverable D01, IST-2000-30155, [Online], Available: http://loveus.intranet.gr/documentation.htm [31 January 2008].

Meng L (2005) Egocentric Design of Map-Based Mobile Services, *The Cartographic Journal,* vol. 42, no.1, pp. 5–1.

Metla (2006) Palvelutarjonta lisää kävijämääriä kansallispuistoissa, Metsäntutkimuslaitoksen tiedote 07.09.2006, (Eng. The Reports of the Finnish Forest Research Institute), [Online], Available: http://www.metla.fi/tiedotteet/2006/2006-09-07-matkailu.htm [31 January 2008], (in Finnish).

MLS (2006) *Mobile Location Services 2006,* May 17–18, Amsterdam, The Netherlands.

MLS (2007) *Mobile Location Services 2007,* May 15–16, Amsterdam, The Netherlands.

Nivala AM, Sarjakoski LT (2007) User Aspects of Adaptive Visualisation for Mobile Maps, *Cartography and Geographic Information Science, Towards Ubiquitous Cartography,* vol. 34, no. 4, pp. 275–284.

Nivala AM, Sarjakoski LT (2003) Need for Context-Aware Topographic Maps in Mobile Devices, in Virrantaus K, Tveite H (eds.), *ScanGIS2003 – Proc. of the 9th Scandinavian Research Conference on Geographical Information Science,* Espoo, Finland, pp. 15–29.

Osman Z, Maguire M, Tarkiainen M (2003) Older Users Requirements for Location Based Services and Mobile Phones, in *Human-Computer Interaction with Mobile Devices and Services, Proceedings of the 5th International Symposium, Mobile HCI 2003,* Udine, Italy: Springer, pp. 325–357.

PARAMOUNT (2002) Public Safety & Commercial Info-Mobility Applications & Services in the Mountains, User Requirements, Deliverable D1, [Online], Available: http://www.paramount-tours.com/ [16 January 2008].

Pospischil G, Umlauft M, Michlmayr E (2002) Designing LOL@, A Mobile Tourist Guide for UMTS, in F. Paterno (ed.), *Proceedings of Mobile HCI02,* Pisa, Italy, pp. 140–154.

Raper J, Gartner G, Karimi H, Rizos C (2007) A Critical Evaluation of Location Based Services and their Potential, *Journal of Location Based Services,* vol. 1, no.1, pp. 5–45.

Reichenbacher T (2004) *Mobile Cartography – Adaptive Visualisation of Geographic Information on Mobile Devices,* PhD Thesis, Verlag Dr. Hut, München.

Research and Markets (2005) Location Based Services (LBS)/Geographical Information Systems (GIS) Market Research 2005, viewing report, [Online], Available: http://www.researchandmarkets.com/reports/306446/ [25 June 2008].

Sarjakoski T, Sarjakoski LT (2005) The GiMoDig public final report, GiMoDig-project, IST-2000-30090, Deliverable D1.2.32, [Online], Available: http://gimodig.fgi.fi/pub_deliverables/GiMoDig-D1231-Final-public.pdf [25 June 2008].

Sayda F (2005) Involving LBS users in data acquisition and update, in Toppen F, Painho M (eds), *AGILE 2005, 8th Conference on Geographic Information Science. Conference proceedings.* Universidade Nova de Lisboa, pp. 33–42.

Schmidt-Belz B, Poslad S (2003) User Validation of a Mobile Tourism Service, in Schmidt-Belz B, Cheverst K (eds), *Proceedings of the Workshop W1 "HCI in Mobile Guides 2003",* in conjunction with Mobile HCI03, Udine, Italy, pp. 57–62.

van Setten M, Pokraev S, Koolwaaij J (2004) Context-Aware Recommendations in the Mobile Tourist Application COMPASS, in Nejdl W, De Bra P (eds), AH 2004, LNCS 3137, Springer-Verlag.

Zipf A (2002) User-Adaptive Maps for Location-Based Services (LBS) for Tourism, *Proceedings of the International Congress on Tourism and Communications Technologies in Tourism (ENTER 2002),* Innsbruck, Austria. Berlin, Heidelberg: Springer.

12 Geo-Identification and Pedestrian Navigation with Geo-Mobile Applications: How Do Users Proceed?

Ioannis Delikostidis and Corné P. J. M. van Elzakker

Department of Geo-Information Processing
International Institute for Geo-Information Science and Earth Observation (ITC)
Enschede, The Netherlands

Abstract

Geo-identification and pedestrian navigation with the use of geo-mobile applications requires a multi source interaction of the users with the environment, the (carto)graphic interface and their mental maps. This interaction is not effectively supported by current implementations that are mostly geared to vehicle navigation. The aim of the research presented in this paper was to look through the ways in which people navigate and orientate themselves in unfamiliar cities or areas supported by mobile map interfaces. In this regard, an experiment with real users and tasks was established in two areas in Amsterdam, based on a research methodology involving a questionnaire, thinking aloud with audio/visual observation and synchronous screen logging and a semi-structured interview. Use was made of a special technical solution that reduces the resources needed for field-based studies and allows for better analysis of the results. The findings of the experiment show the importance of particular landmark types and GPS-independent automatic map orientation for geo-identification and navigation and support further field-based studies on smooth zooming techniques. The research methodology worked well and could be applied to future experiments in order to gain more insight in the mobile users' interactions in real contexts.

Keywords: geo-identification, landmarks, mental maps, smooth zooming, user research methodology

12.1 Introduction

The ever increasing mobility of people pushes the need for effective tools supporting their geographical orientation and navigation. In addition, the wide availability of mobile devices, such as smartphones and PDAs and their capabilities to serve users as a better digital and interactive alternative to paper maps, has opened up an improved potential for mobile orientation and navigation, as well as location based services. There are already a lot of geo-mobile applications with (carto)graphic interfaces available. However, most of them are dedicated to car navigation and are not suitable, for instance, for use by pedestrians. This is not only because of the database contents of such navigation systems, but also because their interface and presentation aspects do not seem to support very well the user's orientation or personal geo-identification. For example, in newspapers we sometimes come across hilarious examples of people who do not know at all where they are after their car navigation system stopped functioning. Therefore, more research is needed in order to make geo-mobile applications a better successor of paper maps, which are still convenient all-round tools for orientation and navigation to many people (Hampe & Elias 2004).

This paper reports on an experiment executed with users (pedestrians) who pay a visit to a city area that is unknown to them and who want to navigate and orientate themselves with two geo-mobile applications that are already on the market. The objective of our research was not to evaluate these existing applications, but to learn more about the ways in which people navigate and orientate themselves with the help of supporting tools with a cartographic interface. As such, this experiment is part of two more extensive research projects in which the authors are involved: PhD research on the usability of geo-mobile applications and a government funded Dutch research project on *Usable (and Well-scaled) Mobile Maps for Consumers* (UWSM2) (URL 1). The PhD research should lead to the prototyping and evaluation of a new and more usable geo-mobile application. To this end a User-Centred Design approach is followed and the experiment reported in this paper can be considered as part of the requirement analysis stage. The UWSM2 project focuses on the development of generalization solutions for mobile maps. This is not only to possibly allow progressive wireless data transfer, but also because zooming in and out is a way of interaction that is required for orientation and navigation. Therefore, in our experiment, specific attention has been paid to this latter requirement.

This paper starts with a brief sketch of the theoretical background of the experiment and the research questions addressed in it. Thereafter, the set-up of the user research and the methodology are discussed. In a separate section specific attention will be devoted to the technical solutions to the desire to obtain synchronous user observation, mobile screen logging and thinking aloud research data for easy analysis later on. The results of the experiment and their analysis constitute the last and major part of the paper before the conclusion.

12.2 Personal Geo-Identification in an Unfamiliar Area

When using geo-mobile applications, visitors to unfamiliar areas interact with information coming from different sources in order to geo-identify (orientate) themselves and navigate. *Personal geo-identification* is the understanding of "where am I?" in geographic space in terms of a mentally translated and identified personal location in the real world. In case of mobile map users, the main information sources are the environment, the representation thereof in terms of (carto)graphics on the interface of the mobile device and the mental maps of the users. In *Figure 12.1,* a pedestrian user of a geo-mobile application who is trying to geo-identify himself is shown. The interaction with the three available information sources can lead to confusion, and as a result, the user can have difficulties in finding proper answers to his geospatial questions.

Imagine a person visiting an unfamiliar city and leaving an underground railway station through one of many exits. Such a visitor may not immediately know where exactly s/he is. But understanding "where am I?" is a necessary very first step before finding solutions to follow-up spatial problems such as in which direction s/he should move in order to reach a particular destination. Several types of *landmarks* and other structural elements may act as common points between the virtual and real worlds available to mobile map users. The importance of landmarks as orientation, navigation and wayfinding aid for this type of users has been reported in many earlier studies (Golledge 1996, Michon & Denis 2001, Millonig & Schechtner

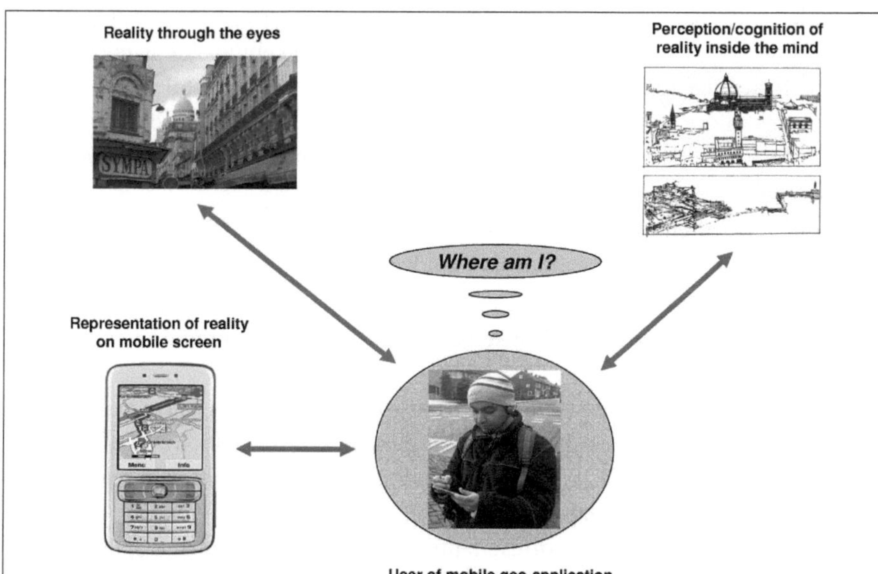

Fig. 12.1. Personal geo-identification through interaction with three different sources.

2005, Sorrows & Hirtle 1999) and has lead to different interesting new ideas, such as active landmarks (Gartner & Radoczky 2007).

In combination with landmarks, and due to the limitations of mobile displays (e.g. screen size and resolution), users frequently use zooming and panning functions of the mobile map in order to have both overview and detailed information about the area that they are (moving) in. The problem is that overview maps lack a lot of information in order to not overload the screen and make the map display illegible, while at the same time zoomed-in views lack the global angle of overviews, one of the main reasons why maps are made (Yammiyavar et al. 2007, Büring et al. 2006). Unfortunately, the change of scale of the mobile map can jeopardize the mental connection between the users' mental maps, the environment and its map representation, leading to disorientation and spatial confusion. In order to deal with this problem, the use of so-called smooth and/or animated zooming, instead of zooming in discrete steps or scale levels, is currently under investigation, for example in our UWSM2 project referred to above (also see Hornbaek et al. 2002, Midtbo & Norvik 2007). Because landmarks are strongly supporting orientation, navigation and wayfinding processes they should be visible in all the used scales so that they support the mental map connection between the real and virtual geographic worlds. Memorization of landmarks and their surrounding less prominent objects is a usual technique that the human mind is often using in order to keep that connection (Harrower & Sheesly 2005, Midtbo & Norvik 2007).

One of the aspects we wanted to investigate in our user experiment is what are the landmarks that users of mobile geo-applications and in particular visitors to unfamiliar cities use in order to orientate, navigate and perform wayfinding. When generalizing map displays, such landmarks may then be kept visible in every scale, so as to foster the relationship between the real world, the mobile map and the mental maps of the mobile users, to help answering their possible geographic questions, even when they are "toggling" between obtaining overview and important spatial details.

From our research objectives and our study of related work (as referred to above) we derived the following questions that guided the set-up of our experiment:

- What is the type of information users of geo-mobile applications are first seeking for in order to geo-identify themselves when they enter an unfamiliar area in a city for the first time?
- What are the landmarks existing in both the users' mental maps and in mobile maps that are important for personal geo-identification and navigation / wayfinding?
- Do users have problems with linking landmarks, as they appear in reality, with those appearing on the map display?
- Are the 'mental' landmarks of users properly linked to the map displays generated by the geo-mobile applications?
- In which ways do users use landmarks when they try to orientate themselves?

- How often and when are users confused about their location and what is the reason for that?
- When users know where they are, do they still make mistakes in deciding which direction to take in order to navigate to a destination? If so, what are the reasons for that?
- Do users benefit from the use of smooth zooming techniques in mobile maps rather than step-wise zooming?
- Do users benefit from 3D map and landmark representations?

12.3 The Experiment

12.3.1 Explorative Research

In answering the research questions listed above it was not the intention to obtain statistically valid quantitative results, nor was it the intention to evaluate the geo-mobile applications used. The objective of the field-based experiment we developed was to provide information about user requirements for personal geo-identification and pedestrian navigation that may be used for future prototype design. For our experiment, two existing geo-mobile applications with different characteristics were selected in order to be used by a group of test persons to perform real world tasks, related to personal orientation and navigation in the context of a visitor to an unfamiliar city. The test persons were observed and their thoughts, behaviour and performance were recorded and analyzed.

12.3.2 Selection of Geo-Mobile Applications

Although today there are many different geo-mobile applications available on the market (*Table 12.1*), most of them are navigation applications for vehicles with only minor abilities for pedestrian navigation (Millonig & Schechtner 2005, Rehrl et al. 2005). The applications investigated were Windows Mobile Smartphone/PPC v.6 compatible, as this was used in the available equipment.

The selection of two applications for the experiment was made through the following criteria:

1. Landmarks presented in 3D
2. Coverage of the study area (Amsterdam)
3. Zooming / panning functions
4. Smooth zooming capability
5. Availability to the researchers

The application that met all the criteria was iGo My way v. 8.0 *(Figure 12.2a)*. Google (mobile) Maps *(Figure 12.2b)* was selected for comparison reasons: it does not offer smooth/animated zooming functionalities and no landmarks on the map display, but we expect that this application will gain a wide distribution in the near future.

Table 12.1. Examples of existing (Windows Mobile 6.0 compatible) geo-mobile applications (numbers refer to the criteria explained in the text).

Web link	Application	1	2	3	4	5
URL 2	agis Navigator	N	N	Y	N	N
URL 3	Alturion GPS Professional v.6.0	N	Y	Y	N	Y
URL 4	amAze v.4.2	N	Y	Y	N	Y
URL 5	Co-Pilot Live v.7.0	N	Y	Y	N	Y
URL 6	Destinator 7	N	Y	Y	N	Y
URL 7	Google (mobile) Maps	N	Y	Y	N	Y
URL 8	Google Navigator v.3.6	N	Y	Y	N	Y
URL 9	iGo My way 2006 plus	N	Y	Y	N	Y
URL 10	iGo My way v.8.0	Y	Y	Y	Y	Y
URL 11	INAV i-Guidance v.4.0	N	N	Y	N	Y
URL 12	Intellinav v.2.0	N	N	Y	N	N
URL 13	Maptech Memory Map v.5.0	N	Y	Y	N	N
URL 14	Marco Polo Mobile Navigator 3	N	Y	Y	N	Y
URL 15	MioMap v.3.2	N	Y	Y	N	Y
URL 16	MioMap 2008	Y	Y	Y	Y	N
URL 17	Navigon Mobile Navigator v.6.0	N	Y	Y	N	Y
URL 18	Nokia Maps (Smart2Go)	N	Y	Y	N	Y
URL 19	Odyssey Mobile 4	N	N	Y	N	N
URL 20	OnCourse Navigator v.6.0 Plus	N	Y	Y	Y	N
URL 21	Pharos Ostia	N	Y	Y	N	N
URL 22	PocketMap Navigator (USA)	N	N	Y	N	N
URL 23	PocketWAW v.3.0	N	N/A	Y	N	Y
URL 24	Route 66 Navigate 7 PPC	N	Y	Y	N	Y
URL 25	Teletype GPS	N	Y	Y	N	N
URL 26	Tom Tom Navigator 6	N	N	Y	N	Y
URL 27	Via Michelin	N	Y	Y	N	N

12.3.3 Test Persons

In order to perform the experiment we were looking for a specific type of user representing a visitor to an unfamiliar city using a geo-mobile application. This visitor is "dropped" at a location unknown to him/her and is supported by the information provided by the mobile interface. Suitable test persons would not have to be deterred by using a mobile application and the selected age range was from 20 to 60 years old. They could have different levels of knowledge in terms of (mobile) maps, use of mobile devices, cartography and orientation and navigation techniques and abilities. The 8 test persons that took part in the experiment were recruited from MSc and PhD students of ITC (the International Institute for Geo-Information Science and Earth Observation). It is true that, compared to ordinary tourists, they may have

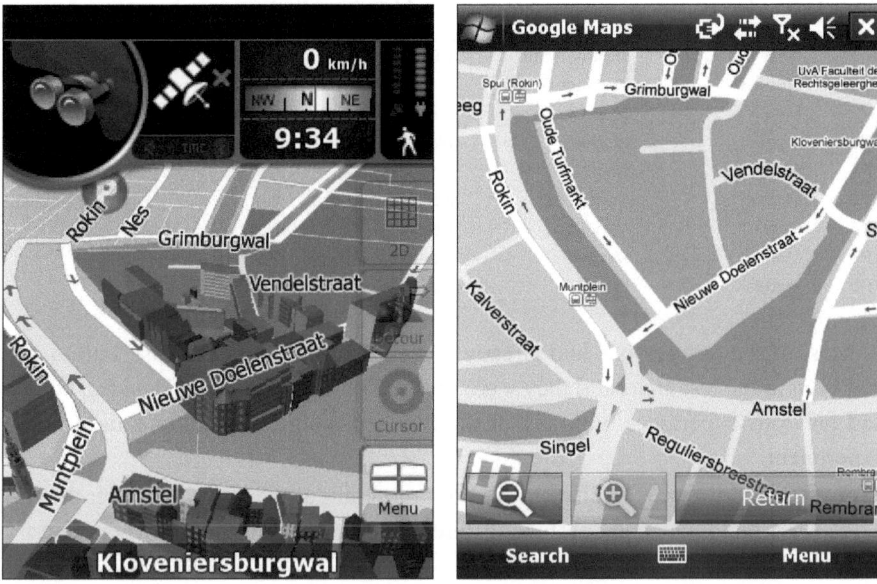

Fig. 12.2. iGo My way v. 8.0 (a) and Google Maps (b) showing the same region of Amsterdam in 3D and 2D respectively.

a more than average affinity with maps and geo-mobile applications, but they do not originate from the Netherlands and were not familiar with the study area. They felt attracted to Amsterdam and free transport to this city was the stimulus for them to participate in the experiment.

12.3.4 User Profiles

Five male and 3 female test persons participated in the experiments, with their age ranging from 24 to 47 years old. They had different countries of origin. The general profiles of the participants are shown in *Table 12.2* and their background in fields related to this research in *Table 12.3*.

Table 12.2. General profiles of the test persons.

Nr.	Age	Gender	Origin	Profession
TP1	34	M	India	Civil Engineering
TP2	27	F	Indonesia	Traffic Management – Transport Behavior
TP3	38	M	Iran	Ecological Modelling - Marine Engineering
TP4	47	M	Iran	Hydrology & Water Resources
TP5	25	M	China	Geographic Information Science
TP6	24	F	China	Geographic Information Science
TP7	29	M	Pakistan	GIS Software Engineering
TP8	26	F	Pakistan	Geohazards Management

Table 12.3. Experience of the test persons in fields related to the research.

Nr.	GPS systems	Digital maps	PDA	Mobile navigation apps
TP1	Modest	Very good	Very good	Very good
TP2	Poor	Modest	Modest	Modest
TP3	Modest	Modest	Modest	Poor
TP4	Poor	No	No	No
TP5	Modest	Very good	No	No
TP6	Poor	Very good	No	No
TP7	Modest	Modest	Modest	Poor
TP8	Modest	No	No	Poor

Most of the test persons had already visited Amsterdam in the past, but no one had been to the test areas and they were thus unfamiliar with them. Besides, no one had previous experience with any of the two geo-mobile applications used in the experiment.

A questionnaire was applied to find out more about the ways in which the test persons normally use and combine different sources of information when they try to orientate and navigate in space:

Most of the participants always or frequently prepare themselves before they go to an unfamiliar city or area and the others sometimes do that. The main sources of information for them are city maps, either printed on paper or consulted on a computer screen. Finding on the map the public transportation stations and the routes to their places of stay, together with points of interest and prominent buildings, are their most important goals. One of the participants also mentioned the need to see pictures of the points of interest in the city so that they can be recognized easily later. Asking other people that have already visited the place is also important for half of the participants, in order to learn about important points of interest that they should visit or can use as landmarks for orientation and navigation purposes.

All test persons sometimes or frequently take the responsibility of orientation and navigation, and the use of paper maps is a common task for them. However, this does not imply that they always complete these tasks with ease, as all of them have sometimes or even always (TP6, TP8) difficulties to orientate themselves in unfamiliar areas with the use of paper maps.

Only one test person (TP2) makes frequent use of mobile navigators when travelling to an unfamiliar place, while the others never do that. This is an interesting outcome of the questionnaire, as half of the participants had answered earlier that they have some experience with mobile navigators.

After their arrival in an unfamiliar city, more than half of the test persons inspects the layout of the city, and the patterns of the streets, in order to try to understand the city better and not to get lost. Besides, half of them are trying to find and locate some easily distinguishable landmarks: churches or other tall buildings (3), train/bus stations (2), big shops, restaurants (1) and rivers (1). When they do get lost, half of them are trying to find their way through street name information, compared to a map, or to find a previously memorized landmark such as a church, a transportation terminal or a big building. Less than half of the test persons would also ask local people to explain to them where they exactly are and to point their position on a map that they may carry. One of the participants (TP1) also uses the sun direction and tries to re-orient the map towards the North.

All test persons find their orientation and navigation abilities improved when they are visiting the city for a second time. According to them, the reason is the memorization of transport stations (2), tall structures, buildings and city centers (3), patterns of the cities (1), previous routes (1), appearance of roads (1) and any other easily distinguishable / unique features (1).

12.3.5 Study Area

Amsterdam was the selected city for the experiment for three reasons. First, it is visited by many people (e.g. tourists) who have not been there before and they often go there by public transport. Secondly, Amsterdam (Department of "Stadstoezicht") takes part in the UWSM2 project as well, as the municipality would like to use a geo-mobile application in relation with parking services. And, thirdly, the train trip from Enschede (the temporary dwelling-place of almost all test persons) allows for interaction between the researcher and the test persons in order to prepare the latter for the experiment and acquire some important additional research data.

For executing the tests, two different areas of Amsterdam with different environmental parameters were selected, in order to investigate how this diversity influences the personal geo-identification and navigation of the test persons. The first (based on the test route) was Wibautstraat Metro station (start) – Krugerplein (1st destination) – Amstel Metro Station (2nd destination) and the second was Dam Square (start) – Begijnhof (1st destination) – Rembrandt House (2nd destination) These can be seen in *Figure 12.3.*The first area has less diversity of features (neighbourhoods mostly comprised of houses) and the second includes many prominent places that could be used as landmarks (churches, governmental buildings, historical places).

12.3.6 Research Methods and Techniques

The testing methodology involved a questionnaire at the start, thinking aloud with audio/visual observation and synchronous screen logging and a semi-structured interview at the end. This combination of methods allows for a deep investigation of the test persons' thoughts and actions, and keeps, at the same time, records of all the test activities. The resulting research materials may be thoroughly and objectively analyzed afterwards (Delikostidis 2007, Van Elzakker et al. 2008).

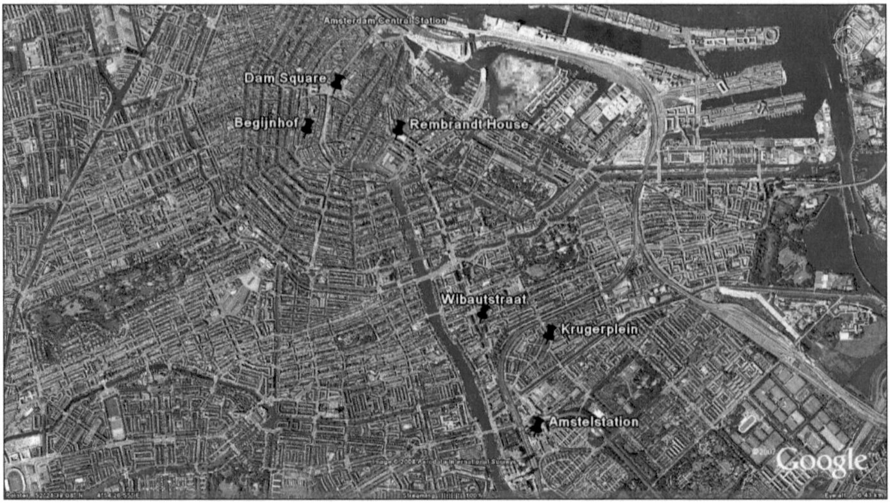

Fig. 12.3. The test areas Wibautstraat-Krugerplein-Amstelstation and Dam Square-Begijnhof-Rembrandt House placed in Google Earth.

The questionnaire had the aim of creating a general profile of the test persons, involving their experience, knowledge and way of acting in fields related to this experiment. Thinking aloud proved to be a very valuable technique for acquiring information about the test persons' thoughts and the reasoning behind their acts, especially in parts where they were confused. During the interview at the end the test persons could also discuss additional issues and could also be asked several questions depending on their previous answers. This was especially useful for clearing things out when there was a noticeable difference between answers of particular test persons in the initial questionnaire and their behaviour and acts during the survey.

The test persons were also asked to draw two mental maps, depicting their place of residence and the route that they followed during the testing. The first mental map was drawn after completing the questionnaire and the second one after finishing the interview. In these drawings they were supposed to include the geographical objects they found important. From the mental map drawings a lot of information can be retrieved, such as the type of landmarks memorized, the points of confusion and the possible reasons for that and the individual's orientation abilities.

The 8 test persons were divided in two groups of 4 people each, one group for each of the two study areas in Amsterdam. The test persons in both groups were asked to use the two selected geo-mobile applications one after the other, performing one navigation task with each application. Each test started by asking the test person to try to orientate with the help of the first geo-mobile application without using street names and then select and follow the shortest route to navigate towards the first destination. After reaching that point, the second application was used in order to execute the same tasks targeting to the second destination. The order of applications was changed every time for comparison reasons.

Although the real contexts of use are dynamic and even unpredictable up to some extent, a set of conditions was applied to the test sessions in order to limit the context diversity as much as possible. The test sessions were only executed during daytime (from 8:00 to 20:00 hrs in the months May and June 2008), in fair weather conditions (cloudy or sunny days, with average temperature and wind speed) and not in highly disturbing instances (demonstrations, national celebrations, road work and the like) in the survey areas.

Before the actual involvement of the test persons, a pilot study took place in the selected areas in order to calculate the required average times for the completion of each task, to test the equipment in real conditions and find out possible problems or limitations that could affect the execution of the tests. After the pilot study, some adjustments to the survey settings had to be made. The initially estimated timings of the different parts of the test were revised and some of the equipment was calibrated in order to work properly with respect to the field's specific parameters (high environmental and electromagnetic noise, possibility of short-time light rain, low GPS signal in narrow streets).

12.3.7 Briefing and Training of the Test Persons

During the transportation of the test persons from Enschede to the test areas by train (two test persons per day) the test persons were first asked to complete the questionnaire and then draw the mental map of their place of residence. Then they were given the mobile device (a PDA HP iPAQ 4700hx and later a PDA-smartphone i-mate Ultimate 9502) running iGo My way 8.0 and Google Maps, with the latter running offline using Google Navigator software (URL 8). The GPS receiver of the mobile device was on while the train was moving, so that the test persons could better understand how each mobile application was working, how they could use the basic functions (zooming, panning) and how the viewing perspective could be changed from 2D to 3D and vice versa. Two city and street name finding tasks involving zooming and panning were also given to the test persons, in order to make them more familiar with the functionalities and interface of the geo-mobile applications. The test persons were informed that the test would start as soon as they had reached the Wibautstraat Metro station (for test persons in the first group), and Dam Square for those in the second group. Additional instructions regarding the execution of the tests and information regarding personal and equipment safety were given as well.

12.4 Mobile Observation and Thinking Aloud in the Field

In order to carry out the experiment, a special technical solution for field-based observation/recording was used. It is based on a system that was already implemented during a previous investigation on methodologies for field-based usability evaluation of geo-mobile applications (Delikostidis 2007, Van Elzakker et al. 2008). This system consisted of several electronic devices, such as two pairs of audio transceivers, three B&W cameras, a laptop, a handheld video recorder, two pairs of video transceivers and a video quad processor. This complicated system was needed in order to reduce the bias from the researcher physically being too close to the test persons, to minimize the human resources required for carrying out the test sessions and to facilitate the analysis of the recorded research materials through synchronization. With this system, the thinking aloud audio signal, the camera observations of the user, the environment and his/her interaction with the mobile device and the logging of the changes on the screen were synchronically recorded with a date/time stamp. The context of use and the participants' activities and expressed thoughts could thus be analyzed later with accuracy, speed and convenience.

For this experiment, the original system was improved and upgraded in order to offer higher reliability, simplicity and performance. Its main parts are a pair of DECT phones connected to headsets, a hard disk-based four-channel mobile video/audio recorder, three high resolution and wide view colour cameras, two pairs of video

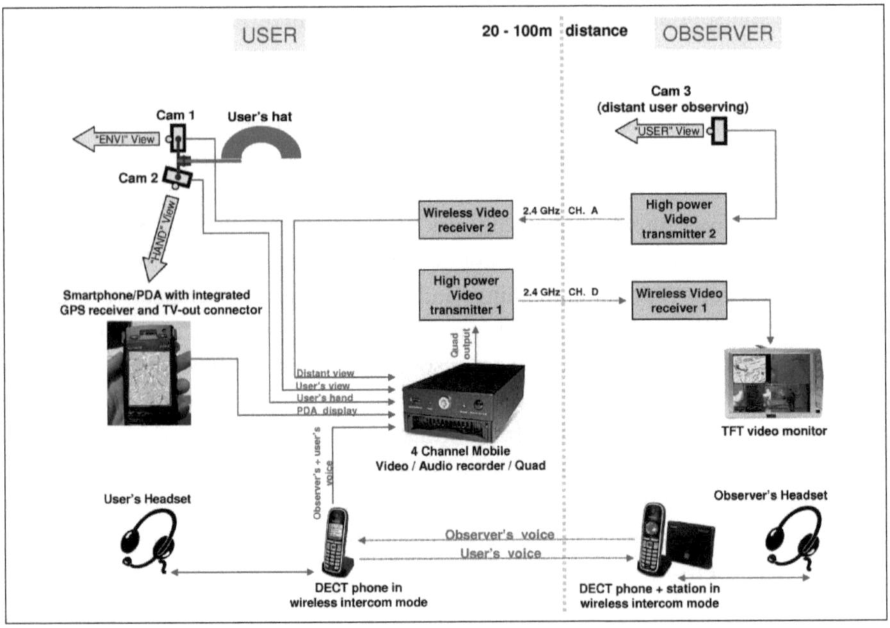

Fig. 12.4. Schematic diagram of the field observation/recording system.

transceivers, a TFT colour video monitor and a mobile device, an i-mate Ultimate 9502 with integrated GPS receiver and video-out capability, running Windows Mobile 6.0 *(Figure 12.4)*. The DECT phones were used as wireless intercoms offering good quality and uninterrupted audio communication between the researcher and the test persons during the sessions. The pair of headsets that both of them were wearing was connected to the DECT phones, through which the thinking aloud could be performed and through which researcher and test person could interact.

The test person wore a hat with two of the colour cameras attached on it. The first one captured his/her interaction with the mobile device and the second one their actual viewpoint. A third camera was carried by the observer, capturing the interaction of the test persons with the environment from a fair distance (20 to 100 meters) and sending this image wirelessly to the user's video receiver. In addition to these inputs, a real-time screen capture of the mobile device display was provided through its integrated composite-type video output.

All the four video signals together with the audio communication were recorded in the 4-channel mobile video/audio recorder, which has enough storage space for many hours of continuous recording. The advantage of using a 4-channel system rather than a single channel one is the higher quality of video per channel, while at the same time there is synchronization between the video/audio channels and date/time stamping which has the benefits described earlier. The researcher could observe all the recorded video signals wirelessly and simultaneously in a quad view (four images in one screen) on the colour monitor that he carried, through a pair of video transmitter/receivers connected to the mobile recorder and the monitor respectively *(Figure 12.5)*.

Fig. 12.5. Sample screenshot of the video recordings of the system.

Fig. 12.6. Equipment checking and final instructions to the test person just before the start of a test session at the Dam Square.

Fig. 12.7. One of the test persons thinking aloud while trying to orientate in front of Begijnhof entrance. The researcher is observing the participant's interactions through the video monitor.

This configuration also overcomes the issue of battery power shortages that came to light in the earlier research. It allows the continuous use of the system for many hours by simply charging a pair of lithium-ion battery clusters for powering the several devices beforehand.

When the researcher/observer and the test persons reached the starting-point, a preparation and installation of the devices of the field observation and recording system was taking place for about 5 to 10 minutes. After that, a few final instructions were given to the test persons and they were able to start the test *(Figure 12.6)*.

During the test sessions, the participants had to think aloud about their thoughts, decisions and confusions. The researcher was frequently reminding them to speak aloud every time they forgot to do so, asking them questions and encouraging them to try finding alternative ways to solve their problems. In order to be able to derive useful conclusions about the interaction between mental maps, reality and mobile maps, the test persons were frequently asked to describe what are the landmarks, features, patterns or any other type of information that they use to orientate and navigate in each situation. It was considered important to inspect what the test persons were looking at during the task execution, even though they were sometimes doing that unconsciously. Questions to the users triggered by these instances provided very interesting and valuable answers. These answers helped in formulating hypotheses about the connections between the real and virtual worlds and in better understanding the process of geo-identification *(Figure 12.7)*.

12.5 Results and Analysis

12.5.1 Task Execution

The test sessions took place between 24 May and 14 June 2008. In general, there were no major problems with respect to the research methodology. There were only a few minor random technical problems with the electronic devices used, which could be expected in this type of research.

The test persons were encouraged to think aloud during the test sessions while trying to orientate and navigate with the use of a geo-mobile application. However, thinking aloud was not always easy for them. In several cases the test persons stopped walking and were trying to say what they think. Outside this apparent verbalization problem, the think aloud method led to valuable results, especially in instances where the test persons were confused about, for example, the selection of a path or direction of movement.

Disturbance by residents or tourists asking us what exactly was the subject of our research was one of the issues we expected to be confronted with. Although that happened a few times during the test sessions, it did not influence their proper execution. Providing a fast and polite explanation was an effective solution that worked in every instance.

12.5.2 Mental Maps

The mental map drawings of Enschede (the town of residence for most of the participants) and those of the test areas, give an overview of what landmarks the test persons find important to be stored in their minds. They also demonstrate differences from person to person in terms of perception of space and easiness of building mental maps of new areas.

Figure 12.8 shows the mental maps of one test area of two different test persons: TP2 and TP5. The park near destination point 2 and the canal are included in both drawings. The main roads are also very important for both of them, and TP2 even remembers the name of one road. Although during the actual testing both of these test persons found the rail track very important to orientate themselves, it was only included in TP2's mental map. TP5, on the other hand, included a big roundabout existing near destination point 3. This roundabout was a point of confusion for that participant during the testing, as it was difficult for him to select which of the surrounding streets was the correct one to follow.

Landmarks and features that most of the participants included in their mental map drawings were main roads, places of residence and work, transport stations, railroads, tall buildings, squares and parks, big shops/supermarkets, center of cities

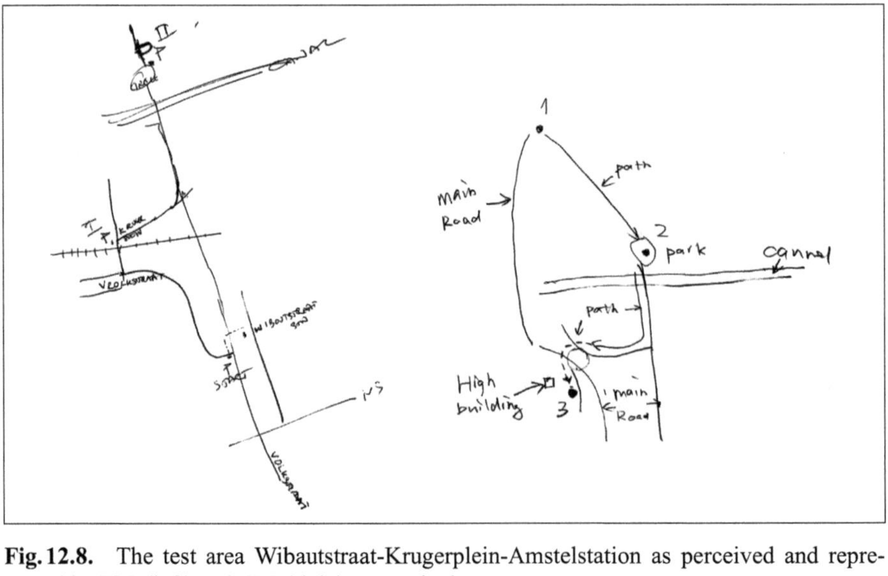

Fig. 12.8. The test area Wibautstraat-Krugerplein-Amstelstation as perceived and represented by TP2 (left) and TP5 (right) respectively.

and canals/rivers. However, a major complaint of the test persons was that they could actually not collect as much information of the environment as they would if they would not use the geo-mobile application continuously. For example, two participants mentioned that they would memorize the location and appearance of shops or they would have noticed several tall buildings along their route if they were using a paper map or no map at all. Apparently, they found it difficult to focus on the virtual and real world at the same time.

12.5.3 Test Outcomes

The results of semi-structured interviews, combined with the answers to the questionnaire, the think aloud protocols and video recordings of the observations and screen logging during the actual tests and the mental map drawings were used to answer the research questions. Only a selection of outcomes can be presented here:

The first information that test persons as visitors to an unfamiliar city searched for in order to geo-identify themselves was their position on the map display through the GPS location arrow. They then linked this to the patterns and sizes of the streets on the map, compared to those in reality, as seeking for street names was not allowed at the starting points. It became clear that for personal geo-identification and navigation with the help of geo-mobile applications, the test persons preferred simple map displays with clear colouring and road sizing related to the actual size of the roads in reality. In this respect they preferred Google mobile maps, but they missed important landmarks on the map displays of this application.

The types of landmarks and features that helped the test persons to orientate and navigate during the tests were the canals, the road patterns and sizes, the street names and the parks/squares and roundabouts. Landmarks that would help them but were not (always) available on the map displays are the bridges, pedestrian paths crossing roads, important buildings, such as municipal offices, or tall buildings that are visible from a distance. Specific landmarks that they expect to come across in order to help them to find their way in an unfamiliar city are big shops and easily distinguishable restaurants, such as fast food branches, churches, noticeable monuments, canals, bridges and parks. Besides, it appeared that the types of landmarks in both the users' mental maps and the mobile maps to support their geo-identification and navigation/wayfinding partly depend on the type of the area. For example, canals and bridges were perceived as very useful landmarks in Amsterdam. Or, in an area with not so many prominent features, a small park or a roundabout gains more importance than in an area with many easily distinguishable buildings. On the map displays, the landmarks should be made more distinguishable (colour, shape, size) or additional information should be provided (photos, text). It was found that applying advanced solutions to the issue of landmark visualization is not always better.

Indeed, as it comes to 3D representation and 3D building models on mobile maps, most of the test persons indicated that it is confusing to have a plethora of 3D buildings on the map and they would prefer to only see important 3D landmarks (which would make the software running faster and smoother as well). One of the participants complained about the perspective view of the 3D maps which is not a human-eye view but a bird-eye one, making it difficult to interpret the image of the map correctly inside the human mind. Indeed, most users preferred to use the 2D rather than the 3D map display in this research. However, in this respect no general conclusions can be drawn, as the test persons could not really get used to the different visualizations in the relatively short time of the experiment. The 3D map interface had more functions than the 2D and many persons confessed that they were afraid to use it as they were getting confused with the totally different way of visualization. They confirmed that the 3D models of the buildings could improve their mental connection with the mobile map, but stated that there should be a careful selection of which types of buildings/landmarks should be included in the map. In iGO 8, for example, they found the screen overloaded with too many 3D buildings making it difficult and slow for them to use the map. At the same time, the similar colours that were applied to the different buildings made it often impossible to understand which one is the building that they were looking at in front of them. Most test persons argued that a pop-up photo of a landmark would be more helpful (and less CPU power-consuming) than their 3D representation. In this regard, photos of corner buildings would be critical for the orientation of TP7, as they would allow fast selection of the correct streets. Continuously visible house numbering would also be a helpful for orientation and navigation, together with a very accurate street size/pattern visualization according to reality. Including railroads, building blocks and pedestrian paths was considered to be important by many test persons as well.

In many cases, test persons could not properly connect landmarks of reality to the mobile map displays as they were either not visible at all on the latter, or they were appearing and disappearing in different zoom levels. Sometimes they were not represented in an easily perceivable form. As it comes to the "mental" landmarks of the users, things were a little bit more complicated. Next to the representation issues already mentioned, the development of their mental maps based on landmarks was decelerated by their looking at the mobile screen most of the time. The majority of test persons argued that if they had used a paper map (or no map at all) they would have developed, combined and memorized more landmarks.

Despite all this, test persons did use landmarks as points of connection between reality and it's graphic representation in the form of mobile maps and they tried to orientate the mobile map towards the position and direction of the real landmarks. This was problematic in areas where there was low diversity of structural elements in the environment. The test persons got confused/disorientated many times when they were trying to rely merely on the position arrow on the map, as this is not a very accurate navigation tool during walking speeds. It appeared that most test persons found the position arrow on the mobile maps very important, and complained about it's inaccuracy in showing the actual direction of movement. They made mistakes during navigation because they tried to follow the arrow, despite the fact that they were continuously informed by the researcher not to rely exclusively on that and try to find other sources of information to understand where they are. The problem is that GPS signal-based position arrows on mobile maps cannot work reliably when the speed of movement is low, as what is happening during pedestrian navigation. Obviously, the test persons would prefer a map continuously rotating towards the direction of their movement and towards their point of view when they are not moving.

A last issue related to personal geo-identification and navigation is the sequential need for overview and detailed map displays. Although iGO 8 has smooth zooming capabilities, none of the participants noticed that during the tests. The possible reason for this finding is that in densely built up areas, such as Amsterdam, the software cannot process the geographic data fast enough in a common mobile device in order to achieve graphically smooth changes during zooming in and out. This technical problem is also addressed in the UWSM2 project.

During the tests, most of the test persons found it easier to keep an overview map of the area in their minds while inspecting a more detailed view. However, they agreed that frequent zooming in and out is required in order not to lose the contact between reality and the maps in their minds. In case of a total loss of the GPS signal, as a consequence of which they would have to find their way through a static mobile map, most test persons would use the road names and the street sizes and patterns in order to first understand where they are and then navigate.

12.6 Conclusions

This paper presented the findings of a field-based experiment investigating the interactions of visitors to unfamiliar cities with the environment, their mental maps and the (carto)graphic interface of two geo-mobile applications. The experiment applied a combination of qualitative research methods and techniques supported by a special technical solution for field-based user research. The results of this experiment contribute to the requirements analysis stage of the user-centred design of a prototype of a more usable geo-mobile application and to finding solutions for automatic mobile map generalization, brought about by the need for zooming, as strived for in the UWSM2 project.

Some of the results that could directly contribute into making guidelines for more usable geo-mobile applications are the determination of the types of landmarks that support user geo-identification and should be included in mobile map interfaces, the representation of these landmarks as a combination of unique icons and popping-up windows showing pictorial and text information. The zoom levels that are mostly used and the landmarks and information that is missing in specific types of points of confusion in current geo-applications can additionally be determined by a more thorough analysis of the existing thinking aloud data.

Further research is needed, though, not only to compare smooth and step-wise zooming, but also in terms of proper visualization of environmental and mental landmarks on mobile (carto)graphic interfaces for pedestrian navigation. Determining the map scales at which particular types of landmarks should remain visible or at which their visual characteristics should be changed depending on the user's requirements and preferences and the context of use is one of the issues to be addressed. Minimizing the need to use the zooming function by properly selecting and generalizing important structural components including landmarks and by applying user-friendly zooming techniques is another one. In this regard, a future field-based research focusing more into the usability of smooth zooming in real contexts, as compared to step-wise zooming, is required. This could be done once the technical problems of geodata transfer have been solved, perhaps by the other results of the UWSM2 project.

In any case, finding proper ways to connect the real and virtual geographic worlds that the user of geo-mobile applications interfaces with, could be one of the keys to developing more usable geo-mobile applications.

Acknowledgements

This work has been executed in the framework of the Dutch research programme Space for Geo-Information (project RGI-233, also referred to as UWSM2). Special thanks to Mr. Fred Harms of project consortium partner Municipality of Amsterdam (Department "Stadstoezicht"), who contributed to the definition of the use case and who suggested the location of the test areas.

References

Büring T, Gerken J, Reiterer H (2006) User Interaction with Scatterplots on Small Screens – A Comparative Evaluation of Geometric-Semantic Zoom and Fisheye Distortion. In: IEEE Transactions on Visualization and Computer Graphics, vol. 12, no. 5, September/October 2006.

Chitaro L (2006) Visualizing information on mobile devices. In: Computer, vol. 39, issue 3, March 2006, pp. 40-45.

Delikostidis I (2007) Methods and Techniques for Field-based Usability Testing of Mobile Geo-applications, MSc thesis, International Institute for Geo-Information Science & Earth Observation (ITC) Enschede, the Netherlands. Available online at: http://www.itc.nl/ library/papers_2007/msc/gfm/delikostidis.pdf

Dilo A, Van Oosterom PJM (2006) Usable mobile maps based on a vario-scale data server. In: Kraak MJ, Meerkerk J (eds.), GIN-RGI Symposium 2006, Ede, pp. 140-14.

Gartner G, Radoczky V (2007) About the Role of Cartographic Presentation for Wayfinding. In: Stefanakis E, Peterson MP, Armenakis C, Delis V (eds), Geographic Hypermedia Concepts and Systems, Lecture Notes in Geoinformation and Cartography, pp 381-398. Berlin Heidelberg: Springer.

Golledge R (1996) Human wayfinding and cognitive maps. In: Golledge R (ed.), Wayfinding Beheaviour, pp. 5-46. Baltimore: John Hopkins University Press.

Hampe M, B Elias (2004) Integrating Topographic Information and Landmarks for Mobile Navigation. In: Gartner G (ed.), Location Based Services & TeleCartography, Proceedings of the Symposium 2004, 28 - 29 January 2004, Vienna University of Technology, Vienna, Austria, pp. 147-155.

Harrower M, Sheesley B (2005) Designing better map interfaces: A framework for panning and zooming. In: Transactions in GIS, vol. 9, no. 2, March 2005, pp. 77-89.

Hornbæk K, Bederson BB, Plaisant C (2002) Navigation patterns and usability of zoomable user interfaces with and without an overview. In: ACM Transactions on Computer-Human Interaction, vol. 9, no. 2, June 2002, pp. 362-389.

Kristoffersen S, Ljungberg F (1999) "Making Place" to Make IT Work: Empirical Explorations of HCI for Mobile CSCW. In: Proceedings of the International ACM SIGGROUP Conference on Supporting Group Work (GROUP'99), Phoenix, USA, 1999.

Michon PE, Denis M (2001) When and Why Are Visual Landmarks Used in Giving Directions? In: Proceedings of the International Conference on Spatial Information Theory: Foundations of Geographic Information Science pp.292-305. London: Springer-Verlag.

Midtbo T, Norvik T (2007) Effects of Animations in Zooming and Panning Operations on Web maps: A Web-based Experiment. In: The Cartographic Journal, vol. 44, no. 4, December 2007, pp. 292-303.

Millonig A, Schechtner K (2005) Developing landmark-based pedestrian navigation systems. In: 8th International IEEE Conference on Intelligent Transport Systems, vol. 8, no. 1, September 2005, pp. 43-49.

Rehrl K, Göll N, Leitinger S, Bruntsch S (2005) Combined indoor/outdoor Smartphone navigation for public transport travellers. In: Gartner G (Hrsg.), Proceedings of the 3rd Symposium on LBS & TeleCartography 2005, Geowissenschaftliche Mitteilungen, Heft Nr. 74, Schriftenreihe der Studienrichtung Vermessungswesen und Geoinformation an der Technischen Universität Wien, 2005, pp. 235-239.

Sorrows ME, Hirtle SC (1999) The Nature of Landmarks for Real and Electronic Spaces. In: Freksa C, Mark D (eds.), Spatial Information Theory, Lecture Notes in Computer Science, 1661, pp. 37-50. Berlin: Springer.

Van Elzakker CPJM, Delikostidis I, Van Oosterom PJM (2008) Field-Based Usability Evaluation Methodology for Mobile Geo-Applications. In: Cartographic Journal, The, Vol. 45, No. 2, May 2008, pp. 139-149.

Van Elzakker CPJM, Wealands K (2007) Use and users of multimedia cartography. In: Cartwright W, Peterson MP, Gartner G (eds.), Multimedia Cartography, Second Edition, Chapter 34, pp. 487–504. Berlin etc.: Springer.

Yammiyavar P, Nainwal S, Banerjee S, Gupta S (2007) Comparison of The Fisheye Technique and Event Horizon Method for Location Based Services. Department of Design, Indian Institite of Technology, Guwahati Assam, India. Obtained on 20/10/2007 from: http://sgupta.newsit.es/projects/fisheye.pdf

URLs

URL 1: RGI-233 research project on Usable (and Well-scaled) Mobile Maps, http://www.rgi-otb.nl/uwsm2/

URL 2: Agis Navigator, http://www.asiagis.com.sg/agis/index.htm

URL 3: Alturion GPS Professional v.6.0, www.alturion.com/products/detailsphp?id=59

URL 4: amAze v. 4.2, http://www.amazegps.com/welcome.php?language=en

URL 5: CoPilot live v.7.0, http://www.alk.com/copilot/

URL 6: Destinator 7, http://www.destinatortechnologies.net/uk/products/

URL 7: Google Maps, http://www.google.com/gmm/index.html

URL 8: Goolge Navigator 3.6, http://www.pdafun.net/index.htm

URL 9: iGO My Way 2006 plus, http://i-go.com/en/news/index.php?aid=114

URL 10: iGO My Way v.8.0, http://www.i-go.com/en/news/index.php?aid=110

URL 11: iNAV iGuidance v.4.0, http://www.inavcorp.com/

URL 12: Intellinav v.2.0, http://www.intellinav.com/products-Software.htm

URL 13: Maptech Memory Map v.5.0, http://www.memory-map.co.uk/software.htm

URL 14: Marco Polo Mobile Navigator 3, www.marcopolo.de/mobilenavigator/en/1.htm

URL 15: MioMap v.3.2, www.mio-tech.be/en/gps-navigation-products-mio-map-v3.htm

URL 16: MioMap 2008, http://www.mio-tech.be/en/MioMap-2008.php

URL 17: Navigon Mobile Navigator v. 6.0, http://www.navigon.com/site/nl/nl/products/mobile_navigation/mobile_navigator

URL 18: Nokia Maps (Smart2Go), http://www.smart2go.com/en/

URL 19: Odyssey Mobile 4, http://www.infogation.com/features/features.html

URL 20: OnCourse Navigator v.6.0 Plus, http://www.oncoursenavigator.com/

URL 21: Pharos Ostia, http://www.pharosgps.com/products/accessories/Ostia_Software/NAV01_EU.htm

URL 22: PocketMap Navigator, http://www.pocketmap.com/navigator.html

URL 23: Pcket WAW v.3.0, http://www.wherearewe.co.nz/pocketwaw.html

URL 24: Route 66 Navigate 7 PPC, http://www.route66.nl/route66/products.php?cid=UK&sec=9&ssec=1&prodid=2240

URL 25: TeleType GPS, http://www.teletype.com/Merchant2/merchant.mvc?Screen=CTGY&Category_Code=Software

URL 26: TomTom Navigator 6, www.tomtom.com/products/category.php?ID=2&Lid=1

URL 27: Via Michelin, http://shop.viamichelin.co.uk/

13 Neo-Cartographic Influence on Map Communication in LBS

Markus Jobst and Jürgen Döllner

Hasso-Plattner-Institute, University of Potsdam, Germany

Abstract

Neo-cartography spans ubiquitous cartography, user participation and considerations for geo-media techniques. This new expansion of multimedia and internet cartography combines the latest Web developments with traditional cartography and imagery research. Therefore the focus within the conceptual communication model shifts and leads to a separate investigation of information-carrier and information-content. At least this separation helps to prepare geospatial data efficiently, especially within virtual 3D presentation methods. But it is also a crucial aspect of geo-media techniques whenever detailed information is put on lower resolution interfaces. This contribution introduces the notion neo-cartography, discusses the mutation of importance within the conceptual communication model and explains its consequences for geo-media techniques in LBS.

Keywords: neo-cartography, communication models, mass-communication, dead information pixels, geomedia-techniques

13.1 Introduction

Ubiquitous cartography, Location Based Services (LBS) and webmapping technologies open up new perspectives for the handling of geospatial information. Whereas geospatial information was accessible only to restricted user groups in former days, this kind of information becomes openly distributed with evolving WWW and LBS technologies. This increase of map distribution shows some impact on mass-communication, where everyone, who is influenced by the media, develops a distorted mental map as well as a distorted individual picture of space (Bülthoff et al. 2001). This effect of mass-communication influences individual decisions, movement in space and knowledge acquisition. Therefore these effects raise urgent questions for possible metamorphosis of existing conceptual communication models, a relocation of communication base items and thus deliver arguments for future directions in cartosemiotic theories, especially for small displays and adopted service-oriented engineering (Hagedorn et al. 2007).

This contribution introduces the notion neo-cartography as well as possible changes of the conceptual communication model, uses these notions to highlight main principles of geo-media techniques for restricted graphical devices (small size), demonstrates the impact on 3D geo-visualization and shows directions of Cartographic Visualization Services for some future work.

13.2 Neo-Cartography

The recording, documentation and analysis of geospatial data were developed throughout centuries within the domain of cartography. On one hand processes of recording, projection schemata and precision formed main elements of cartographic work, on the other hand modifications of these data and partial deformations according to a beautiful artistic visualization mainly shape the "old" use of landscape documentations. Further developments of traditional cartography lead to precise coding mechanisms of recorded data and mostly to its military use. Maps at those times mainly dealt as non-public planning tool for geospatial activities. This exclusive situation of map access changed to a large extent with developing communication technologies and their public accessibility. Main examples are Nasa's World Wind, Google's Earth or Microsoft's Virtual Earth. Basically these applications make use of Web Services, which access geospatial repositories for their rendering of pictures.

Neo-geography is a notion for "new geography", which bases on a public access to geospatial data and participation in geographic applications (Turner 2006). The access to geospatial data is executed via the Internet and various Web Services. A user does not have to load complete datasets to the client computer, but receives simple pictures according to the requests of Web Services that may be used by specific applications (like Google Earth). The participation in geographic applications describes the user's possibility of recording and sharing geospatial data, which have special importance for him/her. One prominent example is openstreetmap, a worldwide initiative for the public recording of street data (www.openstreetmap.org).

In addition to the public recording and exchanging of geospatial data, the notion neo-cartography combines neo-geographic characteristics with ubiquitous cartography and geo-media techniques. Beside a time- and space-independent access to maps and modification of geospatial data, neo-cartography takes the characteristics of transmitting media, the impact of information-content and user needs for the presentation of geospatial information into account. The new aspects of neo-cartography indicate the possibility to access mental imagery directly by using user inputs. The ubiquitous existence of maps and a public participation develop a social imagery of space that should be used for the abstracted and simplified presentation of space.

13.3 Metamorphosis of the Communication Model

The increasing importance of neo-cartography, mass-media influence and public participation in map creation leads to a gradual metamorphosis of conceptual communication models. Although the very simple model with stimulus-object-response (S-O-R) (Kobzina 2006) is used, it is not well suited for the new situation in neo-cartography anymore. Nor was it when mass-communication became efficient with its intense dissemination. By identifying the main components and relations of a general communication process, a process of relocation can be observed. This process then points out some influence of the neo-cartographic framework on the communication model. The metamorphosis of the conceptual communication model can be illustrated by S-O-R, Shannon and Weaver and Westley/MacLean (Jobst 2008).

The S-O-R model focuses on the inter-human communication, whereas individual preferences and settings are considered as central variable within the overall influence process. The general theory of mass-society becomes substituted by concepts of small groups. The individual is neither isolated from society nor an upright part of society, but belongs to groups with specific characteristics (family, friend, work, ...). These groups serve as orientation for individual behaviours (Horsky et al. 1983). The individual behaviour, which underlies an overall influence process (impact of mass-media), can be split in three main components: a cognitive, affective and conative component. The cognitive component concerns perception and imagination, which then builds up new knowledge. The affective component relates to collateral emotions when cognitive tasks are processed. The conative component describes tendencies of behaviour that are activated by perception and imagination of an object. Thus S-O-R models base on a stepwise communication.

Shannon and Weaver formalized communication and built up the mathematical theory of communication (Weaver 1972). The main work lies in the examination of the decoding problem within information theory. For the understanding of Weaver, communication is any kind of contact in a very broad sense (Weaver 1972, p. 3). The principle model of Shannon and Weaver can be exemplified by modalities of telecommunication: a sender uses a transmission line to send information to a recipient. This information needs to be coded in a way that it can be put on the transmission line, e.g. electrical pulses. Consequently this signal has to be decoded again in order to make information accessible. Thus an understandable code at the sender

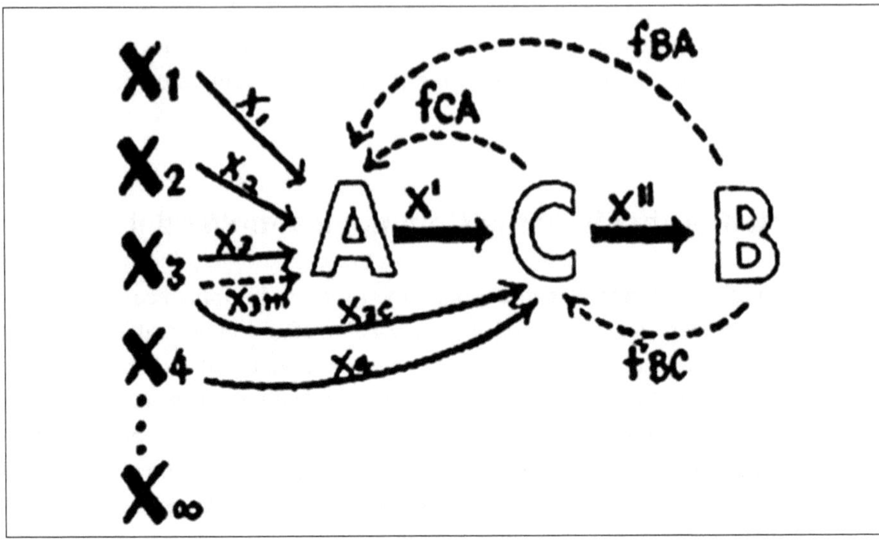

Fig. 13.1. Conceptual model of Westley/MacLean, Lacy 1989, p. 5.

and recipient side is needed to achieve a successful communication. In terms of cartography this simple process has to be expanded to semiotic dimensions, which are formed by syntactic, semantic and pragmatic aspects. For instance the coding of the meaning of an object lies within the semantic dimension, which calls for nearly the same understanding on the sender as well as the recipient side in order to communicate successfully (Petersen 2002, p. 17).

By reason of mass-communication's complexity, information transmission is not as easy as it is explained by S-O-R or Shannon and Weaver, especially when semiotic dimensions have to be considered within the coding/decoding process. The model of Westley/MacLean (Lacy 1989) tries to consider mass-communication, which means that the information reception at the recipient side is divided to a direct, media-based information reception and an environment-based reception. Westley/MacLean assume a ration of 1:4, which means that 20% of the reception are environmental-based and 80% are media-based. Thus this model defines three steps within a mass-communication system that combines actors by feedback processes. The model can be applied to mass-media processes with the overall aim to describe reality on the highest level of objectivity. Then an actor A, who may be a cartographer, describes and prepares reality for specific media C. These processes of preparation underlie various influences X, which disturb an objective description. For example an apparent pollution of water that is obviously disseminated in the news may lead to an extraordinary coding of that affected region in the map (this procedure should only be true for time critical map productions; normally the coding process within a map requires more reliability).

In respect to cartography with a narrow focus on geovisualization and individual visual processing of information, DiBiase stated that "... *visual methods are common and perform a range of functions in scientific research*..." (DiBiase 1990, p. 3). In an idealized visual research sequence the processes span from visual thinking to visual communication and cover four stages. These stages are the exploration of data, confirmation of apparent relationships, generalization of findings and presentation of research results. Based on the bias that abstraction of structural features is the basis of perception, which leads to the beginning of cognition. The main potential of computer-based visualization lies in generating mental images and using these images for new ideas. Therefore visual methods form an important part for cartographic mass-communication and play a central role in neo-cartography.

These given examples for conceptual communication models (even the model of Westley/MacLean) show that they are not appropriate for cartographic mass-communication, because their description of communication is too simple (compared with communication processes in reality). Due to the metamorphosis of communication complexity and structure, these models do not represent the processes best. This situation becomes more complicated with an exchanging role of sender and recipient as it occurs nowadays in neo-cartography. Nevertheless as result of a broad comparison of conceptual communication models, core elements and their relations can be identified.

13.4 Relations and Main Factors in a Conceptual Mass-Communication

A description of mass-communication (and conceptual communication at all) makes use of four core elements, which are heavily related to each other. These elements are the sender, recipient, media and information. Thus six relations can be defined, which are used for expressing importance of the relations.

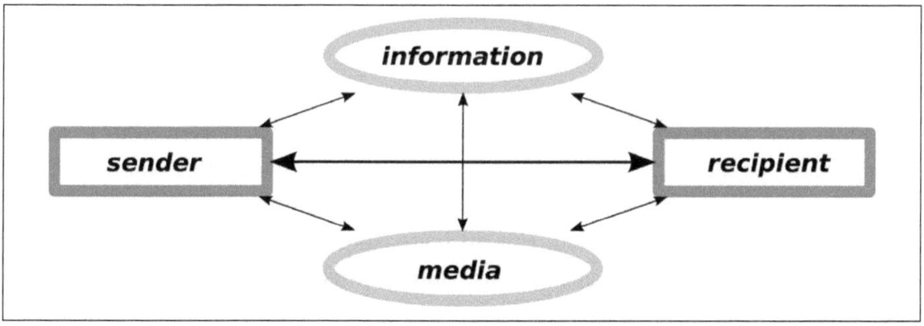

Fig. 13.2. Relations of the main factors in conceptual mass-communication.

- Sender and Recipient: The main characteristic of mass-communication is the indirect and unidirectional processing of communication. This means that the success of communication generally cannot be directly observed and recipients cannot use the same instruments for a response. This technical-based direction does not exclude an alternating relation, especially when appropriate technologies allow spontaneous feedback as it is done in Web 2.0.
- Sender and media: The specific characteristics of media offer concrete possibilities for the design of the interface and selection of content. Following the main aim of a successful communication leads to the significance of media regarding a clear understandable coding and decoding.
- Sender and information: In general the sender packs the message/information according to his/her individual criteria of selection. Thus the sender repairs to a situation of necessity in terms of understanding at the recipient's side: the information forms some kind of "truth" for the recipient. For a graphical design in maps the sender follows the guidelines of an aimed expression or map usage.
- Recipient and media: The importance of the relation of recipient and media lies in diverse characteristics of media and their consequences for information selection, experience and impact. A high support of the human sensual system results in a high value of immersion and thus a "realistic" experience of the presentation. Virtual environments call for a high grade of media immersion and therefore deliver intuitive and direct interactivity for exploring the synthetic world.
- Recipient and information: The relation of recipient and information mainly concerns processes of perception and cognition. The perceptive correct assimilation of information can be understood as selective, projective, signifying and shaping activities. Most of these activities are affected by previous knowledge of the recipient, which leads to an individual interpretation and according to its importance to a first mental storage.
- Media and information: The relation of media and information bases on the two-way dependency of media characteristics and aim of expression. The information has to be adapted to the transmission performance of a selected media. In terms of cartography this adaptation is called cartographic modelling, which includes generalization, interactivity- and content design.

Actual conceptual models of communication predominantly focus on the relations of sender and recipient (Kolacny 1969, Peterson 1995, Brodersen 2001), which is important for the understanding of semantic and pragmatic dimensions. Within the field of neo-cartography, where the role of sender and recipient changes, these models may be restrictedly used. For an increasing amount of examples the role of sender and recipient, the cartographer and map-user, cannot be clearly defined any more (e.g. Openstreetmap Initiative, www.openstreetmap.org). Map-users take their chance and generate maps by defining their areas and content of importance.

Thus the relation of media and information gains importance, although its main focus lies on the syntactic dimension. Supplementary, the pragmatic dimension can be more and more incorporated within the relation of media and information due to increasing issues of usability. So, the relation of media and information leads to a crucial differentiation in the understanding of "media".

13.5 The Differentiation of Media in Neo-Cartography

The notion *media* is widely used for any kind of transmission. The notion per se does neither restrict the transport from sender to recipient nor the usage of deployed resources. It follows classifications from an organisational-sociological point of view by differing transmission sign-systems according to sensual modalities (auditive, visual, ...) (Burkart 2002) or communication-technical aspects by differentiating between physical, communicative, technical and institutional media (Bentele et al. 1994). According to an increasing importance of the media-information relation within the conceptual communication model, a communicational-technical classification should be preferred. This technical orientation then leads to the specific distinction between information-carrier and information-content, which is based on the relation media–information and supports geo-media techniques in many aspects.

The *information-carrier* contains all technical and physical media that are used for an actual transport of information. In exchange all kind of sensual modes, -transmissions and -techniques are used. At least the user independently chooses an appropriate information-carrier that satisfies information needs as expected.

The *information-content* incorporates perceptive manifestations of information. In terms of cartography information-content follows the paradigm of expressive and efficient transmission of geospatial information (Mackinlay 1986). Concerning this matter the perceptive barrier, which is built up by human physiology and characteristics of information-carrier, has to be considered in information-content design. All instruments that are used to understand and successfully transmit information can be subsumed as information-content. By a classical understanding information-content names text, pictures, videos, animations, 3D elements, ... (Schweiger et al. 2001).

The distinction of information-carrier and information-content allows for an expressive quality description of the presentation on a specific transmitting interface. This quality description bases on the relation of information-carrier and information-content, a crucial aspect of geo-media techniques in neo-cartography.

13.6 One Crucial Aspect of Visual Geo-Media Techniques in Neo-Cartography

The user interface variety in neo-cartography complicates a precise information-content preparation for cartographic purposes and possible declarations for its quality. In fact any mobile and fixed device can be used as interface for ubiquitous maps. In many cases these interfaces consist of displays with varying techniques (CRT, LCD, OLED) and resolutions (mobile phones, PDA, MID). Thus the resolution is a key factor in order to either manually or automatically prepare expressive visualizations. By reason that the resolution of the information-content does not necessarily match with the resolution of information-carrier, a relation between these two has to be defined. Furthermore an expansion with the resolution of the eye, which is depending on the viewing distance, provides a value to express semiotic quality. This value can then be used in algorithms, e.g. in case of resolution dependent information-content variation (Jobst 2008).

The main part of this semiotic quality is bases on a "uniqueness relation", which says that the resolution of information-carrier (the transmitting interface) AÜM equals a multiple of the resolution of information-content AIM.

AÜM = Kim * AIM

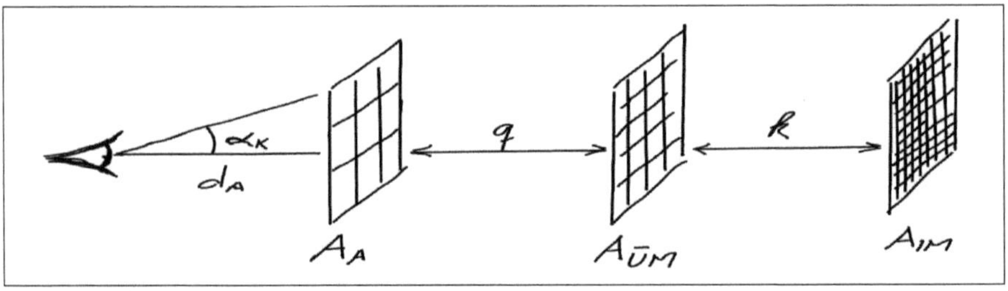

Fig. 13.3. Semiotic quality as relation of the eye's-, interface's- and information-content's resolution.

- *Kim = 1 (case 1):* Describes that the resolution of AÜM equals AIM. In this ideal case one information pixel of the information-content will be represented by one information pixel of the information-carrier. No interferences of the two raster resolutions occur.
- *0 < Kim < 1 (case 2):* In this case one picture element of the information-content will be represented by several pixels of the information-carrier. The more Kim moves to Zero, the more picture elements of the information-carrier will represent one single information-content. As result a rasterization of information-content will be observable. Picture elements of the information-carrier are not efficiently used.

Fig. 13.4. A map on a mobile information-carrier depicting case 2, case 1 and case 3.

- *Kim > 1 (case 3):* In the case that the uniqueness factor is greater than one, several picture elements of the information-content are represented by one single pixel of the information-carrier. This uncontrolled clustering of information generally leads to information loss and often to a disconnection of semantically connected information.

One would argue that this uniqueness relation is only valid for raster to raster projections, when rasterized information-content is used on a rasterized information-carrier (display). In fact this relation is also valid when vector data are rendered on a display. Then case 2 would show no raster artefacts, but in fact it will use much more picture elements of the information-carrier to visualize one information element of the content. Thus the amount of picture elements of the transmitting interface will not be efficiently used. The main principle that each information pixel should have its role for information transmission will not be fulfilled in that case. Beside an enlarging or downscaling of the picture nothing else happened in case 2 and case 3. The information content became not adapted to the resolution of the information-carrier.

13.7 A Common Relevance for Geospatial Communication in Neo-Cartography?

The mentioned crucial aspect of visual geo-media technique is a well known problem in digital cartography, especially when the clear depiction of signs has to be done with digital interfaces (Malic 1998, Neudeck 2001, Brunner 2001). The main problem of resolution remains: in comparison to printed information carrier (paper), digital cartography has to deal with much lower resolution, which means that symbols on standard displays (72–96 dpi) request about three times more space (Lechthaler et al. 2006) in order to be unmistakably understood.

For a general view in neo-cartography this situation becomes even worse: a huge variety of display techniques come along with various resolutions. Whereas the resolution of LCD and LED correspond to standard displays, high resolution OLED's can technically provide nearly printing resolution (300 dpi). Additionally the display extension varies enormously, from 1" to 5" diagonal and more. Thus a useful geospatial content will vary with the resolution and display extension. This aspect is even relevant for syntactical considerations, without thinking of usability or context situation at this stage.

Moreover the graphical and processing performance of mobile devices increases rapidly. This fact and evolving Web Services in geovisualization (Hagedorn et al. 2007) offer new possibilities to incorporate complex computer-graphics on mobile devices. For instance virtual 3D environments can easily be established on PDA or Mobile Internet Devices (MID) with Web Perspective View Services (WPVS). From a communication point of view, virtual 3D environments may enhance geospatial transmission. At least 3D presentation supports the creation of a mental model and knowledge acquisition (Bülthoff et al. 2001, Kirschenbauer 2004). If we want to be sure about the impact of virtual 3D environments on mobile devices, appropriate

Fig. 13.5. In order to avoid "dead pixel values" and use the transmission plane more efficiently, a degressive central projection may deliver better results than the traditional central perspective.

tests should be done. These tests will only deliver useful results for the semantic and pragmatic dimension, if all discrepancies from the syntactical point of view were removed, e.g. the uniqueness relation should be expanded and verified for virtual 3D environments on mobile devices. At least the graphical presentation needs to follow the main premise to efficiently use all information pixels, thus to avoid "dead pixel values". In order to do so, variations reach from deforming projections (progressive and degressive projections) to design-mechanism modifications and appropriate use of graphical variables (Jobst 2008).

A respective example for deforming projections can be given by the degressive central perspective: at the very first glance this example *(Figure 13.5 left)* will make no sense due to the distortion of the ground plate. In fact this distortion helps for an efficient use of the complete presentation plane by syntactical aspects. All information pixels contain relevant information. There are no areas concerned with sky. Although its pragmatic use may be queried, even a possible solution in terms of usability can be given: the "ego-shooter" view in the front allows a high grade of identification within the surrounding, whereas the ground-view in the back allows an overview and short-term planning in movement direction. This situation may be helpful in navigation systems, when maps are the indented information-content.

The variety of user interfaces, their resolutions and future services show the relevance of splitting information-carrier and information-content for neo-cartography. This allows a precise modelling of the syntactic, semantic and pragmatic dimensions of geospatial communication. Furthermore the possibility for modelling the access to various user interfaces becomes more clear from a cartographic point of view. The characteristics of the user interfaces can be considered independent from any information content. This open approach offers a variety of potentials to overcome the massive UI restrictions in terms of overview, resolution or information depth.

13.8 Conclusions

This contribution focused on the new terms neo-geography and neo-cartography, which evolve due to user-participation and opening of "stimulus-object-response" structures. It could be shown that from a conceptual communication point of view, the importance of media and information content grows. Invariably, this leads to a splitting of information-carrier and information-content, which consequently allows a definition of uniqueness relation between both. This relation then enables a more precise description of the transmitting interface and the preparation of information content.

Future developments for virtual 3D environments on mobile devices will rely on the relation of information-carrier and information-content, if it comes to semiotic correct visualizations, as these are requested in cartography. This crucial aspect of visual geo-media techniques defines considerable steps for the 3D rendering process. The development of rendering techniques results in new ways for information enhancing variations of cartographic semiology.

With the conditions of neo-cartography, one direction to go are Cartographic Visualization Services. Beside a ubiquitous access to geospatial applications and data, the information-content has to be adapted to the information-carrier and specific needs of the user. The splitting of carrier and content (media and information) now allows service development. Specific user needs with their situational and cultural based characteristics call for detailed investigations in future.

References

Bentele G, Beck K (1994) Information – Kommunikation – Massenkommunikation: Grundbegriffe und Modelle der Publizistik- und Kommunikationswissenschaft. In: Otfried Jarren (Hrsg.): Medien und Journalismus 1, Opladen, p. 16–50.

Brodersen L (2001) Maps as Communication – Theory and Methodology in Cartography, National Survey and Cadastre Denmark, ISBN 87-7866-308-3

Brunner K (2001) Kartengestaltung für elektronische Bildanzeigen. In: Kartographische Bausteine; Band 19; Technische Universität Dresden.

Bülthoff HH, van Veen Hendrick AHC (2001) Vision and Action in Virtual Environments: Modern Psychophysics in Spatial Cognition Research; in: Vision and Attention; Springer Verlag: New York, Berlin, Heidelberg; ISBN 0-387-95058-3.

Burkart R (2002). Kommunikationswissenschaft – Grundlagen und Problemfelder. 4th. ed., Böhlau Verlag, Wien–Köln–Weimar.

DiBiase D (1990) Visualization in the Earth Sciences. Earth and Mineral Sciences Bulletin 59:2, pp. 13-18. Reprinted in Geotimes 36:7, July 1991, pp. 13–15.

Hagedorn B, Döllner J (2007) High-Level Web Service for 3D Building Information Visualization and Analysis, in: ACM 15th International Symposium on Advances in Geographic Information Systems (ACM GIS), Seattle, WA.

Horsky D, Simon LS (1983) Advertising and the Diffusion of New Products. In: Marketing Science, 2 (1), 1983, p. 1–17.

Jobst M (2008) Ein semiotisches Modell für die kartografische Kommunikation mit 3D, PhD Thesis, Institute for Geoinformation and Cartography, Vienna University of Technology.

Kirschenbauer S (2004) Empirisch-kartographische Analyse einer echt-dreidimensionalen Darstellung am Beispiel einer topographischen Hochgebirgskarte; Dissertation; Technische Universität Dresden; Mensch&Buch Verlag: Berlin.

Kobzina M (2006) Grundlagen der Kommunikationswissenschaft, Universität Wien, Wien.

Kolácny A (1969) Cartographic Information – a fundamental concept and term in modern cartography. Cartographic Journal, 6, 47–49.

Lacy S (1989) The Westley-MacLean Model Revisited: An Extension of a Conceptual Model for Communication Research, School for Journalism, Michigan State University, Michigan, USA.

Lechthaler M, Stadler A (2006) „Cross Media“ gerechte Kartengrafik in einem AIS. In: Proceedings CORP 2006 & Geomultimedia06; Schrenk M (Ed.); Vienna, ISBN 978-3-9502139-0-4.

Mackinlay J (1986) Automating the Design of Graphical Presentations of Relational Information. ACM Transactions of Graphic, 5(2):111–141.

Malić B (1998) Physiologische und technische Aspekte kartographischer Bildschirmvisualisierung; Dissertation; Schriftenreihe des Instituts für Kartographie und Topographie der Rheinischen Friedrich-Wilhelms-Universität Bonn.

Neudeck S (2001) Zur Gestaltung topografischer Karten für die Bildschirmvisualisierung; Dissertation; Fakultät für Bauingenieur- und Vermessungswesen; Schriftenreihe des Studiengangs Geodäsie und Geoinformation; Heft 74/2001; Universität der Bundeswehr München, Neubiberg.

Petersen A (2002) Interpersonale Kommunikation im Medienvergleich. Waxmann Verlag, Münster/New York/München/Berlin.

Peterson MP (1995) Interactive and animated cartography. Prentice-Hall: London.

Schweiger G, Schrattenecker G (2001) Werbung: Eine Einführung. 5. Aufl., Lucius und Lucius Verlag, Stuttgart.

Turner JA (2006) Introduction to neogeography, O‘Reilly Media, ISBN 978-0-596-52995-6.

Weaver W (1972) (orig. 1949) Recent Contributions to the Mathematical Theory of Communication. In: Shannon CE, Weaver W (Eds): The Mathematical Theory of Communication, p. 2–28, University Illinois Press, Chicago/London.

Section III: Location Based Services and Applications

14 The Adoption of Mobile Location-Aware Systems for Field Police Work: Evaluation of a Pilot Initiative at the Dutch Police

Filippo Dal Fiore[1] and Euro Beinat[2]

[1] Austrian Academy of Sciences, Salzburg, Austria and
Massachusetts Institute of Technology, USA
[2] Department of Geography and Geology, Salzburg University, Austria

Abstract

Mobile location aware services for police work are still in an infant phase. In this paper we present the evaluation research conducted for a Dutch police pilot initiative, to identify critical points in the adoption of new mobile and location-aware applications. The service enables officers in the field to receive alerts on activities and events, locate colleagues or police cars and consult police databases. We administered questionnaires and interviewed 20 police officers who had gone through a 10 month testing, after having surveyed daily activities and movements in the field of an enlarged control group before the PDA was introduced. The study shows that the new applications have been unevenly used and liked among officers. The daily activities of neighborhood agents patrolling the streets on foot were the most impacted. The study also shows that technology factors and design choices (such as the frequency of alerts) have a strong impact on utilization. Below a certain level or usability/functionality, which appears to be completely work-process and organization specific, the intention to use the system is very low resulting in poor utilization and little behavioral impact. This paper illustrates the system, the information collected to assess user perception as well as behavioral work changes.

Keywords: mobile LBS, technology acceptance, user study, police field work, PDA

14.1 Introduction

In this paper we present results from a pilot initiative conducted in the Netherlands (city of Groningen), where new location-based applications running on PDA have been tested by 20 field police officers. Our empirical research has been aimed at:

- Researching the extent at which the new applications have been used during the 10 months of testing (April 2007 – January 2008).
- Investigating the reasons why police officers have been used (vs. not used) the new technology, as well as collecting their suggestions on how to improve it.
- Understanding if behavioral change in the field has taken place.

This paper is organized as follows. In *Section 14.2,* we revise the literature on which we grounded our work of scientific evaluators. In *Chapter 14.3* we describe the new applications under assessment, both in their location-based components as well as the rest of services provided through the PDA. In *Section 14.4* we explain how we designed the overall research and the data collection phase, the main findings of which are presented in *Section 14.5.* We conclude with *Chapter 14.6,* where we try to extract the lessons learned through the study.

14.2 Background Literature

Location Based Services are a special class of context-aware systems. From an environmental psychology perspective, context can be defined as a specific set of personal, physical and social aspects of an environment or of a situation (Clitheroe et al. 1998). Lee et al. (2005) made explicit those major elements of context that are relevant to the use of mobile devices:

- An environmental context, including a physical context (characterized by location, distraction and crowding) and social one (interaction, privacy).
- A personal context, encompassing emotion (arousal, pleasure), time (duty, time-on-hand) and movement (static/move).

Accounting for contextual factors is essential at the time of planning an evaluation of location-based services, specially if this is focussed on field police work that by its nature is carried out in a broad variety of contexts. On the contrary, most research in the evaluation of mobile information systems has been carried out through laboratory assessments prioritizing usability (Kjeldskov & Graham 2003).

Different contextual conditions are likely to impact the perceived usefulness and perceived easy of use of mobile devices and their applications, as implied by the technology acceptance model (Davis 1989). It should be also considered that, in most cases, the successful applications are those that best match the tasks that the user must perform (task-technology fit theory, Goodhue 1995). On this regard, a few studies describe in which way mobile work differs from sedentary one, posing greater challenges to IT adoption, since it can be more problematic for mobile workers to execute tasks, plan activities and coordinate with co-workers (Brown & O'Hara 2003, Gruhn & Kohler 2007).

Leveraging on such theoretical background, we devised a research strategy aimed at coupling quantitative observations of police officers' behavior in the field, with qualitative feedbacks on how the new LBS applications have been used. Results confirm the crucial role played by the different contexts of use in leading to the adoption vs. rejection of the new mobile applications and of location based services.

14.3 Description of the New PDA Applications

In the Netherlands, ca 20,000 police officers work in the field. 500 of them operate in the city of Groningen, populated by 180,000 inhabitants and situated in the North of the country. The pilot initiative we evaluated took place in the Korrewegwijk district of Groningen, where the local police department employs 50 uniformed police officers of whom 35 operate in the field. The district hosts 16,700 inhabitants, has a surface area of 1.8 squared kilometres and presents the highest crime rate among the 5 police districts in town.

In early April 2007, 30 field police officers operating at Korregwegwijk were recruited to test a new mobile location-aware system, without any pre-requirement or incentive. 20 of them voluntarily agree to complete a 10 month testing, spanning from April 2007 to January 2008. They were given a PDA (HP IPAQ) coupled with a stand-alone GPS device, while receiving one day of training on how to use the new applications.

The different functionalities supported by the PDA can be grouped in two categories:

1. "p-Info": A bundle of services already consultable through desktop computers.
2. Location-based notification services ("LBNS"): A new set of services, the main focus for the testing in Groningen.

p-Info

P-Info is a collection of regional, national and international police databases, previously available to Groningen policemen on desktop computers. Officers participating into the pilot project can now access them through the PDA. Among others, forms can be queried to retrieve licence plate numbers, names and picture of suspects, relevant addresses and dates. While in the field, police officers can now access information presented during daily briefing sessions at the base station.

Location Based Notification Services (LBNS)

Thanks to this new application, officers participating in the pilot receive notifications on things to be done on the field. The GPS device connected with the PDA allows notifications to get automatically triggered within a 40 meters radius from relevant locations, i.e. the locations to which messages are associated. A sound emitted by the PDA alerts officers, who can visualize the relevant notification in a pop-up screen. Exclamation marks denote notifications, characterized by different colours:

- Red: Notifications about fines to be collected at a given address.
- Black: Notifications about places to patrol (i.e. a park or a building).
- Blue: Notifications coming from daily briefings.
- Green: Notifications from the Chief Police Officer.

Officers can decide to ignore alerts coming from the PDA (by clicking on the icon "X") or read the message associated with them in a pop-up screen (by clicking the icon "I", i.e. Information). If they take action as a consequence of it, they are supposed to click on the icon "A" and eventually input a feedback into the system, choosing among one of the following:

- Nothing was there.
- I did what notified.
- The notification is no longer relevant, please erase it from the system.

In any case, thanks to a small keyboard appearing in the screen police officers can enter comments and complementary information in the form of a written string of text. Also, by means of human icons in the screen, officers can visualize their position in real-time as well as their colleagues', in the same map used to display notifications.

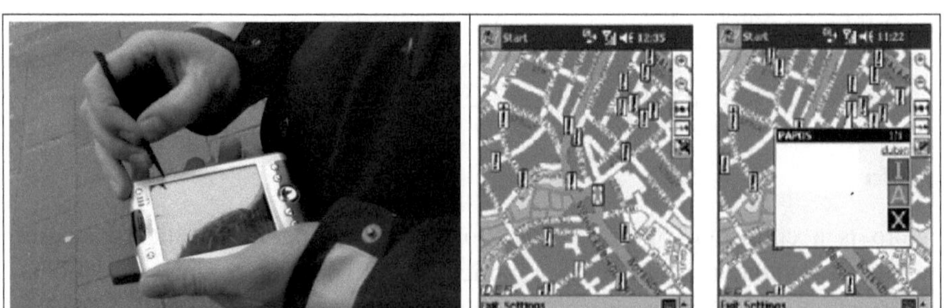

Fig. 14.1. (A) The PDA as it was used in the field. A connected GPS unit is kept by the officer in the pocket (B) Screenshots of the Location Based Notification Services, showing alert icons around the current position (left) with related pop-up information (right).

14.4 Research Design and Data Collection

In simple terms, our evaluation research has been focused on the following questions:

1. To what extent have the new applications been used by police officers during the testing phase?
2. Are the new applications leading to behavioural changes in the field?
3. Which factors drive user's acceptance vs. non acceptance?

On the base of them, we selected quantitative and qualitative methods to be applied for empirical investigation: in-depth interviews, to assess user's acceptance; mobility journals, GPS traces and post-test questionnaires, to quantify behavioural change.

Beyond allowing us to engineer data collection, the three questions were meant to address three problematic issues in the scientific literature:

Use of a new information system. The fact that a new system is available does not necessarily mean that it will actively used; it could simply be "adopted", but then gradually disappear (Davenport & Prusak 1998).

1. Behavioural change. This issue is strictly linked to the reasons why the new information system has been adopted in the first place; therefore, it represents a natural benchmark to assess return on investment (Fitzgerald 1998).
2. Intention to use the system. Intentions do not always translate in actual use. Attitudes and subjective norms play an important role in between, in that subjects value the consequences of performing such behaviour as well the expectations of relevant individual and groups (Fishbein & Ajzen 1975).

14.4.1 Mobility Journals

In March 2007, before commencement of the testing phase, we administered mobility journals to have a better understanding of daily movements and activities of field police officers. Also, these data would have served as a control group to weight the impact of the new applications. For the sole purpose of this research, we designed a paper-based journal by means of which officers would have progressively reported daily activities on the top of police district map. 12 typologies of activities were determined by chief police officers[1], so that, for example, entry (1, A) would demarcate the first activity of the day belonging to typology A.

All field officers operating in Groningen North (N=35) were asked to fill in a mobility journal for five working days, covering their 8 hour shift each time. For each day they were also asked to indicate on the bottom-left side of the diary: [ID, date, shift, transport medium used, type of policing, number of pre-assigned tasks, number of cell-phone/radio notifications received].

[1] These typologies are described in *Section 14.5.1.*

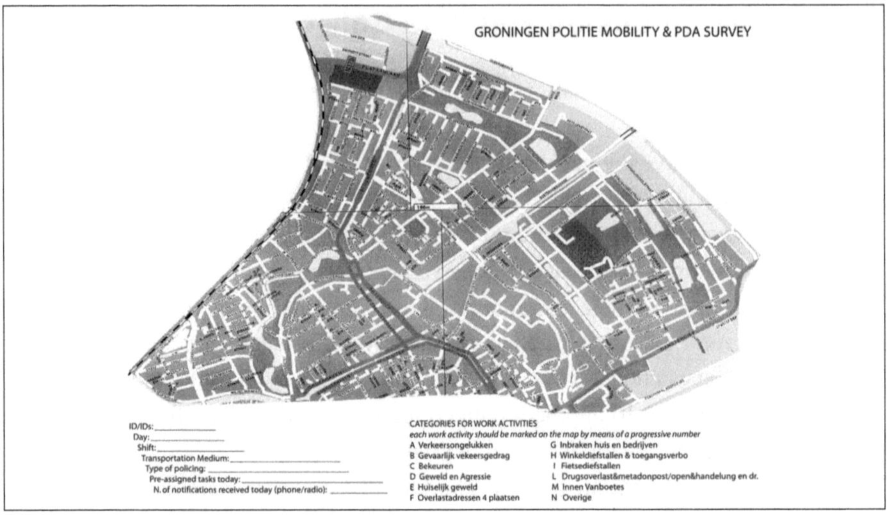

Fig. 14.2. The mobility journal used for data collection in Groningen.

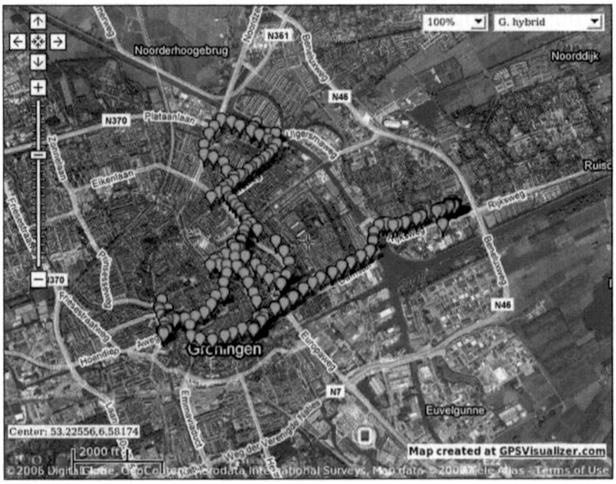

Fig. 14.3. GPS trace for one of the police officers during one work shift.

Initially, we thought we could collect useful data also during the testing phase, by means of the GPS incorporated in the PDA devices that officers were using. Indeed, GPS devices had been programmed to generate a string in a database every 20 seconds, containing the geographical coordinates of police officers' positions. Nevertheless, GPS data would not contain information on activities done in the field, so we had to engineer a way to have officers commenting on them upon return from their shifts. For the resulting operational protocol resulted too complicate and time consuming for police officers, we discarded this idea. *Figure 14.3* shows one of the GPS tracks collected.

14.4.2 Post-Test Questionnaires

In February 2008 – immediately after the test was concluded – we administered a 16 item questionnaire to assess the extent at which the PDA had been used as well as subjects' perception of behavioural change.

14.4.3 Interviews

On completion of questionnaire, we individually interviewed subjects on their motivations for using (or not using) the new applications. On purpose, the interview's script was left as unstructured as possible, to have the chance to follow unbeaten tracks and ask subjects to elaborate more on the comments provided, on a case by case basis. The following precautions have been taken, to ensure objectivity and prevent unconscious manipulation:

- Subjects have been assured of confidentiality.
- Subjects have been assured they can withdraw at any time and that they are not obliged to respond to all questions.
- Subjects have been assured that there are not right nor wrong answers.
- Questions have been formulated in a neutral way, in order not to lead subjects to subsumed answers (e.g. "Why did you use the PDA?" has been preferred to "Did you use the PDA to access richer information?").
- Subjects have been asked for facts before being asked for opinions (e.g. before: "Which applications did you use?"; after: "What do you think of the applications?").

To summarize, in the following table we present a synthetic view of all research activities implemented.

Table 14.1. Research activities and number of participants in each phase of research.

Phase	Research Activity	No. of subjects
Pre-pilot	Mobility journals	N = 35 (entire population)
Post-pilot	Questionnaires	N = 20 (opportunity sample)
Post-pilot	In-depth interviews	N = 20 (opportunity sample)

Officers are aged between 22 and 57; 75% are male; 6 out of 10 report using a computer at home every day. They are all involved in 8 hours shifts (morning, evening and night). 42% operate as response unit specialists (on call for emergency, moving by car in couple), the rest mostly as neighbourhood agents (patrolling a neighbourhood by walking or bike), but also youth agents (presiding schools and youngsters), students (novices) and chief police officers.

14.5 Findings

14.5.1 Journals

76 journals were returned by 29 officers (43% return rate). From descriptive statistics, it emerges that – on average – they visit 5.8 locations per shift (N = 66, sd = 2.6, in the remaining 10 cases they report not having left the police station). This mean value rises to 6.7 locs/shift when considering only response unit officers moving by car (N = 31, sd = 2.8). 6 trips/day is the most recurrent value (15/76).

In the following table, we map out the types of activities carried out while on the field, on the base of the 12 pre-determined categories.

Table 14.2. Types of activities carried out by field police officers.

Type of activity	Number of occurrences (%)
A = management of traffic accidents	27 (6.8%)
B = dangerous traffic behaviour	7 (1.8%)
C = Making fines	64 (16.2%)
D = violence and aggressions	22 (5.6%)
E = domestic violence	15 (3.8%)
F = places with annoyance/abuse	42 (10.6%)
G = burglary of houses or companies	13 (3.3%)
H = shop lifting/trespassing	9 (2.4%)
I = theft of bicycles	7 (1.8%)
L = drug abuse and methadone post	29 (7.4%)
M = collecting fines at people's homes	4 (1%)
N = other	155 (39.3%)

As expected, field police work turns out to be highly dependent on contingencies: only in 18% of cases are activities pre-assigned (typically during briefings at the base station), while in 39% of cases officers are not able to describe them by means of the available categories. Making fines (16.2%), attending places with annoyance/abuse (10.6%) and attending drug abuse and methadone post (7.3%) are the three most recurrent activities. Results from the post-testing questionnaires (presented in the next section) show that field officers still perceive them to be the most recurrent ones also after the 10 month PDA testing, suggesting that the PDA introduction did not bring any major change in the balance of activities carried out in the field.

Implicitly, this also indicates that it is very complex to design an information system to support mobile police work by improving well defined work processes and flows. Processes are hard to describe in precise terms and actual work activities rely significantly on judgment and experience. In this case the concept of process-related added value is problematic to apply, but it is still possible to collect "subjective" evidence by asking directly to participants how they were impacted by the new applications. Indeed, during interviews (whose results are presented in *Section 14.5.3),* several subjects pointed to the fact that their efficiency in carrying out a few operations in the field had increased.

14.5.2 Questionnaires

In the following tables, we present a summary of the findings from data collected in the post-testing questionnaires. The primary roles of the 20 officers who completed it are: chief police officer (2), neighbourhood agent (2), response unit officer (11), biker (2), youth agent (1), other (2).

Table 14.3. Assessment of PDA use and perception of behavioural change

1. How often have you been involved in the following activities in the last 15 days? (scale, min. 0 to max. 6)	median value	min	max
management of traffic accidents	3	1	5
dangerous traffic behaviour	2	1	6
making fines	5	1	6
violence and aggressions	2	1	4
domestic violence	2	1	4
places with annoyance/abuse	3.5	1	6
burglary of houses or companies	3	1	5
shop lifting/trespassing	3	1	4
theft of bicycles	2	1	5
drug abuse and methadone post	4	1	6
collecting fines at people's home	2	1	5
other	1	1	6
2. In the last 15 working days, how many days did you take your PDA with you?			**N**
Less than 3 days			9
3 days			4
7 days			2
10 days			4
15 days			1
3. What did you use the PDA for?			**N**
Checking licence plates or driving licences (p-Info)			19
Checking ID of people or suspects (p-Info)			12
Attending location-based alerts (LBNS)			9
Other			11
4. When using PDA, how important are the followings? (scale, min.0 to max +5)	**median value**	**min**	**max**
I can do more things at the same time	2.5	1	5
The quality of my work has improved	3	1	5
I have more information to do my job well	4	2	5
I know where my colleagues are	3	2	5
My safety	3	2	5
5. Perception of behavioural change (agreement scale, min.-3 to max +3, neutral 0)	**< 0(no) (N)**	**= 0 (N)**	**> 0(yes) (N)**
The PDA changes which things I do in the field	5	3	12
When I use the PDA I change my route in the field	9	8	3
With PDA I do more things in the same amount of time	5	6	9
When I use the PDA, the coordination with colleagues is better	6	7	7

Deriving significant conclusions from these results is problematic, given the small size of the sample. Nonetheless, some considerations can be made. From Question 2 we learn that 13 out of the 20 surveyed officers report having used the PDA only 3 days or less during the last 15 working days. This is a low use percentage. Results from interviews (presented in the next section) indicate that a few technical issues hampered full implementation of the new services and as a consequence full use of them.

Question 3 indicates that the majority of PDA uses concern non-spatial queries. In these cases officers access information from the Police databases (p-Info) and benefit from just-in-time and just-in-case answers. This is not surprising since the location awareness component is useful only for coordination and opportunity-based services (e.g. alerting an officer if in the proximity of a place where s/he can collect an unpaid fine).

Answers to question 4 seem to suggest that efficiency (being able of doing more activities in the same time) is not perceived as the most important goal of the system. Instead, officers assign more importance to the quality of their work and the ability of having relevant information available when needed.

Question 5 indicates that officers perceive that the system changes *what* and *how much* they do in the field, but not *where* they do it. This is understandable given the constraints in terms of patrol areas and zones of competence of the officers, which limit their freedom to operate in different locations, and thus the potential benefit of location-aware information systems. We should also consider that this question is one that easily triggers resistance, since subjects prefer to be the ones in charge of their behavior rather to be passively "guided" by technology.

When segmenting our sample along age and role, we record the following median values in questions 1, 3 and 5.

Table 14.4. PDA use and perceptions of different subgroups (median values).

Sub-sample	1. Days PDA used (1 to 5)	3. Importance (0 to 5, average of 5 medians)	5. Behavioral Change (-3 to 3, average of 4 medians)
Age* < 39 (N=8)	1.5	2.1	- 0.75
Age > 39 (N=7)	4	2.6	+ 0.75
Role: Response Unit Officer (N=11)	1	1.8	0
Role: All other roles (N=9)	2	2	+ 0.25

* 5 subjects did not disclose their age

Younger subjects report having used the PDA less and feeling less impacted by the new applications in their daily behavior. We suspect that this can not only be due to a weaker sense of compliance with organizational goals, but also to the fact that that most of them are response unit officers. Interviews were helpful in understanding why new PDA applications turned out to be unevenly used and liked among different types of officers.

14.5.3 Interviews

Subjects were interviewed after having completed the questionnaire. Once assured about confidentially (i.e. we as researchers could eventually report the content of every comment, but without revealing who provided it), they eventually surprised us by their openness and motivation to provide a wide range of feedbacks during interviews. The overall picture emerging is a fragmented one, suggesting that, beyond common problems, different officers have different reasons to appreciate vs. reject the PDA applications.

The lack of a pre-agreed interview script allowed them to provide a broader variety of information on how the PDA had been used, mirroring the complexity of the issue beyond filters super-imposed by research questions and hypotheses.

130 unique comments have been collected. By means of content analysis, we collapsed them in 69 self-containing issues grouped in the following 10 thematic categories:

- Behavioral and spatial implications of PDA use
- Access to information
- Use contexts
- Relevance and usability of information
- Usability of device
- Technical problems
- Suggestions on how to improve use of PDA
- Suggestions on new functionalities to be supported by PDA
- Suggestions on complementary or alternative devices to be tested
- Preference comments

On average, each officer provided 7 comments. The minimum number of comments recorded during an interview is 2 and the maximum is 13. *Table 14.5* summarizes the comments. Some of them have been rephrased or shortened to facilitate reading.

Table 14.5. Comments collected during individual interviews (number of occurrences).

Comment	*N*
behavioral and spatial implications of PDA use	
Thanks to p-Info, I don't need to call the emergency room so much anymore, so I save time	3
Thanks to p-Info, I have more information on suspects and I end up arresting more	1
When I take care of the location-based alerts, I end up moving differently in the field	1
Rather than leading me to change the things I do in the field, the location-based alerts give me extra things to do	1
When I see a colleague coming on my way (thanks to the GPS functionality incorporated into the LBS) I often decide to take the opposite direction, in order for us to patrol different areas	1
I have to follow a route in the field and along that route I attend the location-based alerts	1
I don't let myself lead by the PDA: I do my job and the PDA helps me in that, but I am the one who looks for problems	2

access to information	
Before starting my shift I look at p-Info to see what happened in the last 24 hours	1
Information on suspects received through the p-Info is more detailed than the one I can get by calling the emergency room	1
I consult the PDA at home, to see what happened (also on Sundays)	2
Thanks to p-Info, now I know in real-time if a person has already been arrested or should be in jail	2
use contexts	
p-Info is very useful during traffic accidents	3
I use the PDA in all contexts: in car, on the street, at people's houses	1
When I sit in the car next to the driver, I use p-Info to check license plates	2
During response unit work, I check driving licenses and license plates through p-Info	1
When I work in schools or I am at someone's house, I share the information in the PDA with families and social workers	1
relevance and usability of information	
The location-based application generates too many alerts, it is distracting and irritating	2
The amount of work is already too much, there is no room for the PDA applications	1
The PDA is too packed with information	1
Using the phone is easier and faster and the info you get access to is the same then the PDA's	3
I don't use the PDA to see where my colleagues are, for I am not certain that all of them are connected to the system	2
If I need support, I simply call for help through the radio transmitter. I don't need to consult the PDA to see where my colleagues are	4
During response unit work, I don't have time to attend the location-based alerts	1
A lot of info in the PDA is not necessary (It does not make me do my job better or faster)	1
I don't attend the location-based alerts because it is not my task	2
I will never quit attending briefing sessions at the base station, since through them I get more info than through the PDA and I can listen to what colleagues have to say	2
usability of device	
The PDA is not handy, specially in cars	4
Having to carry the GPS device and the PDA at the same time is too much, I have already too many things on me	4
I prefer to put the PDA in the front pocket of my jacket, for the PDA case hanging on my belt is not practical	2
The handling of the PDA is problematic, the pen and letters on the screen are too small	2
The PDA applications are not easy to use	2
Charging the PDA everyday is a burden and I often forget doing it	1
If I use the PDA in the field, I can't see in front and around me	2
The PDA is not handy in bikes (I keep it in a pocket of my jacket)	2
If we lose the PDA (for example while running or fighting), a technically savvy person can access all our databases	2
Neighborhood and youth agents are the ones who use the PDA more	1
When I get off the police car, I leave the PDA in there because it is not practical to handle	2
technical problems	
The PDA is too slow in downloading information, sometimes I even need to restart it. When I need information I need it fast	3
Too many log-ins are required to access information	4
Connection to location based services is bad, it often gets abruptly interrupted, specially inside buildings	5
The PDA case is not good: just by unintentional small touches the PDA configuration can be changed	1
Sometimes, when the PDA runs out of battery, I have to install everything again!	1
After clicking on "I" (further information) on the location based alerts, I get lost because I can't recuperate the original message anymore	1

suggestions on how to improve use of PDA	
Instead of attending to the location-based alerts on a case-by-case basis while doing other things, officers should just go a certain area just to clear all alerts there	1
Officers should dedicate one day or half-day per week only to attend alerts	1
More training on how to use the PDA should be provided to police officers	1
Instead of being voluntary, usage of the PDA should be mandatory, we could even try to shut down other systems (for example emergency room for certain kind of information) to incentivize officers to use the PDA	1
suggestions on further functionalities to be supported by PDA	
Taking pictures through the PDA	1
Calling through the PDA	3
Checking e-mails through the PDA (so to save a trip to the base station)	2
Listing the information in the PDA in a way that the first appearing information is the most recent	1
Synchronizing the PDA agenda with the one in our desktop computers	1
Integrate a navigation system (for example, Tom Tom) into the location based services	2
Incorporating an "emergency bottom" in the PDA, similar to the one we have in the radio transmitter	1
Connecting the PDA with our desktop computers, so to allow us to input data from the field through the PDA (so to save a trip to the base station)	1
Expanding the location-based notifications to the entire country (I drive often beyond the borders of the province of Groningen)	1
Being able to locate arrested people in real time (so to save a trip to the base station)	1
suggestions on complementary/alternative devices to be tested	
Testing of on-board computers in cars. They should be fast and incorporated in the dashboard. We should be able to enter data through them (so to save trips to the base station)	8
Testing of home laptops. We should be able to enter data through them (so to save trips to the base station)	2
Testing a of big screen to be put in the base station, with indication of all alerts to be attended in the field (something to take a look at before leaving the station)	1
preference comments	
The PDA is very good to check people (photo and info)	8
The PDA is very good to check cars (license plates)	2
The PDA is particularly good for neighborhood agents	2
The PDA applications and alerts are ok, they tell me what to keep an eye on	1
In the car, the PDA is usable	1
This pilot project is good, experimentation should go on	3
At first I was eager to use something new, then I started hating the PDA	1

These data return a complicated picture. In terms of PDA adoption, a first distinction has to be done on the base of work practices, i.e. contexts of use and the roles of the officer using the PDA services. On this respect, it emerges that response unit officers moving by car have very different windows of opportunity for using the PDA, if compared to neighborhood agents patrolling a given area by walk. Response unit officers also have different priorities, information needs as well as a different amount of time available to use the PDA.

The need of further customizing the new applications on the different contexts is confirmed by the fact that officers don't raise any objection to the location-based notifications themselves, but rather to the way these are delivered to them. Indeed, many complain to receive too many alerts, with too much information delivered at the wrong time. Advanced filtering systems could be considered to address this problem.

A finding in itself is the major role played by ergonomics-related and technical aspects, which may end up making the difference at the time of accepting and adopting the new technology. More than others: logging-in process; size of the characters on the screen; battery duration; wearability of PDA case.

In general terms, most comments related to information relevance and usability of the device are negative, revealing the need for an additional effort to ensure that delivery mechanisms better account for different contexts of use, without interfering or overloading the user. At the same time, and in spite of the evolution of PDAs and other mobile devices, there are clear technical limitations to the capabilities and functionality of these mobile platforms, which may still appear as rudimentary compared to the portability and usability needs of the end users.

On the matter of behavioral change, we found remarkable that several officers pointed to the fact that they would be happy and willing to replace trips to the base station with more time spent on the field, but some of these trips are necessary to use report about field activities on desktop computer. Incorporating this functionality into the PDA (or even in laptop or in-car computers as suggested) would bring a clear benefit.

The overall picture that emerges is one of a general positive attitude towards the mobile location-aware information system, which is expected to improve the quality of police work. Users recognize that this in part takes place, but find that the technology platform needs major improvements to eliminate the overhead introduced by the device and the way information is delivered to users. Taken together, such problems may explain the lower than expected use frequency of the service, and at the same time the openness and willingness to contribute suggestions for improving the system.

It is interesting to compare these results with those reported by Dias (2007). In a study conducted in 2 natural parks, visitors were offered the opportunity of using a traditional information system (a paper information booklet), a PDA-based booklet and a location-aware booklet (a PDA with GPS and real-time personalized alerts). All systems contained the same information, which was delivered in different ways by different tools. The study shows that, for the majority of evaluation metrics, users prefer the paper based booklet over the plain PDA, which contains the same information simply delivered in a digital form. The location-aware service, on the other hand, appears as superior to both other systems.

A plausible explanation is that the delivery of information through a PDA introduces potential benefits, but also a range of usability issues that combined with technical deficiencies may produce a balance against the use of the PDA. This balance can be tilted in the other direction thanks to real-time location-aware information, if this simplifies information access and increases information relevance. It is also interesting to point out that this study was able to measure statistically significant spatial and behavioral changes, in the way visitors with the location-aware PDA experience the park. For instance, these users tended to respect more park indications and environmental protection suggestions.

If we extend these results to the case study described in this paper, it appears understandable that at the present stage of development a few technical and usability factors still weight importantly against a complete utilization of the system, leading to relatively low use and reliance to traditional work tools (phone calls, visits to the office, radio communication). Even if the system has undergone a series of improvements and upgrades during the 10 month test, the anecdotic evidence collected does not suggest a parallel gradual increase of adoption nor satisfaction.

Similarly to what found by Dias (2007), below a certain threshold (which is application and domain specific) users do not likely make the adoption decision and prefer to stick to existing and well known work practices. Below such threshold, there seem to be little proportionality between degrees of device/service sophistication and degree of adoption. The difference can be seen only if usability/functionality threshold is exceeded. It looks like resistance to change combined with service/device deficiencies may only be overcome in full: half ways do not seem to perform well in terms use decisions and behavioral changes.

14.6 Conclusions

In this paper we presented an evaluation study to assess the testing of new mobile LBS for field police work, based on the three research interests made explicit in *Chapter 14.4:* extent to what application had been used; behavioral change triggered; user's acceptance.

Results show that the new applications have been only partially used during the 10 months of testing phase, with ergonomics-related and technical issues playing a major role in users' acceptance. At this stage, qualitative findings suggest that behavioral change affects mostly a few neighborhood agents who walk into the field. They are the ones who have more time to use the PDA for doing things that they would not otherwise done; also they have more freedom in choosing which route to follow in the field and are less dependent on emergencies to be taken care of. Nevertheless, as for all other agents, behavioral change is likely to happen slowly, little step by little step, given initial resistance to modify habits and challenge the status quo, also considered the lack of externally imposed incentives for doing so.

The purpose of this paper was not to elaborate in great detail the technical configuration of the new services, rather to investigate the point of view of final users leveraging on their current work practices. While we deemed our research successful in identifying critical points for adoption, we realize it can still complemented with technically-grounded recommendations on how to improve the current system.

On this regard, a lesson that we can legitimately extract is that there is little room for compromise when dealing with the introduction of new mobile applications. We technologists need to take into account that, specially in the professional realm, user's priorities can be others than adapting to a new device. On the contrary, the technology development and testing phases should be to a certain extent contextual, in order to adapt the new applications to the professional work practices that they aim to support as well as to the future contexts of use. A clear understanding of all types of benefits and costs, as they are perceived by final users, is fundamental to design location-based services that can fulfill their promises.

Acknowledgements

This study would not have been possible without the constant support of Elle de Jonge and Jan Sikkens, respectively IT project manager and Chief Police Officer of the Groningen Police. We have been involved as independent evaluators, in order to help the sponsoring Dutch Ministry of the Interior (Ministerie van Binnenlandse Zaken en Koninkrijksrelaties) to identify critical points before deciding to extend the system to the entire country.

References

Brown B, O'Hara K (2003) Place as a practical concern of mobile workers. *Environment and Planning A*, 35(9), 1565-1587

Clitheroe H, Stokols D, Zmuidzinas Z (1998) Conceptualizing the context of environment and behaviour. *Journal of Environmental Psychology*, 18, 103-112

Davenport TH, Prusak L (1997) *Information Ecology.* Oxford University Press

Davis FD (1989) Perceived usefulness, perceived ease of use, and user acceptance of information technology. *MIS Quarterly,* 13, 3, 319-340.

Dias E (2007) The added value of contextual information in natural areas. Measuring impacts of mobile environmental information. Vrije Universiteit Amsterdam: *PhD thesis*

Fishbein M, Ajzen I (1975) *Belief, attitude, intention, and behavior: An introduction to theory and research*. Reading, MA: Addison-Wesley.

Fitzgerald G (1998) Evaluating information systems projects: a multidimensional approach. *Journal of Information Technology*, 13 (1), 15-27

Goodhue D (2005) Understanding user evaluations of information systems. *Management Science*, 41, 12, 1827-1844

Gruhn V, Kohler A (2007) An Analysis Framework for Mobile Workforce Automation. *Enterprise Distributed Object Computing Conference, EDOC 2007, 11th IEEE International*

Kjeldskov J, Graham C (2003) A Review of Mobile HCI Research Methods. In *Proceedings of the 5th International Mobile HCI 2003 conference*, Udine, Italy, Springer-Verlag, LNCS

Lee I, Kim J, Kim J (2005) Use contexts for the mobile Internet: a longitudinal study monitoring actual use of mobile Internet services. *International Journal of Human-Computer Interaction.* 18, 269-292

15 Enhancing the Experience of the Landscape: The Digital Dowsing Rod

Arend Ligtenberg[1], Ron van Lammeren[1], Martin Goossen[2] and Jandirk Bulens[1]

[1] Centre for Geo-Information, Wageningen University and Research Centre
The Netherlands
[2] Landscape Centre, Wageningen University and Research Centre
The Netherlands

Abstract

This research presents the Digital Dowsing Rod (DIWI[1]) a framework of a Location Based Service (LBS) to explore the cultural heritage of a region. The DIWI consists of a service oriented architecture (SOA), a web-client, a content management system and a mobile client based on smart phones and windows mobile technology. The chosen set-up allows for an easy management of the content and allows for a relative easy extension of the system by additional data-sources.

An extended user test of 150 testers was carried out in the "Grebbelinie" area, the Netherlands. The testers hiked and biked via different routes through that area using the application. Despite a number of technical shortcomings the DIWI was positively valued by the testers. Inspired by the Technology Acceptance Model (TAM) the perceived use, perceived usefulness, and perceived enjoyment were evaluated using questionnaires and detailed user logging.

Keywords: location based services, web map services, service oriented architecture, cultural-historic information, usability testing, technology acceptance

15.1 Introduction

Location Based Services (LBS) are currently widely explored by the leisure industry as an instrument to enrich outdoor experiences (Zipf 2002, Zipf & Malaka 2001,

1 DIWI refers to the Dutch translation of Digital Dowsing Rod: "DIgitale WIchelroede"

Schmidt-Belz et al. 2003, Hinze & Voisard, Dias et al. 2004, Baldzer et al. 2004, Jose & Davies 1999, Raper et al. 2007) and to develop tourism in rural areas (Antikainen et al. 2006). The assumption here is that LBS might enhance access to specific information about landscape, nature, cultural heritage and other tourist amenities, and as such provide additional ways to explore the landscape and give the opportunity to people to learn. Most current LBS applications deliver one-way access to information (D'Roza & Bilchev 2003). Interesting information is pushed to the user based on the location. They do not allow users to add and share own information and experiences. The DIWI is one of the first applications with a two-way access to information through the use of UMTS transmission. It would be interesting to know if this type of interactivity would give an added value to existing LBS approaches (Espinoza et al. 2001).

Until now there is not very much known about the users acceptance of LBS and mobile multimedia services which aims at the enhancement of access to specific information about landscape, nature, cultural heritage and other tourist amenities, and as such provide additional ways to explore and learn about the landscape, with the exception of some studies (Schmidt-Belz et al. 2003, Margherita 2004). To clarify the value of applying LBS in recreational context as sketched above explicit insight is needed in its acceptance. The Technology Acceptance Model (TAM) is still the predominant model for examining factors involved in the acceptance of new technology by users (Davis 1993, Sharp 2007). Its basic concepts are the use, usefulness and enjoyment.

This paper presents the Digital Dowsing Rod, an LBS framework, developed to let users explore the cultural history of a region based on their own preferences. Additionally users can capture, add and share information based on their location. The goal of this research is two-folded. The first goal is to develop a generic architecture together with an accompanying software framework that supports an efficient realization of an interactive location based service targeted to providing cultural-historical information to visitors of landscapes and cities. The second goal is to measure and analyze the acceptance of the Digital Dowsing Rod by carrying out an extensive user test. The hypotheses is that the DIWI will give an added value to an outdoor experience. The second hypotheses is that the application is not accepted for the common citizens but only for specific user groups.

The paper starts with a detailed description of the Digital Dowsing Rod concept, framework, and case study based implementation. Next the results of extended user tests are presented indicating the acceptance of the Digital Dowsing Rod as an instrument supporting users to better explore and understand the 'genius loci' of the surrounding landscape. The test includes the assessment of the cultural-historical contents, and the technical and usability aspects of such LBS. The last section of the paper provides conclusions and discussion about the perceived use, usefulness and enjoyment and the added value of the developed LBS application.

15.2 The Concept of the Digital Dowsing Rod

The base concept of the Digital Dowsing Rod (DIWI) is to reveal the hidden history of a landscape to visitors and provide them with cultural-historical information using broadly available devices. As a location base device, the HTC P3600 has been selected. This smart phone offers GPS, camera and UMTS/HSDPA.

A tailor made user interface has been developed that enables the phone to display routes, points of interests an to push and pull various types of media including text, sounds, pictures and short movie clips. Using the smart phone users can explore a landscape and its cultural history using various levels of freedom. This freedom is important because modern tourists are capable of defining their own needs and preferences. They are not always satisfied with the "travel agent's standard offer". Modern tourists want to have it all, but not at the same time and not at the same spot. Diversity has become a new key word in tourism planning (Goossen et al. 2008). For that reason a DIWI-user can choose one of the following modes:

- Follow predefined trails with tales of history.
- Follow a personalized route based on personal preferences including related tales of history.
- Forage through the landscape, taking for granted what tales of history to encounter.

To provide users with this location based information databases containing maps, routes, and tales of history have been created. The tales of history were initially based on a Dutch national authorized dataset (KICH database)[2]. After a pre-test of the KICH dataset it turned out that these data needed to be adapted to serve a interesting content for the targeted users. For that reason new tales of history have been created by professional storywriters and a multimedia bureau via pictures and multimedia (sound and video) files. The tales of history will be refer to in this paper as collections of Points Of Interest (POI). Additional data sets from various sources were added to the POI database to provide a complete coverage of interesting points. Besides cultural-historical information also POI related to restaurants, pubs and hotel locations were added.

The DIWI application could be used during trips, either by foot or bike, to navigate and to get information. Additionally the device enables people to interact. Users may add their own "stories" about, and experiences around a location. Using UMTS they could record video, audio, picture and text files and uploaded these personal POI's to the server. Once uploaded, these personal stories are available to others passing that location.

2 KICH in Dutch: KennisInfrastuctuur CultuurHistorie (Infrastructure Culture Historic Knowledge)

DIWI is accompanied by a website. Using this website users may:

- Preview predefined routes on a map.
- Construct their own personalized routes.
- View routes they followed and the related uploaded personal POI's.

The concept of predefined routes guides users along an ordered set of points of interest (POI's) which together form a story. During such a trip people are presented short movies, sound-casts and pictures that tell a historical event.

It appears that the "average" walker walks for approximately 1–1.5 hour (Goossen & Ploeger 1997) visiting six POI's (Kramer et al. 2006). Based on this information, story-lines of six, spatially ordered POI's, have been developed. These POI's tell, when visited in the right order, a story about the landscape or a historical event. Following the classical theories of Aristotle and Girot the stories know a tension-line and at least three phases: recognition, personification and empathy. Two types of story-lines have been developed. The first type uses all involved points to slowly build up the tension (type1 in *Figure 15.1)*. It aims on recognition and personification. The second type consists of short, relative autonomous stories, all together explaining a historical event (type 2 in *Figure 15.1)*.

Alternatively users can login into the website and construct their own route based on landscape preferences (forest, heather, open field, etc.), distance preferences (a trip of 3, 6 or 9 kilometers) and activity preferences (walking and cycling). Next a route-engine calculates a list of routes that fit as much as possible to the preferences of the user. The user needs to select the best fitting route. This personalized route shows up after they have logged into the mobile Digital Dowsing Rod client.

Back home people can log into their personal website and explore the actual route they followed (either a predefined route, a personalized route or the route created by strolling around) and their added personal POI's. This is made possible because all trips are tracked on-line via the mobile device and real-time uploaded and stored on the server along with the personal generated content.

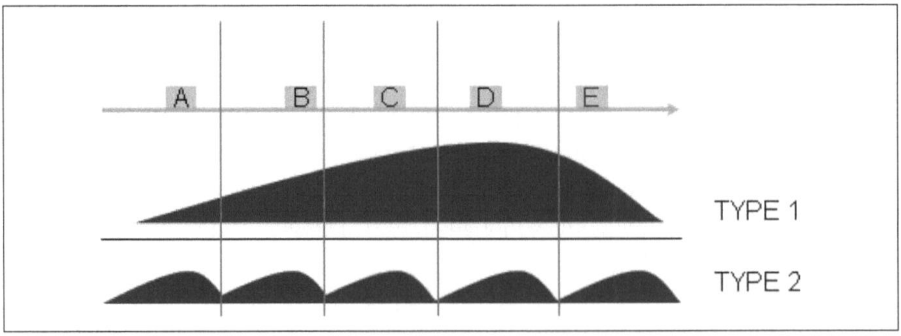

Fig. 15.1. Type of stories implemented for the pre-defined routes (Lammeren et al. 2008).

15.3 Digital Dowsing Rod Framework

15.3.1 Architecture

The concept described above was implemented in, basically, a four-tier application framework. *Figure 15.2* shows the general picture of the DIWI architecture.

At the lowest level distributed databases provide the various data for the LBS, routes, and POI's. Currently the following databases are in place:

- An ArcSDE/Oracle Spatial database providing topographical reference data (1:10.000) and the network data.
- A PostGis database that contains the predefined routes.
- An Oracle database with the POI's embedded in a content management system (CMS) and a database containing the user generated content (UGC).
- A media database serving all the media like photo's and movies.
- A database that stores the tracking information and logs with all interaction data of user with their mobile clients (only for analysis purposes).

To be able to access the various databases a Service Oriented Architecture (SOA) approach is used. In the second tier specialized services are developed that provide access to the geo-information (using OGC standards for map and feature services,

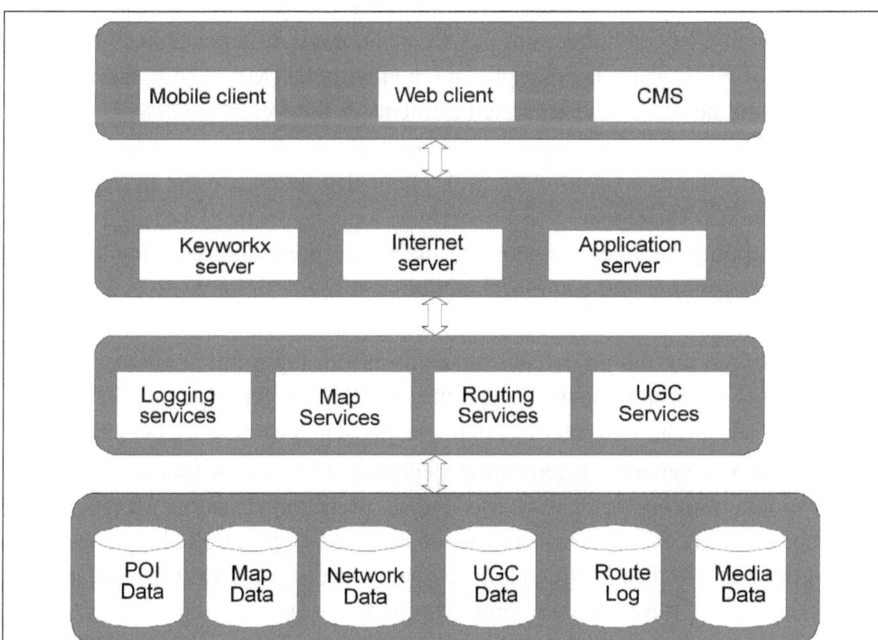

Fig. 15.2. Four-tier architecture of the Digital Dowsing Rod.

WFS and WMS), routing services that selects routes based on user preferences and services that handle the retrieval of poi's and the adding of user generated poi's along with the accompanying media.

To provide a uniform interface to clients an open source application framework called KeyWorx (http://www.keyworx.org/) specially designed to handle location based multi-media information (locative media) has been used. KeyWorx is a software platform that enables developers to develop and deploy multi-user, multi-channel and multi-media applications.

The fourth tier consists of the clients. Three clients are available:

1. a web client that allows users to explore the predefined routes, create their own routes, and access the routes they have followed and the content they have added;
2. the mobile client that offers the LBS and allow users to upload their personal text, photo and video files to the database;
3. a content management system (CMS) that allows administrators and managers to add and manage DIWI-POI's and multi-media, and to moderate and manage the user generated content (personal POI's).

15.3.2 Web Client

To construct of responsive website, minimizing the need for refreshing entire pages, the web client is build using AJAX technology. It depends heavily on the processing of XML using JavaScript combined with dHTML. AJAX user requests can be send to the server and separated elements in the web pages may be updated autonomously.

Figure 15.3 shows a page of the web client that enables users to explore the predefined routes. The OpenLayers client (http://openlayers.org/) was used to access the Web Mapping Services for showing the routes combined with topographical maps. Simple panning and zooming functions allow for an interactive exploration of the region.

Besides exploring the predefined routes users can login into a personal part of the website to construct personalized routes and review the trips they have done. *Figure 15.4* shows the web interface that allows users to generate a personal route.

In the current version of generating personalized routes a user can choose a starting point, walking or cycling trip and its preferred distance. Moreover two landscape preferences can be selected out of six (closed agricultural landscape, half-open agricultural, forest, heath and dunes, natural grasslands/swamp/reed, stream/rivers/pond/lakes). Based on these preferences the system tries to fit a route as close as possible to the users' wishes. This generated route is available on the mobile client after login.

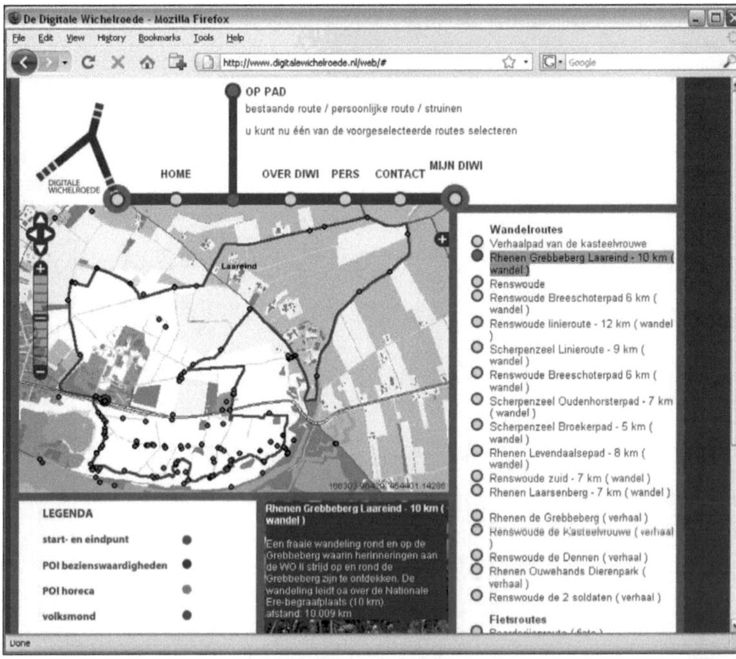

Fig. 15.3. The web client showing predefined routes.

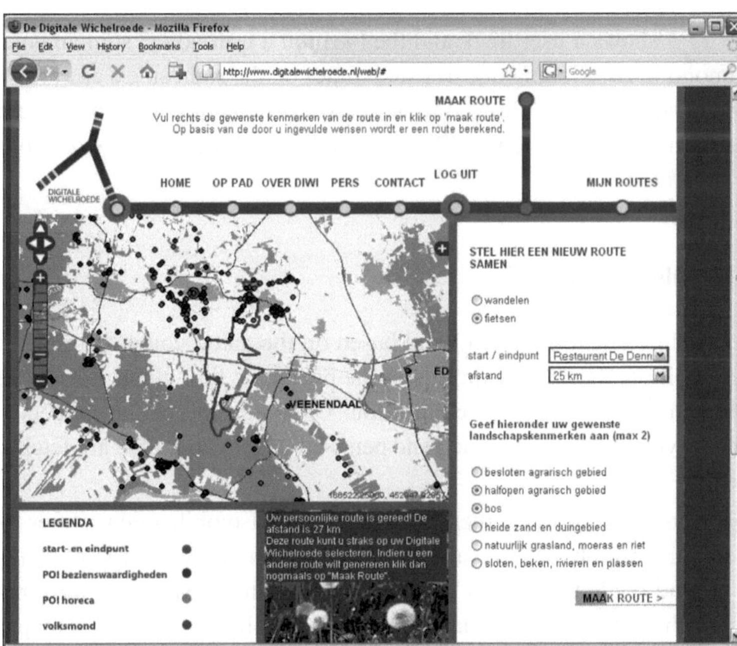

Fig. 15.4. Web interface for generating personal routes.

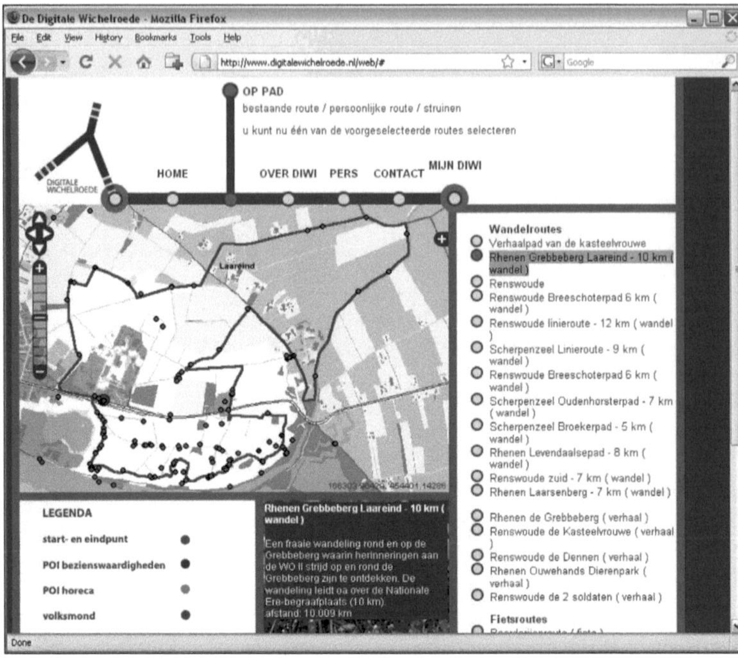

Fig. 15.5. Logged track and user generated content.

Figure 15.5 shows a user track and the location a user added some own content (red "diamond"). By clicking one of listed routes on the right side a crumb-trail will be generated showing the followed route. Personal content is shown on the map as small red diamond shaped icons. Clicking on the "diamond" shows the content of the user generated POI.

15.3.3 Mobile Client

A custom mobile client has been build based on the .net 2 compact framework on Window mobile 5. The application is implemented as thin client fully depending on GPS and minimal a UMTS connection. On logon the client connects to the KeyWorx Server, and the predefined and personal generated routed are fetched from the server.

Figure 15.6 shows some screenshots of the DIWI-mobile client. After login of a user he or she can choose to walk a route (including personal route) or just start to roam around (called "struinen" in the menu). Using WMS route information is combined with a reference base map (currently a detailed topographical map). There is no active navigation system build in. This means a user just tracks himself on the map checking if he or she is still on the route. If a POI is within a certain distance

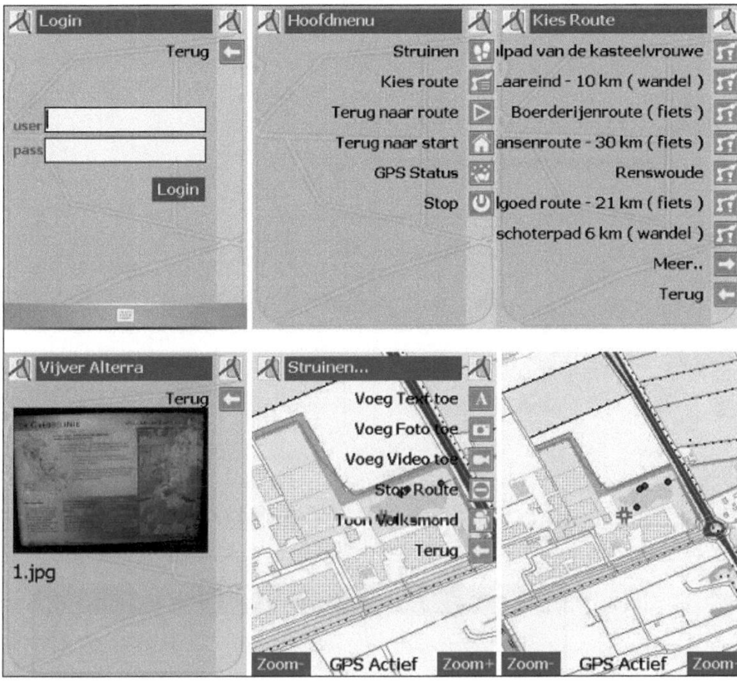

Fig. 15.6. Screenshots of the mobile DIWI client (from left to right: login screen, user modes, predefined routes, personal POI (dummy example), dynamic map and user location and personal records menu).

to the user, he or she will be signaled and is given the opportunity to read, watch or listen to the information. By activating the add photo button (called "voeg foto toe") the build in camera of the Smartphone will be activated. After making the movie shot or photo the results will automatically be uploaded and added to the DIWI database.

The DIWI-mobile client also is equipped with a "go-home" button. This option triggers the system to calculate the shortest path to the starting point. For this service and the personalized route service a dedicated network data set consisting of cycling and walking paths was created.

15.3.4 Content Management

A content management system (CMS) allows for the management of POI's and user content (only accessible for administrators, content providers, and moderators). Via the POI management screen *(Figure 15.7)* POI's could be added by clicking a location on the map or entering coordinates. Besides a description and classification of the POI, media can be linked to the poi by selecting from a media library. Existing POI's can be selected by clicking on the map. All poi's within a user defined range

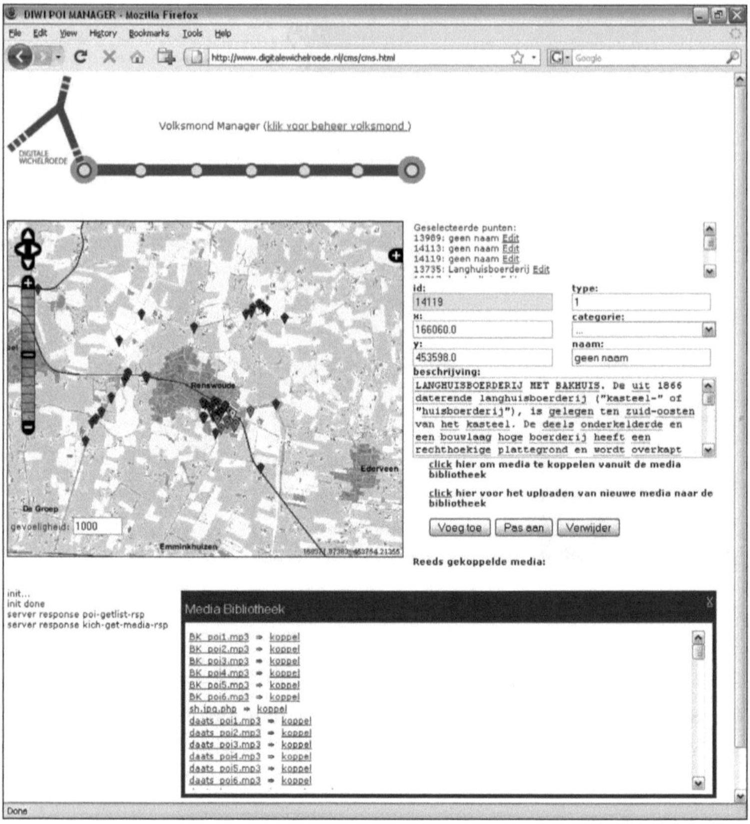

Fig. 15.7. Content management system DIWI: POI edit screen.

will be highlighted and listed by their name. From that list the POI to be edited may be selected. The change of a POI location could be established by dragging the selected POI-object to another location. User generated content might be reviewed and deleted if necessary.

15.4 User Acceptance Test

The test of the use and usability of DIWI took place in the "Grebbelinie" area *(Figure 15.8),* the Netherlands during five subsequent weekends (Friday up to Sunday) in March, April and May 2008.

The "Grebbelinie" was originally created as a military construction consisting of fortifications and areas that could be inundated. It originates from the mid 18th century and was still in use during the 2nd World War. However this 60 kilometers long line of defense was demilitarized in 1940 just after five days of war.

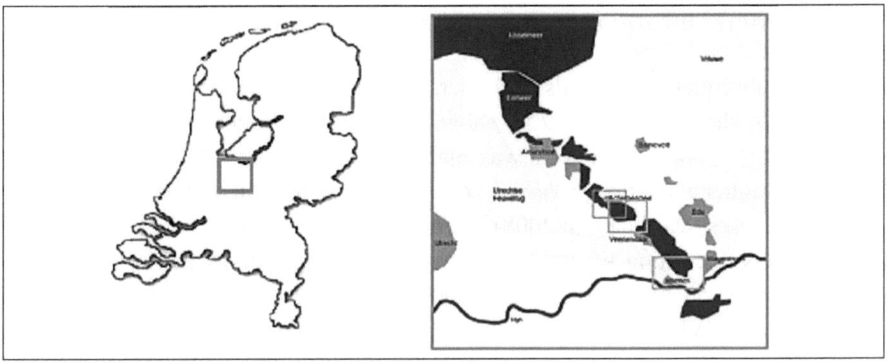

Fig. 15.8. The "Grebbelinie" (dark gray) in the Netherlands and starting points (light gray boxes).

Many remnants of old defense works are still visible today and some of these have been reconstructed. Besides the Grebbelinie area offers beautiful scenery and is known for its high biodiversity.

The long history of heroic actions and battles during the centuries offer many stories to be told. The DIWI user test offers the applicants twenty different routes starting and ending from three different locations. The routes range from 5 to 12 kilometers for walking trips and from 14 up till 30 kilometers for cycling. *Figure 15.8* shows the "Grebbelinie" (dark gray) and the starting points (light gray boxes).

15.4.1 Methodology

A public relation campaign by radio, newspapers and internet directed to an audience that are interested in walking or biking and history of landscapes resulted in 218 persons that register as potential DIWI tester by using the website www.digitalewichelroede.nl. Every tester could select a starting location, and preferred time and date to perform the test. During the on-line registration each applicant had to fill in an intake form with questions about personal interests.

After registration the applicants received a letter with all information about the test and they were informed about the procedure during the test day and the tracking of their trip. During the test day itself they were on site instructed for 10 minutes in the use of the device by assistants. They were explicitly informed about the three user modes and told that the trip should be tracked. Next they started to walk or bike for maximally two and half hours and they had to come back at the starting point again. Next, after their return, they filled in an extensive on-line questionnaire with sets of questions related to the trip, perceived use, usefulness and enjoyment. Finally the research team collected the three main datasets (intake, tracks and TAM questionnaire) and linked them via the user ID as given via the registration.

15.4.2 General Information About the Participants

During 5 subsequent weekends 150 users tested the application. This means 68 users did not show-up. *Figure 15.9* shows the age distribution of the testers. The average age was 48. About 67% was male and 33% female. Most of the testers followed a higher education. *Table 15.1* shows the distribution of the testers by eight mentality classes (Motivaction 2008). Mentality classes represent the variety of the Dutch society according lifestyle and personal attitude. The mentality classes are originally based on a panel of about 95000 internet respondents. The table directly indicates that DIWI is not yet appealing to the common citizen.

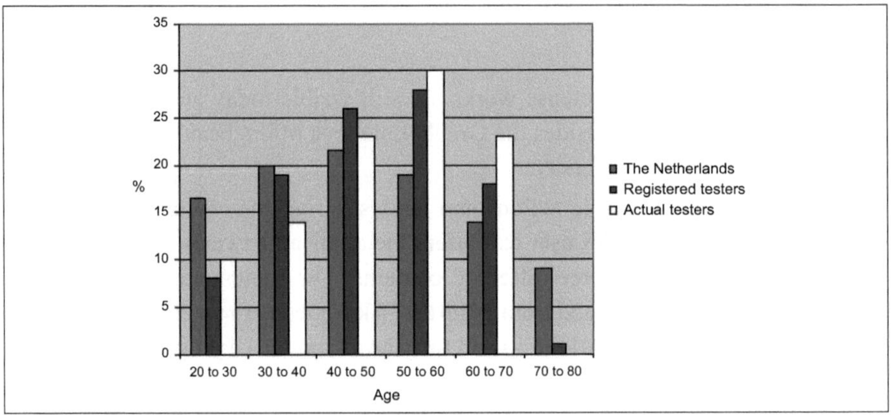

Fig. 15.9. Age distribution of the participants.

Table 15.1. Distribution of the mentality groups (%) for registered participants (DIWI registered), people who really tested (DIWI testers) in comparison with the Dutch society (The Dutch).

Mentality-group	Diwi-registered	Diwi-testers	The Dutch
Traditional bourgeois	9	3	18
Convenience oriented	3	1	9
Modern bourgeois	6	7	22
New Conservatives	5	5	8
Cosmopolitans	18	25	11
Social Climbers	4	3	13
Post Materialists	32	35	10
Post-modern Hedonists	19	20	10
No data	3	1	–

The motives of the DIWI testers to recreate showed up to be different compared to the Dutch in general. *Table 15.2* shows these recreational motives of the testers compared to the rest of the Netherlands

Table 15.2. Recreational motives of the testers (%) compared to inhabitants of the Netherlands.

Motif	%	
	Diwi-testers	The Dutch*
Social interacting	19	27
Break away	22	34
Interest	27	14
Experiencing a different world	14	15
Physical challenges	18	9

* source: Motivaction (Goossen 2008)

15.4.3 Usefulness, Use and Enjoyment

For finding the acceptance rate the Technology Acceptance Model (TAM) (Davis 1989, Davis 1993) and the theory of Hedonic Information Systems of van der Heijden (2004) has been used. The acceptance of the system is explained by the following dimensions:

- Perceived usefulness
- Perceived ease of use
- Perceived enjoyment

Perceived usefulness is "the degree to which a person believes that using a particular system would enhance his or her job performance" while the perceive ease of use can be defined as "the degree to which a person believes that using a particular system would be free from effort". Additionally the perceived enjoyment can be defined as "the extent to which the activity of using the computer is perceived to be enjoyable in its own right, apart from any performance consequences that may be anticipated" (Davis 1993). The three factors together determine a user's attitude towards the DIWI and as an assessment of its actual use.

Perceive usefulness deals with the question if an application provides an additional value while visiting an area. Parts of the concept of perceived usefulness are the benefit, practicality, convenience, helpfulness, efficiency and specialty. For the perceived ease of use the theory of Mayhew describes usability in terms of learning and ease of use (Mayhew 1999). Learning refers to the way user are able to acquaint themselves with the software and the user-interface and the degree they remember how to use it in short, mediate and long periods of intermitted use. The ease of use refers to the way well trained users can carry out tasks with the software and the graphical user interface (GUI). The terms ease of use and ease of learning are somehow contradictory. An interface developed for novice user is not always the most usable for experts and vice-versa. The following aspects are related to the dimension ease of use:

- The complexity of the application, and
- The application behaves as expected.

15.4.4 Perceived Usefulness

About 90% of the testers reported problems using the DIWI, ranging from poor visibility of the screen in sunlight to complete failure of the application. To value the usefulness an inventory of problems encountered by the testers was made. *Table 15.3* shows the distribution for the signaled problems for each of the starting locations and in total.

Table 15.3. Type and distribution of encountered technical problems of the mobile client per starting point in % of the test group (Ntotal=150).

Problem type	Grebbeberg	Renswoude	Scherpenzeel	Ntotal
Need to restart	66.0	51.9	51.7	60.5
Poor visibility of the screen due to sunlight	38.5	55.5	55.2	44.7
GPS went down	45.0	33.3	31.0	40.1
Battery empty	28.6	44.4	34.5	32.7
Long waiting times (downloads)	33.0	25.9	31.0	31.3
Application "freezed"	18.7	29.6	6.9	18.4
Did not receive information about the poi's	17.6	11.1	6.9	14.3
Went lost	5.5	11.1	10.3	7.5
No topographical map	5.5	11.1	6.9	6.8
No sound	3.3	–	3.4	2.7
Problem in typing text	1.1	–	–	0.7
Other	33.0	40.7	34.5	34.0

Based on *Table 15.3* one may conclude that the stability of this mobile client and its screen quality need to be improved. Like other LBS examples battery life is a serious problem. During the Diwi-test on average a battery ran for two hours (by continuous use of back light, gps, data transmission and regular recordings).

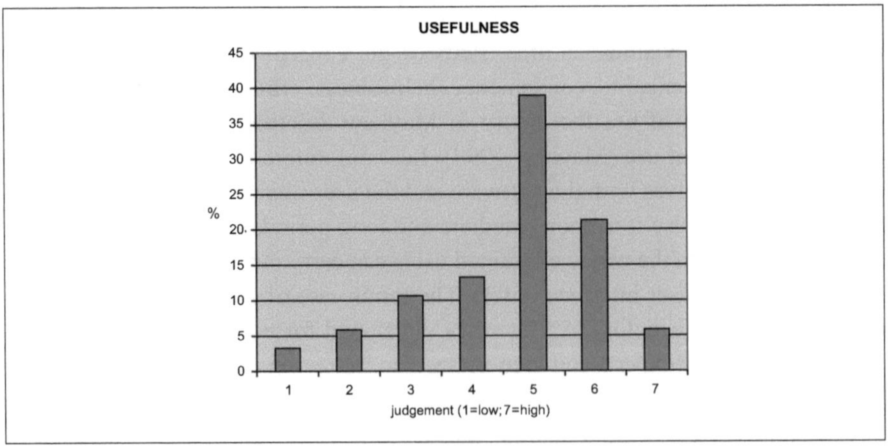

Fig. 15.10. Seven points indication of the usefulness of the mobile client (in %).

Interesting enough all testers had an external battery that easily could be connected when the local battery run out of power, but from the results it curiously turns out that 30% of the test group didn't use that option.

The general opinion about the usefulness was measured using a seven point Likert-scale where 1 represents "disagree very much", 4 "neutral" and 7 "agree very much". About 66% of the testers is positive about the usefulness giving an average judgment of 4.7. The judgments show a skewed normal distribution but with an exception for class five *(Figure 15.10)*.

To gain more insight in the separate dimensions of usefulness *Table 15.4* shows the valuation of it by the testers. From a mean analyses it appeared that most important dimensions are related to benefit, helpful and useful. About 86% of all testers agrees that the mobile client is "special". Almost 75% selects that the client is an aid to explore a region and 60% thinks it is beneficial to the experience of the landscape.

Table 15.4. Average rating for the various dimensions of usefulness on a 7 points scale (1: very much disagree; 7: very much agree) (N = 150).

Dimension	Avg. rating
Useful	4.6
Practical	4.4
Convenient	4.5
Helpful	4.9
Efficient	4.2
Beneficial	5.7

Additionally the testers were asked to value the various types of media offered by DIWI. From *Table 15.5* it appears that testers especially like the diversity in media types but tend to aim for sound and less for text.

Table 15.5. Distribution of the preferences for various types of media in % of N (N = 150).

Type media	%
Text	17
Photo	2
Video	5
Sound	37
Diversity	31
No-pref.	5
Unknown	3

15.4.5 Perceived Use

The perceive ease of use is not uniformly defined. It has been characterized by 7 dimensions: overview, comprehensibility, technical complexity, ease of use, behavioral predictability, functional completeness, physical portability. The seven point

Likert scale has also been used to measure these dimensions. *Table 15.6* presents the average scores for the individual dimensions:

Table 15.6. Average score for the 7 dimensions of use (N=150).

Dimension	Avg. score
Overview	5.0
Comprehensibility	5.1
Requires not much technical knowledge	4.6
Easy to use	4.7
Does what expected	4.2
Need more functionality	4.6
Tiring to carry around	3.6

From a mean analysis it appears that the most important dimensions are comprehensibility and overview. The general opinion is that the mobile client is easy to use. Interesting result is the desire for more functionality. *Figure 15.11* shows the general ratings of the overall perceived use score, which on average (score of 4.7) is above good.

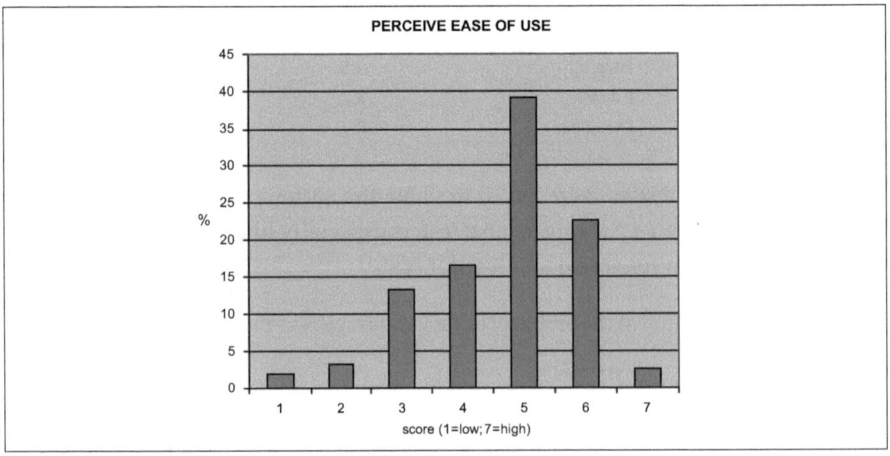

Fig. 15.11. Ratings for the perceive ease of use (% of N).

Additional questions have been asked concerning the design of the mobile client (defined for the test as the combination of the smart phone and the software). *Table 15.7* shows the results. In general the opinion about the design is positive. The readability, in this case related to the screen reflection and contrast under day light conditions, of the mobile client is seen as the most problematic. No significant differences have been observed between experienced PDA or smart phone users and non-experienced testers indicating that the general design of the application is well accessible.

Table 15.7. Average score for the design aspects of the mobile client.

Design aspect	Avg. score
Screensize	4.6
Colors	5.1
Readability	4.3
Weight	5.6
Navigation structure	4.8
Size	5.3
Used symbols and icons	5.0

15.4.6 Perceived Enjoyment

The perceived enjoyment is also specified according a number of dimensions: excitement (1), support for social interactions (2), uniqueness (3), and distraction (4). *Table 15.8* show the questions per dimension and their average scores (7 point Likert scale) for those dimension using partly overlapping questions.

Table 15.8. Average score for the perceive enjoyment dimensions.

Dimension	Avg score
Adventurous (1)	4.8
Boring (1)	2.9
Only fun for those holding the PDA (2)	4.0
Exiting (1)	4.3
Not for groups (2)	3.7
Unique (3)	5.0
Interesting (3)	5.7
Too much of a distraction thus preventing experiencing the landscape (4)	3.5

The general opinion is shown in *Figure 15.12.* Nearly 76% of the testers is rather positive about the perceived enjoyment giving an average score of 5 and higher.

One of the innovative features of the DIWI platform is the ability to add and share personal information by uploading location based content (movie clips, photo's and text). About 54% of the testers added personal content. The testers who did not add content had various reasons: 29% just did not feel the need to add content, 17% experienced it as too complicated and 7% blames the (bad) weather conditions. In total 286 user generated content items were added, an average of almost 4 uploads per tester (taking only into account the testers who did add content). Most of the added content consists of photo's (75%) especially of the landscape scenery, objects along the route, and family snap shots. The additional value of these personal content for the general community is rather limited. Most of it can be classified as too private (not only the family snap-shots but also route notes and corrections).

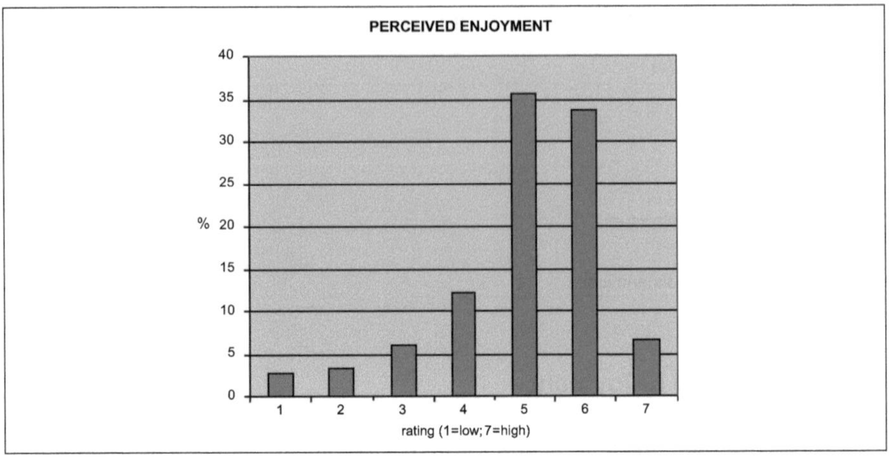

Fig. 15.12. Distribution (% of N) of the ratings for perceived enjoyment.

62% of the testers switched on the user generated personal content on the mobile client to see what others "left behind". Because many of the user generated content was of a too private or specific nature (see for example *Figure 15.6* the personal POI screenshot) the evaluation of the content turns out negatively. The testers do not agree with the proposition that user generated content is interesting and as such an added value to exploring and learning about the region (see *Table 15.9)*. However for purely personal use (for example reviewing the trip on the website) the user generated content does have an additional value, because it offers a personalized location based digital album.

Table 15.9. Average rating of the user generated content.

	Avg. rating
Nice	3.6
Interesting	3.2
Text of other is superfluous	4.3
Photo's of others are superfluous	4.5
Movie clips of others are superfluous	4.5
Provided additional information	3.2

15.4.7 General Acceptance

In general 70% of the testers score the DIWI application with an average score of 4.8. This is a good score despite the fact the various problems were encountered during the tests. The fact that the testers were aware that they are part of a research might explain a possible positive bias and a willingness to accept problems.

Table 15.10. General overview for the general acceptance of the DIWI (n=150).

Reason	# times mentioned
Information	43
Information on the spot	27
Don't loose track/GPS	20
Interactive	13
Replacement of maps	13
Exiting / fun	11
Types of media	11
Made env. More interesting	7

The most often mentioned added values to the use of DIWI are: the possibility to retrieve information on the spot, never loose the way, and uploading of personal content.

Many testers considered the DIWI as a better alternative for paper maps, guides and information panels. 88% of the testers will, once upgraded to a proper product, recommend the DIWI to others. 21% will use it themselves while 61% postpone use due to the costs of the needed data connection (however, currently the average cost of a fast UMTS/HSDPA connection is dropping in the Netherlands). About 11% does not know if to use it in the future and 6% certainly will not use the DIWI. Often the latter is due to extended technical problems faced during use of the DIWI by these respondents. *Figure 15.13* shows the general rating for the perceived additional value of the DIWI compared to not using the DIWI.

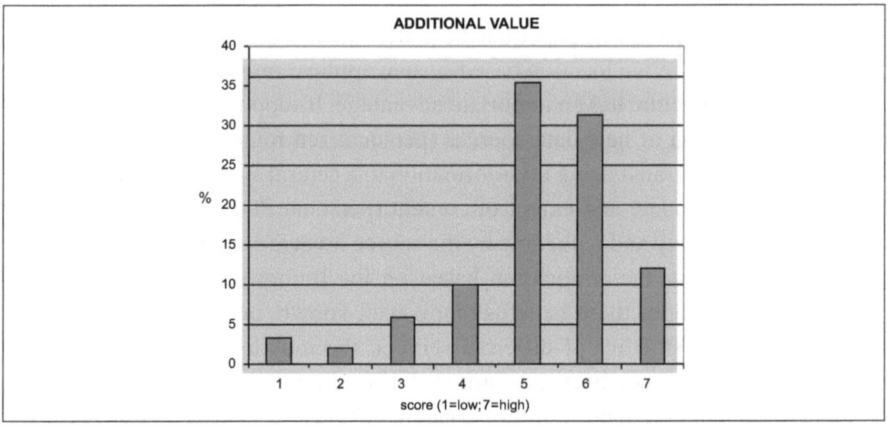

Fig. 15.13. Additional value of the DIWI.

15.5 Spatial Behavior

Besides the use, usefulness and enjoyment it is interesting to explore the actual spatial behavior of the tester in combination with the DIWI application. With spatial behavior we mean the length and type of routes followed and the spatial (landscape) preferences of the testers. The version of the DIWI used by the testers contains 20 pre-defined routes varying from 4 till 26 km. These routes are digitized from various sources. Besides these routes, a complete network of hiking and cycling trails has been digitized. This network is used for the personalized route calculations.

About 25% of the testers used the personalized route generator (73% walking and 24% cycling routes) of the DIWI website. About half (48%) did not follow this route in the field. About 14% followed their personal route completely and 38% partly. From the analyses of landscape preferences most testers preferred forests and water. Also semi-open agricultural landscape was popular. Natural grassland, swamp appeared to be the least popular. The average distance for walking was 5 km and for cycling 25 km.

15.6 Discussion and Conclusions

15.6.1 Regarding the Architecture

The system presented in this paper has been designed to deliver a complete framework for a full interactive location based system application. The approach of using a distributed architecture has an important advantage. It supports an easy on-line and real-time connection of new data sources (personalized routes and personal POIs) without the need of transferring all information to a central system. As such it keeps the maintenance, update and extent of content (personal POIs, DIWI POIs, maps, routes, etc) effectively and realizable by the source owners and the user community. During the user test, the application based on the framework turns out to function quite well. Besides these benefits many, well-known, problems have been run into related to the stability of different servers, the mobile network, and battery life. A crash-test has not been performed during the project but considering the reported failures it is a major point of concern. The test area is known for its variable coverage of UMTS/HSDPA. For one start location the data connection often "degraded" to GPRS connection which appeared to be only sufficient for sending the GPS coordinates to the server and serving the periodic map updates. For downloading of POI media (with the exception of text) GPRS appeared to be insufficient. A (temporal) solution was implemented enabling the client to grab POI media data belonging to a location from a storage-card when the bandwidth was insufficient for timely delivery of media. This option, however, hampers the maintainability of the DIWI because of the personal POI updates.

15.6.2 Regarding the Content

Calculating interesting and reliable routes, requires a comprehensive network based on the description of available cycle tracks and footpaths. Such a network is not yet available, at least not for the whole of the Netherlands. For the test a network was digitized based on routes offered by tourists' offices by maps, topological data and selected and validated gps-tracks.

Additionally creating and compiling POI's, movie, and sound clips requires lot of work. In the DIWI cases professional story writers and voice over's have been contracted to create attractive and understandable content.

15.6.3 Regarding the Test Results

The use, usefulness en enjoyment in general is positively evaluated by the testers despite of a number of technical difficulties experienced during testing. One surprising observation is that the expected added value of content uploaded and shared by fellow testers was not very much appreciated, mainly because of the personal characteristics. The idea of communities forming around similar experiences of landscapes appeared not to be fulfilling a need of users. From additional analysis it appeared that the user generated content could be divided into private data and location based cultural-historic data. The first is not of much value for sharing while the second will be. It would be recommendable to serve the user generated content according various classes, allowing users to only pick those information interesting to them. A community based editor could support such classification. It should be stressed upon that the results of this test is based on a very specific group of testers with a dominance of relatively high educated testers and a-typical social groups (see *Table 15.1).*

15.6.4 Regarding the Soft- and Hardware

Although a reliable performing device, the smart phone used (HTC P3600) suffered from bad visibility when exposed to direct sunlight. Additionally the standard battery did not have enough capacity to support continuous services during the longer routes. Depending on the outside temperature an average battery last for about 2.5–3 hours. Additional, high capacity batteries solved that problem. At the moment of purchasing only very limited smart phones equipped with GPS, UMTS/HSDPA and good camera's running on window mobile 5 were available on the market. Comparison of various devices thus was not possible.

15.6.5 Regarding Future Developments

Since the concept of the DIWI is very appealing (the project received at lot of media attention), there is a lot of interest to have an operational version of it. A number of organizations including local and regional governments already indicated they are interested. On the other hand to "convert" the DIWI in a "product" requires considerable work, especially concerning the content. Having interesting routes accompanied with well though points of interested having professional edited media is essential in providing people an additional experience in the landscape.

Acknowledgements

The authors like to express their gratitude to all staff of the Digital Dowsing Rod project members, Alterra, DS Landschapsarchitects, Galileo Communication Projects, Municipality of Utrecht, KPN, Dutch Cadastre, Province of Utrecht, Foundation "Vernieuwing Gelderse Vallei", Waag Society and Wageningen University, for their participation in, dedication to and support of the DIWI- project. The project has been partly funded by the Dutch National Research program committee on Geo-Information (RGI), the Province of Utrecht and the Dutch Ministry of Agriculture, Nature and Food quality.

References

Antikainen H, Rusanen J, Vartiainen S, Myllyaho M, Karvonen J, Oivo M, Similä J, Laine K (2006) Location-based Services as a Tool For Developing Tourism in Marginal Regions. Nordia Geographical Publications, 35, 39-50.

Baldzer J, Boll S, Klante P, Krosche J, Meyer J, Rump N, Scherp A, Appelrath HJ (2004) Location-Aware Mobile Multimedia Applications on the Niccimon Platform. Informationssysteme fur mobile Anwendungen (IMA'04), Brunswick, Germany.

D'Roza T, Bilchev G (2003) An Overview of Location-Based Services. BT Technology Journal, 21, 20-27.

Davis FD (1989) Perceived Usefulness, Perceived Ease of Use, and User Acceptance of Information Technology. MIS Quarterly, 13, 319-340.

Davis FD (1993) User acceptance of information technology: system characteristics, user perceptions and behavioral impacts. International Journal of Man-Machine Studies, 38, 475-487.

Dias E, Rhin C, Haller R, Scholten H (2004) Adding Value and Improving Processes Using Location-Based Services in Protected Areas: The WebPark Experience. e-Environment: Progress and Challenge Special edition on e-Environment Instituto Politécnico Nacional Mexico, Mexico City, 291-302.

Espinoza F, Persson P, Sandin A, Nyström H, Cacciatore E, Bylund M (2001) GeoNotes: Social and Navigational Aspects of Location-Based Information Systems. Ubicomp 2001: Ubiquitous Computing.

Goossen CM, Ploeger B (1997) Selectie van recreatievormen en indicatoren voor het Beslissingsondersteunend Evaluatiesysteem voor de Landinrichting. Wageningen, DLO-Staring Centrum.

Goossen M, Meeuwsen H, Franke J, Kuyper M (2008) My ideal tourism destination: Personalized destination recommendation system combining individual preferences and GIS data. Journal of ITT (in print).

Heijden Van De H (2004) User acceptance of hedonic information systems. MIS Quarterly, 28, 695–704.

Hinze A, Voisard A (2003) Location-and Time-Based Information Delivery in Tourism.

Jose R, Davies N (1999) Scalable and Flexible Location-Based Services for Ubiquitous Information Access. Proceedings of the First International Symposium on Handheld and Ubiquitous Computing, 52-66.

Kramer R, Modsching M, Hagen Ten K, Gretzel U (2006) Web-based Recommenders for Personalized City Tours: A Usage Analysis. Department of Computer Science. University of Applied Sciences Zittau/Görlitz, Germany, Laboratory for Intelligent Systems in Tourism, Texas A&M University.

Margherita P (2004) Determinants of adoption of third generation mobile multimedia services. Journal of Interactive Marketing, 18, 46-59.

Mayhew DJ (1999) The usability engineering lifecycle, ACM New York, NY, USA.

Raper J, Gartner G, Karimi H, Rizos C (2007) Applications of location–based services: a selected review. Journal of Location Based Services, 1, 89 - 111.

Schmidt-Belz B, Laamanen H, Poslad S, Zipf A (2003) Location-based mobile tourist services-first user experiences. Information and Communication Technologies in Tourism, 115–123.

Sharp JH (2007) Development, extension, and application: a review of the technology acceptance model. Information Systems Education Journal, 5, 1-11.

Zipf A (2002) User-Adaptive Maps for Location-Based Services (LBS) for Tourism. Information and Communication Technologies in Tourism 2002: Proceedings of the International Conference in Innsbruck, Austria, 2002.

Zipf A, Malaka R (2001) Developing Location Based Services for Tourism-The Service Providers View. Information and Communication Technologies in Tourism 2001. Proceedings of the International Conference in Montreal, Canada. Montreal, Canada.

16 MNISIKLIS: Indoor Location Based Services for All

Vassilis Papataxiarhis, Vivi Riga, Vangelis Nomikos, Odysseas Sekkas, Kostas Kolomvatsos, Vassileios Tsetsos, Panagiotis Papageorgas, Stelios Vourakis, Vasileios Xouris, Stathes Hadjiefthymiades and Georgios Kouroupetroglou

Department of Informatics and Telecommunications
National and Kapodistrian University of Athens, Greece

Abstract

MNISIKLIS is an integrated system aiming to provide universal, indoor location-based services focusing on navigation. This paper presents the overall MNISIKLIS architecture and certain implementation details. In the context of the Design for All approach, the system targets to the support of several types of users, including persons with disabilities as well as elderly, by exploiting multimodal interaction. Moreover, the system implements efficient path finding algorithms and provides advanced user experience through highly personalized services. MNISIKLIS adopts Semantic Web technologies (e.g., ontologies and reasoning methods) for representing an d managing application models. Furthermore, MNISIKLIS exploits modern positioning techniques in order to achieve high quality positioning. The paper discusses the algorithms and the models that accommodate the services provided by the system. Additionally, an analysis of the positioning subsystem, the user interaction subsystem and the peripheral infrastructure is given. Hence, a new paradigm in the area of location-based systems is presented.

Keywords: indoor navigation, ontology, RFID, dead reckoning, multimodal interaction

16.1 Introduction

Nowadays, the increasing demand for advanced, personalized, context-aware, intelligent and "always available" applications has led to the convergence of information technology and telecommunication domains. As a result, concepts like pervasive computing, context-awareness and artificial intelligence started to play an essential role to the application development community and researchers, thus, leading to the so-called ambient intelligence paradigm (Aarts et al. 2001).

Location-based services (LBS) constitute a popular domain of context-aware applications. Indoor pedestrian way-finding, in particular, is a very challenging area, mainly due to the unsuitability of the mature and widely established outdoor positioning technologies for use inside buildings. The Global Positioning System (GPS) is an excellent technology that can be used for the determination of absolute location in outdoor environments, but is almost useless indoors.

Furthermore, the vision of Design for All (Stephanidis 2001), which targets at enabling people of all ages and abilities to access services and environments in order to improve their quality of life, becomes more and more popular. The adoption of multimodal interfaces is central to this vision, since Ambient Intelligence aims to enhance human-machine interaction by placing the user at the centre of the computing environment.

This paper presents the MNISIKLIS system that provides real-time, indoor LBS to a wide range of users. The main novelties of the implemented system are listed below:

1. For the first time, to our knowledge, passive UHF Radio Frequency Identification (RFID) technology is used for proximity sensing.
2. Positioning is based on a multi-sensor fusion process, involving Wi-Fi positioning and the Dead Reckoning technique.
3. The system provides a multimodal user interface, thus implementing the Design for All paradigm.
4. The implemented services heavily rely on semantic models and knowledge reasoning techniques. Hence, the overall service logic is highly human-centered.

The rest of this paper is organized as follows. *Section 16.2* discusses related work in the area of indoor location-based systems. A generic architecture of the system and the description of the services supported are provided in *Section 16.3*. *Sections 16.4–16.6* present the main subsystems of MNISIKLIS and give some implementation details. Finally, some concluding remarks are provided in the last section of the paper.

16.2 Related Work

In this section, we present some indicative systems for location-aware applications. Most of them mainly focus on navigation services for either indoor or outdoor environments and, thus, do not support the full spectrum of LBS services as MNISKLIS does. For example, most of them either just implement a navigation service without providing multimodal interaction or do not exploit efficient positioning methods.

iNAV (Kargl et al. 2007) is a navigation framework aiming to providing guidance in indoor environments. It exploits the COMPASS (Kargl & Bernauer 2005) middleware in order to achieve localization and facilitate service discovery. Nevertheless, iNAV mainly targets at typical users, since it does not provide any advanced user interaction features. CoINS (Lyardet et al. 2006) is a context-aware indoor navigation system that involves a complex mechanism for spatial modeling and room detection. With regard to the route selection process, the system exploits an optimized version of the Dijkstra algorithm. However, CoINS does not currently support any multimodal interfaces to support diverse user classes. Another pedestrian navigation and exploration system is presented in (Wasinger et al. 2003). The system exploits GPS, infrared beacons and a magnetic compass as positioning technologies and emphasizes on supporting different modalities. Although, the system involves multimodal interaction, it does not investigate the concept of interaction with multiple devices and it implements only a core navigation service.

IRREAL (Baus et al. 2002) is another indoor navigation system, based on infrared beacons, that adapts the presentation of route directions to the specific device capabilities. The application does not fully support interaction with disabled users. A pedestrian navigation system that investigates complex aspects like multi-criteria path selection and integrated positioning for both indoor and outdoor environments is described in (Gartner et al. 2004). Although the system supports audio guidance, it is not targeting to disabled users. In (Bikakis et al. 2006), the authors exploit Semantic Web technologies in order to develop a context ontology for supporting indoor navigation services. However, this approach does not examine in detail the efficiency of positioning techniques and the presentation of path instructions to the user.

16.3 System Architecture and Implemented Services

The MNISIKLIS platform includes three main subsystems (as shown in *Figure 16.1):*

- *Positioning Subsystem.* It comprises the overall equipment and the algorithms used to estimate the user's position. Specifically, it consists of sensors and positioning techniques, the location fusion component and the interfaces between them.
- *Middleware.* The middleware consists of the services and the navigation algorithms developed as well as the application models. It is also responsible for gluing together the other subsystems.
- *User Interaction Subsystem.* The user interaction subsystem involves the user device (hardware and software), the input/output interfaces and the content selection and representation algorithms.

Apart from these core platform ingredients, a peripheral infrastructure for LBS content provisioning and management has been developed. Such infrastructure includes a GIS system and a Semantic Content Management System (SCMS). In the following sections, we elaborate on the main components of each subsystem.

In the context of the MNISIKLIS project, the following services have been implemented:

- *Static Navigation.* The user asks the system to determine a "suitable" route to a certain destination. The service takes into consideration the application models (e.g., user profile) in order to compute the "best" path and guide the user with the most suitable landmarks.
- *Dynamic Navigation.* An extension of static navigation that periodically traces the user position. In case it detects a significant deviation of the user from the predetermined path, it helps her to find her way by providing more detailed information.
- *Where-Am-I.* The user asks for her current position inside a building. The system responds by providing details about the last known user position. The information about a specific location is organized and presented in different levels of detail.

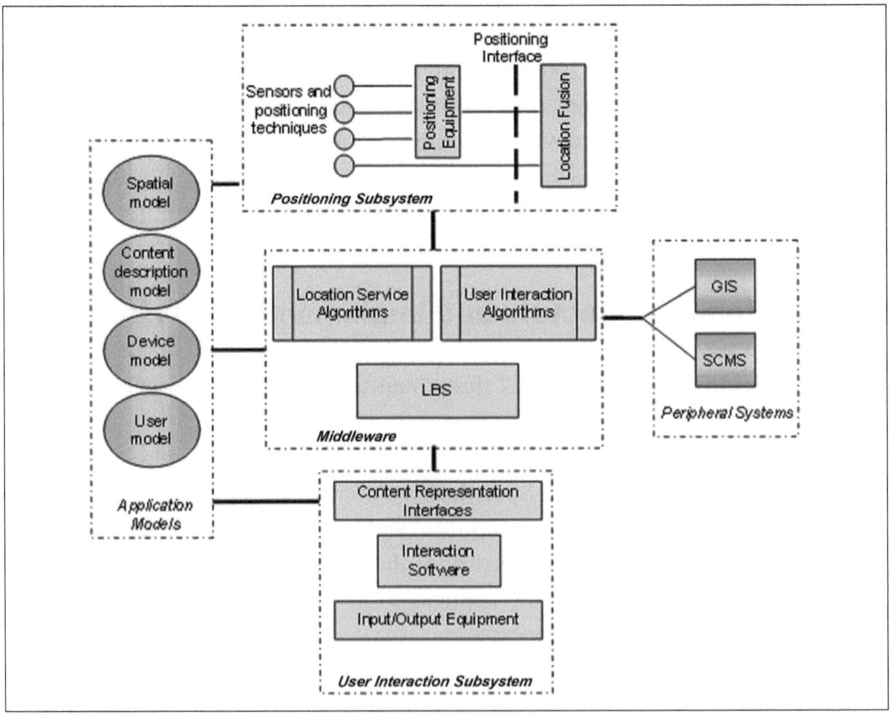

Fig. 16.1. MNISIKLIS Architecture.

- *Exploration.* While the user is moving inside the building, the system provides ("pushes") information about the nearest locations that she may be interested in. Such Points Of Interest (POIs) may have been explicitly stated by the user or not (e.g., significant exhibits in a museum).
- *Nearest POIs.* The system computes the POIs that are closer to the user. The main difference from the exploration service is the push-based nature of the latter. Hence, the system may always return points that are not located close to the user.

16.4 Positioning Subsystem

16.4.1 Sensing Technologies

Accurate indoor positioning can be achieved through two main technology approaches. The first approach is based on radio technology (WLAN, Bluetooth) as well as infrared and ultrasound beacons (Schiller & Voisard 2004). The second approach exploits inertial sensors (accelerometers and gyroscopes) and non-inertial sensors (magnetic compasses) along with appropriate dead reckoning algorithms. Proximity sensing, based on RFID technology has been recently introduced (Koide & Kato 2006) as an alternative. In the MNISIKLIS platform we have adopted three technologies: UHF RFIDs, Dead Reckoning (DR) for pedestrian users (Fang et al. 2005, Dippold 2006) and WLAN Received Signal Strength Indicator (RSSI). This section describes the main sensing technologies that compose the proposed positioning subsystem.

16.4.1.1 RFID Technology

UHF RFIDs (868 MHz) used in the system provide longer ranges (in the order of 1m) with only 50 mW of RF power compared with other approved RFID frequency bands. The RFID reader is the nano-UHF from TAGSENSE, while the RFID tags are manufactured by Texas Instruments and support the EPC1 GEN2 standard for read-write capabilities. In the current implementation an identity has been stored in the tag's EEPROM memory and has been associated with a specific building location in the location server database. The navigated person carries a mobile RFID reader which continuously scans for tags and transfers the specific identity of the proximal RFID to the PDA.

16.4.1.2 Dead Reckoning

In the MNISIKLIS platform a prototype dead reckoning system has been developed, based on a commercially available high performance 3-axis electronic compass

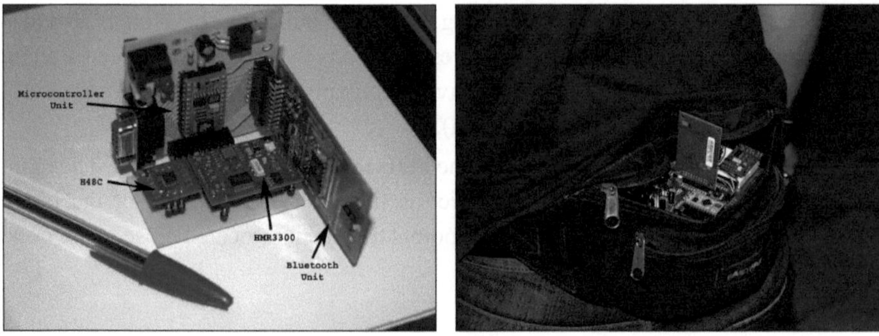

Fig. 16.2. The sensor measurement unit.

(Honeywell HM3300) and a 3-axis accelerometer (Hitachi H48C). The 3-axis magnetometer is used for heading information independently from sensor orientation. For the estimation of distance traveled from a person, various algorithms have been proposed for step detection and step length estimation based on 3-axis accelerometers (Weinberg 2008, Ladetto et al. 2000). The sensor measurements are collected at the sensor unit which is attached to the user's belt (see *Figure 16.2)* and are, subsequently, transferred via a Bluetooth link to a PDA for processing by the dead-reckoning algorithm.

As a first preprocessing step, the raw accelerometer data are low-pass filtered with a moving average filter. The acceleration measurements on the Z (gravitational) axes or from any of the horizontal acceleration axis can be used in the DR algorithm.

The step length, for walking on a flat path, is influenced by the walking frequency and the variance of the accelerometer during one step (Ladetto et al. 2000). Specifically, the predicted step length is computed using the following step model:

$$\textit{step length of } k \textit{ step} = A + B * f_k + C * \mathrm{var}_k \tag{16.1}$$

where

- A, B, C are coefficients estimated through linear regression,
- f_k is the walking frequency at time t_k obtained by the equation $f_k = \dfrac{1}{t_k - t_{k-1}}$, with t_i being the detecting time of the ith step, and
- var_k is the variance of the acceleration in the direction of movement, during the kth step: $\mathrm{var}_k = \sum_{t=k-1}^{k} \dfrac{a_t - mean(a_k)}{N}$, where a_t is the acceleration at time t, $mean(a_k)$ and N is the mean value of the accelerations data and the number of the acceleration samples during the kth step, respectively.

Finally, the walking distance of m steps on a straight flat path is obtained by the equation:

$$walking\ \ distance = \sum_{i=1}^{m} \left(A + B * f_i + C * \mathrm{var}_i \right) \tag{16.2}$$

The position estimations of the dead reckoning algorithm are transferred from the PDA using a WLAN to the location server for further processing (fusion).

16.4.2 Location Server

The location server is the core component of the positioning subsystem. It processes the data received from the mobile device and generates the final estimation of the user's current position. Hence, it is equipped with suitable software which handles the communication with the mobile device, the analysis of collected data and the implementation of inference algorithms *(Figure 16.3).*

The server takes advantage of a database that stores the spatial data and the history of estimations. Below, we provide a more detailed presentation of each component.

16.4.2.1 Communication Component (CC)

The role of this component is the communication with the mobile device. It receives the requests with the collected data and checks their validity (e.g., if the measured value from a WLAN access point is in the range of predefined min-max values).

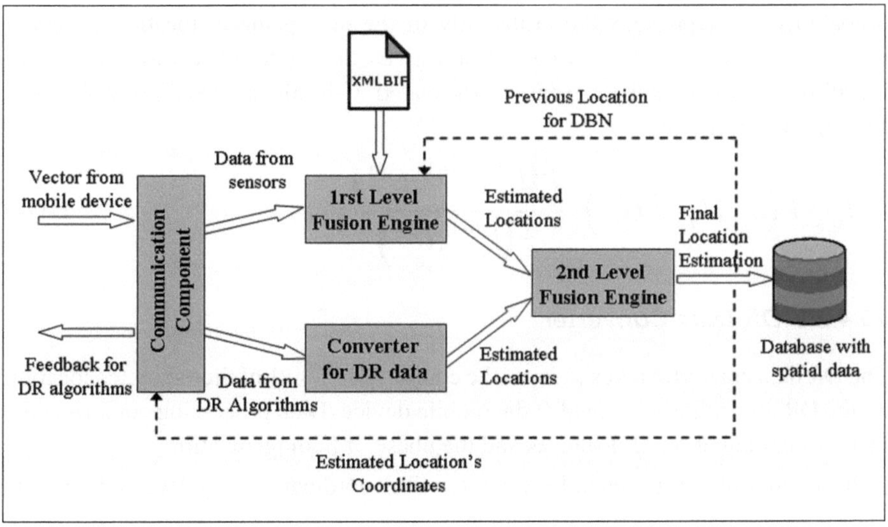

Fig. 16.3. Location Server Architecture.

The form of the data vector is as follows:

$<userId, IE\_Id_1 = value_1, \ldots, IE\_Id_N = value_N, X = x_1, Y = y_1, Z = z_1, Orientation>$

where

- $userId$ is the unique identifier of the user,
- $IE\_Id_i = value_i$ is the pair (Infrastructure Element Identifier, value),
- *X, Y, Z* are the coordinates of the estimated location as they have been computed by the DR algorithms *(Z* denotes the floor level),
- *Orientation* is the compass' measurement for the orientation of user (deviation from the North).

After that, CC proceeds to the quantization of the WLAN RSSI values at N discrete levels: *S1, S2... SN.* If, for example, the value from the access point with IE_Id Id1 is between -30 dBm and -47.5 dBm, the min-max values are -30dBm and -100dBm respectively and N=4, the value *S1* is assigned as the observed value of this access point (assumption of uniformly distributed values).

16.4.2.2 First Level Fusion Engine

This engine uses multiple readings from WLAN access points and RFID tags and it is based on a Dynamic Bayesian Network (DBN), which is used for the location inference (Sekkas et al. 2006). The structure and the required probability distributions for the DBN are loaded from a XMLBIF file (XMLBIF 1998) which is the result of a training phase. The output of this engine consists of a vector of probabilities $< P_{b1}, P_{b2}, \ldots, P_{bN} >$ for each symbolic location $L_1, L_2, \ldots, L_N$ at time *t*. The probability P_{bi} represents the probability of the user being at location L_i given her previous location L_j at time *t-1* and given the values/observations, $O^{(t)}$ of the aforementioned elements/sensors associated with this user. Such probability is calculated as follows:

$$P_{b_i} = P\left(L_i^{(t)} \mid L_j^{(t-1)}, O^{(t)}\right) = \frac{P\left(L_i^{(t)}, L_j^{(t-1)}, O^{(t)}\right)}{P\left(L_j^{(t-1)}, O^{(t)}\right)} \qquad (16.3)$$

16.4.2.3 DR Data Converter

The DR data converter takes as input the coordinates (x, y) of the estimated location by the DR algorithms executed at the mobile device. Data processing comprises the phase of computation of distances and the phase of homogenization.

In the first phase, for each location L_i with coordinates (x_i, y_i) we compute the 2D Euclidean distance from the estimated location, as the DR component cannot locate the user between successive floors.

In the homogenization phase the converter applies a simple transformation rule in order to calculate the probability (P_{ci}) of the user being at location L_i. The distance of the location L_i from the estimated location is not taken into account for the calculation of probabilities when it exceeds a predetermined threshold d_t. For locations of distance d_i greater than d_t or locations found in a different floor than the estimated (x, y) coordinates, we assign zero probability. For the remaining locations we compute the probability $P_{ci}^{'}$ according to the equation:

$$P_{ci}^{'} = \frac{1}{d_i^2} \;,\; i = 1,2,...,N \tag{16.4}$$

which indicates that this probability is reversely proportional to the square distance. The final value of the probability P_{ci} is calculated after the normalization procedure so that the sum of probabilities equals unity:

$$P_{ci} = \frac{P_{ci}^{'}}{\sum_{i=1}^{N} P_{ci}^{'}},\; i = 1,2,...,N \tag{16.5}$$

Thus, the converter produces as output the vector of probabilities $< P_{c1}, P_{c2},..., P_{cN} >$ which feed the 2nd level fusion engine.

16.4.2.4 Second Level Fusion Engine

The second level fusion engine takes as input the probabilities calculated for each symbolic location by the first level fusion engine and the DR converter and produces the final estimation of the current position of the user. Specifically, it uses the probability vectors $< P_{b1}, P_{b2},..., P_{bN} >$ and $< P_{c1}, P_{c2},..., P_{cN} >$ and combines them appropriately so that it calculates the final probability P_i for location L_i. The contribution of probabilities P_{bi} and P_{ci} is determined by the corresponding weights w_b and w_c ($w_b + w_c = 1$) according to the following combination rule

$$P_i = w_b * P_{bi} + w_c * P_{ci} \;,\; i = 1,2,...,N \tag{16.6}$$

The final estimated location is calculated as the location with the highest probability and is stored in the database. Moreover, it is sent to the mobile device as feedback for the DR algorithms and updates the DBN in the first level fusion engine.

16.5 Middleware

16.5.1 Application Models

Four ontologies are the basis for MNISIKLIS: i) the spatial ontology (Indoor Navigation Ontology – INO), ii) the User Navigation Ontology (UNO), iii) the

Device Ontology (DO), and iv) the Content Ontology (CO). The instances of the aforementioned ontologies are connected through semantic relationships in order to provide more intelligent location services.

- *Indoor Navigation Ontology (INO):* The spatial ontology is an extended version of the INO (Tsetsos et al 2006; INO 2008), based on the OWL-DL language (OWL 2004). Specifically, it describes concepts and relationships that correspond to every basic spatial element typically found in indoor environments.
- *User Navigation Ontology (UNO):* UNO (Kikiras et al. 2006) is an ontology that contains the necessary concepts and relations to define the main characteristics and abilities of users, facilitating the provision of highly personalized services. Additionally, UNO design is based on the international standard defined by the World Health Organization (WHO) (ICF 2001).
- *Device Ontology (DO):* Our approach adopts a device ontology in order to represent basic features and the functionality supported by various user devices (e.g., mobile phones, PDAs, headphones). The knowledge captured by the ontology refers to hardware capabilities (e.g., display size, resolution) as well as device supported modalities (e.g., input/output modes).
- *Content Ontology (CO):* Content Ontology describes general categories of content with their properties and relations *(Figure 16.4).* CO includes two main categories of concepts. *Digital Content* is the high level concept including properties related to the general characteristics of the described content. Low level concepts describe each specific content type (e.g., Text, Image, Video).

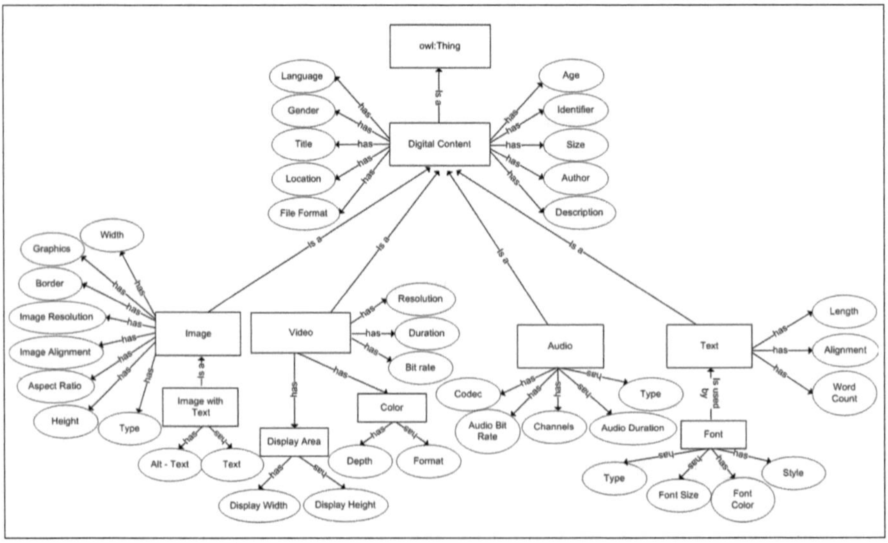

Fig. 16.4. Content Ontology concepts and their properties.

16.5.2 LBS Algorithms

In the context of MNISIKLIS, a hybrid rule-based navigation algorithm has been designed and implemented. The core navigation algorithm (see code below) involves two main steps to compute the best traversable path for each user:

Creation of user-compatible building graph. In this step, a number of path elements incompatible with the user profile (capabilities) are excluded from the building graph. Specifically, several disability rules are applied to the INO instances with respect to the user profile, eliminating the path elements that cannot be traversed by a certain user. These rules are expressed through the SWRL language (Horrocks et al. 2004). For example, all stairway elements should be excluded in case of a wheel-chaired user. The following rule captures that knowledge:

uno:WheelchairedUser(u) $\wedge$ ino:Stairway(s) $\rightarrow$ ino:isExcludedFor(s,u)

Thus, the overall graph of the building is reduced to a user compatible form.

Path computation. This step takes into consideration several metrics in order to compute the best traversable path for each user. Since, in personalized human navigation systems, we expect to have many different constraints (per user) during route computation, we decided not to adopt such a "monolithic" approach. As a result, the algorithm assigns bonus/penalty points to the route elements according to several parameters. The main measures that affect the route computation are:

- *The route complexity.* The complexity of the route instructions is a very important factor in human navigation, since people usually do not follow the shortest but the simplest path. The simplest-path algorithm proposed in (Duckham & Kulik 2003) computes the "easiest-to-describe" path in a graph and has similar computational complexity with a shortest path algorithm.
- *The Euclidean route distance.* Since the distance of a route is usually the main criterion in human navigation, we also take into account the Euclidean distance of a path.
- *The user profile.* User capabilities and preferences play significant role during the path computation process. For example, in the case of a user that prefers to use stairs, the system would penalize paths including the elevator.

```
Navigation(INO, origin, destination, user profile)
  Begin
    Exclude Path_Points incompatible to the user profile by
    applying disability rules
    Create user compatible building graph from remaining INO instances
    Compute the k-Simplest Paths from origin to destination
      Foreach of the k-Simplest Paths
```

```
        Foreach Path Element PE
          Assign bonus/penalty value to PE,
          according to perceptual rules and user preferences
        Endfor
        Compute the total path length
        TotalPathRank = f(path length, bonus vector, penalty vector)
      Endfor
      Return the path with the maximum TotalPathRank
  End
```

The aforementioned core navigation algorithm was developed for the static navigation service. The rest of the services (see *Section 16.3)* were implemented as extensions/modifications of this core algorithm. In this section, we describe in detail the most challenging implementation parts of these services.

Dynamic navigation was the most computationally-hard service since it should spontaneously provide a new route description to user in case the system detects her "far enough" from the initially computed path. It also imposes strict deadlines, since it has to provide useful information to the user at minimal response times. Specifically, the extended algorithm of dynamic navigation does not guide the user back to the initial path, but it computes a totally new path description given the current user trace.

Exploration was also a time critical service, since it had to inform user about nearby location points or regions before she passes them. To achieve acceptable response times, the implemented algorithm takes advantage of the graph structure of the building by examining only the locations that are close to the user position.

Finally, the "K-Nearest POIs" was a memory consuming service, as it computes and manages paths for all POIs of the user. Once again, the graph structure of the building facilitated the service implementation.

16.5.3 Peripheral Systems

16.5.3.1 GIS-Based Ontology Population

In our approach, we use a GIS system in order to define points on the map of a specific building. We categorize points according to the concepts defined in INO. A layered architecture is adopted *(Figure 16.5)*. The blueprints of each floor constitute the lower layer and are used as a reference for the rest of the layers. Based on each floor's map, corridors and other spatial elements are defined. In our approach, 16 layers are used for each floor. There are layers related to navigational or transition points as well as layers devoted to facilitate users' guidance. Moreover, auxiliary

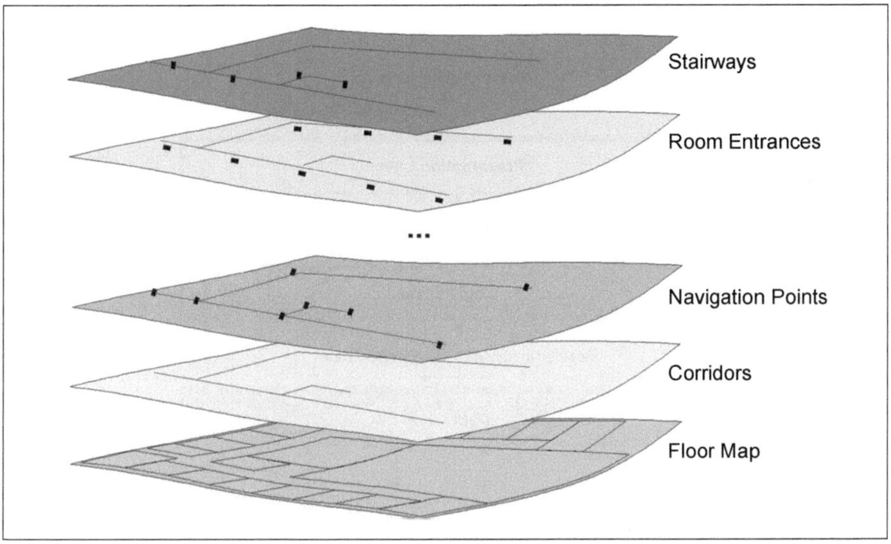

Fig. 16.5. GIS Layers.

points are defined close to transition points (i.e., room entrances) for orientation purposes. Further details on the overall procedure are provided in (Kolomvatsos et al. 2007).

Once the GIS data are in place, one can proceed with the actual ontology population. Each of the aforementioned layers is exported as a shape file (i.e., a popular GIS data storage format) which is subsequently imported in a table of a spatial database. Subsequently, a series of algorithms are used to create the instances based on INO concepts.

16.5.3.2 Semantic Content Management System (SCMS)

Since, the main goal of MNISIKLIS platform is to provide to users services and content according to their location, a Content Management System (CMS) was developed for managing content. We should remind that CMSs are systems used for the creation, organization and the manipulation of digital content (Boiko 2005). Semantic Web technologies can provide efficiency in content retrieval and interoperability. Hence, an ontology used for the annotation of content gives a common view and semantic meaning in order to have effective content retrieval, enabling the exploitation of content in knowledge-based systems and processes (i.e., rule-based path computation). The MNISIKLIS SCMS uses the CO *(Figure 16.4)* in order to semantically annotate content items.

The architecture of our SCMS is comprised by three layers: Data Layer, Logic Layer and Presentation Layer (see *Figure 16.6*).

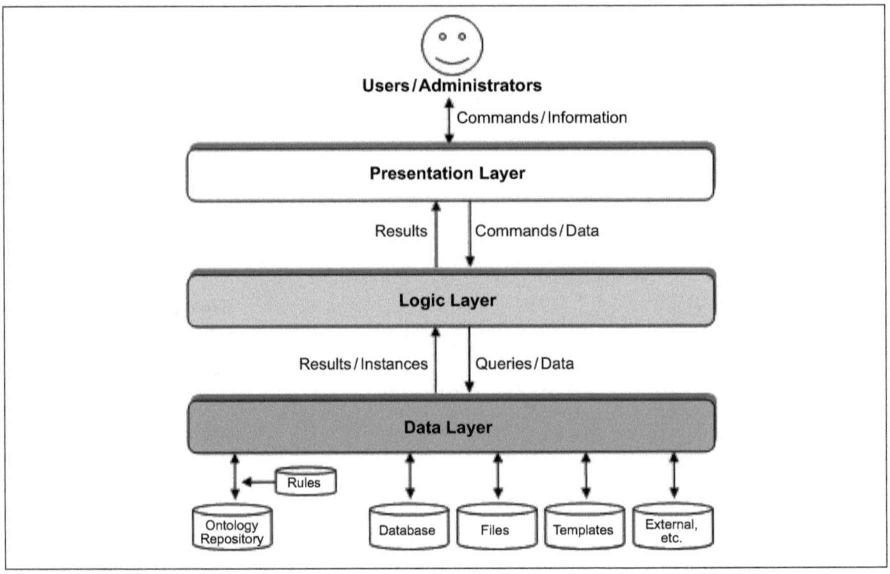

Fig. 16.6. SCMS Architecture.

- *Data Layer:* This layer is responsible for the content storage and retrieval in/from the physical media.
- *Logic Layer:* It deals with the actual administration of the content. Logic Layer processes the retrieved data in order to pass them in the appropriate form to the presentation layer.
- *Presentation Layer:* This layer is used by users in order to retrieve information and by the administrators of the SCMS in order to insert, retrieve and update content and the associated metadata. Content entities can be linked with INO elements through a graphical user interface.

16.6 User Interaction Subsystem

MNISIKLIS, in contrast to other indoor location based systems, targets a wide range of user population. The main user groups include:

1. *Non-disabled users.* This group is comprised of the typical users that do not have any impairment.
2. *Elderly people* (older than 65 years old). Some special features of this group are that they are not fully acquainted with modern technologies. They may also present some degree of disability in perception, memory, vision, hearing and physical movement. Hence, the User Interface (UI) should be as simple and tangible as possible.

3. *People with partial or total vision loss.* If the UI is visual, then it should be as distinguishable as possible (colors, fonts etc). In the case of audio or tactile UI there are no constraints except for the quantity of information given.
4. *People with locomotive disabilities.* Users that use a wheelchair or have very limited ability to walk.

16.6.1 Building an Accessible to All System

One of the main tasks of MNISIKLIS was to create a system Accessible to All. In order to achieve this goal, the design of the UI was based on Design for All principles (Stephanidis 2001). Hence, an analysis of the user groups and their needs was performed from the early stages of the project implementation, mainly based on the analysis of the existing scientific literature. To better define those groups, the International Classification of Functioning, Disability and Health (ICF 2001) was used.

16.6.2 MNISIKLIS Devices

In the context of the MNISIKLIS system, four types of user terminal devices are used by the target groups for accessing the LBS: Tablet PC, Smart Phone, PDA (Personal Digital Assistant) and Mobile Phone. The basic criteria for the selection of those devices were ergonomy and economy. Of high importance was also their compatibility to some assistive technology and their characteristics, such as processor speed, screen size and networking. Depending on the user profile, some input/output peripherals may be connected wirelessly (e.g., Bluetooth) or through cables to some devices, such as earphones, head-mounted screens and Braille displays. *Table 16.1* presents the possible combinations of peripheral devices with the user device.

Table 16.1. Combinations of portable terminals with peripheral devices.

	Head-mounted LCD screen	Braille Display	Bluetooth Earphones
Tablet PC			√
Smart Phone		√	√
PDA	√	√	√
Mobile Phone			√

16.6.3 Multimodal User Interfaces

The UI of MNISIKLIS is multimodal. It adopts three modalities for input/output: visual, audio and haptic. In the haptic mode the user can interact using the keyboard, special buttons, the touch-screen, or the Braille display. The user can also interact orally, through a speech-based dialogue subsystem (Fellbaum & Kouroupetroglou 2008) using a Mobile Phone. In the visual mode, the output is presented through a common graphical UI (GUI). Combinations of the above described modalities are also supported.

Visual Modality: In the case of a GUI, the output data may be in the form of text, image and map. Specifically, the user has at her disposal a textual menu consisting of buttons and selection lists to make a service choice. When the user has requested a specific navigational service, she receives an output consisting of textual instructions, images of the nearby POIs, landmarks and corridors and a map depicting the current floor (see *Figure 16.7).* The user is navigated on a turn-by-turn basis using textual instructions that are given dynamically according to her position and notifications, whenever necessary. The map, in this case, depicts the segment of the path/ itinerary that is already covered by the user and the remaining part left to travel. When the user asks for an informational service such as Nearest POIs, Where-Am-I or Exploration, then she obtains three layers of proliferating information in the textual instructions. She also obtains the current map depicting her position and the position of the Nearest POIs.

The user can choose among five different text font sizes, two font colors (black or white) and five colors for the drawing on the maps. The maps are represented in

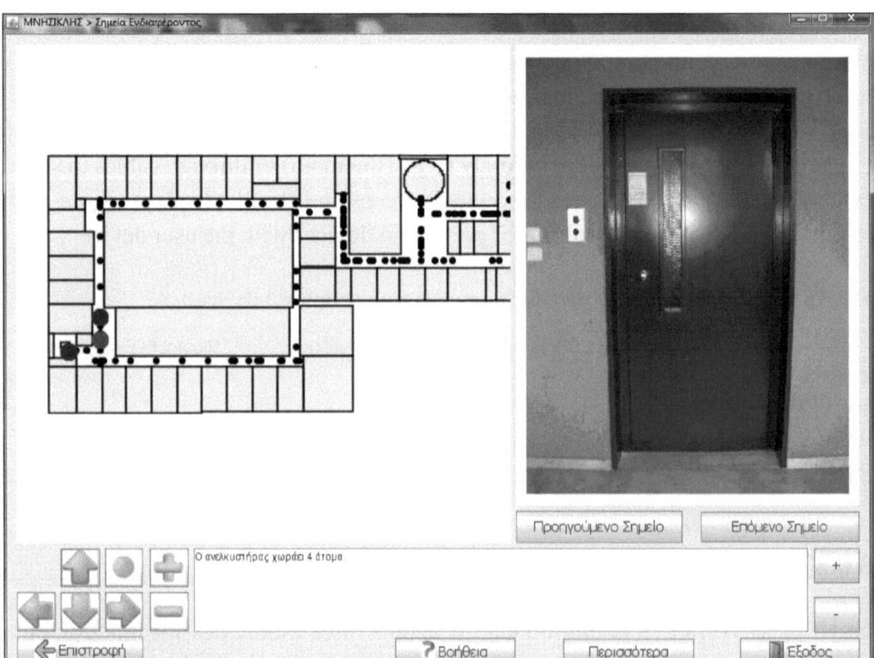

Fig. 16.7. Screenshot of the graphical UI.

Scalable Vector Graphics (SVG) format and can be transposed and zoomed-in and -out by the user without any loss of detail.

Audio Modality: A person without hearing loss is able to obtain audio output in the form of synthetic speech or in combination with the visual output. The design of the GUI ensures support of traditional Interactive Voice Response (IVR) mode. For the case of the blind users full spoken-based dialogue interaction (Speech-only User Interface) is supported (Freitas & Kouroupetroglou 2008).

Haptic Modality: In the case of haptic interaction, the user makes use of her hands to provide input (touch screen, buttons) or/and obtain output (Braille display). The output on the Braille display is just the same as if the user was using audio output.

16.6.4 Architecture of User Interaction Subsystem

The User Interaction subsystem of MNISIKLIS adopts a client/server architecture. The client part is installed on the users' terminal. It constitutes a cross-platform software component compliant with a variety of devices. Specifically, the client was installed on a Tablet PC, a PDA, a Smart Phone and a speech interaction server. The server side of the subsystem, on the other hand, receives the requests of the client and, in collaboration with the subsystems of Positioning Services, Content Management and GIS, composes and returns the output to the user, regardless of the device used. The architecture of the User Interaction subsystem is depicted in *Figure 16.8.*

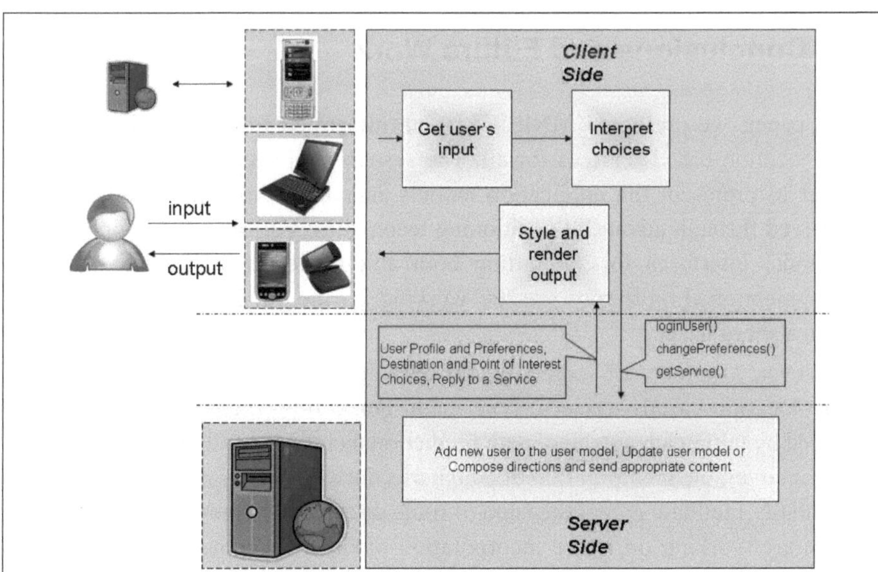

Fig. 16.8. User Interaction Subsystem Architecture.

The client is developed using Java and other open source technologies. Java was selected because of its interoperability and platform neutrality. Moreover, there are various tools available to support the technologies integrated in MNISIKLIS.

Maps are visualized through SVG and are displayed using the Batik open-source framework (from Apache Foundation). This caters for efficient map handling (zoom, rotate, drawing of POIs and paths etc). On the PDA, the Tinyline SVG toolkit is used to perform the same tasks.

16.6.5 Evaluation and Field Trials

Currently, a number of trials with real users are underway. Users belong to all target groups addressed by MNISIKLIS. The field tests take place in the building of Department of Informatics and Telecommunications – National and Kapodistrian University of Athens and a university museum. Users are asked to select/execute a certain service (for example, follow the instructions of navigation). All the users' choices, significant intermediate results, outputs and positions are logged for post-processing and analysis. Furthermore the users are asked to change their profile, preferences and selected POIs, resulting in changes to UNO and INO ontologies. Finally, after having used the system, the users are asked to fill in some questionnaires. The results of these tests are expected to bring forward directions for further research and improvement of MNISIKLIS.

16.7 Conclusions and Future Work

In this paper, we presented MNISIKLIS, an integrated framework that provides indoor location-based services for all. The system exploits semantic technologies in order to represent the application models and the estimation of user position is achieved through advanced positioning techniques. Furthermore, it implements multimodal interfaces for supporting both able-bodied and disabled users. As already mentioned, currently, we are working on the validation of MNISIKLIS through a set of trials.

However, a number of issues remain open for further research in the area of indoor location-based services. The path generation methodology could be enhanced by taking advantage of path prediction techniques and historical information. Moreover, the incorporation of landmarks during the navigation process could substantially facilitate comprehension of route directions by the user. Another issue that we are working on is the incorporation of other technologies and elements (infrared beacons, ultrasound sensors) in the positioning system, thus enhancing its accuracy and robustness. Furthermore, different methods and techniques (e.g.,

Evidential Reasoning) will be examined in order to improve the results of the second level fusion. Additionally, Kalman and particle filtering will be utilized for improving the dead-reckoning and the sensor fusion position estimations (Evennou & Marx 2006), as well.

Acknowledgements

This work has been partially funded by the Greek General Secretariat for Research and Technology (GSRT) under grant PABET 2005 – Project Code: 282.

References

Aarts E, Harwig R, Schuurmans M (2001) Ambient intelligence. In *The Invisible Future: The Seamless Integration of Technology into Everyday Life*, McGraw-Hill Professional.

Baus J, Krüger A, Wahlster W (2002) A Resource – Adaptive Mobile Navigation System, *International Conference on Intelligent User Interfaces IUI02*, San Francisco, USA.

Bikakis A, Patkos T, Antoniou G, Papadopouli M, Plexousakis D (2006) A Semantic-based Framework for Context-aware Pedestrian Guiding Services, *2nd International Workshop on Semantic Web Technology For Ubiquitous and Mobile Applications (SWUMA)*, Riva del Garda, Trentino, Italy.

Boiko B (2005) *Content Management Bible*, 2nd Edition, Wiley Publishing Inc., 2005.

Dippold M (2006) Personal Dead Reckoning with Accelerometers, *Third International Forum on Applied Wearable Computing, IFAWC2006*, Bremen, Germany.

Duckham M, Kulik L (2003) "Simplest" Paths: Automated route selection for navigation, in: W. Kuhn, M.F. Worboys, S. Timpf (Eds.), Spatial Information Theory: Foundations of Geographic Information Science, Lecture Notes in Computer Science, vol. 2825, Springer, Berlin, 2003, pp. 182–199.

Evennou F, Marx F (2006) Advanced Integration ofWiFi and Inertial Navigation Systems for Indoor Mobile Positioning, EURASIP Journal on Applied Signal Processing, Volume 2006, Article ID 86706, Pages 1–11.

Fang L, Antsaklis PJ, Montestruque LA, McMickell MB, Lemmon M, Sun Y, Fang H, Koutroulis I, Haenggi M, Xie M, Xie X (2005) Design of a Wireless Assisted Pedestrian Dead Reckoning System – The NavMote Experience, IEEE Transactions on Instrumentation and Measurement, vol. 54, pp. 2342–2358.

Fellbaum K, Kouroupetroglou G (2008) Principles of Electronic Speech Processing with Applications for People with Disabilities, J. Technology a Disability, Vol 20, No 2, 55–85.

Freitas D, Kouroupetroglou G (2008) Speech Technologies for Blind and Low Vision Persons, Journal *Technology and Disability*, Vol. 20, No 2, pp. 135–156.

Gartner G, Frank A, Retscher G (2004) Pedestrian Navigation System for Mixed Indoor/ Outdoor Environments – The NAVIO Project. In: Schrenk, M. (Ed.): Proceedings of the *CORP 2004 and Geomultimedia04 Symposium* (pp. 165–171). Vienna, Austria.

Horrocks I, Patel-Schneider PF, Boley H, Tabet S, Grosof B, Dean M (2004) SWRL: A Semantic Web Rule Language Combining OWL and RuleML W3C Member Submission. [Online]. Available from http://www.w3.org/Submission/SWRL/ [6th June 2008].

ICF, International Classification of Functioning, Disability and Health (2001) World Health Organization, Geneva. http://www3.who.int/icf/icftemplate.cfm [6th June 2008].

INO (Indoor Navigation Ontology) (2008) [Online]. Available from http://p-comp.di.uoa.gr/ont/INO.owl [Accessed: 6th June 2008].

Kargl F, Bernauer A (2005) The compass location system. In: Strang T, Linnhoff-Popien C (eds.) *LoCA 2005. LNCS*, vol. 3479, Springer, Heidelberg.

Kargl F, Gessler S, Flerlage F (2007) The iNAV Indoor Navigation System, *Lecture Notes In Computer Science*, Springer.

Kikiras P, Tsetsos V, Hadjiefthymiades S (2006) Ontology-based User Modeling for Pedestrian Navigation Systems, *ECAI 2006 Workshop on Ubiquitous User Modeling (UbiqUM)*, Riva del Garda, Italy.

Koide S, Kato M (2006) 3-D Human Navigation System with Consideration of Neighboring Space Information, *2006 IEEE International Conference on Systems, Man, and Cybernetics*, (pp. 1693-11698), October 8–11, 2006, Taipei, Taiwan.

Kolomvatsos K, Papataxiarhis V, Tsetsos V (2007) Semantic Location Based Services for Smart Spaces, in Proc. of the *2nd International Conference on Metadata and Semantics Research (MTSR)*, Corfu, Greece.

Ladetto Q, Gabaglio V, Merminod B, Terrier P, Schutz Y (2000) Human Walking Analysis Assisted by DGPS, GNSS, Ediburgh, 1–4 May.

Lyardet F, Grimmer J, Muhlhauser M (2006) CoINS: Context Sensitive Indoor Navigation System, Proceedings of the *Eighth IEEE International Symposium on Multimedia*, pp. 209–218, 2006, IEEE Computer Society Washington, DC, USA.

OWL (Web Ontology Language) Guide (2004) [Online]. Available from http://www.w3.org/TR/owl-guide [Accessed: 6th June 2008].

Schiller J, Voisard A (2004) *Location-Based Services*, Morgan-Kaufmann Publications.

Sekkas O, Hadjiefthymiades S, Zervas E (2006) Enhancing location estimation through data fusion, in Proceedings of the *17th IEEE International Symposium on Personal, Indoor and Mobile Radio Communications (PIMRC '06)*, pp.1–5, Helsinki, Finland.

Stephanidis C (2001) *User Interfaces for All – Concepts, Methods, and Tools*, Mahwah, NJ: Lawrence Erlbaum Associates.

Tsetsos V, Anagnostopoulos C, Kikiras P, Hadjieftymiades S (2006) Semantically Enriched Navigation for Indoor Environments, *International Journal of Web and Grid Services (IJWGS)*, Vol.4, No.2,Inderscience Publ.

Wasinger R, Stahl C, Krüger A (2003) M3I in a Pedestrian Navigation & Exploration System, Proc. of the *Fourth International Symposium on Human Computer Interaction with Mobile Devices*, pp. 481–485, Pisa, Italy.

Weinberg H (2008) Using the ADXL202 in Pedometer and Personal Navigation Applications, *Application Notes AN-602, Analog Devices*, [Online]. Available from http://www.analog.com/UploadedFiles/Application_Notes/513772624AN602.pdf , [6th June 2008].

XMLBIF, The Interchange Format for Bayesian Networks (1998) Available from http://www.cs.cmu.edu/afs/cs/user/fgcozman/www/Research/InterchangeFormat/ [6th June 2008].

17 geoHealth Monitoring – Real-Time Monitoring for Action Forces During Disaster Operations

Michael Lippautz, Manuel Kohlmann, Christian Sternig, Irene Piralli, Renate Rauchenschwandtner, Peter Haber and Simon Kranzer

Information Technology and Systems Management
University of Applied Sciences Salzburg, Austria

Abstract

This paper presents a new type of monitoring environment for action forces during any indoor or outdoor mission to increase efficiency and manageability. The major benefit of the system is to improve short time planning of staff commitment so that staff can be retrieved or exchanged as required by the circumstances. The system itself consists of two main components: i) Data collection via an embedded device and a web server involving and standardized data transmission (SensorML, Observations) to a web server ii) Data visualization on a central or mobile device (computer or mobile phone).

This paper illustrates the system by describing how two essential vital signs – pulse rate and oxygen saturation – are used to monitor action force members under adverse conditions and how this data is graphically displayed for the control unit staff so that they can immediately intervene as necessary. The components of the system and their interaction are explained in detail as well as the implementation and the results of a first test phase. The innovative approach of this specific system lies in the combination of the user-friendly data display based on the collected data in near real-time and the geographical location information. An additional advantage is the flexibility and scalability of the system as it runs on a variety of devices both mobile and fixed.

Keywords: monitoring environment, geographic information, SensorML, soft real time system, command and control client

17.1 Introduction

Due to recent developments in embedded systems in general, systems for monitoring and improving the security of endangered occupational groups have gained importance. In the context of the project *geoHealth Monitoring* at the Salzburg University of Applied Sciences a completely new system has been designed to support action forces, such as fire brigades and rescue teams during difficult operations. The system which is now working efficiently using a soft real-time operating system can document not only the history of vital signs over a longer period of time but also display the location of each action force member. The advantages of the system are the fast processing of the data, the standardized communication between the components with SensorML (URL 12), and the clearly structured layout of the command and control unit. Application planning can be done more conveniently as described in detail in the following chapters. (URL 12)

17.1.1 System Description

The system can be split into three separate parts (see *Figure 17.1)*. First there is the input of the data, which is done by the sensor and the processing of the information using the Gumstix board (1).

Second a merging point stores all the data providing access for all the other components (2). Finally a visualization and analysis interface is provided through a combination of a Microsoft Silverlight implementation combined with Microsoft Maps (3). (URL 13)

During a mission, data is read in by a sensor – in this project a finger respectively an ear clip sensor has been used, which could, however, easily be replaced

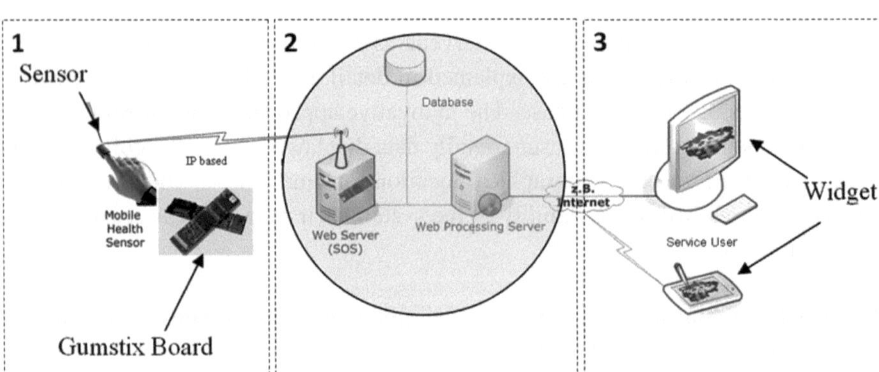

Fig. 17.1. System Overview.

by any other sensor type. Specific vital signs are measured continuously, in this context pulse rate and oxygen saturation were used. The sensor is connected via a serial interface to an embedded device where the data is then processed. This embedded device (Gumstix board) provides a WLAN and a GPRS interface, which enables a direct, standardized and less time consuming data communication with the server. Here the data is stored in a database. Then the third component of the system comes into play the so called command and control client, realized as a widget. This unit accesses the data and rehashes it and visualizes it in an appropriate and user-friendly manner. This widget is either executed by a mobile device or a central PC. Any member of the operational command center is now able to react on any critical incident such as an extremely low value of oxygen saturation or excessive pulse.

The innovative approach in the interaction of the different system components provides a sophisticated monitoring environment that combines vital sign alerts with location information. The following chapters describe the three subsystems in more detail.

17.2 Intra System Communication

The system uses different ways of communication through three interacting units. The first unit is the mobile health sensor device. This portable sensor is connected with the mobile embedded device by a serial interface, for example, an RS232 cable. The second unit is the web processing server, which processes the received data and stores it. This server is the merging point of all sensor data. Between these two units data is transferred via W-LAN or GPRS. Another option for data transfer across short distances would be Bluetooth, which can easily be installed by exchanging one of both additional boards. In the near future UMTS as transfer medium has been considered.

The third unit constitutes any device used by the command and control action forces. Such devices can be static or portable, depending on their purpose and operation field and can be a PC, PDA or mobile phone. The internet is used for data transmission between the web processing server and command and control unit.

The protocol used partly depends on the transfer medium. The whole transfer chain is based on IP, whereas the transported data is packed into the container protocol SensorML (URL 5) (URL 12), standardized by the Office of Geospatial Intelligence Management (OGM). At first the sensor sends the received data to the embedded device, from where it is sent to the Web Processing Server via this protocol. Then the third unit receives the vital sign data, wrapped in this SensorML, parses the vital sign data and visualizes the received information.

17.3 Sensor

The sensor is responsible for the collection of the health parameters, pulse and oxygen saturation of a participant. With the help of these two parameters a basic medical diagnosis about the current health status of a person can be made.

As the selection of a suitable sensor for the measurement of pulse and SO2 saturation was an important decision, 6 suitable developing kits and sensors were evaluated (URL 1, URL 2, URL 3, URL 4). The evaluation criteria for the selection were the interfaces, the functions, the flexibility and the price. A prerequisite was also that there should be an I2C or RS232 interface on the sensor to transfer the data to the LINUX board and from there as standardized SensorML combined with GPS information to a server or database for further processing.

After a comparison of the advantages and disadvantages of the two sensors that fulfilled all criteria, Nonin OEM III and Nonin Xpod, the Nonin OEM III developing kit with 8000Q ear clip and 8000SM finger clip sensor was chosen.

The Nonin OEM III Module provides a simple way to incorporate exceptional pulse oxygen. Its compact design, power-efficiency and expanded output capabilities provide maximum flexibility in a small module. With the help of the finger clip and ear sensor, measurement can be done in two different ways.

To provide an optimal and standardized output of the measured health parameters of a participant in SensorML conform language (URL 5), several output format options must be considered and assessed.

17.3.1 Serial Output Formatting Options of Nonin OEM III

The Nonin OEM III developing kit provides 3 different output formats, which can be set using a onboard switch with several pins (URL 1).

The format for serial output data is determined by the amount of resistance present between the serial data format switch (J1, pin 9) and ground (J1, pin 15).

The selected serial transmission rate for all data formats is as follows:

Table 17.1. Serial transmission rate.

Bits per Second	Data Bits	Parity	Stop Bits	Flow Control
9600	8	None	1	None

The serial interface is configured with these parameters, so that the transmission from the sensor to the serial interface works properly.

It was decided to use output format #1, because it was sufficient to get the output information of 3 bytes once a second.

So it is possible to get measured values of pulse and oxygen every second. This is important to guarantee data in a soft real time.

Table 17.2. Description of the data format. (URL 1)

Byte 1 – Status							
BIT7	**BIT6**	**BIT5**	**BIT4**	**BIT3**	**BIT2**	**BIT1**	**BIT0**
1	SNSD	OOT	LPRF	MPRF	ARTF	HR8	HR7

* Note: Bit7 is always set

Byte 2 – Heart Rate							
BIT7	**BIT6**	**BIT5**	**BIT4**	**BIT3**	**BIT2**	**BIT1**	**BIT0**
0	HR6	HR5	HR4	HR3	HR2	HR1	HR0

* Note: Bit7 is always clear

Byte 3 – SpO2							
BIT7	**BIT6**	**BIT5**	**BIT4**	**BIT3**	**BIT2**	**BIT1**	**BIT0**
0	SP6	SP5	SP4	SP3	SP2	SP1	SP0

* Note: Bit7 is always clear

Explanation of bit states (see *Table 17.3):*

Table 17.3. Error Messages. (URL 1)

SNSD	Sensor Disconnect	Sensor is not connected to oximeter or sensor is inoperable.
OOT	Out of Track	An absence of consecutive good signals.
LPRF	Low Perfusion	Amplitude representation of low signal quality (holds for entire duration).
MPRF	Marginal Perfusion	Amplitude representation of medium signal quality (holds for entire duration).
ARTF	Artifact	A detected pulse beat didn't match the current pulse interval.
HR8–HR0	Heart Rate	Standard 4-beat average values not including display holds.
SP6–SP0	SpO_2	Standard 4-beat average values not including display holds.

With the help of this output information, a program in C (see *Code Snippet 17.1)* has been designed, that converts the sensor data into readable information for pulse and oxygen saturation.

```
#include <stdio.h>
int main(int argc, char* argv){
    FILE* port;
    unsigned char data(URL 5);
    int i;
    short hr=0;
    port = fopen("/dev/ttyS0","r");
    if(port == NULL) { printf("Couldn't open serial interface");
        return 1; }
while(1) {
    fread(data,1,5,port);
    hr=0;
    for(i=0;i<5;i++){
        if(data[i] > 127){
            if( (data[i] & 0x40) != 0) printf("Sensor disconnected!\n");
```

```
            if( (data[i] & 0x20) != 0) printf("Sensor Out of Track!\n");
            if( (data[i] & 0x04) != 0) printf("Artefakt!\n");
            hr=(short)data[i+1];
            if( (data[i] & 0x01) != 0)
             hr+=128;
            if( (data[i] & 0x02) != 0)
             hr+=256;
            if( (data[i] & 0x64) == 0) {
                printf("Sp02: %d\n",data[i+2]);
                printf("Pulse: %d\n",data[i+1]); }
                printf("\n");
                break;
            } //end if
    } //end for
} //end while
    fclose(port);
    return 0;
}
```

Code Snippet 17.1.

If a person is now connected to the ear clip or finger clip sensor, the measured data is transmitted from the OEM III developing kit to the serial interface of a computer or a real time LINUX Board where the information is displayed in this form:

Pulse: 95

SpO2: 90%

In combination with the OEM III module this program can be used on every Linux based operation system and works absolutely reliably with every kind of sensor. Once a second the parameters of the current pulse and SpO2 of a participant are available.

17.4 Embedded System

As the project required a wearable health sensor device, it was necessary to reduce the size of the hardware as much as possible, to ensure ease of use and support of various communication standards. For that reason, the Gumstix system was chosen because it is Linux based, small, powerful and pliable through various expansion boards.

Furthermore the Gumstix system is modular so it is for example possible to replace the medical SpO2 sensor with a temperature sensor without the necessity to rebuild the whole system. Another big advantage is the ability to create customized firmware images through the open-source build root environment with the help of an external PC as host-system.

17.4.1 Real Time Operating System

The project has to compute important values of vital signs. Therefore a big challenge was to implement a Real-time Operating System. For that reason the hardware has to maintain specific CPU processing deadlines. While an RT patch for the Linux vanilla kernel is available an attempt to patch this kernel to the adapted Gumstix 2.6.21 kernel failed.

The intended implementation of real time linux on the Gumstix board was deferred as the ARM kernel would only be supported with next update as mentioned on Real time Linux Road Map at osadl.org. (URL 14)

Fig. 17.2. Gumstix Boards. (URL 7)

The table below points out the most important features of the board used.

Table 17.4. Board Features. (URL 6)

Processor Marvell® PXA270 with XScale™ **Speed** 400MHz **Memory** 64MB RAM 16MB Flash **Features** – Bluetooth(TM) communications (includes u.fl antenna @ 2.4 GHz) – USB host signals – CCD camera signals **Connections** 60-pin Hirose I/O connector 120-pin MOLEX connector 24-pin flex ribbon	**Features** – GPS – Audio in/out – LCD – USB client – 4 GPIO **Connection** 60-pin I/O header	**Features** – 10/100baseT Ethernet * – microSD storage capabilities** – 802.11(b) and 802.11(g) wireless communications*** **Connection** 120-pin bus header

The current device used is the Gumstix Verdex XM4-bt board with 400 MHz ARM CPU and two additional expansion boards. The main expansion board is the „GPSstix" which is responsible for receiving and processing GPS signals through the u-blox LEA-4H receiving module. As second expansion board the 'netwifimicroSD-EU' is used to provide Ethernet and WLAN-access as well as a microSD card interface to expand the built-in flash memory.

An assembled Gumstix system consists of the motherboard and two expansion boards. The expansion boards have either a 60-pin Hirose or a 120-pin MOLEX connector, which limits the combination possibilities of expansion boards.

The GPSstix is needed in every case but for international usage it is feasible to replace the netwifimicroSD EU by a netwifimicroSD FCC to provide WLAN access in the USA.

17.4.2 Firmware

The Gumstix system uses a Linux based firmware with several kernel patches provided by Gumstix Inc. So it is possible to create ones modified firmware images to ensure future compatibility and modularity. For building firmware images, it is necessary to use the build root environment which is also open source and easy to use. Build root gives users the ability to modify the kernel and the software packages to integrate the firmware images as they are used in common Linux distributions. So the learning curve is very flat and it is possible to produce stable firmware images within a short period of time. (URL 15)

Figure 17.3 shows the software architecture of the project. The two main daemons, the GPS daemon (GPSd) and the Sensor daemon (sensord), are responsible for processing the raw data into one standardized XML file. This XML file is made available through a web server for remote access or can be transmitted to the central web server via the sensord itself and stored there.

The current image has all used hardware drivers compiled direct into the kernel to avoid potential module loading errors. The GPSd, the sensord and the web server are started automatically. To ensure reach ability, a running DCHP client for the Ethernet interface is installed. So using the system does not need any specific computer knowledge.

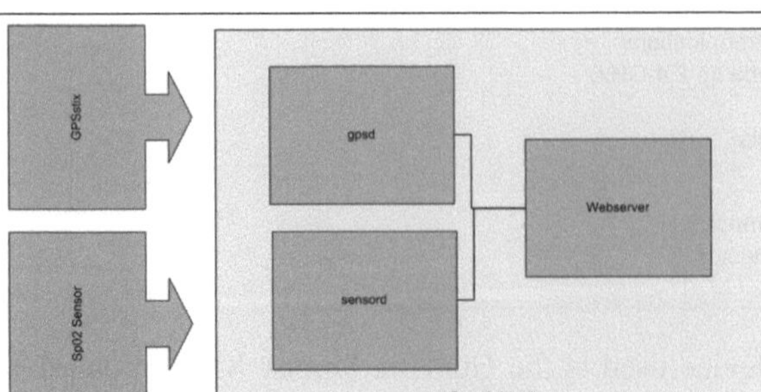

Fig. 17.3. Software Architecture.

17.5 Command and Control Client

As stated above the main purpose of this project was to observe vital signs of persons under pressure and exposed to adverse physical influences. This part of the paper deals with the monitoring of these people, commonly referred to as the command and control client or *geoHealth Monitor*.

To suit the actual needs, as specified below, this software application was built as widget, which is a browser based program that can be executed in various different browsers. (URL 13)

17.5.1 Purpose

This widget mainly serves the following purposes:

- Providing *visualization* of the SensorML standard (URL 5, URL 12).
- Being *as portable/mobile* and
- As *compatible* with other media as possible.

The visualization part uses the SensorML standard as data source and builds up data graphs to display the information included in the monitoring. The monitoring standard provides different information types. Firstly it includes data with a timestamp. Data can be anything that can be measured by a sensor. In the example below information about the pulse of a person is transmitted with the standard. As the sensors used are usually connected to digital signal processing boards, they are capable of attaching GPS information. Again these values can be found in the SensorML standard and therefore require some form of visualization. Important values have been highlighted in the XML monitoring as exemplified in *Code Snippet 17.2.*

```
<?xml version="1.0" encoding="utf-8"?>
<om:Observation xmlns:gml=http://www.opengis.net/gml
                xmlns:om="http://www.opengis.net/om/0.0"
                xmlns:swe=http://www.opengis.net/swe/0.0
                xmlns:xlink="http://www.w3.org/1999/xlink"
                xmlns:xsi="http://www.w3.org/2001/XMLSchema-instance"
                gml:id="VitalSensor1">
<gml:name>urn:iSPACE:observation: VitalSensor1</gml:name>
    ...
    <swe:field name="time">
        <swe:Time definition="urn:ogc:phenomenon:time:iso8601" />
    </swe:field>
    <swe:field name="longitude">
        <swe:Quantity definition="urn:ogc:phenomenon:location:EPSG:4326:longitude">
    <swe:uom code="deg" />
    </swe:Quantity>
    </swe:field>

    <swe:field name="latitude">
```

Code Snippet 17.2.

17.5.2 Global Architecture

The system is based on a web processing server, which basically serves the website holding the visualization client. There are no specific requirements for the software since the whole software client can run as a client side browser plug-in.

The website layout is based on XHTML 1.0 and therefore provides the best possible compatibility for most browsers. The content of the site, which is the visualization monitor, is basically divided into two major parts, a graphical user interface as Microsoft Silverlight browser plug-in and a Virtual Earth Map plug-in.

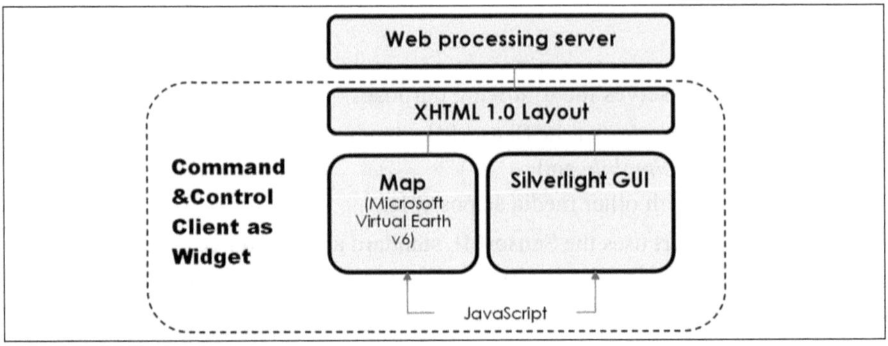

Fig. 17.4. Global Architecture.

As the purpose was to maximize the compatibility between browsers and therefore the Silverlight application is embedded as a self-identifying object. Due to the fact that there is currently no user control that provides mapping services implemented in Silverlight, the map is embedded as an additional part in the website.

Prerequisite for the Microsoft Silverlight plug-in and the Microsoft Virtual Earth plug-in is a supported browser.[1] Additionally JavaScript is needed and has to be enabled.

Basically any map such as Google Maps, Yahoo Maps could be used, but in this case Microsoft Virtual Earth was selected to minimize the number of different frameworks used. As a result, a communication bridge was needed to make interaction with the mapping service and the visualization tool possible. The bridge is written in JavaScript since this is the only language that is supported by both, the Virtual Earth map and the Silverlight browser plug-in.

[1] Silverlight supported browsers: http://www.microsoft.com/silverlight/overview/faq.aspx
Virtual Earth supported browsers: http://www.microsoft.com/virtualearth/product/faq.aspx

17.5.3 Silverlight GUI Architecture

The architecture of the Silverlight visualization plug-in can be split into three parts as shown in *Figure 17.5:*

- Initialization
- Time-based actions
- Event-based actions

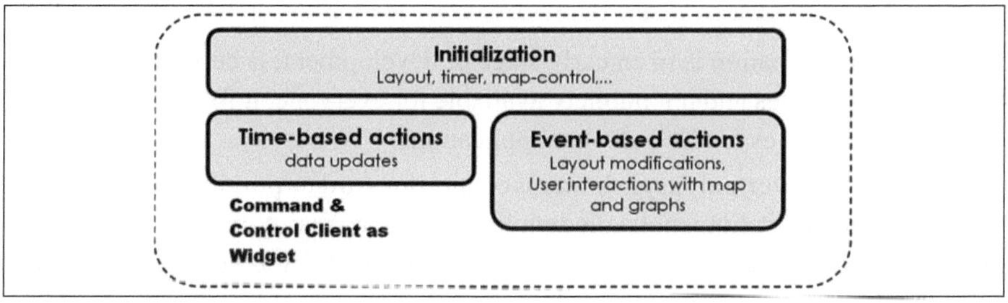

Fig. 17.5. Silverlight Plug-in Architecture.

17.5.4 Initialization

The initialization process first generates the software application layout and starts the timer for the time-based update actions. Secondly it also sets up the options for the map plug-in using the JavaScript communication bridge. Although the process itself is quite fast, the first time the application starts, the whole package has to be downloaded and therefore the first software start up can take some time. In order to give the user feedback about the process, a loading progress bar has been added. After this initialization, no more loading is needed.

17.5.5 Time-Based Actions

After the timer is started by the initialization process, the command and control client periodically fetches the data XML files, if a client is selected. This timer is important as it regulates the update frequency of the data. A timer set for every 10 minutes or even rarely, is probably suitable for offline monitoring, whereas a timer set every 5 seconds generates nearly real time monitoring which perfectly meets the conditions of this project. As stated above the monitoring XML files are organized as SensorML (URL 12) standard containing a timestamp, data values, and the corresponding latitudes and longitudes. If no client is selected, the timer has no effect and the application is in an idle state. The resources needed while executing the update process, heavily depend on the amount of data that is parsed by an event that handles the storing of the information inside the widget.

17.5.6 Event-Based Actions

Event-based actions are responsible for user interaction with the application. These interactions include the setting of options, the interaction with the data graphs, but also map customizations.

17.5.7 Robustness

Although the application is in an early stage of development, is currently only used for research purposes and not publicly available for everyone, it follows some basic robustness guidelines for C# software applications.

Currently, however, only asynchronous connections to data providers are used. This ensures that the application is not stalled while loading new data into the internal database. This internal database is an object structure (C# objects) that holds the information while the program is running. (URL 10)

17.5.8 Layout

The layout results from the following requirements recommended by the customer, including the need for:

- A visualization of certain parameters in form of a line graph.
- An area containing information about the geographic location of the measured data.
- The space for specifying certain options to customize the whole application.

Due to the fact that most modern displays have wide screen resolutions, the areas are arranged horizontally.

In order to provide the best flexibility the layout of the software consists of one workspace (2), a map container (3) and a container allowing some customization with option fields (1).

The map container and the customization pane can be hidden to enlarge the workspace. The workspace is organized as a whiteboard with several windows containing all the data graphs. Each window has the possibility to host different types of plug-ins. Currently, only a graph and a text plug-in are included. While the text plug-in can be used to show different types of messages, the graph plug-in is used to visualize the information. Layout customization to the layout can be made by editing a configuration XML file to be found under /ContentResources/config.xml. Adjustments include colors used in the plug-ins as well as icons being displayed in each plug-in window. Additionally, the whiteboard can also display debugging information which can be used to track errors and application failures.

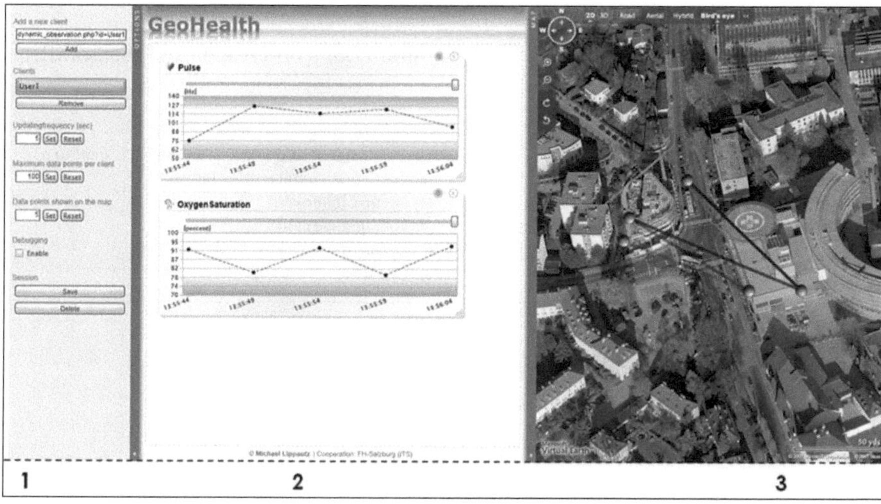

Fig. 17.6. Layout/Design of the *geoHealth-Monitor.*

17.5.9 Features

In order to focus on the different parts of the application, the features have been divided into different classifications. (URL 13)

17.5.10 Basic Graph Features

As stated above each window can host different plug-ins. Usually this will be the graph plug-in to display the data graphs, which have been fetched in the background. The data graphs are displayed as x/y graphs and adopt their scaling to the data visualized. The abscissa holds the time information and is self-organized by the application. Depending on the number of points that should be displayed, the time values are only scaled down and then printed on the axis. The ordinate holds the scaling of the data depending on its type. The ranges of the specific information types are stored in an extra XML file that is used for configuring the application. Only information types that have a valid entry in this file can be displayed. The types that are needed for the configuration are identical with the types used in the SensorML (URL 12) observation file.

Additionally, each graph can display a range of critical values by adding a colored gradient that indicates non standard values. Furthermore the graph window can be resized by the user. The scaling of the axis is then adjusted dynamically by the application.

Example:

Table 17.5. confix.xml

```
<Phenomenon id="Pulse">
 <MinValue>50</MinValue>
 <MaxValue>140</MaxValue>
 <UpperCrit>80</UpperCrit>
 <LowerCrit>120</LowerCrit>
```

Table 17.6: observation.xml (SensorML)

```
...
<swe:field name="Pulse">
        <swe:Quantity       definition="urn:x-
ogc:def:phenomenon:OGC:Pulse">
 <swe:uom code="Hz" />
 </swe:Quantity>
</swe:field>
```

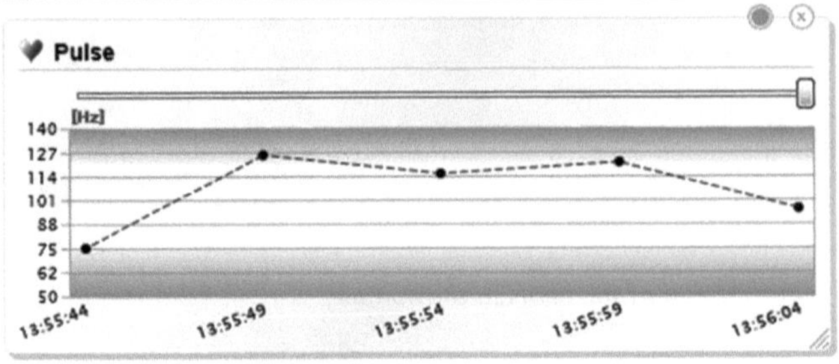

Fig. 17.7. Example for pulse monitoring.

In the above example the ordinate holds the scales from the config.xml *(Table 17.4)* file which values are ranging from a minimum of 50 to a maximum of 140. The unit of the measurement is defined in the observation.xml *(Table 17.5)* file and is Hertz. The critical areas are marked as color gradients and can be found in the config.xml file. The abscissa is self organized and displays the time information. At the top of the graph plug-in, a slider is visible which makes navigating to past values possible. Additionally, a heading is placed in the upper corner. This heading basically is the last part of the swe:Quantity definition of the observation. As regards customizability, an icon and a base color can also be defined in the configuration.

17.5.11 Customizability/Options

This section holds options that can be used to adjust the displaying of the data graphs. Despite these customizing possibilities, there are also options that adopt the update timers. As a result, the application can be adjusted to be more real time for important data and less real time for data that only has a low alteration rate. Concerning graph adoption, options such as "maximum points per client" and "points shown on the map" have been added.

Additionally the graph plug-in includes a history option, where the past data points are stored, and if required displayed. This option can be used to analyze the gradients after an observation has been made to determine possibly wrong decisions when planning a route of evacuation or operational areas. This makes the program being usable for real time observations, but also for the analysis after an observation incident has been completed.

17.5.12 Map Plug-In

A map service is the second displaying feature besides the graph visualization. This service displays two categories of information.

First it shows a route of the points where the observing person has been. The data is again fetched from the observation file. The route is not only displayed but additionally a colored gradient has been added above it to indicate the movement direction.

Second the information data (pulse, oxygen saturation) can be shown by moving the mouse cursor over the markers on the map.

17.5.13 Map-GUI-Interaction

This section covers features that deal with map to GUI interaction or vice versa. *geoHealth Monitor* offers some highlighting features such as data point highlighting and map focus. Whenever a data point is selected in the graph visualization, the corresponding point is highlighted in the map plug-in. Conversely whenever a user selects a point marker in the map, the graph plug-in not only focuses on the point, but also highlights it.

Again the customizability of the widget offers the possibility to include alerting in different colors. This can be used to indicate where the possible vital failure has happened.

17.5.14 Real Time Relevance

In terms of exact traditional definitions the application can neither be classified as hard real time, nor can it be classified as soft real time. The reason for this is that the components used cannot guarantee any processing times and execution delays. In order to guarantee various delays one would have to build a field test and check the various response times of the different components.

Components to be tested would be the protocol of the sensor which fetches the basic data, the communication between the sensor and the embedded system, but also the internet as the transport medium and the widget as the visualization tool.

Additionally, one would have to bear in mind that not only the transport medium but also the operating systems and used software have to fulfill several criteria regarding execution delays.

Despite this fact the command and control client can be seen as a tool for visualizing geographic vital data in near real time. Every calculation which has to be made is performed during usual execution of the applications. The result is that it can be used to track data with a delay of some seconds. The exact determination of the delay is not possible due to the implementation in a high level programming language.

Another feature which usually is typical of real time applications is that there is only one loading process which interrupts the normal usage behavior. This is the initial start up process. After the start up process every data update is run in the background without interrupting the observing user.

17.6 Advantages / Disadvantages

Overall, the whole system is implemented as a lightweight core system containing the Gumstix board and is easily expandable with different sensors and network technologies, such as Bluetooth, GPRS, WLAN and UMTS. Another benefit is the low power consumption of 2.5W with active and 1.5W with non active WiFi.

Furthermore, the standardized SensorML and the system's modularity guarantee compatibility with every type of hardware. Another advantage is the autonomy of the components which ensures the easy setup of the infrastructure in any environment. Furthermore the command and control client provides an overview of the geographical position and the health condition of the action force members. The system can be managed and controlled easily form a geographically independent command and control centre.

The major disadvantage is that it is not possible to display a vast amount of data points in the map area, as its size is limited. Currently the options limit the amount of data points to a value of five, which could, however, be expanded to ten. Any further increase would not be feasible, as the map would be overcrowded with data points.

Additionally, there is another aspect that limits the information content displayed in a map. Points with a great distance between them, for instance sensor with high travel speed, force the map to increase the scale value and make exact positioning impossible. The longer the geographic distance between the values, the greater the loss of accuracy. The same problem occurs when the sensor measures values not in line with the expected trend of the vital sign parameter. In the case of such outliers the visualization of the data loses accuracy until the data points are out of the scope.

17.7 Conclusions

Within the scope of the project *geoHealth Monitoring* the design and implementation of a completely new system to visualize geographical and vital signs within a map have been carried out. The innovation of this project does not lie in a single component, but in the cooperation of individual components within the whole system. As a result of the standardization of the single modules and their interconnections, a high degree of flexibility and modularity is guaranteed. The currently used recording device can easily be replaced due to the usage of the SensorML standard. The low power-consuming data processing is an additional advantage of the Gumstix board besides its small size, the standardized interfaces as well as the real time functionality. Thus the system developed in this project could not only meet all original requirements, but additionally a highly dynamic data visualization component, executable in every browser – the command and control client – has been developed as a combination of Microsoft Silverlight and Microsoft Maps. The whole system is now fully functional and ready for implementation but has not been tested under real life conditions. A large scale field test would therefore provide an opportunity to potentially fine-tune the system.

References

URL1: Nonin (2008) OEM III Specifications, viewed 15 January 2008 http://www.nonin.com/documents/OEM%20III%20Module%20Specifications.pdf

URL2: Polar (2007) Product Specifications, viewed 10 November 2007 http://www.polar-austria.at

URL3: Guger Technologies (1999–2007) Product Specifications, viewed 12 November 2007 http://www.gtec.at

URL4: BioMedical Instruments Inc., Product Specifications, viewed 10 November 2007

URL5: Open Geospatial Consortium, SensorML, viewed 27 June 2008 http://www.opengeospatial.org/standards/sensorml

URL6: Gumstix Inc (2004–2008) Motherboard Specifications, viewed 17 June 2008 http://Gumstix.net/wiki/index.php?title=Motherboard_specifications

URL7: Gumstix Inc (2004–2008) Product Image, viewed 3 February 2008 http://Gumstix.com/store/catalog/index.php?cPath=27&osCsid =d52de4bf688ac7c5a18f5b7da7f4564d

URL8: Microsoft (2008) Working with XML Data in Silverlight, documentation release, 2008, viewed 3 February 2008 http://visualstudiomagazine.com/columns/article.aspx?editorialsid=2085

URL9: Wagner B (2007) Write Robust Exception-Handling Code, media release, June 2007, viewed 1 January 2008 http://visualstudiomagazine.com/columns/article.aspx?editorialsid=2085

URL 10: Microsoft (2008) Accessing Web Services in Silverlight, documentation release, 2008, viewed 3 February 2008 http://msdn.microsoft.com/en-us/library/cc296254(VS.95).aspx

URL 11: Wagner B (2007) Write Robust Exception-Handling Code, media release, June 2007, viewed 1 January 2008 http://visualstudiomagazine.com/columns/article.aspx?editorialsid=2085

URL 12: Open Geospatial Consortium (1994–2008) OCG Sensor Web Enablement: Overview And High Level Architecture, 2006, viewed May 2008 https://portal.opengeospatial.org/files/?artifact_id=26667

URL 13: MSA (2008) DecBOS.Command II Software für den Katastropheneinsatz, 2008, viewed November 2007 http://www.msa-auer.de/uploads/tx_comacatalog/Schnellere_Krisenkommunikation_Feuerwehr_3-2008.pdf

URL 14: OSADL eG (2008) Realtime Linux, viewed March 2008 http://www.osadl.org/Realtime-Linux.projects-realtime-linux.0.html

URL 15: Gumstix Wiki (2008) Buildroot, viewed January 2008 http://docwiki.Gumstix.org/index.php/Buildroot

18 Modelling Home and Work Locations of Populations Using Passive Mobile Positioning Data

Rein Ahas, Siiri Silm, Erki Saluveer and Olle Järv

Department of Geography, University of Tartu, Estonia

Abstract

The article introduces a model that uses passive mobile positioning data to determine respondents' home and work anchor point locations. Passive mobile positioning data is secondary data concerning the location of call activities or hand-overs in network cells, which is automatically stored in the log files of service providers. This data source offers good potential for monitoring of the short-term mobility of populations, since mobile phones are widespread, and similar standardised data can be used around the globe. We developed the model and tested it with 12 months' data collected by Estonia's largest mobile service provider EMT, covering more than 0.5 million anonymous respondents. Modelling results were compared with population register data; this revealed that the developed model described the geography of the population relatively well, and can hence be used as a quantitative source in the study of population geography or for developing location-based services.

Keywords: population geography, mobile positioning, location-based services, anchor points, modelling

18.1 Introduction

The assessment and analysis of the location of individuals and populations in today's mobile world requires new methods and approaches. Traditional census and population registers are solid sources for long-term processes. For the short term and everyday mobility, more flexible methods such as various registers and indirect databases (Raymer et al. 2007), satellite-based methods (Chen et al. 2006) or modern sensing technologies (Kwan 2000, Eagle & Pentland 2005, Shoval 2007) are needed. One of the proposed methods for developing such a monitoring tool is mobile positioning or mobile telephone tracking (Mountain & Raper 2001, Spinney 2003), which has also been called a social positioning method (Ahas and Mark 2005) or a cellular census (Reades et al. 2007).

Mobile positioning is often considered to be a novel and exciting source of information for the studying of the spatial dynamics of human society. The most positive aspect of mobile positioning is that nowadays more than half of the population of the world have mobile phones; it was estimated in November 2007 there were 3.3 billion mobile telephone users in the world (Reuters 2007). Is also significant that the differences in mobile phone penetration between developed and developing countries are smaller than the variance of other socio-economic statistics. This enables the implementation of interesting worldwide studies on population monitoring and mobility assessment.

In order to conduct population and mobility studies based on positioning data, a great deal of preparatory work is necessary, because data needs to be processed and assessed, and the appropriate methods must be developed.

The objective of this paper is to model the geographical location of Estonian homes and workplaces using passive mobile positioning data and compare their geographical distribution. Passive mobile positioning data is a secondary source that is automatically stored in the memory files and logs of mobile operators (Ahas et al. 2008). The model uses a 12-month anonymous dataset from all subscribers of Estonia's largest mobile operator, EMT.

Passive mobile positioning data and the anchor points model we present herein are used in several urban studies in Estonia by Positium LBS and Department of Geography University of Tartu. For example, we have studied the urban sprawl in the Tallinn and Kuressaare areas, commuting and the growing transportation demand in rapidly developing parts of Estonia (Ahas et al. 2007b, Silm et al. 2007). This has been used quite extensively to assess the areas of summer cottages, e.g. for the investigation of the feasibility of water piping and sewer system projects (Silm and Järv 2007). We have also helped town planners understand the time issues and social and geographical dimensions of the use of the city centre areas of Tallinn and Tartu. The most interesting application of our anchor point model was as part of the project for the renewal of Estonia's largest highway (Tallinn-Tartu) (Järv et al. 2007).

18.2 Theoretical Framework

Mobile positioning is the tracing of the location coordinates of mobile phones. There are different frameworks for positioning, for instance handset-based, network-based or GPS-based. In order to locate phones, radio waves are used, and positioning is done using various methods, such as cell ID, triangulation with direction angle and/ or distance from an antenna. Different positioning methods are used because of the different network standards (GSM, CDMA, 3G), and there are different purposes for positioning (Zhao 2002). Mobile positioning can be divided into active and passive positioning.

Active mobile positioning is used for mobile tracking in which the location of the mobile phone is determined (asked) with a special query using a radio wave (Ahas et al. 2007a). In order to track certain phones for research projects, a special permit from the phone holder is required. In addition, a questionnaire about travel behaviour is conducted with respondents, in order to obtain more information about their travel behaviour.

The passive mobile positioning used in this study is data that is automatically stored in memory or log files (billing memory or the hand-over between network cells) of mobile operators (Ahas et al. 2008). Passive mobile positioning data do not require personal contact with the people involved, and yield a large amount of anonymous data.

The simplest method for passive mobile positioning is "a billing log" that is recorded for call activities. Any active use of a mobile phone (call and SMS messages in and out, GPRS, etc) is deemed to be call activity. Passive mobile positioning data is normally collected to the precision of one network cell. Every cell has a certain geographical coverage area and unique identity code, and therefore this method is called Cell ID. The maximum distance from a handset to an antenna in the GSM network is less than 35 km. Amplified antennae that cover greater distances are used in GSM networks in less inhabited or coastal areas. Mobile operators can aggregate anonymous geographical data from log files, such as location points or movement vectors, and researchers can use this in surveys for scientific purposes.

For the analyzing of passive mobile positioning data that consists of hundreds of location points recorded for every telephone (random ID, person) in Estonia, we need to set up a theoretical framework. Frequent use of certain mobile phones in specific locations gives us the possibility to learn about important or meaningful places for this phone (person) (Nurmi and Koolwaaij 2006). In order to calculate such meaningful locations, we need a conceptual framework, which has been developed in travel behaviour research under the name of anchor points. Anchor points are locations where people regularly stay (Golledge & Gärling 2001, Golledge & Stimson 1997). "Common anchors" are significant places in the environment that are commonly recognized and used as key components of cognitive maps (Dijst 1999). "Personalized anchors" are related to a person's activities, and they mark a specific workplace or home-base (Golledge 1990). In this study we developed a methodology for the calculation of personalised or personal anchor points of home and work-time places for mobile telephone owners in Estonia. The calculated location of home and work-time anchors was compared geographically.

18.3 Data and Methods

In this study we use passive positioning data from EMT, Estonia's largest mobile operator, which has more than 500,000 active phone users in its network. The database contains the locations of all "calls out" (calls initiated by the respondent) with the precision of one Cell ID over a period of one year. EMT uses network hardware and positioning middleware from Ericsson Ltd: GPRS/EDGE and WCDMA/3G networks and the MPS 9.0 mobile positioning platform. The entries for every outgoing call include: a) the time of the call; b) a random ID number for the phone; c) the cell ID. Because of privacy protection, we do not possess any personal information about the respondents, just a randomly assigned ID. The data is gathered by the company Positium LBS, which has a contract with the 2 major mobile operators in Estonia regarding the use of LBS data.

In order to calculate anchor points, an 8-step model that finds the home location, work-time location and secondary anchors for every ID was developed by Positium LBS (E. Saluveer, O. Järv), and is designed to work with huge databases in the PostgreSQL database manager. As a result of the modelling, the location of more than 500,000 persons for every 12 month in the period 1.11.2006–31.10.2007 was calculated. The average number of calls per month is 65.3 million; there is an average of 100–120 calls per ID per month.

In the first step, the model determines points of regular cells and separates them from random cells (1). The next stage of the model removes from the database respondents with too high or too low a number of calls (2). If the number of calls made is too low, it is not possible to calculate anchor points. The reason for there being too many calls is an organised call procedure (service centre etc) or a technical device using a GSM network. The third step is to determine home and work-time anchor points (3). Home and work-time anchor points are determined using regular cells, based on the average start time of calls (the average of all calls made during a 24 h day) and the standard deviation of call commencement times. The fourth step is to consider the neighbouring relationship in the case of 2 home or 2 work-time anchor points (4). Then we assessed the proportion of days spent at an anchor point (5) and determined the missing home or work-time anchor point by the addition of a third point (6). The seventh step is to classify an anchor point as the missing home or work anchor point (7). The last step is to format everyday and secondary anchor points.

Terms used to describe the model and the results:

- *Anchor point:* An anchor point or personal anchor point is a regular cell, which according to the number of calls has a significant meaning in the everyday life of the respondent. Regular cells are referred to as anchors when the model gives meaning (home, work-time, multifunctional) to them.

- *Regular cell:* A network cell regularly visited by one respondent, and from which the respondent has made calls on at least 2 different days a month.
- *Random cell:* A network cell from which one respondent has made calls on only one day a month.
- *Everyday anchor point:* Anchor point at which the respondent has spent time on most days, and which has thus been designated as a home or work place.
- *Secondary anchor point:* Anchor points that have lower visiting regularity than everyday anchor points.
- *Home anchor point:* An everyday anchor point which, based on the model, is the probable location of the respondent's home.
- *Work-time anchor point:* An everyday anchor point which, based on the model, is the respondent's probable work-time location. The anchor is called a work-time location because it is not possible to differentiate between work, school and other activities in the place where a person regularly and most often spends time in business hours during a month.
- *Multifunctional anchor point:* An everyday anchor point in which the home and work-time anchor points are located in the same network cell and cannot be separately identified.

In this study we analyse home anchors, work-time anchors and multifunctional anchors. There is a linear correlation between the size of municipalities in the modelled dataset (home anchor points) and the number of inhabitants in municipali-

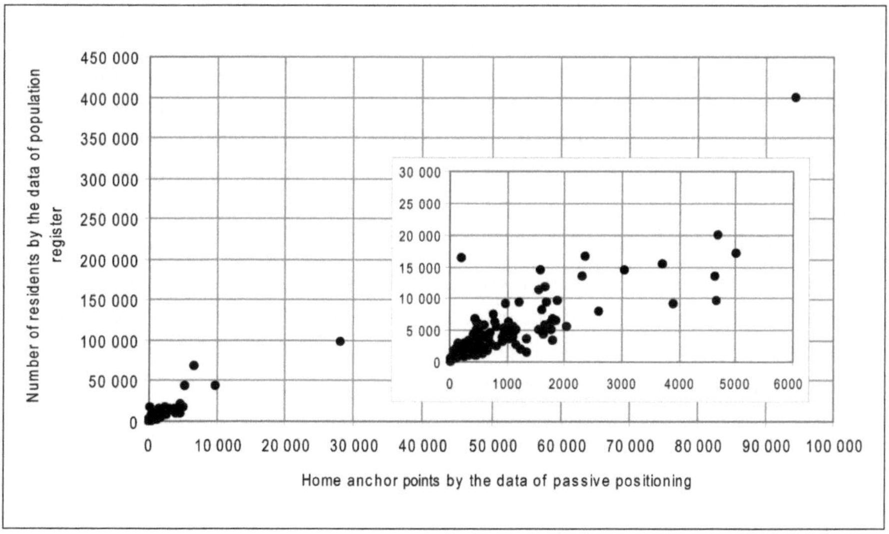

Fig. 18.1. Correlation between the size of Estonian municipalities by modelled home anchor points (mobile positioning data from 1. 11. 2006 to 31. 10. 2007) and the population register as of 1. 1. 2007.

ties. The greatest differences occur in larger towns and north-eastern Estonia, places which have more Russian speaking inhabitants, and former Soviet industrial towns *(Figure 18.1)*.

18.4 Results

18.4.1 Distribution of Home, Work-Time and Multifunctional Anchors

The modelled distribution of home and work-time anchors follows quite closely the distribution of inhabitants by the Population Register in Estonia. The number of home as well as work-time anchor points is higher in Harju County (over 30%) *(Table 18.1),* which is the county that contains the capital of Estonia, Tallinn. Of anchors in Harju County, 72% of home and 79% of work-time anchor points are located in Tallinn. The next county with anchor points, which is noticeable apart from others, is Tartu County, which has 14.4% of home anchors and 14.7% of work-time anchors in Estonia.

The most home places are in the city of Tallinn and in other larger towns and neighbouring municipalities *(Figure 18.2)*. The number of home anchors is highest in Tallinn (about 94,000), followed by the city of Tartu (about 28,000). In other municipalities the number of home anchor points is less than 10,000. The density of home anchors is higher in the city of Tartu (726 anchor points/km^2), followed by Tallinn (593 anchor points/km^2).

The distribution of modelled work-time anchor points is also similar to the location of homes. The distribution of multifunctional anchor points, i.e. the people whose home and work-time anchor points are in the same network cell, is more equal than it is at home and work-time anchors. The number of multifunctional anchor points is higher in larger towns and their surroundings, because of the higher number of inhabitants or network cells. The number of multifunctional anchor points is smaller on the islands and in the east and southeast of Estonia. It is important to estimate the distribution of multifunctional anchors because is shows which regions allows individuals to work quite near their homes. This is described in greater detail in the capture, which analyses distances between home and work-time places. It also shows the worst coverage of network cells.

Table 18.1. Distribution of modelled home, work-time and multifunctional anchors in Estonian counties.

	Home anchors			Work-time anchors			Multifunctional anchors		
County	No	Density (no/km²)	%	No	Density (no/km²)	%	No	Density (no/km²)	%
Harju County	131,883	30.4	46.1	137,221	31.6	47.8	50,087	11.5	26.7
Tartu County	41,169	13.8	14.4	42,188	14.1	14.7	18,829	6.3	10.0
Ida-Viru County	20,508	6.1	7.2	20,446	6.1	7.1	18,249	5.5	9.7
Pärnu County	16,912	3.5	5.9	16,932	3.5	5.9	13,651	2.8	7.3
Lääne-Viru County	12,737	3.5	4.5	12,363	3.4	4.3	11,371	3.1	6.1
Viljandi County	11,120	3.3	3.9	10,706	3.1	3.7	10,271	3.0	5.5
Rapla County	7438	2.5	2.6	5471	1.8	1.9	8495	2.9	4.5
Saare County	7327	2.5	2.6	7212	2.5	2.5	7165	2.4	3.8
Võru County	7186	3.1	2.5	7061	3.1	2.5	8301	3.6	4.4
Põlva County	6714	3.1	2.3	6036	2.8	2.1	8964	4.1	4.8
Jõgeva County	5893	2.3	2.1	5344	2.1	1.9	8619	3.3	4.6
Valga County	5618	2.8	2.0	5221	2.6	1.8	8651	4.2	4.6
Järva County	5410	2.2	1.9	4988	2.0	1.7	7505	3.1	4.0
Lääne County	4619	1.9	1.6	4307	1.8	1.5	5190	2.1	2.8
Hiiu County	1464	1.4	0.5	1405	1.4	0.5	2145	2.1	1.1
Total	285,996	6.6	100.0	286,900	6.6	100.0	187,491	4.3	100.0

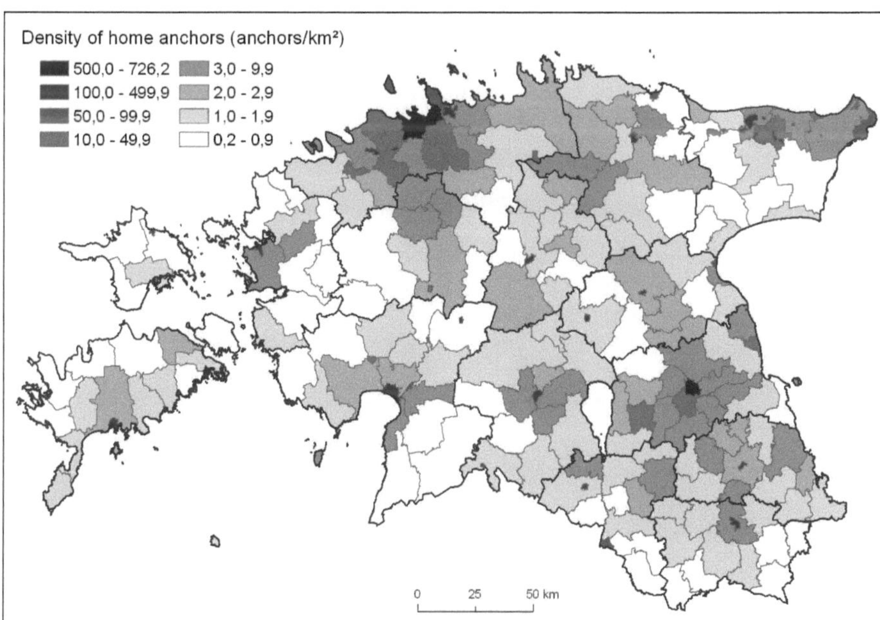

Fig. 18.2. Geographical distribution of modelled home anchor points in Estonian municipalities.

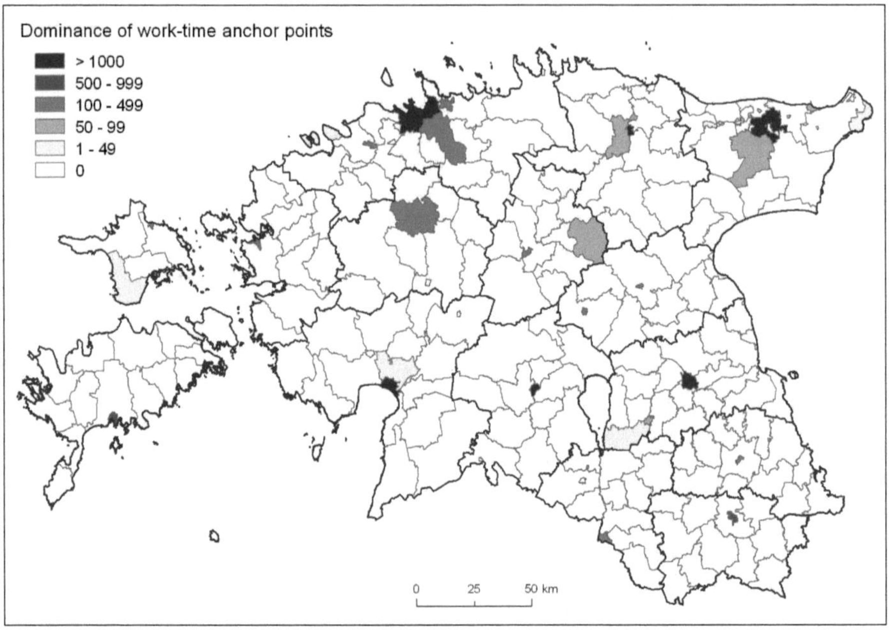

Fig. 18.3. The Difference in the distribution of work-time and home anchor points in Estonia; the dominance of work-time locations over homes.

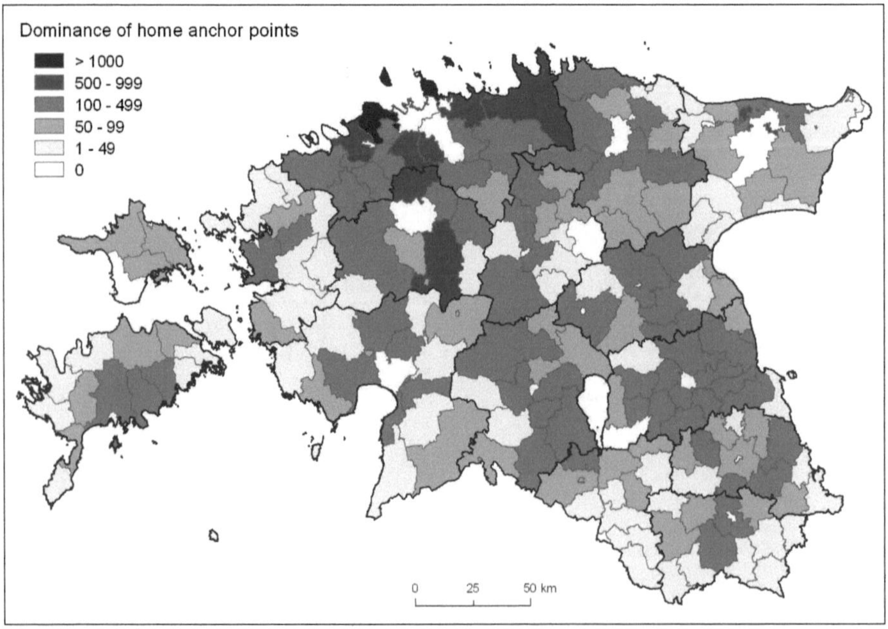

Fig. 18.4. The difference in the distribution of work-time and home anchor points in Estonia; dominance of home locations over work-time places.

18.4.2 Differences in the Geographical Distribution of Home and Work-Time Anchors

Despite the general similarities in the distribution of modelled homes and work-time places, there are geographical differences in the distribution of locations of modelled homes and work-time anchors. To a greater extent than modelled homes, work-time anchors are concentrated in towns, especially larger towns. In Tallinn, for example, there are 94,566 home anchors and 108,579 work-time anchors, and in Tartu 28,190 home anchors and 32,348 work-time anchors. The majority of towns have more work-time anchors than homes: there is a higher number of work-time anchors in most towns *(Figure 18.3, Table 18.2):* Pärnu (2062), Jõhvi (1700), Viljandi (1249) and Rakvere (1005). Võru and Kuressaare have 500-1000 work-time anchors more than homes, and some suburban communities in the Tallinn urban region, such as Rae parish, also have a greater number of work-time locations than homes. In smaller towns the difference between the number of modelled homes and work-time places is smaller but still noticeable.

Homes dominate over work-time places in suburban regions near larger towns, as the majority of municipalities near Tallinn. The number of home anchors is higher

Table 18.2. Municipalities with the greatest difference in home and work-time anchor points in Estonia.

No	Municipality	Dominance of work-time anchors	Municipality	Dominance of home anchors
1	Tallinn town	14,013	Harku parish	1763
2	Tartu town	4159	Viimsi parish	1519
3	Pärnu town	2062	Kohila parish	861
4	Jõhvi parish	1700	Saku parish	722
5	Viljand town	1249	Keila parish	670
6	Rakvere town	1005	Kohtla-Järve town	632
7	Kuressaare town	945	Jäelõhtme parish	599
8	Võru town	713	Kuusalu parish	533
9	Paide town	331	Loksa town	533
10	Keila town	323	Kiili parish	521
11	Haapsalu town	322	Kehtna parish	511
12	Maardu town	321	Ülenurme parish	450
13	Rapla parish	301	Raasiku parish	448
14	Jõgeva town	298	Paikuse parish	428
15	Põlva town	272	Konguta parish	416
16	Rae parish	209	Tartu parish	406
17	Valga town	170	Toila parish	361
18	Põltsamaa town	139	Anija parish	353
19	Kärdla town	112	Saue parish	346
20	Mäetaguse parish	97	Saue town	332

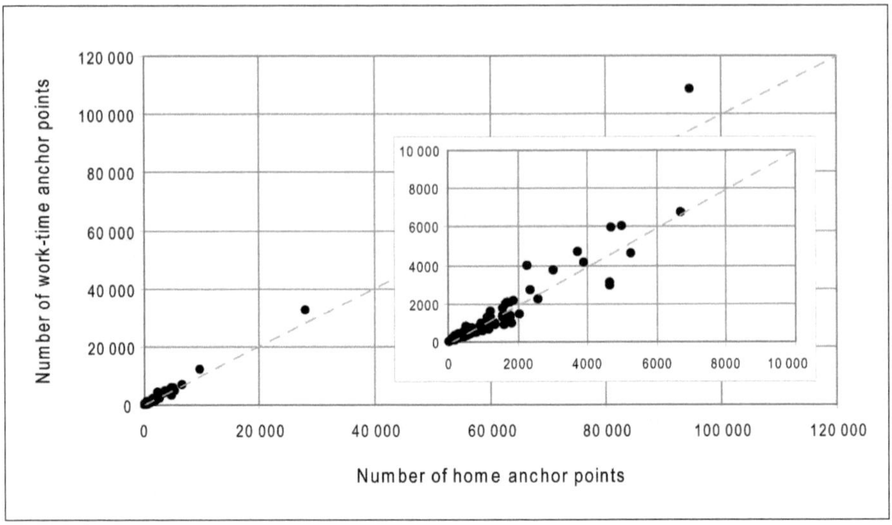

Fig. 18.5. Correlation between home and work-time anchor points.Distance between home and work-time anchors.

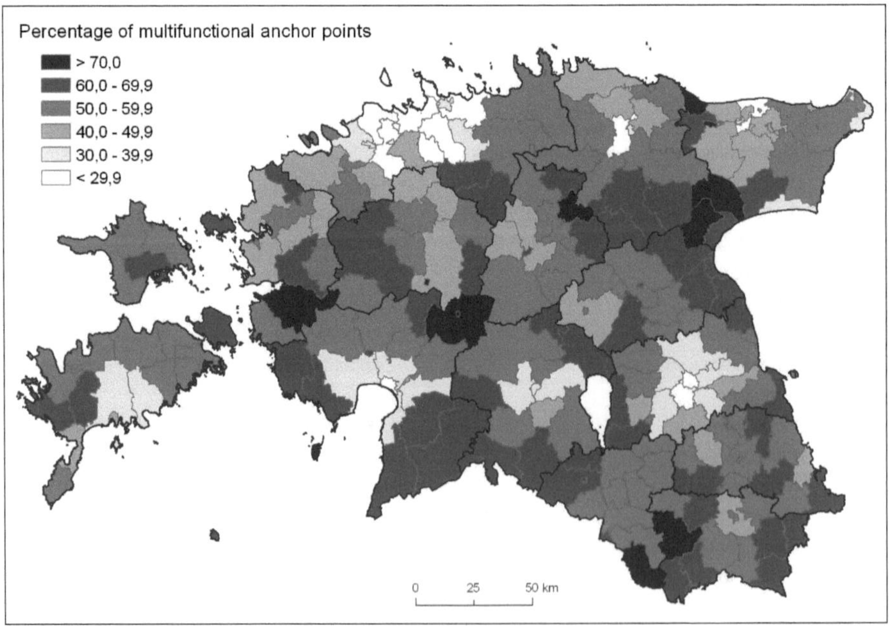

Fig. 18.6. The percentage of multifunctional anchors, of whole everyday anchors (home, work-time and multifunctional anchor points).

than work-time anchors in the following municipalities: Harku (1763), Viimsi (1519), Kohila (861), Saku (722) and Keila (670) *(Figure 18.4, Table 18.2)*. Areas where home anchors dominate over work-time anchors in limits over 100 units are found in the Tallinn hinterland, and near the cities of Tartu, Pärnu, Viljandi and Kuressaare. The difference between the number of home and work-time anchors is smaller in municipalities of south-eastern Estonia.

As the correlation between home and work-time anchor points shows, the work-time anchors are dominant municipalities that have more inhabitants *(Figure 18.5)*. There is no such trend in municipalities with a smaller number of home and work-time anchor points.

The percentage of multifunctional anchor points - the number of people who live and work in the same network cell area – is higher in municipalities that are farther from larger towns *(Figure 18.6)*. The percentage of multifunctional anchors is smaller in and near urban municipalities. Less than 25% of people have home and work-time anchor points in the same network cell in the cities of Tartu and Tallinn and in parishes surrounding Tallinn – Viimsi and Saue parish. Other larger towns (Pärnu, Viljandi) and municipalities surrounding these are also noticeable.

18.4.3 Distance Between Home and Work-Time Anchors

The percentage of multifunctional anchors is higher in rural areas, because the network cells are larger there or the people's spatial behaviour differs from urban areas.

People who have home and work-time anchors in the same network cell are most likely pensioners and housewives who stay at home during the day, as well as people who work quite close to where they live.

The distance between a home and work-time anchor point is shortest in Tartu county (10.2 km), which is followed by Harju County (10.4 km) *(Table 18.3)*. The distance is longest for people who live in Hiiu County (21.3 km). Based on the county in which the work-time anchor point is situated, the distance between home and work-time anchor is shortest for people whose work-time anchor is in Rapla County (9.9 km) and longer than the home anchor in Hiiu County (18.3 km). For people working in Harju County, their home anchor is on average 12.1 km away from their workplace.

By municipalities, longer distances between home and work-time anchor points are found on the islands and in areas near the border *(Figure 18.7)*. In the Tallinn metropolitan area, the distance between home and work-time anchors is longer near the border of the metropolitan area, i.e. over 20 km. The distance is shortest from a home anchor point in Tallinn (average 9.1 km), followed by smaller towns (Maardu, Keila, Saue) and surrounding parishes.

Table 18.3. Average distance between home and work-time anchor points in counties proceeds from home and work-time anchors.

County	Average distance from home anchor point to work-time anchor point	Average distance from work-time anchor point to home anchor point
Tartu County	10.2	10.6
Harju County	10.4	12.1
Ida-Viru County	11.2	11.1
Viljandi County	11.4	10.0
Järva County	12.3	10.7
Võru County	12.4	10.7
Pärnu County	12.4	11.8
Lääne-Viru County	12.5	11.2
Jõgeva County	12.5	10.8
Põlva County	12.6	10.3
Rapla County	13.5	9.9
Valga County	14.5	12.9
Lääne County	16.3	12.9
Saare County	16.5	14.7
Hiiu County	21.3	18.3

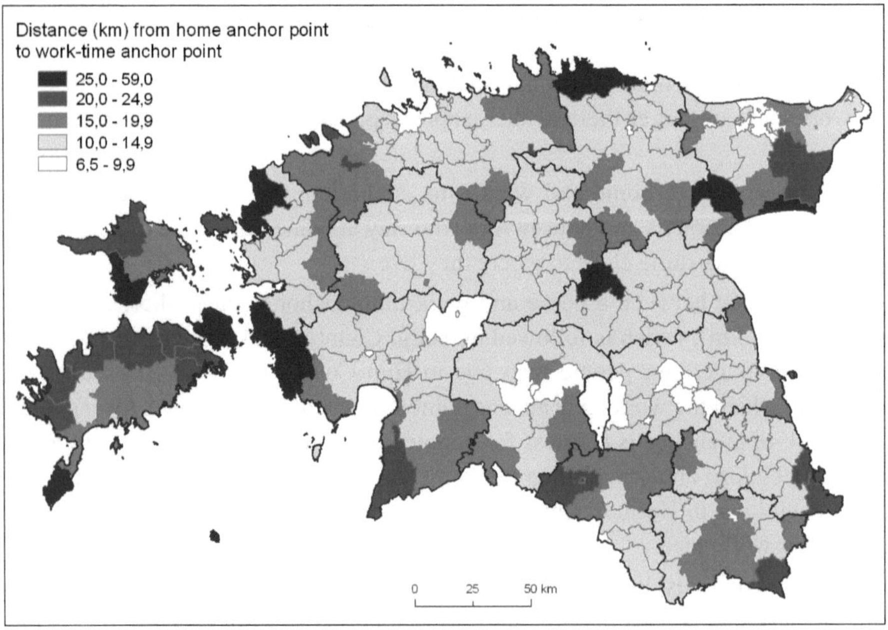

Fig. 18.7. Average distance between home and work-time anchor points by municipalities, from home anchor point.

18.5 Discussion and Conclusions

The objective of the current study was to compare the geographical location of homes and work-time places in Estonia. For the analysis, we used a new method based on a passive mobile positioning database. We used an anonymous dataset from EMT, Estonia's largest mobile operator, which has more than 500,000 subscribers, which is approximately 45% of Estonian citizens. This method was selected as it is necessary to develop a new methodology for population monitoring in rapidly developing countries such as Estonia. Rapid suburbanisation near the largest towns is ongoing in Estonia (Tammaru et al. 200x). The population is also changing in rural areas and post-Soviet industrial towns in north-eastern Estonia. Because of those rapid changes, mobile positioning data can be an alternative source for population monitoring, as the population register is not functioning properly and the last (2000) population census data is out of date.

Our results and comparison with the population register data showed that the method we used to monitor the locations of homes and work-time places was accurate. The modelled locations of homes and work-time places followed their logical distribution. The number of modelled work-time locations was higher in towns. As there are no other sources available that show geographical statistics of jobs, our discussions and comparisons with different databases showed that this result is correct. Jobs are concentrated in town centres.

Homes were predominantly found in suburban regions near bigger towns. This trend is also logical, as the rate of suburbanisation in Estonia is quite rapid, and thousands of families have moved out of towns (Leetmaa et al. 2009). Our studies of new suburban settlement areas show that the inhabitants of new suburbs are highly educated (Silm et al. 200x, Ahas et al. 2007a), and most of their jobs are located in town centres. Therefore we can say that suburbanisation is increasing transportation demand in Estonia. The second important result involves the location of jobs that are still concentrated in towns. As homes move out of towns and the labour market is increasingly mobile throughout Estonia, there is a considerably longer distance between home and work. Even if Estonian municipalities plan to create new workplaces in the suburbs, as is the case in the urban region of Tallinn, highly qualified jobs happen to be far away from homes (Ahas et al. 2007b). It has also been studied in other western countries that more educated persons have a longer distance from home to work than persons with simple jobs (Kwan 2000). As our anonymous database does not include respondents' social status, we cannot measure this relationship with the current dataset, but comparing suburbanisation patterns in Estonia and education level, we can assume that this is also the case in Estonian suburbanisation (Tammaru & Leetmaa 2007, Silm et al. 200x).

There are a number of issues to discuss concerning our unusual data and method. The preliminary results of our investigation of the locations of homes and the comparison of these data with the population register shows that our method is accurate. We also compared our personal activity spaces with the modelled ones for 26 cases, and the results were reliable. Problems are connected with rural locations where the mobile network is less dense, and many home and work-time anchor points are in the same network cell. This result is still useable and interesting, but it shows the level of precision of our method. There is also a problem with the privacy and anonymity of the data, which must be handled properly, as the accuracy of future research depends on it. Today's precision and anonymity of data is not really a problem in this area. In the near future, however, the quality of data is increasing rapidly, as A-GPS telephones have been in operation in the EMT network since 2007, with an accuracy of a few metres. This is a milestone where the discussion about spatial privacy is becoming truly important, and may be an important discussion point for geographers in the near future.

References

Ahas R, Aasa A, Roose A, Mark Ü, Silm S (2008) Evaluating passive mobile positioning data for tourism surveys: An Estonian case study. *Tourism Management* 29:469–486.

Ahas R, Aasa A, Silm S, Aunap R, Kalle H, Mark Ü (2007a) Mobile Positioning in Space-Time Behaviour Studies: Social Positioning method Experiments in Estonia. *Cartography and Geographic Information Science* 34(4):259–273.

Ahas R, Jauhiainen J, Silm S, Nuga M, Tähepõld A, Saluveer E, Kivi K (2007b) Tallinna ja ümbritsevate omavalitsuste koostöövõimalused ja perspektiivid valglinnastumise kontekstis. Tartu, University of Tartu, Institute of Geography.

Ahas R, Mark Ü (2005) Location based services – new challenges for planning and public administration? *Futures* 37:547–561.

Chen XL, Zhao HM, Li PX (2006) Remote sensing image-based analysis of the relationship between urban heat island and land use/cover changes. *Remote sensing of environment* 104(2):133–146.

Dijst M (1999) Two-earner families and their action spaces: A case study of two dutch communities. *GeoJournal* 4(3):195–206.

Eagle N, Pentland A (2005) Reality mining: Sensing complex social systems. *Personal and Ubiquitous Computing* 10(4):255–268.

Golledge RG (1990) Cognition of physical and built environments. In *Environment, cognition and action; a multidisciplinary integrative approach*, ed Garling T, Evans G. New York: Oxford University Press.

Golledge RG, Gärling T (2001) Spatial behaviour in transportation modeling and planning. In *Transportation and Engineering Handbook*, ed Goulias K.

Golledge RG, Stimson RJ (1997) *Spatial behavior: a geographical perspective.* New York: The Guilford Press.

Järv O, Saluveer E, Ahas R (2007) Analysis of traffic distribution and frequencies in Kose-Võõbu and Võõbu-Mäo units of E263 Tallinn-Tartu-Luhamaa highway (E263 Tallinna-Tartu-Luhamaa maantee Kose-Võõbu ja Võõbu-Mäo lõikude liiklusuuring mobiilpositsioneerimise abil). Tartu: Positium LBS.

Kwan MP (2000) Gender differences in space-time constraints. *Area* 32(2):145–156.

Leetmaa K, Tammaru T, Anniste K (2009) Urban actors and residential suburbanisation in the Tallinn metropolitan area. In *Special issue: Neighbourhood change in European cities: new developments in the context of the changing role of the state.*, ed Van Kempen R, Murie A. Tijdschrift voor Economische en Sociale Geografie.

Mountain D, Raper J (2001) Modelling human spatio-temporal behaviour: a challenge for location-based services. *Proc. of the 6th Internat. Conference on GeoComputation.* Brisbane, Australia: University of Queensland.

Nurmi P, Koolwaaij J (2006) Identifying meaningful locations. *3rd Annual International Conference on Mobile and Ubiquitous Systems: Networks and Services (MobiQuitous).* San Jose, CA, IEEE Computer Society.

Raymer J, Abel G, Smith PWF (2007) Combining census and registration data to estimate detailed elderly migration flows in England and Wales. *Journal Of The Royal Statistical Society Series A-Statistics In Society* 170(4):891–908.

Reades J, Calabrese F, Sevtsuk A, Ratti C (2007) Cellular census: Explorations in urban data collection. *IEEE Pervasive Computing* 6(3):30–38.

Reuters (2007) Global cell phone use at 50 percent. 29 November 2007 http://www.reuters.com/article/rbssTechMediaTelecomNews/idUSL2917209520071129

Shoval N (2007) Sensing Human Society. *Environment and Planning B* 34:191–195.

Silm S, Ahas R, Nuga M (200x) Gender differences in space-time mobility of dwellers from new residential areas in the suburbs of Tallinn. *Europe Asia Studies.* (Forthcoming).

Silm S, Järv O (2007) Summerhouses, users and usage frequencies in Saku county (Saku valla suvilate kasutuse uuring). Tartu, Positium LBS.

Silm S, Järv O, Ahas R (2007) Urban sprawl in Kuressaare Urban Region (Valglinnastumine ja sellest lähtuvad mõjud Kuressaare linnaregioonis: esmane analüüs ja ettepanekud planeerijatele). Tartu: Positium LBS.

Spinney JE (2003) Mobile positioning and LBS applications. *Geography* 88:256–265.

Zhao Y (2002) Standardization of mobile phone positioning for 3G systems. *IEEE – Communications Magazine* 40(7):108–116.

Tammaru T, Leetmaa K (2007) Suburbanisation in relation to education in the Tallinn metropolitan area. *Population, Space and Place* 13(4):279–292.

Tammaru T, Leetmaa K, Silm S, Ahas R (200x) New residential areas in the Tallinn Metropolitan Area. Urban studies (Forthcoming).

19 A Geographical Information System for Real Estate (GEOVAL)

Efstratios Stylianidis, Themistocles Roustanis and Nikolaos Karanikolas

School of Urban-Regional Planning and Development Engineering
Faculty of Engineering, University Campus, Thessaloniki, Greece

Abstract

The basic aim of the project is the design of a geographical system which has the ability to provide all the benefits of geoinformatics on issues and projects of land valuation and land management. By nature, real estate market is spatial market and thus GIS can provide, in one hand, several very powerful tools for land value estimation, management of real estate and on the other hand land promotion. These three procedures are the central components for the design of GEOVAL system. It is well known to real estate firms that the use of GIS can offer lots of benefits for them. These firms, which did use GIS, reported that GIS is important or very important to their overall mission (Fryrear et al. 2001). GIS can also provide more accurate measurements on land and contributes to productivity as well. For example, GIS can provide a superior location variable relative to the traditionally used straight-line distance assumption (Rodriguez et al. 1995). Mobile GIS provides also powerful tools for real estate firms, especially due to field work and the actual need for transportable solutions. Valuation is also a fieldwork-based process which is based on geodata collection in order to provide the best accuracy to the estimation of land values. All these advantages can lead to a major reduction of real estate firms costs. GIS can also provide important tools for storing a large amount of data with spatial relationship. GEOVAL system is a tool for many spatial functions like other GIS systems for real estate. The spatial functions are in general terms geographic selection, manipulation, exploration and finally confirmation (Anselin 1998).

Keywords: mobile geographic information system, real estate, property valuation, mobile GIS

19.1 Introduction

The basic aim of the project is the design of a geographical system that provides all the benefits of geoinformatics on issues and projects relative to land valuation and land management and promotion as shown on *Figure 19.1*. These three procedures are the basic components constitute the GEOVAL system:

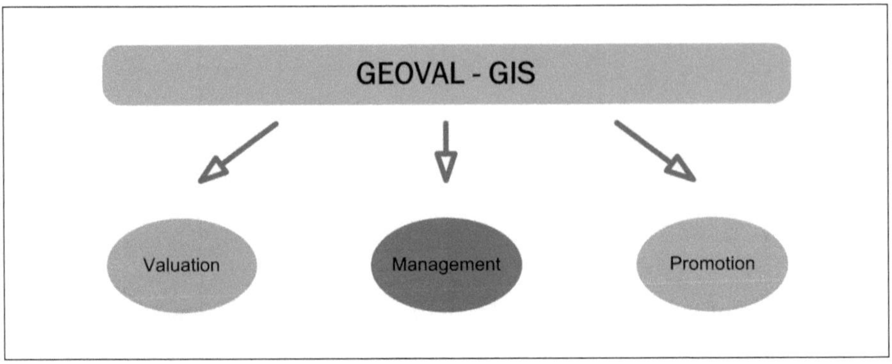

Fig. 19.1. The three main components of GEOVAL system.

The tools offered through GEOVAL system are grouped on three levels as illustrated on *Figure 19.2*. To begin with, a tool that is workable on a mobile PC and constitutes the mobile GIS of GEOVAL system. This mobile component is a valuable tool for the valuators during both the surveying procedure and autopsy. Secondly, a desktop system, which can be used for a holistic data analysis; this is the main tool for the concluding valuation of the real estate. Finally, there is a web-

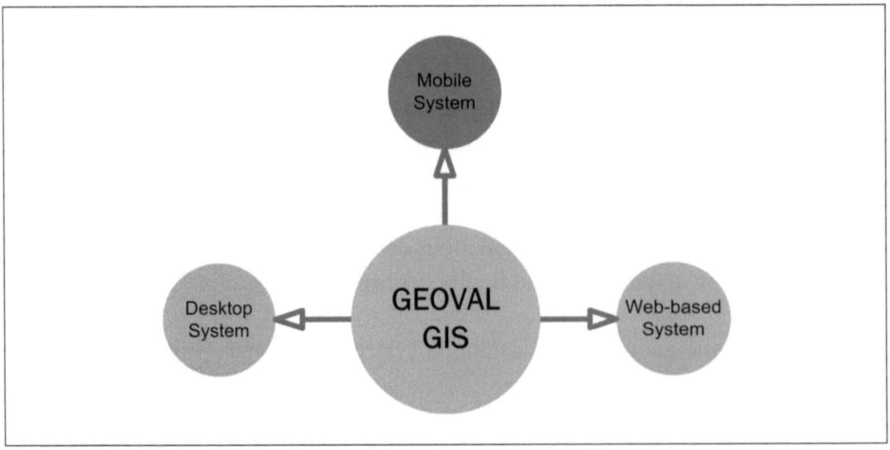

Fig. 19.2. Tools of the system.

based system acting as a communication tool for the mapping and promotion of real estate according to actual needs and firm's strategy.

The main advantage of the system is the configuration and transformation ability according to user's needs. The primary concern of the system is the respect on sensitive information as well as to the directions of International and European Valuation Standards. From the literature, it is well known that property value is a function of locational, physical, legal and economic factors (Wyatt 1997). *Undoubtedly, the principal factor that affects land values is location* (Goodall 1972). Consequently, the valuation technique always requires the valuator's expert knowledge for locality and inter–locality relations. For these reasons, a GIS model may contribute to the provision of all the spatial measurements of property values. In addition, the GIS model can also present the graphic evidence of spatial relationships. The map can be the source of evidence and is capable of leading from the "evidence-based thinking" to "using graphics not just to present a conclusion but also to reach it" (Zachry & Thrall 2004). There is also Bertin's statement that "graphics is the visual means of resolving logical problems" (Bertin 1983).

19.2 GIS Applications for Real Estate

For real estate, *location* is one land characteristic of high importance for economic analysis. As a matter of fact, location is the crucial spatial attribute on which estimation and analysis of land value can be transformed and create different types of study. Prior to GIS period, the location determinant impact on land values was difficult to be succeeded due to the existing accuracy levels on positioning. Nowadays, GIS offers techniques in tens to facilitate the analytic spatial reasoning (Arbia 1989, Tomlin 1990, Huxhold 1991, Star & Estes 1990). Today also, the cost of a GIS system, the revolution on computer technology, the availability of digital maps and geographic data on internet, the business applications promotion through GIS (Fryrear et al. 2001) are the critical factors for the increased use of GIS in real estate market.

GIS can also play an important role in real estate research (Rodriguez et al. 1994). For instance, in the areas of spatial interaction models (Haynes & Fotheringham 1984) and spatial diffusion models (Morrill, Gaile & Thrall 1988), GIS is of great importance for formulating accurate spatial models. Spatial interaction models, which can also referred to as gravity models, forecast traffic models, investment analysis and shopping centre revenue and should be used to identify optimal site locations. Clapp & Rodriguez (1995) also show the importance and improvement of the real estate market analysis using GIS.

Thrall in his works (Thrall 1984–1988), contains many relevant works that bridge the gap between geography, real estate and urban economic disciplines.

Other literature covers such topics as central place theory (King 1984), point pattern analysis (Boots & Getas 1988), and spatial autocorrelation (Odland 1988). Point pattern analysis is of interest in that most real estate research does not require the specification of precise boundaries surrounding a parcel. As such, the precise location of a parcel is represented by only a single point contained somewhere within the boundaries of that parcel of land. Nevertheless, given that each location on the surface of the earth is influenced by other locations, econometric analysis must be concerned not only with the possible errors introduced by the problem of time autocorrelation but must also address the errors that can be introduced by the problem of spatial autocorrelation. The real estate literature has not yet dealt adequately with these issues.

19.2.1 Geographic Characteristics of Real Estate

All geographers take into consideration a number of basic views when they perform spatial analysis. The lack of geographic thinking can lead to problems when analysis is taking place. Many non-geographers, who work in real estate market, are capable of avoiding mistakes by taking into consideration some necessary geographic tools. Geographic scale, absolute and relative locations, geocoding, characteristics of continuous and non-continuous space, spatial autocorrelation, geographic accessibility, proportional and absolute geographic measurements, geographic surveying are some of the tools that real estate analysis must always consider.

For instance, geographic scale can be a common error in data analysis. There is always a basic consideration that GIS produce successful new datasets if data are collected in high resolution. For example, statistical census collections are able to provide data for households' income by census tract because the information exists due to households. When real estate analysts are working with data and creating new datasets (Shilton & Stanley 1995, Bible & Hsieh 1996) sometimes they confuse geographic scale of the existing data and at the same time are proposing solutions on higher resolution even though initial data are collected in lower resolution levels. The analysis on geographic scale and geographic position shows up a dependency which is known as the "modifiable area data problem" (Arbia 1989).

Is space continuous for all geographic phenomena? Although in physical geography many phenomena are considered and treated as continuous – for example elevation is a continuous phenomenon which alters from place to place in smaller or bigger changes – the situation is slightly more problematic when it comes to land values. All locations do not have land values. Even though a contour program in GIS maps surfaces giving a value everywhere, apparently there is not good accuracy in such a mapping. On the contrary when contours of land values are illustrated, this does not mean that land value is a continuous phenomenon. For instance, there are no values in the centre of a lake or along a river. Although algorithms creating surfaces are mainly based on mapping surfaces from spatially continuous measurements, common real estate characteristics are discrete phenomena. Real estate analysts can make use of software to create maps of land values but they must always consider how results can be interpreted and used.

Another fundamental problem is spatial autocorrelation. In this case, values of 2006, for instance, must occur before values of 2007 in the relevant used model and therefore the neighbourhoods are expected to have more similar values than distant estates. Spatial autocorrelation is a well-recognized problem by geographers (Norcliffe 1977, Silk 1979, Cliffe & Ord 1981, Goodchild 1986, Odland 1988, Anselin 1988, Cressi 1991). A few methods are capable of detecting spatial autocorrelation as well as other procedures are used to correct spatial autocorrelation.

Measuring externality in real estate analysis is difficult. For instance, migration in western Thessaloniki, Greece gave increase on land values during the last years. Suspend of migration and arrival of new habitats has already stopped the increase of values and at the same time reduced the demand for new houses. General theory of land values and uses would suggest that the general trajectory of land values would be downward in areas impacted by negative externalities (Thrall 1982, 1988).

One of the most common errors in real estate analysis is distance measurement between two points of interest. The distance measured in many GIS programs is the so called as the crow flies method of measurement. It is a straight line between two points in the map. This can be a mistake in real estate analysis where accessibility is a basic analysis factor in many cases. Geocoding is a powerful tool, which can properly assist the real estate analysts and valuators. Collection of real estate data can be of great accuracy, in a really fast way, with the use of geocoding tools.

The phenomenon of "ground truth" is a critical problem for real estate analysts. It is never obsolete to visit locations and study the way the model in paper describes reality in space. The data analysis must always be ground truth.

19.3 The GEOVAL System

GEOVAL system is the extract of the gained expertise throughout the last years working in real estate, cartography and GIS. It is an information system which is based on the modern views of GIS technology. The basic structural elements of the system as shown on the GEOVAL architecture *(Figure 19.3)* are described like the following:

- *Metadata database:* A system, which stores the information related to the structure and operation mode.

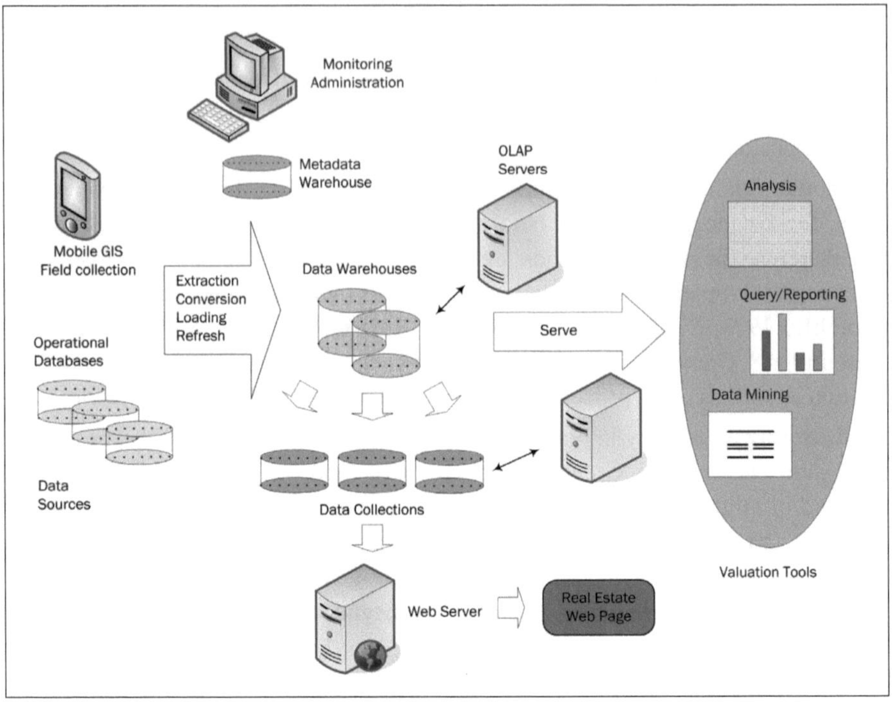

Fig. 19.3. The architecture of GEOVAL system.

- *Administrator:* An application that offers the appropriate tools for managing the system completely.
- *Data sources:* Various data sources that support or complement the central database of the real estate. External data sources such as information provided by other information or GI systems, as well as application files and text of media files.
- *Converters:* Applications which execute specific procedures for transferring data from the data sources to the data warehouse, like for instance, from the mobile devices and the mobile GIS to the central system.
- *Data warehouse:* The main system where all data are stored in a predefined mode in order to be offered to the users.
- *Data collection:* Data groups that are exporting on demand from the system in order to be used, for instance, from the web server.
- *Applications:* Applications, which are accessing the data warehouse and have the ability for particular analysis.

The GEOVAL system consists of tools that handle data extraction from the different operational databases. Besides, the system incorporates tools for handling field data, cleaning, transformation, incorporation and data transfer in the warehouse.

The system tools are also use for the data periodical renewal in order to record the updates in the primary sources and finally for the liquidation of the warehouses. Apart from the main data warehouse, various regional data groups may exist. The information on data warehouses as well as the data on regional warehouses are stored and managed from one or more central computers. This is a multidimensional approach and various tools are disposable: querying tools, reporting, analysis tools and restoring data tools. Finally, there is a dedicated data warehouse for storing and managing metadata and tools for reporting and storing the system.

19.3.1 Data Sources

The regional system for field data collection is useful during fieldwork. A land valuator during the field campaign has the ability to collect data through a pocket PC *(Figure 19.4a)* which incorporates a mobile GIS with all functionalities and tools. The collectable data could be the following:

- The location of the real estate through the GIS/GPS mobile device. The location is stored in a spatial database with all relevant attributes. The positioning process is essential for the automatic data transfer in the main system as well as the real estate geometry in the main spatial database.

Fig. 19.4. (a) Field data collection through mobile GIS and (b) The satellite image representation in the field GIS application.

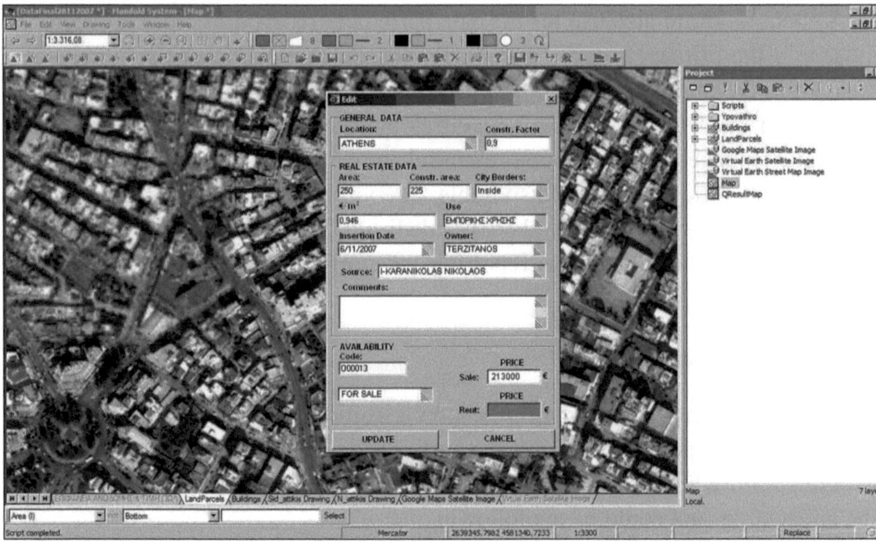

Fig. 19.5. The GIS desktop application.

- Simultaneously, with the real estate location, the municipality code and other descriptive regional data are recorded automatically in the local (mobile) spatial database, based on the recorded location.
- At the same time, the valuator has the opportunity for recording the address, numbering and other information regarding the recorded location, through the user friendly mobile GIS interface.
- The ability of recording the distance of the real estate from the centre of the proper municipality. There is also the capability for the automatic recording of the time-distance from any chosen centre.
- Finally, the option of multimedia recording, like sound (voice recording), video and photographs is also provided. Such means are useful to transfer the real view of the real estate situation.

The mobile GIS has all the appropriate tools regarding mainly spatial and other data acquisition and secondly data management during field campaigns. For instance, it could be used satellite images through the mobile GIS application *(Figure 19.4b)* in order to guide the land valuator during the field campaign.

19.3.2 The Regional System for Processing Field Data

After data collection during the field campaign, the pocket PC is connected to the main system for downloading collected data. The downloaded records are imported in the system, however only in a temporal space. This means that data are not abso-

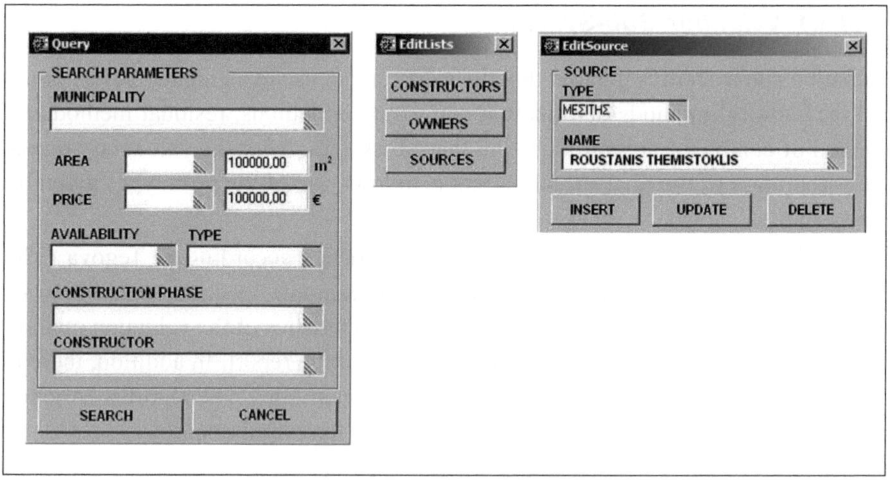

Fig. 19.6. Querying and editing tools.

lutely incorporated in the system unless an authorized user selects the real estate category: (a) for rent, (b) for sale and (c) for valuation.

Furthermore, additional information regarding the real estate has to be completed by the user through the GIS desktop application dialog boxes as shown on *Figure 19.5.*

The GIS desktop application offers also querying and editing tools *(Figure 19.6)* providing the necessary support to the user for acquiring any kind of information from the databases.

19.3.3 Regional System for Updating Additional Information

The GEOVAL system has the ambition to play a role over than real estate valuation, management and promotion. The broader plan is to enrich the system with additional and valuable information, further that the real estate like for instance: employment, demography, economy, environment etc. This additional information is considered in combination with the initial information and other essential spatial and descriptive data for the analysis and final estimation of the real estate.

There are many theories developed on the interactivity between land values and demographic or social data. For example land values in suburban zones followed upward trajectories because of the increase of the demand for new homes. New families who were leaving centres of the cities and choose these zones for their residence changed the demography of these areas. Social level of new habitats is one of the main factors of affecting land prices on a residence zone.

19.3.3.1 Valuation Sub-System

Valuation sub-system is the tool that the valuator can use to estimate market value. All the classical methods of valuation comparative methods, residual method and method of contractor are introduced in the sub-system. After the valuation process the preparation of the final valuation report follows one of the prototypes given by the system.

The final report follows the wanted characteristics according to Tegova (The European Group of Valuers' Associations). For example location, description of estate, urban and legal identity of the estate, characteristics of the valuation method and finally valuation, constitute the main content of the report. In addition, there is also information about the valuator and field work activities. The report is enriched could be enriched with remote sensing images and terrestrial images.

19.3.3.2 Sub-System for Exporting Internet-Publish Data

The system throughout easy use and friendly designed processes gives the opportunity for exporting selected information regarding real estates in the Internet in order to help publicity purposes and sales marketing. Site selection is critical for planning a real estate development project (Li, Yu & Cheng 2005). The dedicated webpage represents the geographical position of the real estate and retrieves the relevant information from the databases according to the user-define query.

19.4 Conclusions

GEOVAL system can be a great tool for real estate analysts. This technology can increase analyst's productivity and accuracy. GEOVAL was designed as a holistic spatial approach to the problem of land management and valuation. The designers of the system do anticipate that this could be the beginning of a discussion of land managers and geographers for the land sustainable development through real estate approaches. The system has two main advantages, accuracy and speed. Thus, it can be used in many different fields like town planning and development, land use etc. The continuous upgrading of the system is a natural selection.

The system introduces innovative elements, which mainly make users' life easier and the production line more constructive. GEOVAL core system remains unchangeable while at the same time is dynamic. This option gives the opportunity to change the shell and adapt the interface in actual user needs. A user friendly interface always assists the users to exploit all system functionalities.

Furthermore, all system functionalities, like for instance, valuation, analysis etc. can be transferred from the desktop system to the web-based system. This development is a great advantage for the GEOVAL while gives the possibility to offer such services through the internet.

Finally, by using GIS technology for spatial analysis and mobile GIS/GPS for field work during real estate recording and valuation, the user transfer portion of the central database to the PDA. This database during field work activities is a powerful tool for the valuator. Not only is a special aid in valuator hands for data collection (e.g. auto-completion due to GPS positioning) but also for having all instruments and information for a draft valuation.

GEOVAL system has been designed and developed based on the actual ideas and experiences from the real estate market. Initiating from the theory of Geography, Cartography, GIS and IT, the accumulated knowledge was combined with the ideas rising from the real estate market. The system has been developed in a step by step process, following the market rules and the actual needs of the users. The goals for developing GEOVAL system has been mainly introduced by the existing and potential users.

The system is useful because has a real estate market placement. At the moment, the system has been installed in two real estate firms. Even though the system is workable in two real estate firms, this period is useful in order to refine the system. The comments and suggestions from the users are acting as "think-tank" for improving system performance and maximising its capabilities and functionalities for the users' benefit.

References

Arbia G (1989) Spatial Data Configuration in Statistical Analysis of Regional Economic and Related Problems. London: Kluwer.

Anselin L (1998) GIS Research Infrastructure for Spatial Analysis of real Estate Markets. Journal of Housing Research, Volume 9, Issue 1, Fannie Mae Foundation.

Anselin L (1988) Spatial Econometrics: Methods and Models. Boston: Kluwer.

Bertin J (1983) Semiology of Graphics, University of Wisconsin Press.

Bible DS, Cheng-Ho H (1996) Applications of Geographic Information Systems for the Analysis of Apartment Rents. Journal of Real Estate Research 12(1), 79–88.

Boots B, Getas A (1988) Point Pattern Analysis, Vol. 10 in G. I. Thrall, editor, Scientific Geography Series, Newbury Park, Calif.: Sage Publications.

Clapp J, Rodriguez M (1995) Using a GIS for Real Estate Market Analysis: The Problem of Grainy Data, Working paper, University of Connecticut.

Cliffe AD, Ord JK (1981) Spatial Processes: Models and Applications. London: Pion.

Cressie N (1991) Statistics for Spatial Data. New York: Wiley.

Fryrear R, Prill E, Worzala ME (2001) The Use of geographic Information Systems by Corporate Real Estate Executives, Journal of Real Estate Research, Volume 22, Nos ½.

Goodall B (1972) The Economics of Urban Areas (Oxford: Pergamon Press).

Goodchild, M (1986) Spatial Autocorrelation. Norwich, Eng.: GeoBooks.

Haynes KE, Fotheringham AS (1984) Gravity and Spatial Interaction Models, Vol. 2 in

Huxhold WE (1991) An Introduction to Urban Geographic Information Systems. Oxford and New York: Oxford Univ. Press.

King LJ (1984) Central Place Theory, Vol. 1 in G. I. Thrall, editor, Scientific Geography Series, Newbury Park, Calif.: Sage Publications.

Li H, Yu L, Cheng EWL (2005) A GIS-based site selection system for real estate projects. Construction Innovation, Volume 5, Number 4, pp. 231–241(11). Emerald Group Publishing Limited.

Marks A, Stanley C, Thrall GI (1994) Criteria and Definitions for the Evaluation of Geographic Information Systems Software for Real Estate Analysis. Journal of Real Estate Literature 2(2), 227–241

Morrill R, Gaile G, Thrall GI (1988) Spatial Diffusion, Vol. 10 in G. I. Thrall, editor, Scientific Geography Series, Newbury Park, Calif.: Sage Publications.

Norcliffe GB (1977) Inferential Statistics for Geographers. New York: Wiley.

Odland J (1988) Spatial Autocorrelation, Vol. 9 in G. I. Thrall, editor, Scientific Geography Series, Newbury Park, Calif.: Sage Publications.

Odland John (1988) Spatial Autocorrelation. Vol. 9. Scientific Geography Series. Thousand Oaks, CA: Sage.

Rodriguez M, Sirmans CF, Marks A (1995) Using Geographic Information Systems to Improve Real Estate Analysis. Journal of Real Estate Research Vol. 10(2), 163–174.

Shilton L, Stanley C (1995) Spatial Filtering: Concentration or Dispersion of NCREIF Institutional Investment. Journal of Real Estate Research 10(5), 569–582.

Silk J (1979) Statistical Concepts in Geography. London: George Allen &Unwin.

Star J, Estes J (1990) Geographic Information Systems: An Introduction. Englewood Cliffs, NJ: Prentice Hall.

Thrall GI (1998) Common Geographic Errors of Real Estate Analysts. Journal of Real Estate Literature 6, 45-54.

Thrall GI (1998) GIS applications in Real Estate and related Industries. Journal of Housing Research 9,1, 32-59.

Thrall GI, Marks A (1993) Functional Requirements of a Geographic Information System for Performing Real Estate Research and Analysis. J. of Real Estate Literature 1(1), 49–61.

Thrall GI, Thrall SE (1990) A Computer-Assisted Decision Strategy for Evaluating New Satellite Hub Sites for a Local Utility Provider. Computers, Environment, and Urban Systems 14, 37–48.

Thrall GI, Thrall SE (1993) Business GIS Data, Part Two: Highend Tiger/Line. Geo Info Systems 3(9), 66–70.

Thrall GI, Thrall SE (1994) Business GIS Data, Part Three: ZIP Plus 4 Geocoding. Geo Info Systems 4(1) (January), 57–60.

Thrall GI, Fandrich J, Thrall SE (1992) The Location Quotient: Descriptive Geography for the Community Reinvestment Act. Geo Info Systems 5(6) (June), 18–22.

Thrall GI, McClanahan M, Thrall SE (1995) Ninety Years of Urban Growth as Described with GIS: A Historic Geography. Geo Info Systems 5(4) (April), 20–27.

Thrall GI, Thrall SE, Ruiz M, Sidman C (1993) Using GIS to Analyze and Visualize Spatial Data. Geo Info Systems 3(5) (May), 59–65.

Thrall GI (1979a) A Geographic Criterion for Identifying Property Tax Assessment Inequity. Professional Geographer 31(3), 278–283.

Thrall GI (1979b) Spatial Inequities in Tax Assessment: A Case Study of Hamilton, Ontario. Economic Geography 55(2), 123–134.

Thrall GI (1982) Public Goods, Externalities, and the Consumption Theory of Land Rent. Papers, Regional Science Association 50 (Montreal), 132–149.

Thrall GI (1988) Land Use and Urban Form. London: Routledge/Methuen.

Thrall GI (1996) Surfer (A Review of a Three-Dimensional Surface Modeling Software Program). Journal of Real Estate Literature 4(1) (January), 73–75.

Thrall GI (1997) GIS and Business Geography: A Retrospective. Geo Info Systems 7(6) (June), 46–52.

Tomlin CD (1990) Geographic Information Systems and Cartographic Modeling. Englewood Cliffs, NJ: Prentice-Hall.

Wyatt PJ (1997) The development of a GIS-based property information system for real estate valuation. International Journal of Geographical Information Science, 11, 435–450.

Section IV: Positioning and Sensor Fusion

20 Quality Assurance / Quality Control Analysis of Dead Reckoning Parameters in a Personal Navigator

Shahram Moafipoor[1], Dorota A. Grejner-Brzezinska[1], Charles K. Toth[1] and Chris Rizos[2]

[1] Satellite Positioning and Inertial Navigation (SPIN) Laboratory
Department of Civil and Environmental Engineering and Geodetic Science
The Ohio State University, Columbus, USA
[2] School of Surveying and Spatial Information Systems
University of New South Wales, Sydney, Australia

Abstract

The personal navigator (PN) prototype, developed at The Ohio State University Satellite Positioning and Inertial navigation (SPIN) Laboratory, integrates GPS, tactical grade inertial measurement unit (IMU), digital magnetometer compass, digital barometer, and uses a human locomotion model to support dead reckoning (DR) navigation for rescue work, security and emergency services, police safety and military applications. The human locomotion model is represented here by the step length (SL) and step direction (SD). In the absence of GPS signals, the SL is predicted by a knowledge-based system (KBS) in the form of artificial neural network (ANN) and fuzzy logic (FL), while the SD is directly measured by the magnetometer and gyro IMU and modeled by a special module of a Kalman Filter, referred to as DR-KF. If the duration of a GPS outage is prolonged, the gyro and to some extent the magnetometer sensor errors will increase due to lack of updated calibration parameters that may result in an unacceptable level of navigation error.

The current target accuracy of the system is 3–5 m circular error probable, 50% (CEP), where the navigation performance depends predominantly on the quality of SD estimation. In this paper, a quality assurance/quality control (QA/QC) mechanism is introduced to test methodologies that predict SL and SD parameters and to monitor their integrity during DR navigation. The QA/QC process proposed here includes fully automated data processing for verification of the measurement and other data content acquiescence and detecting outliers, essential for rejecting incorrectly attributed DR parameters; all these processes are performed in real-time. If the QA/QC analysis rejects the quality of the predicted DR parameters, particularly the SD value, the system may cease operation and issue a warning message requiring the recalibration of sensors. To test the algorithm, in this paper, a performance analysis in the indoor environments is discussed.

Keywords: personal navigator, quality assurance / quality control, knowledge-based system, data validation

20.1 Introduction

Personal navigators (PN) have been studied for about a decade in different fields, such as visual surveillance, rescue work, security and emergency services, police safety and military applications, with the main objective of providing position/heading information for the individual users in various environments. The PN prototype, developed at The Ohio State University Satellite Positioning and Inertial navigation (SPIN) Laboratory, includes GPS and a range of self-contained sensors. The GPS is primarily used to navigate in the open-sky environment and to calibrate the self-contained sensors, using the Extended Kalman Filter (EKF) algorithms. Once all sensors are calibrated, they can be used to support the estimation of parameters required for DR navigation, including step length (SL), step direction (SD), and altitude.

The sensor design of the proposed PN is shown in *Figure 20.1.* The PN backpack configuration includes a dual-frequency Topcon Legacy GPS receiver, a tactical grade Honeywell HG1700 IMU that replaced a Crossbow MEMS IMU used in the early prototype, the HMR3000 magnetometer (inclinometer/compass) that replaced the Azimuth 1000 compass (Grejner-Brzezinska et al. 2006a). The high accuracy Vaisala PTB220 digital barometer (1.5 m height accuracy (1 sigma)) provides the vertical coordinate measure, and a set of four GPS time-synchronized micro-switches, located in the shoe soles (at heels and toes), are used to sense step events, i.e., the instances when the operator's shoes hit and come clear of the ground.

The adaptive EKF module, which includes multisensor calibration functionality, as illustrated in *Figure 20.2,* provides the navigation solution, depending on the quality of the IMU/barometer/compass sensors used, as well as facilitates the GPS measurement modeling (pseudorange, carrier phase, or their linear combinations). The proposed adaptive EKF is used to improve the navigation solution by better evaluating the system and measurement noise covariances. The adaptivity scheme is based on fuzzy logic rules (see, e.g., Sasiadek et al. 2000, Moafipoor et al. 2008).

In *Figure 20.2,* the autocalibration of the magnetometer is an initial calibration procedure designed to calibrate the magnetometer compass with respect to the ferrous materials in the local magnetic field (Moafipoor et al. 2007b). If the

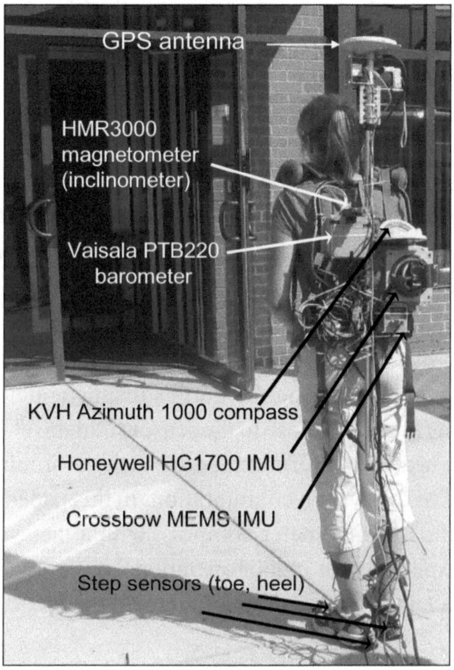

Fig. 20.1. PN sensor configuration.

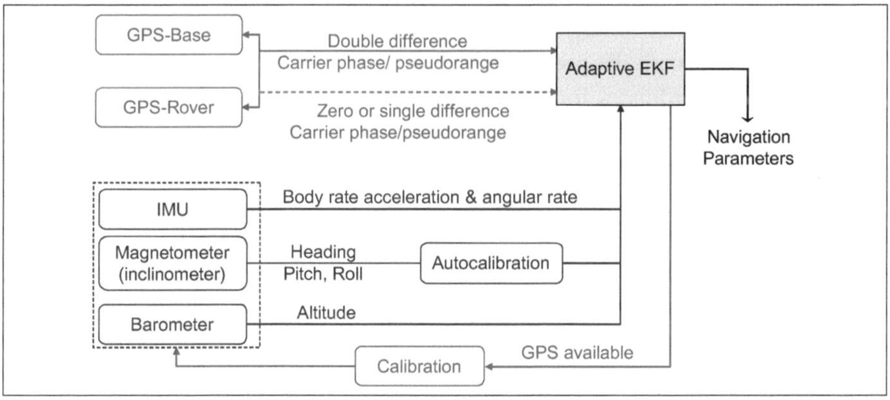

Fig. 20.2. Adaptive EKF and sensor calibration.

magnetic flux density of the operational environment is comparable to the autocalibration environment, the magnetometer heading is likely to be reliable. Otherwise, the magnetic disturbances can be detected using an empirical threshold; any disturbance below the threshold level introduces an error in heading estimation (Ladetto & Merminod 2002).

The approach chosen in the system to determine the SL is based on a human locomotion model (Ladetto et al. 2002, Beauregard 2006, Grejner-Brzezinska 2006b-c and 2007a-b). The human locomotion model is focused on analyzing measurable parameters of gait, as well as their interpretation, such as human dynamics (step interval, step frequency, locomotion pattern, etc.) from their gait. To properly capture the SL of a mobile user in a variety of environments and dynamics, two different versions of the adaptive knowledge-based system (KBS) based on Artificial Neural Networks (ANN) (Grejner-Brzezinska et al. 2006b & 2007b), and Fuzzy Logic (FL) (Toth et al. 2007, Moafipoor et al. 2007a) have been implemented.

The heading solution, proposed to determine the SD, is derived from the combined magnetometer compass and gyro measurements. The magnetometer compass operates based on sensing and measuring the Earth's magnetic field; therefore it is very sensitive to any ferrous materials close to the sensor (Ladetto et al. 2002). To improve the reliability of the heading determination in such environments, a high-quality gyro should be added to the system configuration. In this paper, the heading solution is mainly based on the magnetometer compass, with a tactical grade gyro used to correct the heading especially for indoor operational environments, where the entire system is subject to a substantial amount of ambient magnetic perturbations.

20.2 Introduction to QA/QC Analysis

Starting from a known point, with heading, SD, measured by the gyro/magnetometer in association with the pre-calibrated SL, predicted by the KBS, the DR horizontal position coordinates for the next point can be calculated. However, the DR trajectory reconstruction based on this approach can easily bias the determination of the subsequent positions over time due to factors such as imprecise SL prediction and uncalibrated magnetometer/gyro SD determination. The accumulation of these errors would eventually cause the system to go off the target accuracy specifications. In order to help constrain the growing DR error, different levels of integrity monitoring and reliability assessment are required (Mehra & Bayard 1995, Bhatti 2007). Integrity is an indicator of the trust that can be placed on the precision of the data supplied by the entire system. This includes, in general, the capability of a system to warn the users when the system should not be used for the intended operation. In the DR navigation mode, the warning message generally targets the integrity of the SD and SL parameters, which define the position errors. For this purpose, a validity range should be defined for each parameter separately.

By definition, reliability describes the probability that a system will operate without failure, under given conditions, for a specified period of time. The reliability analysis is mainly considered with two aspects: the potential of detected outlier (inner-reliability), and worst effect of undetected outlier (outer-reliability).

To properly address the reliability and integrity requirements, a quality testing mechanism has been designed to assess the estimated/predicted SL/SD values before sending them into the DR trajectory reconstruction module. In general, there are two basic elements of a system quality testing: Quality Assurance (QA), and Quality Control (QC). The term QA describes all the planned and systematic actions necessary to assure that a parameter will satisfy the specified requirements. The QC, on the other hand, is a set of procedures designed to evaluate the system's output to detect outliers in the delivered measurements, i.e., SL and SD parameters in this case (Juran & Godfrey 2000). Because of different functionality, the QA functions are independent of the QC functions.

The rest of the paper is structured as follows: *Section 20.3* explores the QA/QC functions' design. The comparative performance results, compatible with the new QA/QC processes, are discussed in *Section 20.4.* Finally, the summary and conclusions are presented.

20.3 QA/QC Analysis

The QA is a set of procedures designed to ensure that the predicted and/or estimated process is adequate to meet the PN system objectives. Therefore, the QA activities focus on the process elements and ensure that the process is defined and appropriate, while the QC is a set of procedures designed to evaluate a developed system output (Juran & Godfrey 2000). The QC in the PN focuses on finding outliers in the sensor measurements. To complete the chain of the QA/QC process, a third component, called testing, should also be used. Testing is the process of executing the QA/QC procedures with the intent of finding outliers in the controlled environment. This process includes test planning prior to the execution of the test cases. Thus, testing the performance of the QA/QC algorithms is accomplished by an off-line testing where artificial outliers are introduced, and some metrics defining the level of success is evaluated. Both QA and QC activities are generally required to assure successful system operations.

The QA/QC analysis involves all procedures required to investigate the accuracy (uncertainty), precision, and consistency of the measurements. The most promising system for the purpose of quality testing is offered by accurate reference GPS data. However, under GPS-denied conditions, detection of failures in the PN system requires external information. In addition, because of low redundancy in the DR navigation proposed here, many standard methods for outlier detection (Snow 2002, Cothren 2005), and integrity monitoring (Hewitson & Wang 2006, Bhatti 2007), cannot be implemented. Accordingly, the QA/QC procedure adapted and proposed for the DR navigation in the PN includes, as a minimum, six key elements: (1) assessment of sensor calibration, (2) determination of the upper and lower limits of the sensors' responses, (3) determination of the acceptable accuracy and precision levels, (4) customization of the prediction procedures, (5) training the QA/QC procedures, and (6) sensor data validation. More details on these components are provided next.

20.3.1 Assessment of Sensor Calibration

An assessment of sensor calibration for a PN system in DR navigation mode with three primary outputs, SL, SD, and altitude, depends primarily on defining three criteria: sensor noise, sensor drift, and sensor repeatability (Retscher 2004, Elkaim & Foster 2006). Sensor noise is defined as the internal random errors affecting the sensor measurement. To account for this kind of error in measurements, a proper stochastic model should be developed for the system errors. The sensor drift is the instability of steady state values of the sensor for a constant process value, resulting typically in a departure from the correct (known) measurement values over time. The sensor repeatability is the ability of the sensors to output the same values every time the hardware is placed in the same conditions. *Table 20.1* shows the stochastic error characteristics, repeatability, and white noise introduced for the PN sensors. Notice that the numbers provided in *Table 20.1* are determined for a "generic" sensor, and an empirical derivation of the stochastic models for particular sensors is recommended for better reliability and error handling (Yi & Grejner-Brzezinska 2005, Yi 2007).

The IMU-derived position and attitude information, barometric height, magnetometer (inclinometer) attitude, and GPS data (carrier phase and/or pseudorange measurements in the double difference mode, undifferenced pseudorange or ionosphere-free linear combination of pseudoranges) are integrated together in the tightly coupled adaptive EKF (Grejner-Brzezinska et al. 2006c). The basic idea of the proposed adaptive EKF is to use the innovation vector to self-calibrate the noise covariance of the system or the observations. The innovation vector is defined as a vector of the difference between the measurements and the estimates of the measurement. This vector can be used to keep track of the performance of the filter and adjust its process noise covariance matrix. In this application, the innovation vector comprises the differences between the position and heading estimates obtained from GPS/barometer/compass observations, and their predicted values. In the optimal properly tuned EKF, the innovation vector should have zero-mean Gaussian white noise values (Sasiadek et al. 2000). Deviation of the innovation vector from a zero-mean vector by more than a chosen threshold indicates a reduction in filter performance. So, the performance of the EKF is tracked by monitoring the innovation vector values. Excessive deviation of the innovation vector could be corrected by updating either the state prediction, or the process and measurement noise covariances (Ding et al. 2000).

Table 20.1. Stochastic error characteristics and repeatability for multi-sensor error sources; (mg) stands for 10^{-3} *g*, (μg) stands for 10^{-6} *g*, and *g* is the gravity constant; 1 hpa=1 mbar.

Sensor	Error Sources	Stochastic Error Model	Repeatability	White noise
HG1700 accelerometer	Bias	Random walk	1 mg	20 μg √hr
	Scale factor	Random constant	120 ppm	0
Hg1700 Gyroscope	Bias	Random walk	1 deg/hr	0.125 deg/√hr
	Scale factor	Random constant	10 ppm	0
PTB220 Barometer	Bias	Random walk	0.08 hpa	0.1 m √hr
	Scale factor	Random constant	1 ppm	0
HMR3000 magnetometer	Bias	Random walk	0.3 deg	1 deg √hr
	Scale factor	Random constant	1 ppm	0

20.3.2 Determination of the Upper and Lower Limits of the Sensors Response

The upper and lower limit boundaries of the KBS-based SL prediction and KBS-based SD determination are usually determined during the GPS signal reception. Once these boundaries are created, the actual measured data must be within these ranges, to be considered as valid.

In order to show the way these boundaries are created and knowledge acquired by self-contained sensors, this section displays one particular example, locomotion pattern, representing the desired input and output values. As shown in (Moafipoor et al. 2007a), the locomotion pattern is predicted by using several sensor measurements, where each of the sensors is capable of sensing a response. Then, the KBS module receives these responses and predicts a locomotion pattern for each possible combination of the responses, wherein each of the sensor measurements can be considered to be acceptable or unacceptable. *Figure 20.3* shows an example of the upper/lower limits, represented here as membership function, for the maximum acceleration during a gait cycle (Moafipoor et al. 2007a). To design the membership functions used here, an expert analyzed 24 sample data sets, collected by three operators in a number of field trials. Different experts would perhaps produce a different collection of membership functions; however, due to the tolerance of the fuzzy systems with respect to approximation, the systems would eventually yield similar results, if all the experts have captured the main points of interest (Sasiadek et al. 2000).

During these trials, the operators were walking trajectories with different dynamics and at various speeds, in different environments. As a result, the practical contributions of the membership function presented here are based on a behavioral analysis of the collected data, and therefore subject to better tuning by gaining more information from additional data sets, in which different operators move in a variety of conditions.

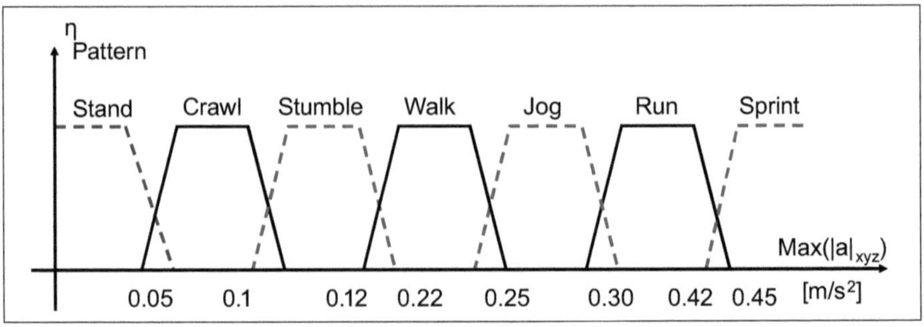

Fig. 20.3. Locomotion pattern membership functions based on maximum acceleration, Max ($|a_{xyz}|$).

In *Figure 20.3,* the vertical axis corresponds to the membership degree that indicates to what level the value of locomotion pattern is in a set, and the x-axis indicates the boundaries defined for each pattern, respectively. This variable is defined based on measured 100 Hz HG1700 IMU accelerations. If the accelerometer drifts over time, and for a predefined locomotion pattern the maximum accelerometer data is not matched with the boundaries defined in *Figure 20.3,* it can be concluded that the sensor calibration no longer holds.

20.3.3 Determination of the Acceptable Level of Accuracy and Precision

To determine the acceptable level of accuracy for SL and SD parameters, a Taylor-Karman Structure is used here (Schaffrin 2006). A Taylor-Karman structure can be used to determine the acceptable level of accuracy for SL and SD estimations for a given target accuracy at specified trajectory length. For example, for a desired accuracy of 3–5 m CEP 50% with a pre-defined trajectory of 500 m, the implementation of the Taylor-Karman structure shows that the SL and SD for each step should be estimated with a precision better than ±5 cm and ±3°, respectively. For longer trajectories, greater precision is required for the DR input parameters.

20.3.4 Customization of the Prediction Procedures

Hausdorff et al. (2001) showed that human motion demonstrates specific characteristics which make the definition and the identification of generic models of human motion feasible. To this end, it may be possible to customize a prediction model for KBS-based SL and SD values (Moafipoor et al. 2008). For example, if the locomotion pattern does not change during the movement, the SL fluctuations are in the range of random white noise with a predefined variance. If the locomotion pattern changes gradually, e.g. from normal walking to fast walking, a Gauss-Markov process can be applied. As the locomotion pattern switches from walking into running, the kinematics of locomotion change abruptly, and therefore an uncorrelated noise can be considered. Applying such knowledge to the DR navigation not only assures correlation between successive SL values, but also improves the system's performance by rejecting the outliers in the predicted SL. Similarly to the framework applied to SL prediction, the SD parameters can also be supervised (or controlled) on the basis of generic models of human motion. Applying such knowledge to improve the integrity of the system produces a framework which exploits statistical models which characterize the human locomotion model. This framework also provides a probabilistic model to combine DR parameters, predicted SL and SD, and to introduce *a priori* knowledge related to the locomotion pattern of operators to detect and track SL and SL parameters during DR navigation.

20.3.5 Training the QA/QC Procedures

Once a structure to predict the DR parameters is created, an independent subsystem is required to implement the QA/QC procedures. Thus, a KBS in the form of a Kalman filter, separate from the navigation/calibration EKF, referred to as DR-KF, is introduced to model the DR parameters. The DR-KF is used for not only implementing/updating the predicting model for SL and SD parameters but also reconstructing the DR trajectory. The DR-KF includes five unknowns: SL, heading rate, SD, and horizontal position coordinate.

The flexibility of the DR-KF is mainly achieved by assigning different prediction error models as a function of the locomotion pattern. Using the knowledge about the mechanism of walking in terms of SD and SL can be useful in adjusting the estimations, particularly SD, and in smoothing the trajectory reconstruction. The observation model, including SD and SL measurements with the corresponding error models, updates the prediction model. The recursive structure of the DR-KF also estimates the covariance matrix, indicating the quality of the reconstructed trajectory. Furthermore, this scheme converts the point-to-point DR navigation into a dynamic navigation. The fundamentals of this method are discussed in (Moafipoor et al. 2008).

20.3.6 Sensor Data Validation

The goal of sensor data validation is to ensure a high degree of certainty for sensory data at all times, and to ensure that the system runs according to the predetermined specifications (Juran & Godfrey 2000). A real-time sensor data validation procedure in the PN can be developed on the basis of assigning a level of validation that is an empirical number in the range, for example, 0 to 3 as shown in *Table 20.2,* indicating the degree of assurance in the data. These numbers do not have any physical meaning but they allow for a rough assessment of the level/content of the sensory data before using them in the DR navigation mode.

Table 20.2. DR navigation parameters at each data validation level.

Level	SD	SL	Altitude
0	– Raw gyro data – Raw magnetometer data	– Untrained locomotion pattern – Raw IMU/barometer data	– Raw barometer data
1	– Magnetic anomaly – Applying last calibration parameters to the gyro	– Untrained locomotion pattern – Applying last calibration parameters to the IMU/barometer	– Change of initial temperature and pressure
2	– Magnetic flux stable – Applying last calibration parameters to the gyro	– Trained locomotion pattern – Applying last calibration parameters to the IMU/barometer	– Applying last calibration parameters to the barometer
3	– Calibration of magnetometer compass – Calibrated gyro	– Trained locomotion pattern – Calibrated IMU/barometer data	– Calibration at a known level, or known point

Level 0 is essentially assigned when the PN system operates without calibration for a long time. Because no valid sensor calibration is available, the data output can be considered as raw data, and thus, have the lowest validity level for reconstructing the DR trajectory. In contrast, if GPS-based calibration exists, current biases and scale factors are known, the system enters the top validation level, which involves quality checks for calibration of the PN sensors; quality flags are assigned to each sensor to indicate its calibration quality. Data are considered at level 3 after all calibration of the PN sensors is completed. While level 3 validation for SD and altitude is determined by the validation of the corresponding sensory data, the validation of the SL modeling must also be confirmed by the locomotion pattern; if the SL KBS module is not trained for the current locomotion pattern, then a lower validity level should be assigned to SL modeling.

Levels 1 and 2 are generally considered for validation of operation in confined environments. Under GPS-denied conditions, the PN performance depends on how well the last recorded calibration parameters represent the status of the various sensors. For the magnetometer compass, the validation is considered in relation to the stability of the magnetic field. If the magnetic flux density of the operational environment is not comparable to the initial calibration environment, the magnetometer heading is likely to be unreliable (level 1). Similarly, the barometer performance is strongly affected by the temperature in the changing environment and pressure changes due to varying airflow. If the barometer is moving in areas with different pressure and temperature, its re-calibration may not be feasible, and the barometer data can be used only in relative terms. Provided that the environmental conditions, such as temperature, pressure and magnetic density of the initial calibration and DR navigation areas remain largely unchanged, a level 2 validation can be assigned to the data. Without any absolute update, however, the accumulation of the errors would exceed the target accuracy over time, and thus, for longer trajectories, an additional mechanism is required to help constrain the growing DR error, such as the recalibration of the PN sensors using other independent data, such as known waypoints, direction, or level surface assumptions. This aspect is an ongoing investigation task.

20.4 Performance Analysis

This section provides a performance evaluation of the prototype, with a special emphasis on QA/QC techniques applied in the DR navigation. Several data validation and integrity monitoring experiments (explained in the previous section) were conducted in order to verify the system performance, as demonstrated by the quality of the navigation solution for a trajectory in a confined environment.

For this purpose, a series of datasets were collected on August 21, 2007 at the Center for Mapping (CFM), inside a single-story building located on The Ohio State University campus. After performing outdoor calibration (all outdoor calibration demonstrations in this paper are based on double-difference carrier phase observations) and tuning the filter as well as updating the KBS (Grejner-Brzezinska et al. 2007b), the system was ready to perform navigation in the confined environment. After sufficiently long outdoor calibration and training performed as a part of "regular" navigation tasks, the user entered the building and walked a part of the hallways of the CFM for about 78 m in 2.5 minutes. En route, the person followed the control points marked in the hallways. The floor plan of the building was previously acquired by classical surveying methods, and several control points were established in the hallways with an accuracy of around 1 cm, to provide the reference trajectory. An example data set, which includes a few steps on a staircase along the operator's path, was selected for a preliminary assessment of the QA/QC procedures that are currently under development.

In accordance with the QA/QC functions, described in *Section 20.3,* several tests were performed. The main objective was to identify the potential failure situations in the SL, SD, and altitude modeling, as well as in their integration to reconstruct the DR trajectory. The process began with analyzing the prediction of these variables. First, the operator's motion was parameterized for each step by a set of variables, derived from the PN sensor outputs. According to the type of model (ANN/FL), the corresponding KBS was used to process the information and to predict the SL modeling.

The method used to determine the SD value was based on the integration of gyro and magnetometer sensors, because in the absence of GPS signals, both the gyro and, to a lesser extent, the magnetometer suffered from many environmental limitations. Nevertheless, gyro/magnetometer integration may aid accurate heading determination over a longer period of time. For this purpose, in the approach presented here, the two sensors were integrated through the same EKF that was used for multisensor integration. However, depending on the duration of the GPS gap, different weights were assigned to the gyro and the magnetometer measurements. For example, in the current implementation, if the gap duration is within a few minutes, the noise variances of the gyro and the magnetometer compass heading observations are set to $(0.1^{o})^{2}$ and $(1^{o})^{2}$, respectively. For larger gaps, the gyro's weight is decreased, and therefore, in such cases, the integrated solution will be governed by the magnetometer compass measurements.

The first step in QA analysis of the PN is the assessment of the quality of the sensor calibration. The appropriate indicator, as explained in *Section 20.3.1,* is the innovation sequence. *Figure 20.5* shows a sample innovation sequence derived on the basis of the double-difference carrier phase GPS/IMU/barometer/magnetometer integration. The horizontal axis in *Figure 20.4(a)* represents the number of GPS observations, and in *Figure 20.4(b),* the horizontal axis represents the innovation sequence for four other measured variables, including barometer observations and magnetometer (inclinometer) attitude observations. Since in an optimal, properly tuned EKF, the innovation vector should have zero-mean Gaussian white noise values, *Figure 20.4* confirms that the EKF is properly tuned, and also that stochastic error characteristics, repeatability, and white noise observed in *Table 20.1* are acceptable.

The second step of the QA/QC analysis was carried out using a scheme for global and local checking of DR parameters, which aims at determining if the predicted values are valid or not, and how a local outlier can be identified. The global checking is briefly addressed in *Table 20.2,* where different data validation levels are presented. For local checking, in an off-line testing process, attempts were made to identify several artificial outliers which were introduced in the controlled

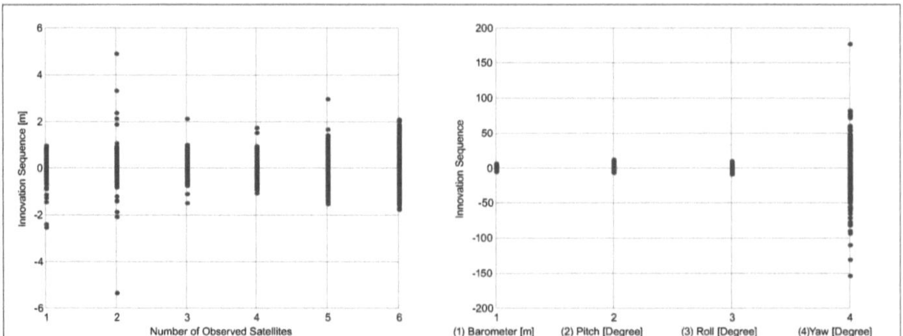

(a) GPS double difference carrier phase data (b) Barometer, pitch, roll, and yaw data.

Fig. 20.4. Innovation sequence and double-difference carrier phase GPS/IMU integration.

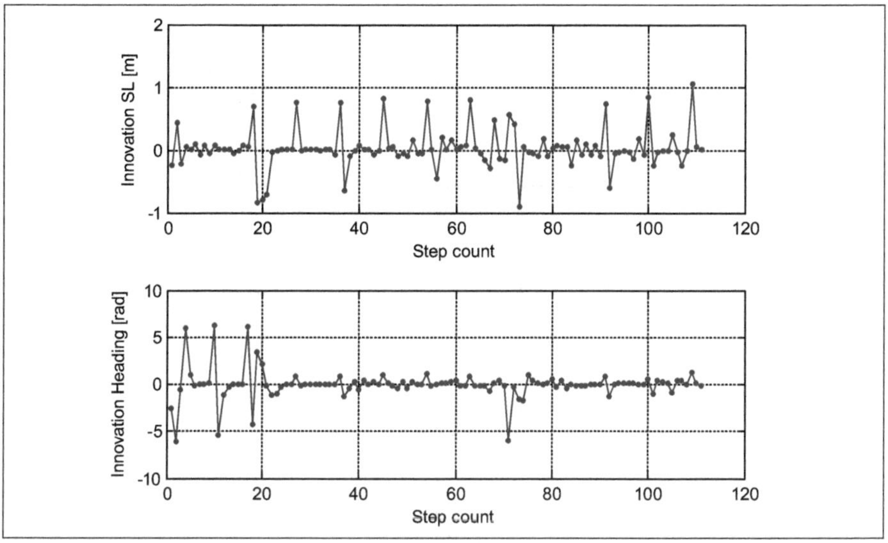

Fig. 20.5. The potential failure modes and the innovation sequence.

environment. Accordingly, random Gaussian errors, with zero mean and standard deviations of about 0.5 m for SL values and 20 degrees for SD values were simulated and introduced into sufficient number of random points. To detect the simulated SL/SD outliers, the innovation sequence, obtained from the DR-KF structure, was used. As shown in *Figure 20.5,* the residuals higher than 0.5 m and 20° should represent the outliers. The performance of the QC functions was tested for outlier detection, and a 90% success rate was achieved for SL and 60% for SD.

The difference in the success rate between SD and SL data is caused by the fact that SL/SD change factors from one step to the next are not necessarily the same. Since the locomotion patterns of the simulated points have not been changed, a comparison between the prediction (designed on the basis of previous and current locomotion patterns) and observation models can identify the outlier observations. However, for the SD parameter, defining a threshold is more complex and requires more redundancy and environmental information, such as the curvature of path, to determine the outlier observation.

Once the potential failure modes and SL/SD models were tested (inner-reliability), the next validation step was applied to the predicted SL and SD values, and their integration to reconstruct the DR trajectory. Instead of many integrity monitoring systems where the bad observations are simply rejected in the DR navigation, due to a low redundancy, an attempt is made to correct the "suspect" values. Thus, in the next stage, it must be decided whether the "fixed" data are sufficiently reliable to reconstruct DR navigation trajectory for a particular period (outer-reliability). In the case of outer-reliability test failure, the DR navigation is terminated.

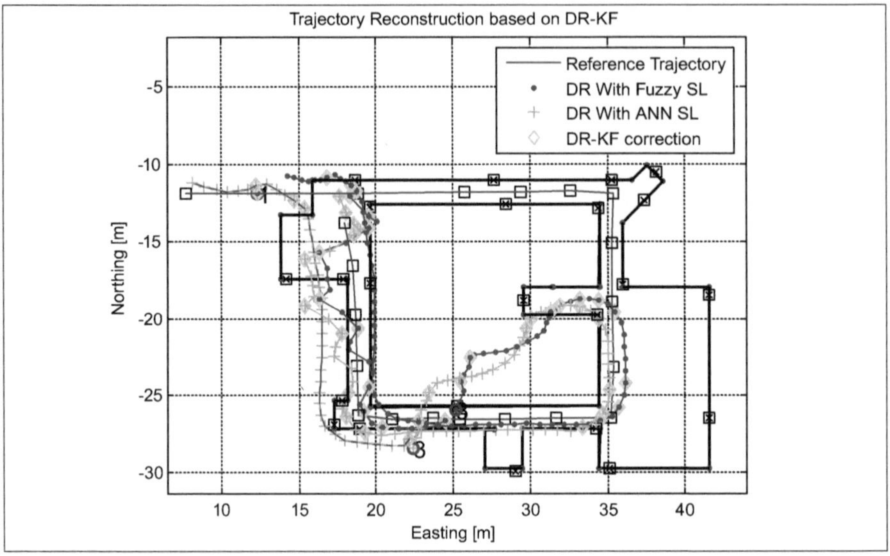

Fig. 20.6. CFM floor plan and DR-KF trajectory reconstruction based on FL SL (dot symbols) and ANN SL (plus symbols) integrated with gyro and magnetometer heading, adjusted by DR-KF module (diamond symbols).

Figure 20.6 shows the DR navigation trajectory, reconstructed using FL-based and ANN-based SL modeling with integrated gyro and magnetometer sensor data. The square symbols in *Figure 20.6* represent the ground control points that were followed by the operator, representing the reference trajectory. The DR trajectories reconstructed by SL-Fuzzy and SL-ANN and the SD are plotted in dot and plus symbols. The diamond symbols represent steps where inconsistencies were detected by the QA/QC mechanism; these inconsistencies were successfully fixed in the DR-KF structure later.

On the DR navigation solutions shown in *Figure 20.6,* several comprehensive analytical QA/QC procedures were carried out with the objective to maintain the integrity of the DR navigation results. These analyses were conducted for ANN/FL-based SL modeling, magnetometer compass heading solution, and gyro heading solution. One of the challenges in this experiment was to properly model the SL while walking the stairs. Despite the expectation that a new kind of locomotion pattern for walking stairs would be required, it was finally concluded that it is a walking (or stumbling) activity with displacements in both horizontal and vertical directions. It was, therefore, properly handled by the FL-based approach. On the other hand, the ANN method, designed to perform input-output mapping, requires training sets for staircase data which were not available prior to this test. Consequently, inadequate prior ANN training resulted in outliers in the test results. Similarly to the study performed for potential failure modes, to detect the ANN-SL outliers the innovation sequence obtained from the DR-KF structure was used. As shown in *Figure 20.7,* the residuals higher than 0.5m represent the outliers. Once the suspect data were identified, they were flagged for additional review and corrective action, as necessary. To fix the outliers, the prediction value, corresponding to the locomotion pattern, can be used.

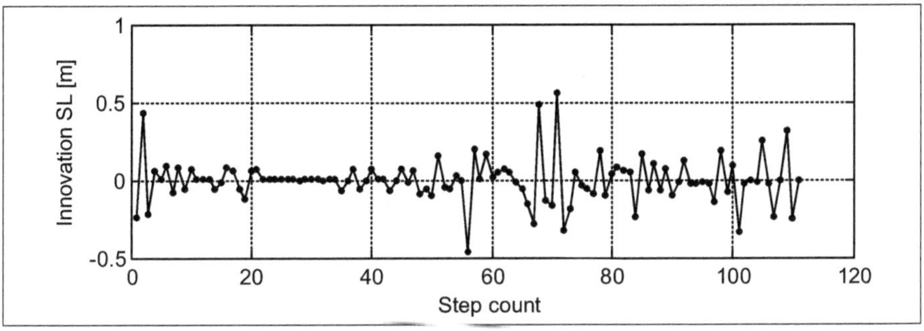

Fig. 20.7. Outlier detection and the innovation sequence for ANN-SL.

The performance of the magnetometer compass in the confined environment was not as good as in the open environments. The validation of the magnetometer compass heading is based on testing the magnetic anomaly. For example, the difference in magnetic flux density between the autocalibration area (outside) and the confined environment (inside the CFM building) was of more than 50 milligauss, causing several tens of degrees of heading deviation. The experiments showed that the magnetic anomaly less than 10 milligauss (empirical threshold) was a suitable threshold to validate the magnetometer heading. If the magnetometer heading was to be used in reconstructing the DR trajectory, its integrity had to be monitored on the basis of generic models developed in the DR-KF. *Figure 20.8* shows the innovation sequence, as the difference between the prediction heading and the observation heading.

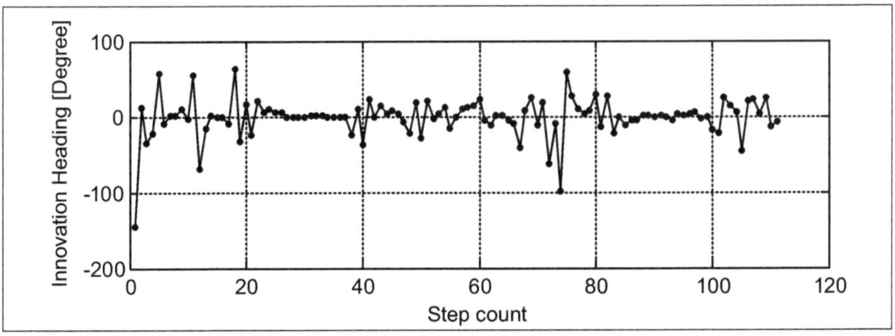

Fig. 20.8. Outlier detection and the innovation sequence for magnetometer heading.

Several large residuals in *Figure 20.8* showed the epochs where severe magnetic anomalies were observed, the biggest one corresponds to the time the user entered the building. Large residuals were also observed when the operator approached the steps on the way, where the support person had to perform a complex maneuver in order to let passing the portable recording device on the steps; the logging unit included the cable, connections between the self-contained sensors and the main computer with power supplies. After that, only a few outliers were detected, and even after applying corrections in the attempt to fix the heading value, the resulting values were still unreliable.

In this experiment, the gyro solution validation was set to level 2, as illustrated in *Section 20.3.6.* Thus, the reliability of the gyro heading is a function of time. For a few minutes after calibration, and given moderately unchanged locomotion dynamics, the gyro heading was sufficiently reliable to hold the attitude. In order to increase the reliability, the DR-KF heading was used in this experiment to determine the SD value. The solution is predicted based on the walking locomotion pattern and is updated based on the observation values obtained from the integration of gyro and magnetometer.

On the basis of the QA/QC strategy, if the difference between the actual measurement heading and the predicted measurement is more than a predefined threshold, this observation will be considered an outlier. The empirically selected threshold ranges between 3° and 5°, depending on the locomotion pattern. This threshold is applied to the DR-KF results in outlier rejection and observation replacement by the predicted measurement values, especially for the data collected at the entrance to the building, where many magnetic anomalies were observed that impacted the measured SD value. In general, the DR-KF was successful in trajectory reconstruction and adequate QC for both SD and SL. This can be observed in the statistical results shown in *Table 20.3,* and in the trajectory points fixed by this method, marked in *Figure 20.6* with cyan and green diamond symbols.

Table 20.3. Statistical fit to reference trajectory of DR trajectories generated using SL predicted with FL and ANN, and the integration of gyro and magnetometer compass heading adjusted by the QA/QC module.

Test data set	SL model	Mean [m]	Std [m]	Max [m]	End Misclosure [m]	CEP (50%) [m]
Trajectory length: 78 m	FL	2.97	1.73	3.05	1.43	1.57
	ANN	2.82	2.71	4.24	3.64	1.58

These results indicate a difference in the performance of the SL modeling and trajectory reconstruction performed by the ANN KBS and FL KBS. However, the 78m trajectory was reconstructed in the DR-KF module with a CEP (50%) less than 2m. Note that the CEP (50%) is defined here somewhat loosely. Since a sufficiently large number of test trajectories is not available at this point to define the

CEP based on the misclosures of a representative sample of trajectories, in these tests each navigation step is treated as a "sample," and used to compute its residual with respect to the reference trajectory. These samples are subsequently used to determine the 50% threshold that is reported here. The position failure is indicated whenever the difference between the true position and the derived position exceeds the target accuracy of the system.

20.5 Conclusions and Future Work

The QA/QC mechanism for a PN supported by an adaptive KBS has been designed and is currently being implemented and tested. The QA/QC is the essential component in designing any navigation system, to assure data integrity and navigation reliability over time. The QA/QC analysis involves several procedures required to investigate the accuracy (uncertainty), precision, and consistency of the system performance. In order to monitor the integrity of the PN system, several factors were defined, including (1) the number of redundant observations accessible for SL, SD, and altitude, (2) the inner-reliability and the potential failure modes, (3) the size of acceptable error, and (4) the quality of the data validation.

The preliminary performance of the QA/QC algorithms, currently under development, was tested by analyzing the system in a challenging environment. It was determined that during indoor navigation, where the user walked 78 m in about 3 minutes, the DR performance tested by the QA/QC mechanism still met the required specifications, with the CEP within the required 3—5 m range. More tests are required to properly validate the QA/QC algorithms along more complex paths, including longer stairways, multi-storey building, etc. These tests are currently underway.

Acknowledgements

This research is supported by the National Geospatial-Intelligence Agency 2004 NURI grant.

References

Beauregard S (2006) A Helmet-Mounted Pedestrian Dead Reckoning System. Mobile Research Center, TZI Universität, IFAWC2006, March 15-16, Bremen, Germany, pp. 79-89

Bhatti UI (2007) Improved integrity algorithms for integrated GPS/INS systems in the presence of slowly growing error. (http://www.cts.cv.ic.ac.uk/documents/theses/BhattiPhD.pdf) (accessed 3 July. 2008).

Cothren J (2005) Reliability in Constrained Gauss-Markov Models: An Analytical and Differential Approach with Applications in Photogrammetry. Geodetic Science and Surveying Report number 473 (http://www.ceegs.ohio-state.edu/gsreports/), The Ohio State University, Columbus, OH.

Ding Z, Leung H, Chan K (2000) Model-set Adaptation using a Fuzzy Kalman Filter, Proceedings of the 3rd International Conference on Information Fusion, Paris, France, pp. 3-9

Elkaim GH, Foster CC (2006) MetaSensor: Development of a Low-Cost, High Quality Attitude Heading reference System, Proceedings of the ION GNSS Meeting, September 26 - 29, Fort Worth, Texas, USA, pp. 1124 - 1135.

Grejner-Brzezinska DA, Toth CK, Moafipoor S, Jwa Y (2006a) Multi-Sensor Personal Navigator Supported by Human Motion Dynamics Model, Proceedings of the 3rd IAG Symposium on Geodesy for Geotechnical and Structural Engineering/12th FIG Symposium on Deformation Measurements, May 2006, Baden, Austria, CD ROM.

Grejner-Brzezinska DA, Toth CK, Moafipoor S, Jwa Y, Kwon J (2006b) A Low Cost Multi-Sensor Personal Navigator: System Design and Calibration. Presented at IEEE/ION PLANS Meeting, April 25-27, San Diego, CA, USA.

Grejner-Brzezinska DA, Toth CK, Jwa Y, Moafipoor S (2006c) Seamless and Reliable Personal Navigator. Proceedings of the ION Technical Meeting, January 18-20, Monterey, CA, USA, CD ROM, pp. 597-603.

Grejner-Brzezinska DA, Toth CK, Moafipoor S (2007a) Adaptive Knowledge-based System for Personal Navigation in GPS-Denied Environments, Proceedings of the ION National Technical Meeting, January 22-24, San Diego, CA, USA, CD ROM, pp.517-521.

Grejner-Brzezinska DA, Toth CK, Moafipoor S (2007b) Pedestrian Tracking and Navigation using Adaptive Knowledge System based on Neural Networks and Fuzzy Logic, Journal of Applied Geodesy, vol. 1, No. 3, pp. 111-123, 2007, invited.

Hausdorff JM, Ashkenazy Y, Peng CK, Ivanov PC, Stanley HE, Goldberger AL (2001) When Human Walking Becomes Random Walking: Fractal Analysis and Modeling of Gait Rhythm Fluctuations, PHYSICA A, Vol. 302, pp.138-147.

Hewitson S, Wang J (2006) GNSS Receiver Autonomous Integrity Monitoring (RAIM) with a dynamic model. Accepted for publication in the Journal of Navigation. (8 Nov. 2006)

Juran JM, Godfrey AB (2000) Juran's Quality Handbook, McGraw-Hill International Editions: Industrial Engineering Series.

Ladetto Q, Merminod B (2002) Digital magnetic compass and gyroscope integration for pedestrian navigation. 9th Saint Petersburg International Conference on Integrated Navigation Systems, Saint Petersburg, Russia, May 2002.

Ladetto Q, Seeters J van, Sokolowski S, Sagan A, Merminod B (2002) Digital Magnetic Compass and Gyroscope for Dismounted Soldier Position & Navigation, in NATO-RTO meetings, Istanbul, Turkey, October 2002.

Mehra R, Bayard D (1995) Adaptive Kalman Filtering, Failure Detection and Identification for Spacecraft Attitude Estimation, Proceedings of the 4th IEEE Conference on Control Applications, September 28, Albany, NY, USA, pp.176-181.

Moafipoor S, Grejner-Brzezinska DA, Toth CK (2007a) A Fuzzy Dead Reckoning Algorithm for a Personal Navigator, Proceedings of the ION GNSS Meeting, September 2007, Fort Worth, Texas, USA, CD-ROM, 2007, pp.48-59, in review for NAVIGATION.

Moafipoor S, Grejner-Brzezinska DA, Toth CK (2007b) Adaptive Calibration of a Magnetometer Compass for a Personal Navigation System, Proceedings of the International

Global Navigation Satellite Systems Society (IGNSS) Symposium, the University of New South Wales, December 2007, Sydney, Australia, CD ROM.

Moafipoor S (2008) Adaptive Kalman Filtering for Personal Navigation System, Presented at the 2nd Annual Workshop of the Consortium of Ohio Universities on Navigation and Timekeeping (COUNT), April 7-8, The Ohio State University, USA.

Moafipoor S, Grejner-Brzezinska DA, Toth CK (2008) Multi-Sensor Personal Navigator Supported by Adaptive Knowledge Based System: Performance Assessment, Proceeding of the IEEE/ION PLANS 2008 Meeting, May 5-8, in Monterey, California, USA, CD ROM.

Retscher G (2004) Multi-sensor Systems for Pedestrian Navigation, Proceedings of the ION GNSS 17th International Technical Meeting of the Satellite Division, September 21-24, Long Beach, CA, USA, CD-ROM.

Sasiadek JZ, Wang Q, Zeremba MB (2000) Fuzzy Adaptive Kalman Filtering for INS/GPS Data Fusion, Proceedings of the 15th IEEE International Symposium on Intelligent Control (ISIC 2000), July 17-19, Rio Patras, Greece, pp. 181-186.

Schaffrin B (2006) GS765: Network Analysis. Lecture notes, The Ohio State University, unpublished.

Snow KB (2002) Applications of Parameter Estimation and Hypothesis Testing to GPS Network Adjustments, Geodetic Science and Surveying Report number 465 (http://www.ceegs.ohio-state.edu/gsreports/), The Ohio State University, Columbus, OH.

Toth CK, Grejner-Brzezinska DA, Moafipoor S (2007) Pedestrian Tracking and Navigation using Neural Networks and Fuzzy Logic, Proceedings, IEEE International Symposium on Intelligent Signal Processing, Alcala De Henares, October 2007, Madrid, Spain, CD ROM, pp. 657-662.

Yi Y, Grejner-Brzezinska DA (2005) Non-Linear Bayesian Filter: Alternative to the Extended Kalman Filter in the GPS/INS Fusion Systems, Proceedings of the ION GNSS, September 13-16, Long Beach, CA, pp.1391-1401.

Yi Y (2007) On Improving the Accuracy and Reliability of GPS/INS-based Direct Sensor Georeferencing, Geodetic Science and Surveying Report number 484 (http://www.ceegs.ohio-state.edu/gsreports/), The Ohio State University, Columbus, OH.

21 Efficient Overlay Mediation for Mobile Location-Based Services

Amine M. Houyou and Hermann De Meer

University of Passau, Faculty of Computer Science and Mathematics
Passau, Germany

Abstract

Data communication in mobile and cellular environments is still in its infancy. Data traffic generated by mobile users lacks the diversity of the fixed Internet. On the one hand, the limited bandwidth available for mobile users is still a major hurdle for some Internet applications. On the other hand, a universal inexpensive mobility solution built to support the different wireless architectures while offering service ubiquity is still missing. Despite this, some applications such as location-based services (LBS) have found a natural application domain by mobile users. Data traffic, in such applications, is generated by users accessing information linked to their movement context. To improve access to these services while limiting the effects of mobility, communication should occur near the edge, while staying localized. In this paper, a distributed approach is proposed for mediating location information between heterogeneous geographic information severs in a peer-to-peer manner. A design methodology is proposed to ensure the localized impact of mobility-lead communication overhead. Both numerical and simulative analyses are carried out.

Keywords: mobile LBS, vertical handover, overlay, range queries, clustering

21.1 Introduction

Data services in wireless cellular networks still have not reached the extent of the fixed Internet, partly due to lacking bandwidth and to a costly mobility management. Cellular networks track the user's position in order to limit interruptions due to handover, which is an essential quality of service requirement for voice services, and also to deliver calls initiated by remote correspondent nodes. With the development of new wireless techniques such as cooperative and mesh networking, more bandwidth is offered at the wireless edge, but so is the diversity which new mobility management techniques have to deal with. In such scenarios vertical handover could lead to improvement of access to bandwidth but might introduce lengthy interruptions.

Mobile or ubiquitous location-based services (Rao & Minakakis 2003) are an example where mobility is a major incentive to communication. Normally, we talk about LBS (Schiller & Voisard 2004), even for services accessed through fixed Internet in a centralized manner, if the service on offer is associated to a geographic location (e.g. online access to a database or geographic information servers (GIS)). A mobile LBS, in contrast, allows mobile users to ubiquitously and adaptively access different content and information which is tailored to their changing location. The location of the user is taken into account to deliver timely, relevant content and information.

In this paper the transition between wireless networks is not of focus, but rather placing the LBS in a distributed manner as near to the mobile as possible. This is done thanks to an overlay-based architecture, which allows efficient and distributed range queries while taking into account the distributed nature of GIS servers and their heterogeneity. The overlay reorganizes the data scattered among the different GIS on a new plane to allow quick readapting to the status of the managed objects and retrieval of data with minimal communication overhead. The P2P overlay offers in this case a type of middleware that mediates content information back to the user. Mobility management is intrinsic to the overlay solution, which takes the user's location and movement into account.

The paper is structured as follows. The related work is presented in *Section 21.2*. *Section 21.3* takes an example application with hard requirements such as decentralized GIS, timeliness of data retrieval, and a dynamic status change of the retrieved LBS data. *Section 21.4* presents the analytic analysis of the range query cost. The structure of the overlay is further optimized as a result of the analysis. Simulation results, in *Section 21.5,* also confirm the structure choices of the overlay system. *Section 21.6* concludes the paper.

21.2 Related Work

Most mobile LBS assume a mobility solution at the lower layers. While mobility research has matured regarding cellular networks and even 3G and beyond scenarios (Akyldiz et al. 2004), major efforts are still needed in WMNs and Ad-Hoc networks (Akyldiz et al. 2005). Using overlays to manage mobility or as a support for mobility has also resulted in several proposals including earlier work by Stemm and Katz (1998) supporting vertical handovers. I3-based protocol (Zhuang et al. 2005) followed as a large scale routing framework replacing mobile IP. Similar to the home and visited location registries the Palma P2P protocol manages the user's location (Sethom et al. 2005). The work presented in this paper looks at supporting the mobile user itself in identifying the wireless diversity offered in 4G scenarios. The overlay manages mobility similarly to i3 delivering query results to whichever network the user is. An important aspect of the proposed use of overlay is the support of continuous location-based range queries. The mobile user, and based on its location, can query the decentralized LBS (Houyou et al. 2006) offered by the overlay network.

Range queries in P2P systems have addressed several issues such as non uniform partitioning of the object space. In this type of range queries, data ranges are neither continuous nor uniformly partitioned. This is the case for dictionary entries or names of people, where given ranges abruptly change. For this type of queries, several proposals have been made including Kademlia (Maymounkov et al. 2002) and more systematically P-Grid based overlay (Datta et al. 2005). P-Grids are structured as prefix hash trees (PHT) (used in traditional database research). This suggests organizing the overlay as a data trie (Datta et al. 2005), which addresses nodes according to a binary tree structure. For instance, a peer addressed with "001" would be on the route to all nodes or objects whose addresses share this prefix. In order to extend the scope of each peer at the lower part of the binary tree, these latter edge peers include pointer to other far off edge peers. Combined with search algorithms at the edge, the proposal in Datta et al. (2005) is well suited for fragmented keyword based range queries. It is though worth noting that once at the edge the search algorithms are closer to flooding.

Another family of solutions is based on the multi-dimensional DHT CAN (Ratnasamy et al. 2001). In Sahin et al. (2002) a geographic coding is proposed. Although offering a better mapping between the multiple dimensions needed to describe a geographic zone, CAN based solutions partition the key space in multiple dimensional large zones, leading to flooding at the edge once reaching the zone. The dimension conversion is very restrictive, since once set, it is hard to encode anything else apart from zones, and therefore even for exact match query a range query is used (Datta et al. 2005).

For this a Hilbert-based transformation function used in this paper has the advantage of offering both the possibility to convert any multi-dimensional attribute space to a one-dimensional hash value. It also allows an exact match query as well. Knoll et al. (2006) have demonstrated the clustering property of the Hilbert Curve when applied to modelling geographic coding. Moon et al. (2001) have made a pioneering study on the clustering property of the Hilbert-curve. And in this paper, the same space filling curve is used as a main cornerstone to design location-based distributed overlay systems.

21.3 Overlay-Based LBS Design

21.3.1 Application Example: LBS Support for Vertical Handover

Mobile LBS require service provision which takes user movement into account. The service delivered changes with the movement and location of the user. Movement tracking is usually done with the help of a tracking hardware like GPS, which makes the user aware of its location.

Assuming that the user can use the movement tracking capabilities, mobility management, especially in 4G scenarios could be supported by an LBS. Based on the location of the user, a query is started to search for available heterogeneous wireless cells belonging to various operators. The discovery of layout of wireless cells allows the mobile node to select a wireless node that most suits its context. This latter problem has been approached in (Balasubramaniam & Indulska 2004) as multiple objective optimization decision process. Vertical handover is seen as the process of selecting the cell that suits the QoS needs of the user among several alternatives. Each alternative cell is evaluated according to several objectives like (i) minimizing interruption time, (ii) maximizing bandwidth, (iii) longest connectivity possible, (iv) minimizing price of communication, (v) minimizing jitter, etc. The score each prospective cell achieves is combined in a utility function, which is used to select the best handover alternative between overlapping cells or next in line cell.

In another simpler approach, the mobile node simply looks up nearest free of charge wireless LAN access point (similar to war driving) the user can move towards that place using a handheld navigation system.

Such a query has to be constrained by the following system requirements:

- Look up heterogeneous wireless cells among different operators.
- Look up only those wireless cells near the movement path of the user.
- Look up those wireless cells that the end device is capable to support.
- Limit the query path in the network for timeliness considerations (i.e. a result of a query is only relevant if it is sent back before the user has moved out of that cell).
- A wireless resource status could change which means that the query result is more likely to be invalid if the query occurred too long before the position has changed.

21.3.2 Data Model

The data items[1] represent an element managed and provided by a specific GIS. The data item is represented with a description file (XML file), listing the different

[1] Data items refer to objects, resources, XML descriptions, or service instances managed by the overlay and are the goal of the query or search process

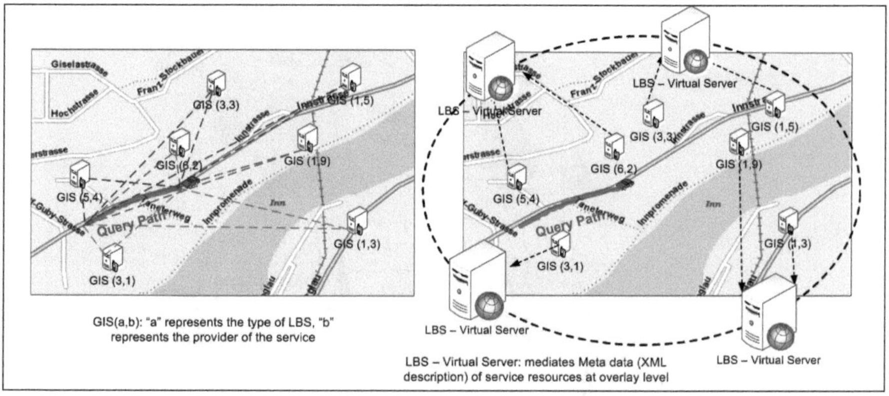

Fig. 21.1. Left: current situation of fragmented LBS among operators and type of services. Right: use of a single management plane to coordinate heterogeneous GIS via an overlay.

attributes and functionality of the geographically bound objects. Each description includes a list of (i) static attributes, describing (for our application example) a wireless network with its link technology, frequency channel/modulation, location, coverage geometry, and access rights (AAA) and possibly IP-address prefix; (ii) dynamic attributes, which could include the number of connected users, whether the cell is available for handover or not, or possibly cost of access. Other types of content/services could be integrated in the decentralized LBS as long as they are geographically addressable.

Figure 21.1 shows on the left hand side a naïve approach which requires a separate query to each GIS system which might contain data items linked to the same movement path. An overlay constructs a virtual network capable of routing to reach virtual nodes (virtual servers) as well content.

A virtual server gathers pointers and description files of the underlying GIS it represents. Further SQL query in a virtual server allows integration of information in the reply message. A single response message is generated by each virtual node back to the mobile node.

21.3.3 Overlay Node Organization and Interactions

The overlay connects virtual LBS servers. Its structure is formed with the help of a DHT protocol. The virtual nodes (or peers) build up a structured P2P network.

The overlay offers the following functionalities:

- Automatic resource discovery: content is pushed by GIS with a soft-state, but any other peer can discover available content once part of the overlay.
- Translation between service providers: the overlay offers an abstraction from the different content providers, also to the different services offered.

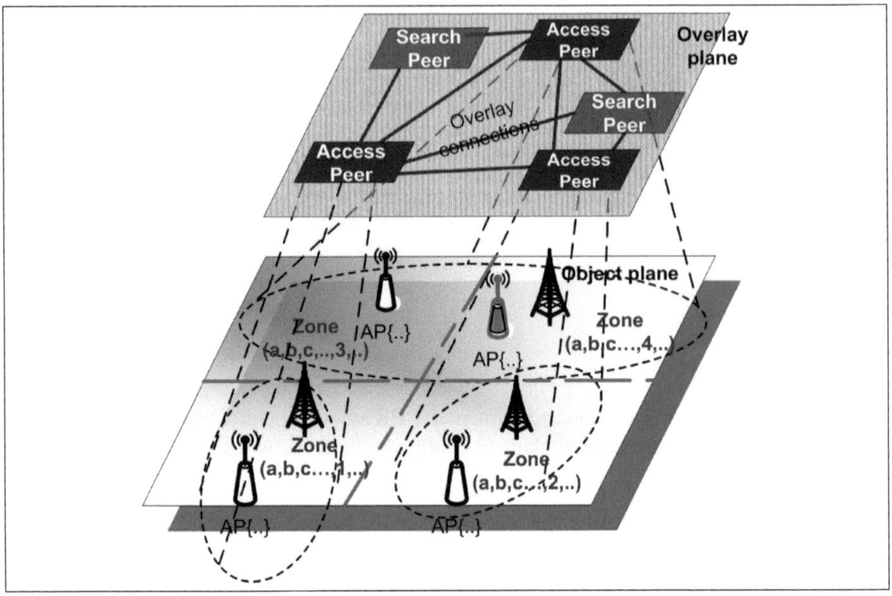

Fig. 21.2. Geo-coding on overlays.

- Reorganization of LBS: despite the separate GIS along service types or operator boundaries, the overlay re-structures the combined LBS according to location only. GIS contributors are free to push content with some access restrictions (e.g. AAA, pricing).

Besides the virtual-LBS servers, which represent the single entities in the larger distributed LBS. For the overlay management, two types of nodes are identified: (i) access peers, and (ii) search peers (see *Figure 21.2).*

Access peer: represents a virtual-LBS server in the overlay. So an access peer is a node capable of storing and managing content values. It can be reached by other peers based on some P2P protocol (like Chord, Kadelmia, etc). An access peer maintains routing tables for the P2P network as well. The access peer also plays the role of forwarding search requests as well as filtering out the information sent back as a response.

Search peer: This is a bootstrapping node contacted by mobile terminals, or users of the distributed overlay, in general. These nodes are not responsible for managing content neither for storing routing information to reach the content managed by the access peers. Instead, they keep a location trigger (e.g. care of address) to the mobile users and entry to known access peers. They additionally define the search parameters and are capable to start searches in the overlay as well as responding to the user (in which ever network he/she moves to). In other words, this node manages user mobility.

21.4 Analytic Study to Query Overhead

21.4.1 Choice of Overlay Protocol

Mediation or routing in a structured overlay like Chord (Stoica et al. 2003), allows a node to reach any other node (and its successor IDs) in $O(log(N))$ hops, given N the number of participating nodes. For this, each node stores routing information to $O(log(N))$ other nodes (or fingers). The random partition of object values and their corresponding keys, result in the loss of any semantic relationship that exists between the objects gathered in a node. Although all content or objects could be reached via the overlay (given their known value), the $log(N)$ effort has to be repeated for each object. Instead of looking up each object separately, clustering items along the ring, so that their semantic neighbourhood is preserved by the addressing scheme, reduces the mediation effort. The $log(N)$ search effort should be repeated for each cluster instead of each object.

W number of objects, require $O(W \cdot log(N))$ search effort using the analytically well studied Chord. With a range gathering w objects, the search effort is reduced to $O((W/w) \cdot log(N))$ messages. If a range is split among several direct neighbour peers M, then the overhead becomes:

$$O((W/w) \cdot [log(N)+M]) \quad (21.2)$$

In Houyou et al. (2008), the number of ranges W/w called clusters (*Table 21.1.* notation $N_2(k, k+n)$)and M are analyzed through the use of the Hilbert curve.

Figure 21.3 demonstrates the role of the overlay as an abstraction layer reorganizing the separate GIS in order to optimize the query process itself. A range query requires a distribution of content or items among the peers in a way to preserve their semantic neighborhood as well as their overlay neighborhood. In order to achieve this, the following functions in a Chord-like DHT need to be modified:

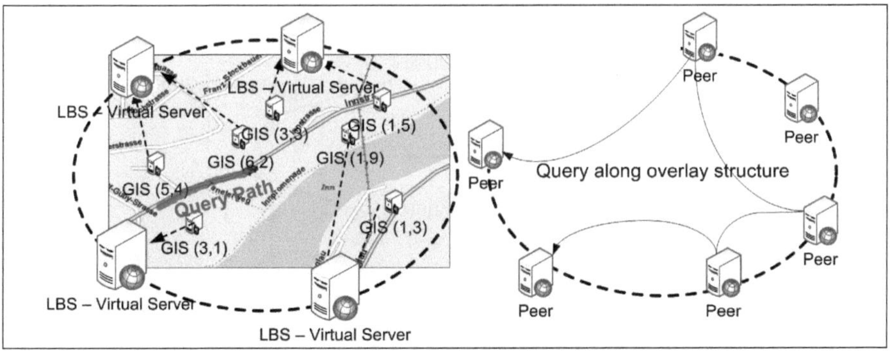

Fig. 21.3. Overlay efficient querying: reach an LBS – Virtual Server instead of searching each GIS; follow overlay structure to reach a cluster of objects (recursively), then iteratively hop to closest successor nodes to cover the cluster.

- Inserting an object in the overlay requires a coding function from the two-dimensional geographic coordinates to a single coordinate which preserves neighborhood.
- Query definition transforms a geographic range into the set of keys which should be grouped as clusters.
- A set of neighboring keys in the Chord space are easily converted back into a geographic range.

For the above functionalities to be fulfilled, a space filling curve could be used.

21.4.2 Hilbert Space Filling Curve

The Hilbert curve is a special space filling curve with the geometric form illustrated in *Figure 21.4.* In general, a *d*-dimensional Euclidean space with finite granularity can be filled with the γ-approximation of a *d*-dimensional Hilbert space-filling curve ($\gamma \geq 1$ *and* $d \geq 2$), which maps an integer set $[0, 2^{\gamma \cdot d}-1]$ into a *d*-dimensional integer space $[0, 2^{\gamma}-1]^d$.

Notation: For γ*, let* H_γ^d *denote the* γ *order approximation of a d-dimensional Hilbert space-filling curve, which maps* $[0, 2^{\gamma \cdot d}-1]$ *into* $[0, 2^{\gamma}-1]^d$*.*

In *Figure 21.4* the 3rd approximation of a *2*-dimensional Hilbert curve is taken to model the earth surface. This means that given that the earth surface is covered by a $2^3 \times 2^3$ mapping (shown on the left), any point of the earth could be addressed with the coordinates of the box in which this point exists. The Hilbert curve mapping converts the 2D *(x,y)* coordinates to a 1D coordinate (i.e. a single integer). Example shows in *Figure 21.4* how the coordinates of the zones (1,1), (3,2), and (5,1) are converted to integer IDs 2, 9, and 56 accordingly. The mapping used to divide a

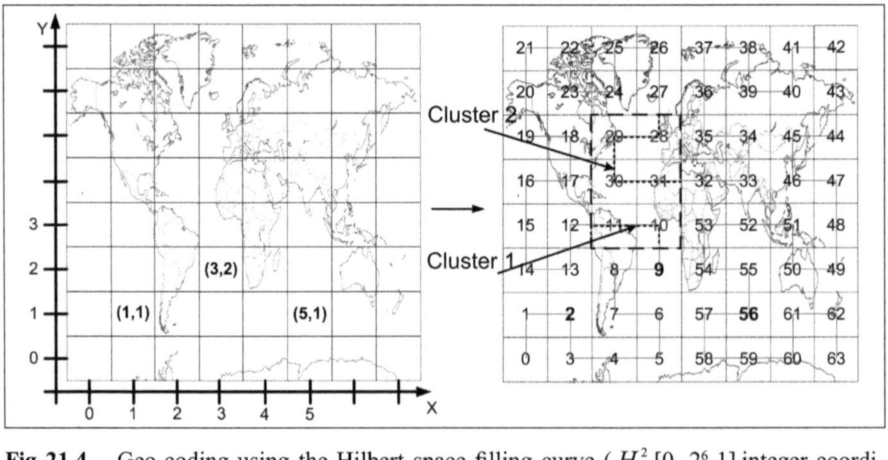

Fig. 21.4. Geo-coding using the Hilbert space filling curve (H_3^2 $[0, 2^6\text{-}1]$ integer coordinates into the 2-dimensional grid mapping covering the earth surface $(x,y) \in [0,2^3\text{-}1]^2$).

space in a $2^{\gamma} \times 2^{\gamma}$ grid requires a H_{γ}^{2} Hilbert curve, which converts the $2^{2\gamma}$ possible *(x,y)* coordinates into one of the $2^{2\gamma}$ integer IDs. Such an integer space requires 2γ-bit long ID. The addressing of each geographic zone inside a single cell results in a binary address which depends on (i.e. the approximation γ) and the *(x,y)* coordinates. The same transformation could be done in reverse.

Clusters are groups of grid points inside a query box that are consecutively connected by a mapping (or a curve). An example is given in *Figure 21.4,* where a random query box is selected. Two clusters exist in that query box [10, 11] and [28, 31].

The average number of clusters within a square query box of size $2^k \times 2^k$ in a $2^{k+n} \times 2^{k+n}$ region, covered by a H_{k+n}^{2} Hilbert curve is given in (Moon et al. 2001) as $N_2(k,k+n)$:

$$N_2(k,k+n) = \frac{(2^n-1)^2 2^{3k} + (2^n-1)2^{2k} + 2^n}{(2^{k+n} - 2^k + 1)^2} \tag{21.2}$$

The complexity of the overlay search in a single square search box in *equation 21.1* becomes:

$$O(N_2(k, k+n)[\log(N_p) + M]) \tag{21.3}$$

Where *Np* is the number of nodes part of a Chord ring and *M* the number of nodes covering the same cluster (see *Table 21.1).*

Table 21.1. List of variables used in the asymptotic study.

Variables	Definition
$\gamma = k + n$	The granularity of the modeled space given as the maximum index of both x and y coordinates where $(x,y) \in [0, 2^{\gamma} - 1]^2$
k	Binary index of a square search window, i.e. an area of $2^k \times 2^k$
H_{k+n}^{2}	A 2-dimensional Hilbert Curve filling a $2^{k+n} \times 2^{k+n}$ square grid
$N_2(k, k+n)$	The average number of clusters inside a $2^k \times 2^k$ query box in a $2^{k+n} \times 2^{k+n}$ closed space filled by a H_{k+n}^{2} Hilbert Curve
M	The number of peers covering a single cluster (worse case)
Np	Total number of peers taking part in the Chord network

21.4.3 Analytic Optimization of Query Overhead

Figure 21.5 shows the effect of choosing the right Hilbert curve degree on the overhead of a range query. The earth surface could be modeled with different cell-sizes translating in different Hilbert curve degrees ($\gamma = k+n$). The cases shown in *Figure 21.5* are those of *(k+n) = 19, 18, 17, 16,* resulting in a Chord key space *of 38, 36, 34, and 32* bits respectively.

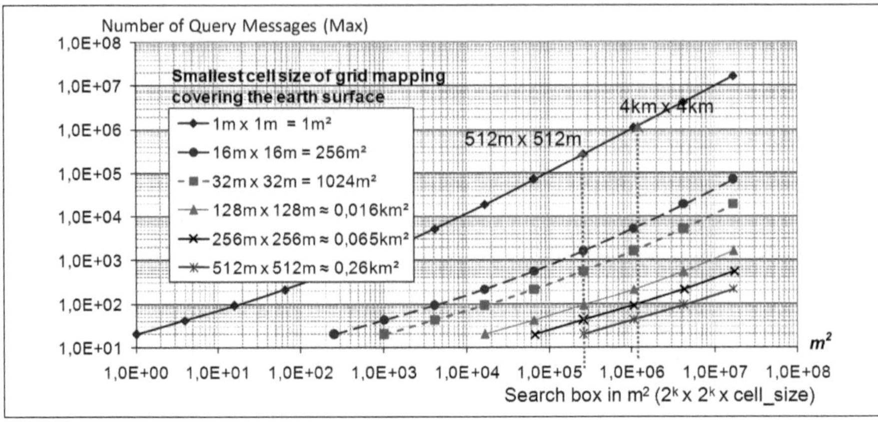

Fig. 21.5. Complexity estimation: varying query box for different Hilbert granularities (log scale).

The geographic size of each grid cell is given in the legend of *Figure 21.5.* The query box (see *Table 21.2)* is varied from $k = 0$ to $k = 12$, resulting in a different geographic range depending on the grid size used. The more granular the Hilbert curve is, the larger the overhead becomes. This is partly due to the fact that each object (in our case a wireless cell) of let say *500m* $\times$ *500m* size requires *2500* Chord keys when using $\gamma = k+n = 19$, and only *one* Chord key when using $\gamma = k+n = 15$. The problem with $\gamma = 15$, though, is that a smaller cell of let say *50m* $\times$ *50m* will be encoded as well in a *500m* $\times$ *500m* cell adding a significant loss of information about the stored data in the Chord ring. In our modelling of data items, it is most efficient to limit the number of keys per data item, but not to loose the granularity of let's say heterogeneous coverage (such as the case of wireless cells). In fact in our simulation scenarios, $\gamma = k+n = 15$ is found to offer a granularity of *16m* side distance, which means a *50m* radius WLAN cell would surround a square cell of surface $2*50^2\ m^2$ and require about 16 squares of area $19^2\ m^2$. In other words the 16 Chord keys are required to represent the granularity of the data item.

Table 21.2. Analytic evaluation of complexity for different Hilbert curve degrees.

Parameter	Geographic Interpretation	Chord Interpretation
Earth Surface	300 000 km²	Total modeled surface
$2^\gamma \times 2^\gamma$	Number of cells in a grid covering the earth surface the size of each cell is $\text{cell\_size} = \frac{3\times 10^{11}}{2^{2\gamma}}\,\text{m}^2$	Hilbert Curve H_γ^2 the Chord key length is $(2\cdot\gamma)$-bit long
$Np = 2^m$	$2^{2\gamma}$ grid cells through 2^m peers	Each peer manages $2^{2\gamma-m}$ possible keys, where $2\cdot\gamma > m$
Query Box $2^k \times 2^k$	**Objects:** Depending the modeled earth grid, each query box covers $2^{2k} \times$ cell_size	Each query box requires $N_2(k, k+n)$ searches

In addition to the number of clusters, the Hilbert curve mapping or partition of the space also influences the Chord overhead. In other words, the structure of the overlay does not require a fully meshed Chord, where each node has a *O(logNp)* fingers, but rather requires *O(logNp')* fingers, where *Np'* is a subset of *Np* whose key range covers that of an urban geographic subspace.

21.4.4 Partition of Objects Among Peers

The partition of the subspaces among the peers reflects the fractal nature of the Hilbert curve. Assuming a $2^{\gamma} \times 2^{\gamma}$ mapping of a 2D space is filled by the γ approximation of a Hilbert curve, if each cell in that grid is divided into 4 subsquares, each resulting cell requires 2-bit longer ID, while sharing the same 2γ-bit long prefix (see *Figure 21.6).* This fractal property could be applied to tree structure node scope and connectivity as follows:

1. Geographic urban groups require well meshed peers, whereas far away nodes are less important in a finger table.
2. Transition between large geographic zones need to be supported, to allow ubiquity of the system and support moving between cities or in rural subspaces.

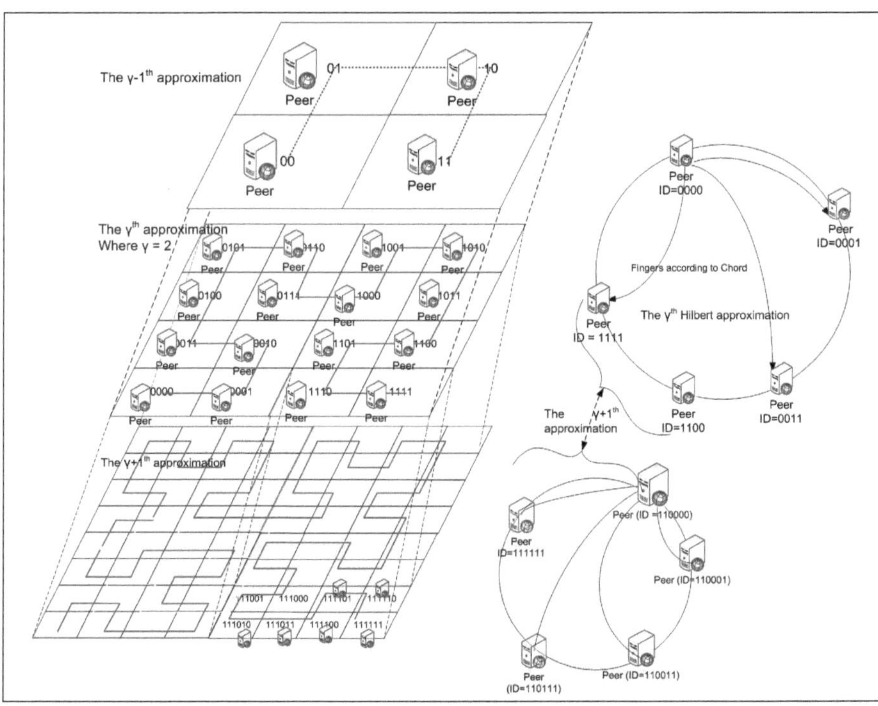

Fig. 21.6. Hierarchical partition of peer IDs following the Hilbert *1st, 2nd, and 3rd* approximation.

Figure 21.6 demonstrates the addressing principles and scope of each node in the overlay. The construction procedure of the overlay relies on the granularity of the object IDs which are managed. The virtual LBS servers are bootstrapped similarly to the Chord protocol. Each node advertises the minimum and maximum Hilbert ID, they manage. Based on several iterations, zones could be identified in the sense that those nodes sharing the longest prefix can be grouped to form a subspace representing an urban agglomeration. This subspace is a single Hilbert cluster starting from a min ID and finishing at a max ID. The keys in the latter cluster are associated with a predecessor peer, keeping a load-balancing rule requiring each node to be responsible for a similar number of keys.

In the urban subspace, a fully meshed Chord ring is built. The participating peers within that zone can elect a single node which is used as a gateway to access another ring at a higher hierarchical level. This node takes as an address the common prefix shared by all the nodes and objects at a lower level and can be part of a smaller sized Chord ring (at a higher level).

The efficiency of the multiple Chord hierarchy is proved in the following section. In fact, a fully meshed Chord ring is used while the scope of the queries is analyzed to find out how many nodes are involved in the query.

21.5 Simulation Scenarios

The simulation relies on a Java based overlay discrete event simulator. The overlay network is constructed allowing all nodes to store fingers (or routing information) to any other nodes with a global view of the whole network. A demonstration of the scope of the queries and the required fingers is sought.

21.5.1 Simulation Assumptions

The simulation model assumptions are summarized in *Table 21.3:*

Table 21.3. Network Simulation Model.

Elements	Values
Network model	Chord nodes are simulated at the overlay level, i.e., fully reliable nodes and no packetization or delays
Overlay nodes	Both access and search peers are simulated. Chord is implemented. Node IDs are obtained using the Hilbert curve partition of the key space
Simulated object space	Using a fixed grid cell size of $16 \times 16m^2$, i.e., 2^{38-8} possible keys or 30-bit key space $(\gamma = k + n = 15)$
Urban area	268 km^2 large, where access networks and users are located, e.g. Passau in Germany is about 69km^2 large
Peer distribution	2^{10+p} of which, 2^p responsible of the 268 km^2 city and the remaining 2^{10} for the rest of the earth
p	is varied between 10, 9, 8, 6, and 4

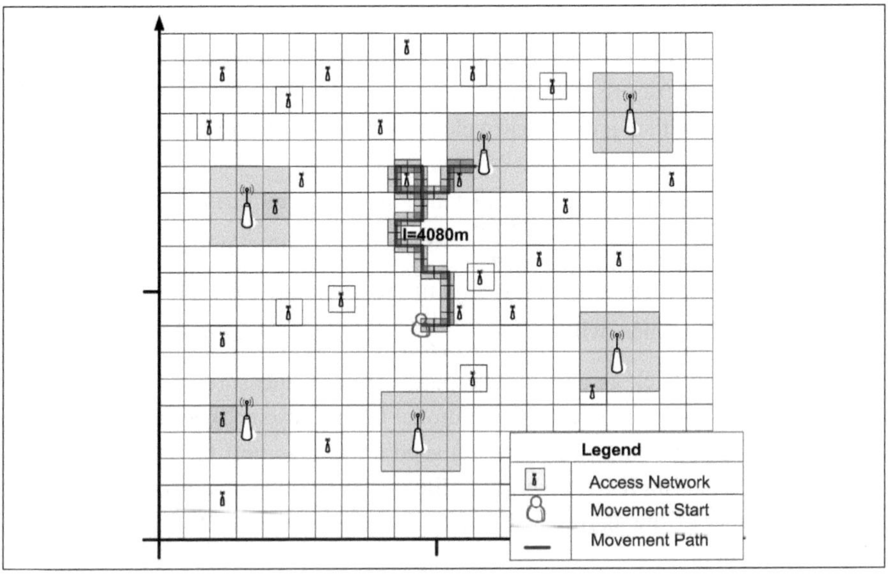

Fig. 21.7. A portion of the simulated environment with example query boxes.

Given the geographic area of a medium-sized city, heterogeneous wireless cells are modeled as randomly distributed cells. Similar to Zdarsky et al. (2006), 50 wireless cells are distributed randomly for each *1 km²*. That makes about 13400 access networks for the whole *268 km²* modeled city (see *Table 21.3).* Since wireless cells are heterogeneous in nature, we consider different cell sizes. Macro cells whose radius is up to 128m result in grid area of $256m \times 256m \equiv 16m \cdot 2^s \times 16m \cdot 2^s$, *where* $2^{2 \cdot s}$ *r*epresents the number of keys required to encode this cell in the Chord space, where *"s"* is varied uniformly between 1 and 4.

A graphical illustration of part of the urban city is shown in *Figure 21.7.* Further to the object model, the query is also modeled according to some movement pattern assumptions.

- We assume that the user follows a random walk through a Manhattan type of street model composed of a grid street system with junctions each *256 m*, where the user randomly selects the next direction with probabilities: *0.5* for continuing straight on and probability *0.25* for turning either right or left.
- The travelled distance to *4080 m*, the movement direction can change by each street junction, i.e. every ($\tau = \frac{256m}{velocity}$) seconds.

The resulting query boxes could be said to have the following properties.

- Assuming smaller query boxes as the full length of the traversed path, query boxes are emitted by the user recursively as he/she progresses along the movement path.

- A query box cannot be generated as long as the user has not moved out the area where his previous query has be generated, this excludes for slow users to generate large queries to frequently.
- A user moving with a given linear velocity crosses each single Chord key every *16m.*
- Each side of a query box covers a distance of $2^k \cdot 16m$ requiring $k_{street} = \log_2 \frac{velocity \times \tau}{16m} = \log \frac{256m}{16m} = 6$ to cover a street portion.
- To cover the full distance of about *4 km* movement path, for each *k* the query has to be repeated Q times, where $Q = \frac{4080m}{2^k \cdot 16m}$ (plotted in *Figure 21.9).*

The messages sent back as part of a response are also measured as both message units (where each unit represents an additional Chord key sent back).

It is worth noting that a peer does not send a message unit for each key found, but rather constructs an XML description file, for instance, with a list of found keys. The exact format of the aggregated results depends on the XML format, which is not of interest here.

21.5.2 Simulation Results

21.5.2.1 Search Overhead vs. Query Results

The number of peers covering the urban zone is initially set to 2^{10}, i.e. $p=10$ in *Table 21.3.* For the same distance of *4080m,* the number of query boxes needed *Q* is plotted in *Figure 21.9.* The effect of the velocity has been investigated in Houyou et al. (2008). This could be summarized in two points:

1. The size of the query box and velocity of movement affect the frequency at which the queries are started.
2. The data throughput required per range query could be formulated as follows: $$Query\ Throughput = \frac{query\ messages\ count \times velocity}{distance}$$

In *Figure 21.8* the summation of both query messages and response message units shows the effect the selected query box for the selected context. The ideal query size would be in that case $k=3$ or $k=4$. The query messages decrease with increasing *k*, whereas the bulk response information size increases (for the whole path).

In *Figure 21.9,* the number of response messages (XML format) actually decreases since fewer *Q* queries are started, but the total information size (plotted in *Figure 21.8)* increases. According to our application scenario, only those results near the movement path are of interest here, making the large amount of data response for larger *k* require further filtering. This is an indication of the scalability of the whole approach. Other mechanisms such as Caching (at the search peers for instance) could be used to re-use similar query results generated from different users.

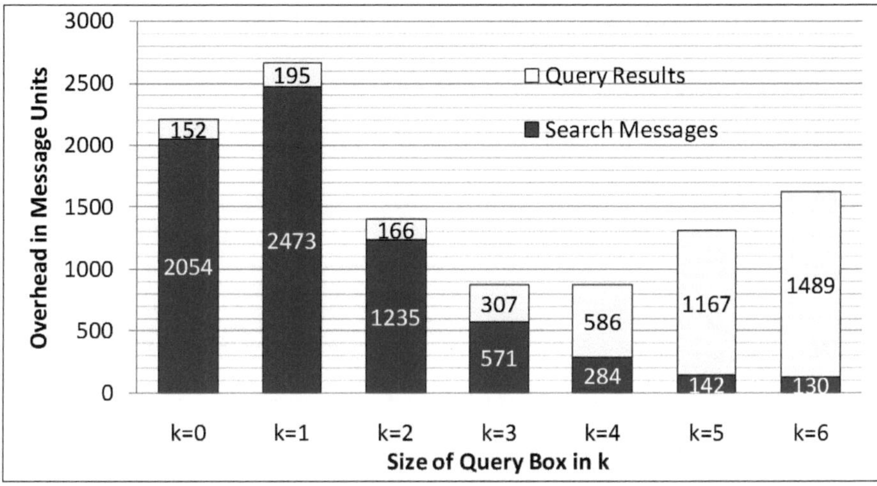

Fig. 21.8. Communication overhead in message units (search & response units).

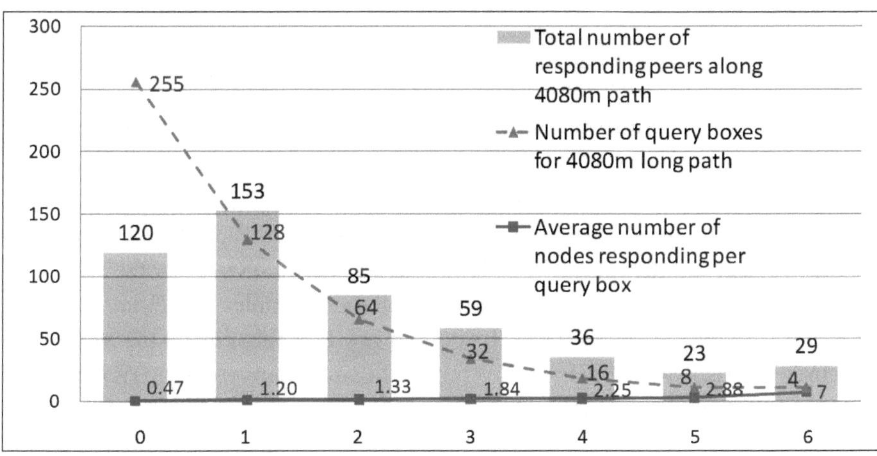

Fig. 21.9. Overlay behavior of responding peers.

Each time a peer responds with a full aggregated number of keys along the movement path, a message unit is counted (for each *k*) (see *Figure 21.9*). When compared with the amount of the expected query boxes (*Q*), the average number of nodes responding per query box grows from less then 0.47 for *k=0* to 7 responding peers for *k=6*. This number corresponds to a combination of the number of clusters in each single query box, and the partition of the clusters among the predecessor peers *(M)*. If no objects are found along those clusters, then there is no response sent. This depends on the density of objects inserted in the data base system.

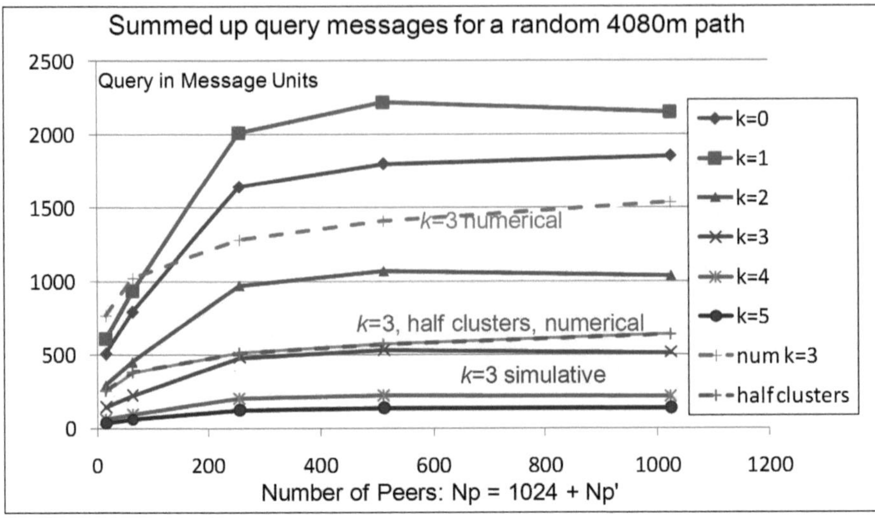

Fig. 21.10. The effect of number of urban peers (Np').

21.5.3 Comparing the Simulation Results with the Numerical Ones

In *Figure 21.10* the number of peers that manage the modelled urban zone is varied. The number of peers, indicated in *Table 21.3* by the parameter $p = 4, 6, 8, 9, 10$ corresponds to $Np' = 16, 64, 256, 512, 1024$ nodes respectively.

For a smaller query boxes ($0 \leq k \leq 2$), the query overhead is mostly affected by the number of the peers. For larger query boxes ($3 \leq k \leq 4$) the effect of the number of peers is minimal. Now recall our complexity *(Equation 21.3)* $O(N_2(k, k+n) \cdot \log_2(N_p) + M)$ for each single query box that covers the movement path. In comparison with the simulation results the complexity (for $k = 3$) is compared using the following equation:

$$O(Q \cdot [N_2(k, k+n) \cdot \log_2(N_p) + M]) \tag{21.4}$$

Where Q is the number of search boxes required for the same path.

According to the results in Houyou et al. (2008), M turns out to be at most one additional hop. The number of clusters also turns out to be less than the theoretic average $N_2(k, k+n)$. In the case of $k = 1$, the clusters' partition among peers has lead to a slight decrease in the search cost (from $Np' = 512$ to $Np' = 1024$). The average number of hops (given in Chord) is actually 0.5 log(Np). When taking Np' instead and half the clusters, the complexity line still bounds the simulative results.

The number of hops required to reach a cluster (Chord hops) $k = 3$ results in the histogram shown in *Figure 21.11*. The trials are repeated for the original scenario with $Np' = 1024$ nodes, 100 times. In each trial another random path is taken of the same distance resulting in a comparatively similar search boxes but with different clusters in each search box. The search being started by the search peer (as a proxy of the user), the same orginator ID is assumed. The number of the search boxes is $100 \times \frac{4080}{2^3 \times 16} = \mathit{3200}$. Whereas the number of clusters with objects on them has been 1577 clusters. In order to reach the first peer responsible of all clusters, Chord hops are required. These are plotted below *(Figure 21.11)*. The distribution of the number of hops is shown in the histogram along the cummulative distribution function of the maximum number of hops. In over 90% of the cases the number of hops is less than 8 which is about $\log_2$(256 peers). The worse case occurs in a few cases where the number of hops reaches 10 (i.e. $\log(Np')$, where $Np' \ll Np$).

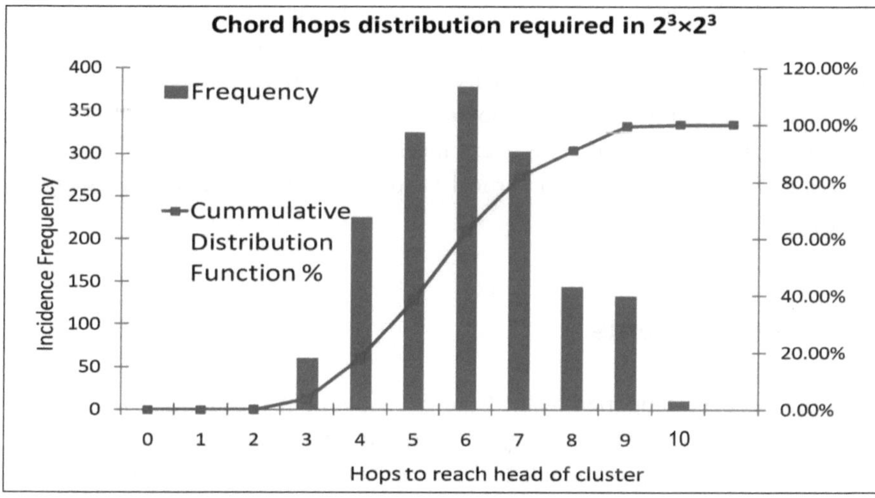

Fig. 21.11. Number of Chord Hops averaged over 100 trials.

The resulting complexity could be re-formulated (from *Equation 21.4*) based on the finding above as follows:

$$\min O(Q \cdot [0.5\, N_2(k, k+n) \cdot 0.5 \log_2(N_p{}') + M]) \qquad (21.5)$$

$$\max O(Q \cdot [N_2(k, k+n) \cdot \log_2(N_p{}') + M])$$

21.6 Conclusions

In this paper a design strategy is offered to build efficient mobile LBS based on overlays. An overlay offers a global ubiquity and mobility management which allow locating both mobile nodes and the sought content or composite services in a purely location-based manner. The geographic-centered semantic structure of the information and data is used to address network nodes which then structure a global overlay network. The analytic and simulative results show that the mobile-range queries can benefit from the overlay structure. In a reverse process, the overlay is also structured to limit the scope seen by each node is a global network. The routing aspect is therefore optimized thanks to the self-similarity and fractal addressing scheme offered by the Hilbert space filling curve. The curve can address geographically dense environments with higher granularity, while keeping a less granular knowledge about less dense areas. This work concentrated on demonstrating the feasibility and scalability of the proposed work especially in urban environments. An example LBS, which offers a query mechanism for heterogeneous collocated wireless cells to allow predicted vertical handover, is demonstrated and analyzed in this paper.

Acknowledgements

This work has been supported by the Euro-NF network of excellence, EU grant IST 216366, and specifically the subproject CAMYM. The authors would like to thank Alexander Stenzer for providing valuable support for the simulation work.

References

Akyildiz IF, Wang X, Wang W (2005) Wireless Mesh Networks: A Survey. Computer Networks Journal, Elsevier. 47(4), pp. 445–487

Akyldiz IF, Xie J, Mohanty S (2004) A Survey of Mobility Management in Next Generation all-IP based Wireless Systems. IEEE Wireless Communications. 11(4), pp. 16–28

Balasubramaniam S, Indulska J (2004) Vertical Handover Supporting Pervasive Computing in Future Wireless Networks. Computer Communications. 27(8), pp. 708–719

Datta A, Hauswirth M, John R, Schmidt R, Aberer K (2005) Range Queries in Trie-Structured Overlays. In: Proc. of Fifth Peer-to-Peer Computing Conference, pp. 57–66. IEEE Press, Konstanz, Germany (31 Aug.–2 Sept. 2005)

Houyou AM, De Meer H, Esterhazy M (2006) P2P-based Mobility Management for Heterogeneous Wireless Networks and Mesh Networks. In Cesana, M., Fratta, L. (eds) Wireless Systems and Network Architectures in Next Generation Internet, Menaggio Como, Italy, July 13–15, 2005. LNCS, vol. 3883, pp. 226–241. Springer, Heidelberg

Houyou AM, Stenzer A, De Meer H (2008) *Performance Evaluation of Overlay-based Range Queries for Mobile Systems*. In Llorenç Cerdà-Alabern(eds.) Wireless Systems and Mobility in Next Generation Internet, Barcelona, Spain, January 16 – 18, 2008. LNCS, vol. 5122 – Springer, Heidelberg

Knoll M, Weis T (2006) Optimizing Locality for Self-organizing Context-Based Systems. De Meer, H., Sterbenz, J. P. G. (eds), Self-Organizing Systems, First International Workshop, IWSOS 2006, Passau, Germany, September 18-20, 2006. LNCS, vol. 4124, pp. 62—73, Spinger, Heidelberg

Maymounkov P, Mazieres D (2002) Kademlia: A Peer-to-Peer Information System Based on the XOR Metric. In: Goos, G., Hartmanis, J., van Leeuwen, J. (eds) 1st International Workshop on Peer-to-Peer Systems (IPTPS), Cambridge, MA, USA, March 2002, LNCS, vol. 2429, pp. 53—65. Springer, Heidelberg

Moon B, Jagadish HV, Faloutsos C, Saltz JH (2001) Analysis of the Clustering Properties of the Hilbert Space-Filling Curve. In IEEE Transactions on Knowledge and Data Engineering. 13(1), pp. 124–141, IEEE Press

Rao B, Minakakis L (2003) Evolution of mobile location-based services, Communications of the ACM, v.46 n.12, pp. 61–65

Ratnasamy S, Francis P, Handley M, Karp R, Schenker S (2001) A scalable content-addressable network. In SIGCOMM'01: Proceedings of the 2001 conference on Applications, Technologies, Architectures, and Protocols for Computer Communications, pp. 161–172. ACM Press, New York

Sahin OD, Gupta A, Agrawal D, El Abbadi A (2002) Query Processing Over Peer-to-Peer Data Sharing Systems. Technical Report UCSB/CSD-2002-28, University of California at Santa Barbara

Schiller J, Voisard A (2004) Location-Based Services. Morgen Kaufmann/Elsevier, San Fransisco, CA, USA

Sethom K, Afifi H, Pujolle G (2005) Palma: A P2P based Architecture for Location Management. In 7th IFIP International Conference on Mobile and Wireless Communications Networks (MWCN 2005), Marrakech, Morocco, September 19-21

Stemm M, Katz RH (1998) Vertical Handoffs in Wireless Overlay Networks. In Mobile Networks and Applications. 3 (4), pp. 335–350. Springer, Netherlands

Stoica I, Morris R, Liben-Nowell D, Karger DR, Frans Kaashoek M, Dabek F, Balakrishnan H (2003) Chord: a Scalable Peer-to-Peer Lookup Protocol for Internet Applications. IEEE/ACM Transactions on Networking. 11(1), pp. 17–32

Zdarsky FA, Martinovic I, Schmitt JB (2006) The Case for Virtualized Wireless Access Networks, De Meer H, Sterbenz JPG (eds) Self-Organizing Systems, First International Workshop, IWSOS 2006, Passau, Germany, September 18-20, 2006. LNCS, vol. 4124, pp. 90–104, Spinger, Heidelberg

Zhuang S, Lai K, Stoica I, Katz R, Shenker S (2005) Host Mobility Using an Internet Indirection Infrastructure. In Wireless Networks. 11(6), pp. 741–756

22 Three-Dimension Indoor Positioning Algorithms Using an Integrated RFID/INS System in Multi-storey Buildings

Kefei Zhang[1], Ming Zhu[1], Günther Retscher[2], Falin Wu[1] and William Cartwright[1]

[1] School of Mathematical and Geospatial Sciences, RMIT University, Australia
[2] Institute of Geodesy and Geophysics, Vienna University of Technology, Austria

Abstract

Location based services (LBS) require a reliable, accurate and continuous position determination of mobile users. This is particularly true in indoor environments where the widely used Global Positioning System (GPS) is not available due to its signal outages. One solution is to integrate different techniques in a multi-sensor positioning system to overcome the limitations of a single sensor. In this chapter an approach is described using a three-dimensional Radio Frequency Identification (3D RFID) location fingerprinting probabilistic approach with map-based constraints in order to provide reliable positions in indoor 3D environments. An Extended Kalman Filter (EKF) is used to integrate 3D RFID positioning method with an Inertial Navigation System (INS) in order to produce an accurate and continuous positioning estimation.

The multi-storey experiments conducted at RMIT University, Australia, show that the 3D RFID positioning method can determine the mobile user's movements in a kinematic mode to meter-level by using the fingerprinting probabilistic approach. The smoothing method and the RFID/INS integration can both improve the positioning accuracy by tackling the RSS instability problem. Besides the 100Hz updating rate, the RFID/INS integration method can provide more reliable estimation based on the mobile user's kinematic characteristics rather than simply smoothing the estimations. The results also show that the algorithms for the Integrated RFID/INS indoor positioning system developed can satisfy the requirements for personal navigation services.

Keywords: 3D indoor positioning, RFID positioning, INS, Extended Kalman Filter (EKF)

22.1 Introduction

To date, Global Positioning System (GPS) has become a widely used technique for positioning and tracking. However, signal jamming, obstruction, interference and the relatively low update rate are still the major limitations of the low-cost GPS receivers. For personal navigation services, the capability to provide reliable and accurate position is required and sometimes as well as the capability of continuous positioning and high updating rate (e.g. for blind people guidance and athletes tracking). Current research suggests that the standalone techniques are difficult to satisfy these requirements. One solution is to integrate different positioning techniques to overcome the limitations in a single sensor (Retscher & Kealy 2005, Retscher & Thienelt 2004).

The integration of GPS and Inertial Navigation System (INS), to some extent, can overcome the problems caused by GPS outage and the update rate can be significantly increased by using the low-cost, high update rate INS (Godha & Cannon 2007) However, INS is prone to significant instrument biases and drifts due to the noise, device instability and operational environments. For a long period of GPS outage, the INS accuracy can be greatly degraded due to its drifting problem. Under this situation, other high accuracy positioning techniques during the period of the GPS signal outage are required.

The Radio Frequency Identification (RFID) system is a radio-frequency-based system widely used for identification applications (Finkenzeller 2003). Since an active RFID system is so powerful as it can propagate the signal through thin walls, this technique is regarded as a promising method to provide absolute positions indoor. Potential applications of RFID systems have been studied and some positioning systems have been implemented, such as *LANDMARC* (Ni et al. 2004) and the RFID positioning system along roads (Chon et al. 2004). Nevertheless, these systems are mainly for 1D or 2D positioning and do not provide continuous position estimations.

However, for the personal tracking applications in a multi-storey building, a 3D continuous positioning service is preferred. This paper deals with algorithmic developments of 3D RFID positioning and RFID/INS integrated positioning. The RFID location fingerprinting probabilistic approach is adapted for the 3D positioning algorithm using RFID. The strapdown INS navigation framework and an EKF are used in the algorithm for integration positioning in order to provide a continuous and reasonably accurate indoor positioning service for the mobile users moving between different levels in a multi-storey building.

22.2 3D RFID Positioning Algorithm

The 3D RFID Positioning Algorithm is based on the location fingerprinting method. This method was first implemented in the *RADAR* system (Bahl & Padmanabhan 2000) which is an RF-based indoor tracking system developed by *Microsoft Research*.

This positioning method achieves the best positioning results by two stages: the off-line training phase (off-line phase) and the on-line position determination phase (on-line phase). In the off-line phase, the received signal strength (RSS) from all the detectable transmitters and other relevant information including the physical coordinates of the receiver, which are measured by other methods, such as using total stations or tapes, in the local coordinates, are collected at predefined reference points and stored in a database. The set of the database is called fingerprint. During the on-line phase the mobile user samples the RSS patterns and searches for similar patterns in the off-line database so as to find its most possible position.

22.2.1 3D Off-Line Database

Generally, the 3D off-line database is a combination of the conventional 2D off-line databases in the different layers along Z-axis and in each layer, the RSS is measured and stored with four horizontal orientations (e.g. 0°, 90°, 180°, 270°) in order to represent the gain patterns of the RFID antenna. According to Li (2006), the accuracy of the fingerprints can be improved if more reference points are measured. However, in reality, large number of reference points leads to the heavy workload in the off-line measurements. One alternative method to generate a fine-scale off-line fingerprints database, based on the limited number of reference points, is interpolation, since the RSS and its variation are position-related variables in static propagation environments (without moving obstructions).

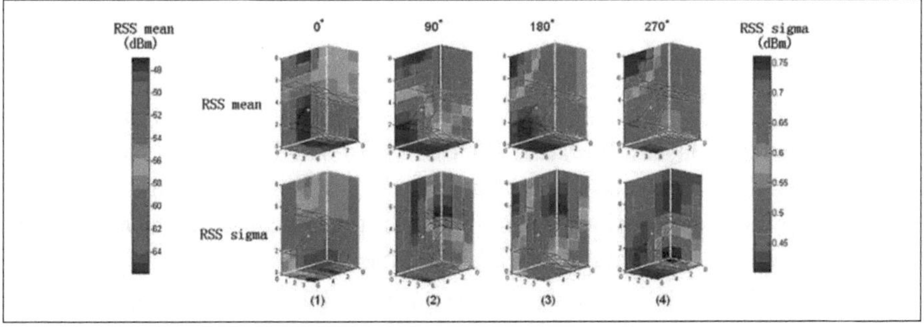

Fig. 22.1. An example of the interpolated RSS grids of the tag 200168210 located at (5.85, 2.8, 3.85) in the 3D RFID off-line database (The RSS mean and its corresponding sigma grids with the horizontal orientations of 0 °, 90 °, 180 °and 270 ° are shown in this figure respectively).

In order to generate the RSS mean and sigma grids with one-cubic-meter resolution in the operation area from the limited number of off-line reference points, the Kriging interpolation method (Krige 1951) is applied. The interpolated results are stored in the database with the location, horizontal orientation and tag ID.

22.2.2 Probabilistic Approach of the 3D RFID Positioning

The on-line phase can be conducted through two different methods, the probabilistic method and the deterministic method. The former one calculates the mobile user's position according to the conditional probabilities of the location under certain RSS values. This method is relatively accurate in positioning but requires a large off-line database not only with the RSS but also the information of the RSS probabilistic distributions. In contrast, the later method determines the mobile user's position according to the shortest distance between the rover-sampled RSS vector and the off-line measured RSS vectors. This method does not require a large database whereas in most cases, it is not as accurate as the probabilistic method. In order to achieve high accuracy, the probabilistic approach of the location fingerprinting method is applied in this study.

The estimated position of the mobile user ($\hat{L}_r$) is defined as the weighted average of cells' positions in the operation area grid.

$$\hat{L}_r = \frac{\sum_{i=1}^{n}[P_i(L_r|M) \cdot L_{ri}]}{\sum_{i=1}^{n} P_i(L_r|M)} \tag{22.1}$$

The weight, $P(L_r|M)$, of the cell i's position L_{ri} is defined as the probability of the mobile user at the position L_r under the condition of the RSS vector M, and

$$P(L_r|M) = \frac{P(L_r) \cdot P(M|L_r)}{P(M)} \tag{22.2}$$

according to the Bayes Rule (Li, 2006), where $P(M|L_r)$ is the conditional probability of the RSS vector M at the position L_r, $P(L_r)$ refers to the probability of the mobile user at the position L_r and $P(M)$ is the probability of the mobile user measuring the RSS vector and achieving the value M.

The term $P(L_r)$ is sometimes determined by the prediction based on the previous kinematic variables and/or the maps indicating the possible tracks of the mobile user. In *Equation 22.2*, $P(L_r)$ can be determined as a constant if the prior knowledge is not accessible. $P(M)$ is a constant independent on position. Consequently, $P(L_r|M)$ can be simply treated as $P(M|L_r)$.

$$P(L_r|M) \sim P(M|L_r) \tag{22.3}$$

where "~" refers to the equivalence relation.

Then $P(M|L_r)$ can be determined by the marginal probability distribution of $P(RSS_{Tag_i} = s_i)$ which is the probability of the measured RSS (s_i) from the tag (Tag_i) at the position (L_r).

$$P(M|L_r) \sim \prod_{i=1}^{n} P(RSS_{Tag_i} = s_i) \tag{22.4}$$

The experiments measuring the RSS in an unchanging environment (without people movements and furniture relocation) show that the RSS can generally follow the normal distribution. Based on the assumption of the normal distribution of the RSS, $P(RSS_{Tag_i} = s_i)$ can be estimated according to the following equation.

$$P(RSS_{Tag_i} = s_i) = \Phi_{\mu(L_r,D_r,Tag_i),\sigma^2(L_r,D_r,Tag_i)}(s_i)$$

$$= \frac{1}{\sigma(L_r,D_r,Tag_i)\sqrt{2\pi}} \int_{s_i-0.5}^{s_i+0.5} \exp\left(-\frac{(u-\mu(L_r,D_r,Tag_i))^2}{2\sigma^2(L_r,D_r,Tag_i)}\right) du \tag{22.5}$$

where $\mu(L_r,D_r,Tag_i)$ is the mean of the RSS from the tag (Tag_i) measured at the position (L_r) with the horizontal orientation (D_r), $\sigma^2(L_r,D_r,Tag_i)$ is the corresponding variance and u is the integral variable.

22.2.3 Map-Based Constraint

In reality, the assumption of measuring the RSS in an unchanging environment is not quite reasonable. Consequently, the normal distribution of the RSS will be disrupted by the changing environment and the *Equation 22.5* will not be fully satisfied. This error usually leads to a few meters error of the positioning estimation by probabilistic method. Some constraints can be applied in order to tackle this problem. One of the constraints is the map or the floor plan, which indicate the possible routes of the mobile user. This is similar to the map matching methods (Syed & Cannon 2004).

According to the floor plan and the assumption that the mobile user must move along the stairway when walking between different levels, a map-based constraint to the RFID positioning is applied. This constraint defines a grid of probabilities which equal to the zero-mean normal distributions with the distances away from the center line of the stairway, $\Phi_{0,\sigma^2}(Dist(L_r))$. The variance of the normal distribution, assigned to satisfy the requirement that the 98% cumulated probability, is within the width of the stairway (see *Figure 22.2)*. The probability of the mobile user at the position, $P(L_r)$, can be determined by this constraint and the weights

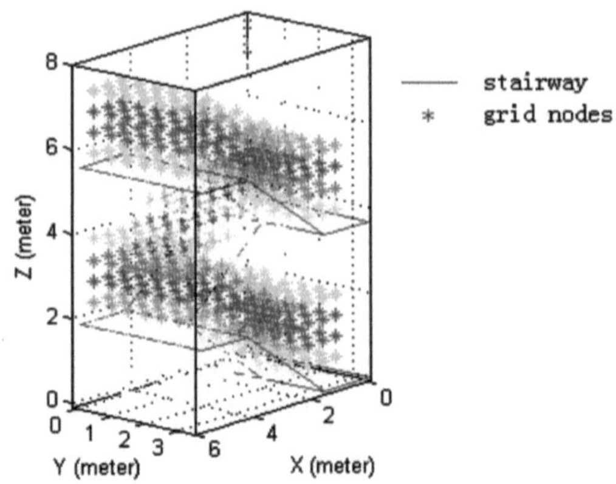

Fig. 22.2. The map-based probabilistic along the stairway (The grid nodes represent the probabilities of the mobile user's position. The lighter the note, the less probability of the user appearing at that position, so the dark nodes present as a line along the stairway indicating the user's possible route).

used to calculate the mobile user's position can be determined by *Equation 22.7.* In order to balance the effects from the constraint and the RFID fingerprinting estimation, two weights, w_L and w_{RFID}, are introduced and assigned arbitrarily (see *Equation 22.8).*

$$P(L_r) = \Phi_{0,\sigma^2}(Dist(L_r)) \tag{22.6}$$

$$P(L_r|M) \sim P(L_r) \cdot P(M|L_r) \tag{22.7}$$

$$P(L_r|M) \sim P(L_r)^{w_L} \cdot P(M|L_r)^{w_{RFID}} \tag{22.8}$$

22.3 Integrated RFID/INS Positioning Algorithm

In order to improve the positioning performance and increase the update rate of the positioning system, an INS is integrated with the RFID positioning system. On one hand, the INS is a self-contained, high update rate device, which does not suffer the signal propagation problems, such as the multipath effects and obstructions, and can produce more continuous positioning estimations. On the other hand, the RFID positioning method is drift-free so it can be used as the reference in the INS error models. An Extended Kalman Filter (EKF) based algorithm is developed to integrate these two systems.

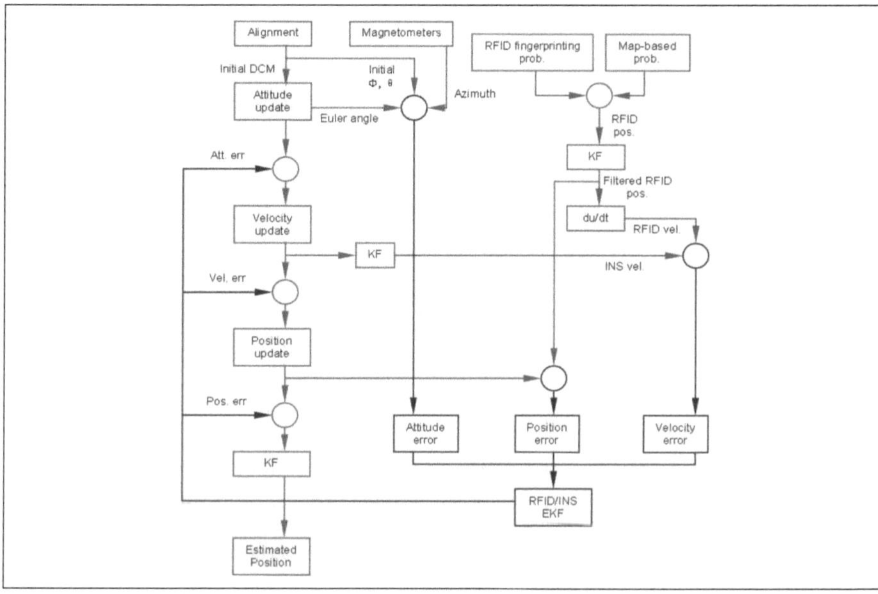

Fig. 22.3. The structure of RFID/INS positioning algorithm.

22.3.1 Structure of the RFID/INS Positioning Algorithm

In the algorithm, the strapdown INS local geographic navigation frame mechanisation (Titterton & Weston 2004) is combined with the tri-axis inertial error model (Brown & Hwang 1997) and the RFID location fingerprinting method to produce accurate and continuous positioning estimations (see *Figure 22.3)*. Some linear KFs are used in the process to smooth the outputs.

22.3.2 EKF for the RFID/INS Positioning Algorithm

A basic 9-state dynamic model (Brown & Hwang 1997) is used as the RFID/INS EKF model. In this model, the state vector $\tilde{x}$ contains three position errors, Δx, Δy and Δz, three velocity errors, ΔV_x, ΔV_y and ΔV_z, and three Euler angle errors, $\Delta\phi$, $\Delta\theta$ and $\Delta\psi$ (see *Equation 22.9)*. The state transition matrix $\widetilde{\phi}$ represents the tri-axis inertial error model as shown in *Equation 22.10.* The observation vector $\tilde{z}$ equals to the INS estimations and the reference measurements. In this case, the RFID positioning and velocity estimations are used as the position and velocity references. The initial pitch and roll angles from the alignment stage and the azimuth estimated by the tri-axis magnetometers are used as the references of the Euler angles since the mobile user itself does not contain significant changes in pitch and roll (see *Equation 22.11)*.

$$\tilde{x} = \begin{bmatrix} \Delta x \\ \Delta V_x \\ \Delta\phi \\ \Delta y \\ \Delta V_y \\ \Delta\theta \\ \Delta z \\ \Delta V_z \\ \Delta\psi \end{bmatrix} \tag{22.9}$$

$$\tilde{\phi} = \begin{bmatrix} 1 & \Delta t & 0 & 0 & 0 & 0 & 0 & 0 & 0 \\ 0 & 1 & g\cdot\Delta t & 0 & 0 & 0 & 0 & 0 & 0 \\ 0 & \Delta t / R_e & 1 & 0 & 0 & 0 & 0 & 0 & \omega_x \cdot \Delta t \\ 0 & 0 & 0 & 1 & \Delta t & 0 & 0 & 0 & 0 \\ 0 & 0 & 0 & 0 & 1 & g\cdot\Delta t & 0 & 0 & 0 \\ 0 & 0 & 0 & 0 & \Delta t / R_e & 1 & 0 & 0 & \omega_y \cdot \Delta t \\ 0 & 0 & 0 & 0 & 0 & 0 & 1 & \Delta t & 0 \\ 0 & 0 & 0 & 0 & 0 & 0 & 0 & 1 & 0 \\ 0 & 0 & 0 & 0 & 0 & 0 & 0 & 0 & 1 \end{bmatrix} \tag{22.10}$$

where g is the vertical component of the gravity, R_e is the radius of the Earth and ω_x and ω_y are the pitch and roll angular rates respectively.

$$\tilde{z} = \begin{bmatrix} x - x_{RFID} \\ V_x - V_{RFIDx} \\ \phi - \phi_0 \\ y - y_{RFID} \\ V_y - V_{RFIDy} \\ \theta - \theta_0 \\ z - z_{RFID} \\ V_z - V_{RFIDz} \\ \psi - Ori_{mag} \end{bmatrix} \tag{22.11}$$

where x, y and z are the position estimations of the mobile user respectively. V_x, V_y and V_z are the velocity estimations of the mobile user respectively. ϕ, θ and ψ are the pitch, roll and yaw angles estimates of the mobile user respectively. x_{RFID},

y_{RFID}, z_{RFID}, V_{RFIDx}, V_{RFIDy} and V_{RFIDz} position and velocity estimations by RFID positioning respectively. ϕ_0 and θ_0 are the pitch, roll estimated in the alignment stage respectively. Ori_{mag} is the azimuth estimated by the tri-axis magnetometers.

The velocity estimated by RFID, V_{RFIDx}, V_{RFIDy} and V_{RFIDz}, are assigned as the average velocity between the current RFID position estimation and the previous RFID position estimation. It may not represent the mobile user's current velocity accurately but, these values provide a significant contribution in eliminating the INS drifts. That is because the kinematics of the user is relatively low (with the speed of less than 10 m/s) and the measurements by RFID are drift-free even though they contain errors caused by the RFID positioning errors and the arbitrary assignment of the RFID velocity estimations.

$$\widetilde{H} = \widetilde{I}_{9\times 9} \tag{22.12}$$

The EKF is implemented and provides the estimations of the errors in position, velocity and attitude determinations according to the referencing measurements generated from the RFID positioning system, magnetometers and the alignment. These estimated errors are then used to correct the INS positioning estimations.

22.4 Experiments and Analysis

The 3D RFID/INS positioning system, which is carried by a mobile user, includes a *MinimaxX* module which contains an Micro-Electro-Mechanical Systems (MEMS) INS and tri-axis magnetometers developed by the Satellite Positioning and Navigation (SPAN) research group at RMIT University in collaboration with the Australian Institute of Sport (Wu et al. 2007, Zhang et al. 2004, Zhu et al. 2007), an *IDENTEC SOLUTIONS*' Intelligent Long-range RFID system and a laptop (see *Figure 22.4)*.

The INS which is used in *MinimaxX* is a MEMS strap-down INS (SINS) including three gyroscopes and one tri-axis accelerometer. The gyroscopes use the *Analog Devices ADXRS300* chips which can measure ±300°/s angular rate. The accelerometers use the *Kionix KXM52* chips. Their measurement ranges are ±2 g. The magnetometers used are the *Hitachi Metals HM55B* chips. These chips can measure the magnetic field between -200 μT and 200 μT. The RFID system provides 100-meter reading range in free space but in indoor environments, due to the obstructions, the range varies from 10 to 30 meters depending on the environments. The RFID reader can measure the received RSS and this value can be accessed from a build-in function provided by the RFID system driver.

The indoor multi-storey positioning experiments are conducted between level 9 and level 11 in Building 12, RMIT University city campus. The local positioning

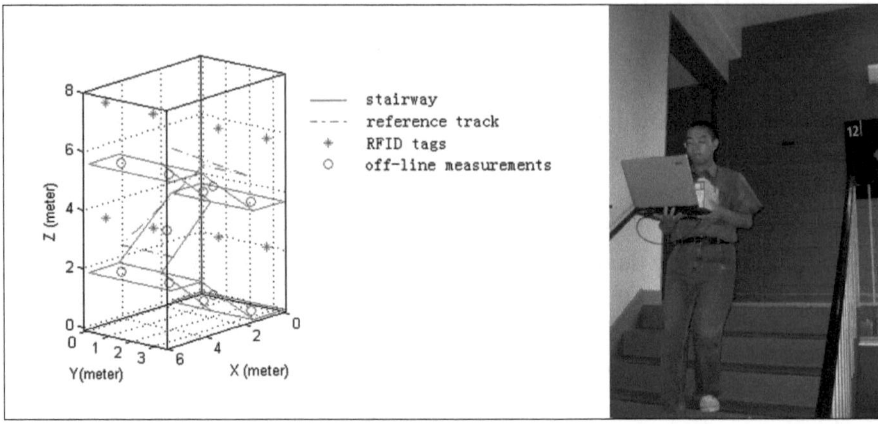

Fig. 22.4. The experimental area settings and the prototype 3D RFID/INS positioning system used in the experiments (left: a schematic plot of the experimental area, right: the integrated positioning system).

coordinate system is defined as follows. The origin is set in the northwest corner of the intermediate level between level 9 and level 10. The X-axis points horizontally to the south parallel with the stairway, the Y-axis points horizontally to the east, perpendicular with the stairway, and the Z-axis points to the vertical up. Eight RFID tags are mounted on the walls 2 m above the floors. The two tags on the same level are 2.05 m away from each other and the distances between the tags on different levels are over 4 meters. 11 RFID fingerprinting off-line reference points are chosen along the central line of the stairway from level 9 to level 11 (see *Figure 22.4*). The off-line database contains the RSS and the corresponding sigma with one-cubic-meter resolution grid, interpolated by the Kriging method according to the measurements of the 11 reference points.

In the on-line phase, the 3D RFID/INS positioning system is carried by a mobile user moving along the referenced track which is from the entrance on level 10 to the entrance on level 11 along the central line of the stairway. The entire experiment takes approximately 40 seconds.

Figure 22.5 shows the three axes positioning results with three different positioning methods, including the RFID fingerprinting method with the map-based constraint, the filtered RFID fingerprinting method with the map-based constraint and the RFID/INS positioning methods. The 3D RFID fingerprinting positioning method can roughly indicate the mobile user's movements with large errors, up to 5 m, mainly caused by the unstable RFID RSS. In the most cases, this signal instability is related to the change of the surrounding environments, especially the movements of the obstacles between the transmitters and the receivers. The filtered RFID fingerprinting method used a linear Kalman Filter to smoothing the significant positioning noise by using RFID stand alone then applied with the map-based

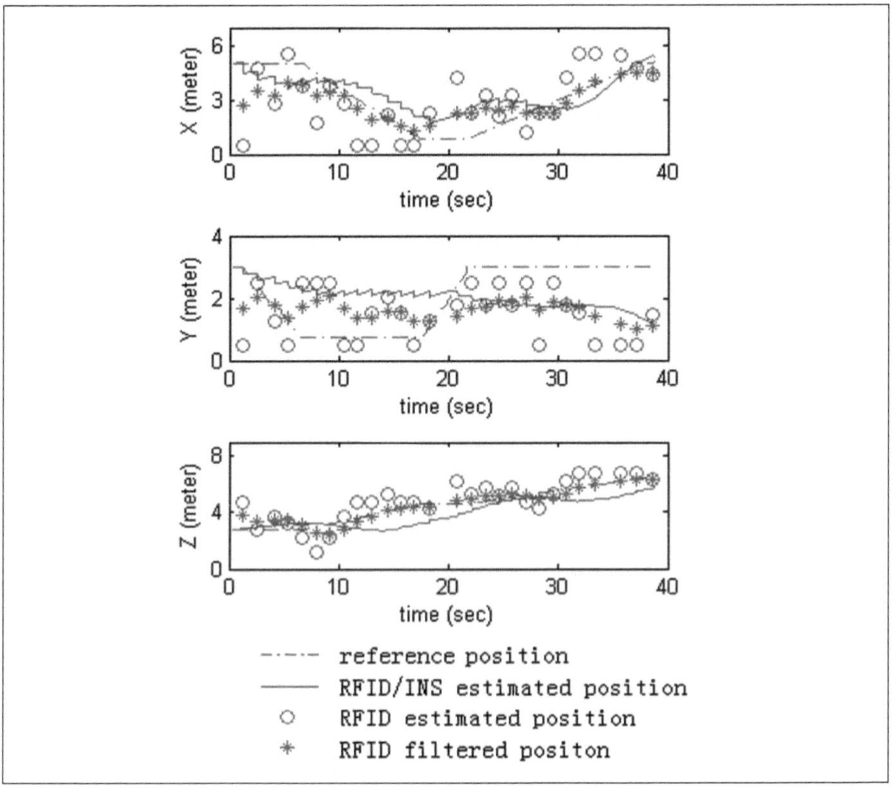

Fig. 22.5. The positioning results with three different methods.

constraint. The state vector of this linear Kalman Filter is the estimated coordinates, *x, y* and *z,* of the mobile user. The results indicate that the linear KF and the constraint applied to the RFID fingerprinting positioning method can greatly improve the RFID positioning performance, especially in the X direction and Z direction. The Y direction positioning performance does not improve as much as the other two directions. This is mainly caused by two reasons, the width of the Y-axis and the geometry of the RFID tags' grid. Firstly, the width of the stairway is only 1.8 m. Considering about the RFID standalone positioning accuracy of 2.4 m (see *Table 22.1),* it is very easy to fall into the wrong stairway sections by the constraint. Secondly, the different separations of the tags along each axis will indicate the different positioning performance along different axis. The distance between tags in Y-axis is only 2.05 m, which is the shortest distance among three axes and without any permanent obstructions. It means that the RSS from any two RFID tags in the same level will be similar and the effects of changing environments on the RSS instability are significant. Consequently, the positioning accuracy along the Y-axis is degraded.

The INS positioning method is numerically compared with the above three methods in *Table 22.1.* The comparison shows that the significant drift in INS, which is 8.6 m in 40 seconds, is greatly eliminated by the integrated RFID positioning system. The filtered RFID positioning method, which uses a linear KF for smoothing the RFID positioning estimations, provides the lowest positioning error RMS, i.e., 1.6 m, among four methods but with a low update rate of 1 Hz. The RFID/INS positioning method can increase the update rate to 100 Hz and maintain a similar positioning accuracy.

Table 22.1. Numerical comparisons of positioning performance and update rates of four different positioning methods.

	X error RMS (m)	Y error RMS (m)	Z error RMS (m)	Positioning error RMS (m)	Update rate (Hz)
RFID positioning	1.7	1.4	0.9	2.4	1
Filtered RFID positioning	1.0	1.1	0.4	1.5	1
INS positioning	5.1	6.3	2.8	8.6	100
RFID/INS positioning	0.9	1.2	0.8	1.7	100

22.5 Conclusions

A 3D RFID fingerprinting positioning method is developed with the constraints based on the prior knowledge of the mobile user's possible positions. The experiments indicate that the RFID probabilistic positioning approach with the map-based constraint can produce less accurate position estimations to indicate the mobile user's movements in a kinematic mode. In theory, the accuracy of the positioning estimation should be dominated by the separating distance of the tags (in this case 2.05 m) but for RFID positioning, the instability of the RSS always degrades the accuracy dramatically from this level. Consequently, the constraints, smoothing methods and integration of the external sensors are always required to improve positioning performance.

By involving a linear KF to smooth the estimated results, the performance can be improved to 1.6 m in terms of the total error. Experiments also indicate that the RFID fingerprinting positioning performance is largely influenced by the tags' grid geometry. In these experiments, the performance in Y-axis, which is the shortest axis, is poor since the stairway is a long and narrow area along X-axis and the tags along Y-axis are too close to each other so that the RSS measured from different tags on the same level are similar along this axis and the RSS instability caused by the changing environments is significant. This degrades the positioning accuracy accordingly.

A new algorithm of integrating the RFID positioning and the INS estimations has also been developed. This algorithm uses an EKF combining the INS local geographic navigation frame mechanisation and the 3D RFID fingerprinting positioning probabilistic approach together. It improves the positioning update rate from 1 Hz to 100 Hz and maintains the positioning accuracy in the range of 1.7 m. Theoretically, the RFID/INS integration method can provide more reliable estimation based on the mobile user's kinematic characteristics rather than simply smoothing the estimations.

It has demonstrated that the 3D positioning algorithms for integrated RFID/INS indoor positioning system can satisfy the requirements for accuracy, reliability and continuity for personal navigation services, especially for those who need the continuous positoning information or moving in a relatively high kinematic mode. Our future research efforts will be focused on investigating the relationship between the positioning accuracy and the RFID tags' grid geometry in order to improve the RFID positioning performance, and the INS error models to improve the EKF performance.

Acknowledgements

This research is partially supported by the Department of Innovation, Industry, Science and Research of Australia through its International Science Linkages Programme (Australia-China) – "Intelligent gas disaster early-warning, robust emergency response and rescue systems for coal mining based on geospatial information technologies" endorsed to a team led by Professor Kefei Zhang. Mr Ming Zhu is financially supported through RMIT University International Research scholarship.

References

Bahl P, Padmanabhan V (2000) *RADAR: An In-Building RF-Based User Location and Tracking System.* Paper presented at the 19th Annual Joint Conference of the IEEE Computer and Communications Societies, Tel Aviv, Israel.

Brown RG, Hwang PYC (1997) *Introduction to Random Signals and Applied Kalman Filtering* (3rd ed.). New York: John Wiley & Sons, 484 pages.

Chon H, Jun S, Jung H, An S (2004) Using RFID for Accurate Positioning. *Journal of Global Positioning Systems, 3*(1–2), 32–39.

Finkenzeller K (2003) *RFID Handbook: Fundamentals and Applications in Contactless Smart Cards and Identification* (2nd ed.). Hoboken, NJ: Chichester.

Godha S, Cannon ME (2007) GPS/MEMS INS Integrated System for Navigation in Urban Areas. *GPS Solutions, 11*, 193–203.

Krige DG (1951) A Statistical Approach to Some Basic Mine Valuation Problems on the Witwatersrand. *Journal of the Chemical, Metallurgical and Mining Society of South Africa, 52*, 119–139.

Li B (2006) *Terrestrial Mobile User Postioning Using TDOA and Fingerprinting Techniques.* Unpublished PhD thesis, The University of New South Wales, Sydney, Australia.

Ni L, Liu Y, Lau Y, Patil A (2004) LANDMARC: Indoor Location Sensing Using Active RFID. *Wireless Networks, 10*, 701–710.

Retscher G, Kealy A (2005) *Ubiquitous Positioning Technologies for Intelligent Navigation Systems.* Paper presented at the 2nd Workshop on Positioning, Navigation and Communication (WPNC'05) & 1st Ultra-Wideband Expert Talk (UET'05).

Retscher G, Thienelt M (2004) NAVIO – A Navigation and Guidance Service for Pedestrians. *Journal of Global Positioning Systems, 3*(1–2), 208–217.

Syed S, Cannon M (2004) *Fuzzy Logic Based-Map Matching Algorithm for Vehicle Navigation System in Urban Canyons.* Paper presented at the ION National Technical Meeting, San Diego, CA.

Titterton D, Weston J (2004) *Strapdown Inertial Navigation Technology* (second ed.). Stevenage: The Institution of Electrical Engineers, 574 pages.

Wu F, Zhang K, Zhu M, Mackintosh C, Rice T, Gore C et al (2007) *An Investigation of an Integrated Low-cost GPS, INS and Magnetometer System for Sport Applications.* Paper presented at the the 20th International Technical Meeting of the Satellite Division of the Institute of Navigation ION GNSS 2007, Fort Worth, Texas.

Zhang K, Deakin R, Grenfell R, Li Y, Zhang J, Cameron WN, et al (2004) GNSS for Sports – Sailing and Rowing Perspectives. *Journal of Global Positioning Systems, 3*(1–2), 280–289.

Zhu M, Zhang K, Wu F, Cartwright W, Retscher G (2007) *An Investigation of Pedestrian Positioning Algorithms Based on Integrated Low-cost INS, GPS and Magnetometer.* Paper presented at the International Global Navigation Satellite Systems Conference, Sydney, Australia.

23 Prediction of GNSS Availability and Accuracy in Urban Environments – Case Study Schiphol Airport

Frank Kleijer[1], Dennis Odijk[2] and Edward Verbree[1]

[1] Delft University of Technology, OTB Research Institute for Housing
The Netherlands
[2] Delft University of Technology, Department of Earth Observation and Space Systems, The Netherlands

Abstract

Because of the increased call for positioning in urban areas, the performance of GNSS is analyzed under conditions with a decreased satellite visibility caused by buildings blocking the lines of sight. With GPS and Galileo almanacs and two city models visibility fingerprints of time–location combinations are computed. Outdoor availability and accuracy are predicted for both city models: Real-life data of Schiphol Airport, and an imaginary urban canyon model (located in the same area as Schiphol) with variable parameters for street width, street length and building block height. The accuracy is predicted based on dilution-of-precision values. GPS, Galileo, and a combined constellation are considered for single-frequency single-epoch positioning.

Even for mildly difficult urban environments 95% availability is not reached for GPS (or Galileo) close to buildings or in streets in north-south directions. In the combined GPS-Galileo constellation a substantial improvement of 30%–50% availability can be reached as compared to GPS only for those locations where GPS availability is already more than ~10% (under severe conditions such a high increase cannot be achieved). At Schiphol Airport, however, conditions are very mild since there are only a few enclosed areas, surrounded by more than one building block and in most vital areas 95% availability can be reached even with GPS only. The accuracy close to the buildings and piers is, on average, decreased however, up to a factor of about three very close to buildings. This effect is even larger on the south-west and south-east side of buildings because there the high satellite density in north-east and north-west directions cannot be exploited.

Keywords: Global Navigation Satellite Systems (GNSS), line of sight (LOS), dilution of precision (DOP), availability, Schiphol Airport

23.1 Introduction

With the advent of the Global Positioning System (GPS) positioning has become daily practice for many vehicles as well as increasingly so for pedestrians. Ever more new location-based services (LBS) are being developed based on GPS. Advantages of GPS over other techniques are that it is operational, global, 3D, accurate, and relatively cheap. Besides, in principle, no additional infrastructure of reference points, transmitters and/or receivers is needed by providers of location-based services, although they may chose to provide such an infrastructure, e.g., by assisted GPS. Nevertheless, accurate and independent positioning at any time and at any place in urban environments is still problematic. In urban environments, where positioning needs are often most present, the GPS signals are being blocked or attenuated and often are received via multipath signals. This means that both availability and accuracy of GPS are seriously harmed. With availability (Swan et al. 2003, Verbree et al. 2004) we mean the percentage of time that a sufficient amount of satellites are visible, or, in other words, a sufficient amount of satellites have unblocked direct lines of sight (LOSs). In combination with Galileo, the future European Global Navigation Satellite System (GNSS), both availability and accuracy will increase.

In this paper we study to what extent the shielding of signals by buildings degrades the availability and accuracy of both GPS and Galileo single-point positioning as well as the combination of both systems (G2). For this we used two models: A simple but very instructive urban canyon model with variable parameters that we could easily construct ourselves, and a real-life model in the Netherlands of Schiphol Airport with rooftops of terminals, piers, parking garages, and other buildings that we received from the Schiphol Group as Autocad data based on terrestrial surveys. Unlike previous work (Verbree et al. 2004, Tiberius & Verbree 2004, Liu et al. 2006, Taylor et al. 2007) related to our topic we were therefore not concerned with gathering data and how to do that. Also we were not concerned with LOS computations given a certain digital terrain model because we assumed flat streets in the urban canyon model and a flat terrain at Schiphol Airport. In the latter model, a flat terrain was close to reality for most of the area.

With the urban canyon model we focus mainly on understanding GNSS availability under different urban conditions, taking into account street width, building heights and receiver location. It has a typical shape where GNSS signals are blocked by buildings on two sides of a street. Accuracy was predicted for one set of city model parameters.

The other model is more open: LOSs are at most areas blocked from one side only, and in some areas at three sides. Schiphol has been an interesting case study because it is often considered a small city where 24 hours a day activities take place, and because it has a high potential for LBS. Many applications are under consideration by the Schiphol Group, both indoors and outdoors. This paper is a report of a partial project where we only consider the GNSS outdoor case; the GPS indoor case is treated by Odijk and Kleijer (2007) and Odijk and Kleijer (2008).

Outdoor positioning at Schiphol may be done for navigation, tracking & tracing, or just for recording coordinates. Procedures can be improved by positioning of personnel, vehicles, passengers, luggage, materials, and incidents. Personnel that may benefit from personal positioning are, e.g., security personnel, authority officers, firefighters, and emergency services. These personnel will have a need for positioning in vehicles as well. Vehicle positioning is important for basically all vehicles on the runway, but also for many vehicles on the platform and on landside. One may think of luggage trolleys, airplane lorries, grit sprinklers, street cleaning cars, mowing machines, etc. Registration of incidents is important to improve compliance: Reports of incidents are made for analyses to detect trends. For these reports coordinates need to be recorded accurately and on the spot with simple devices, e.g. in case of possible environmental pollution. From interviews with employees of the Schiphol Group it became clear that for many applications a 10 m accuracy is sufficient. For some special cases, however, higher accuracy is advisable.

Our approach to LOS computations is given in *Section 23.2*. The GPS and Galileo availability are studied in *Section 23.3* for both data sets. In *Section 23.4* the attainable accuracy in the two models is studied by means of dilution-of-precision (DOP) values. *Section 23.5* concludes this paper.

Table 23.1. GPS and Galileo system characteristics.

GNSS	GPS	Galileo
Status	Operational	Planned
Number of orbits	6	3
Number of satellites	24	27 + 3
Inclination	55°	56°
Semi-major axis	26560 km	29600 km
Period	11 h 58 min	14 h 4 min 42 s
Ground track repeat	10 days / 20 orbits	10 days / 17 orbits

23.2 Line-Of-Sight Computations

With GNSS almanac data and a given 3D city model one can predict which satellites are visible at a certain receiver position in that city model. In other words, if there is a building blocking the LOS, one can predict that no signal or a much weaker signal will be received. Almanacs contain information about which satellite is where at which time. For this study a software tool was created with which we did LOS computations based on both GPS and Galileo almanacs. Some characteristics of both systems are given in *Table 23.1*. For the GPS LOS computations we used a Yuma almanac (http://www.navcen.uscg.gov/gps/default.htm) of GPS week 328, December 2005. For the Galileo LOS computations an almanac was used that was created (Verhagen et al. 2002) based on a 27/3/1 Walker constellation (Walker 1984). In these almanacs there were 29 GPS satellites, slightly more than the nominal constellation, and 30 Galileo satellites. Unless stated otherwise, for the computations of sections 3 and 4, satellite coordinates were determined for 96 epochs with a separation of 2.5 hours spanning 10 days. This choice was made to have a homogeneous spread of satellite constellations for GPS, Galileo, as well as for the combination (the ground track repeat differs for GPS and Galileo, see *Table 23.1)* without the need of high computing power (and therefore long calculation times). In the LOS computations we only considered satellites with an elevation cut-off angle above 10° because in real-life circumstances such a cut-off angle would make sense as well to avoid multipath signals.

All visibility information of satellites at a certain epoch for a certain receiver location is considered to make up a GNSS fingerprint of an epoch–location combination. For each of the 29+30 satellites the result of a LOS computation is either:

1. The LOS is not blocked.
2. The LOS is blocked.
3. The LOS is below the elevation cut-off angle.

Satellite coordinates are computed in earth-centred earth-fixed (ECEF) WGS-84 coordinates (X_s, Y_s, Z_s). However, we found it easier to do the LOS computations in a local 3D Cartesian coordinate system. Horizontal coordinates (x, y) of the Schiphol data set were available in the Dutch RD (Rijks Driehoeksmeting) system. These are Cartesian map coordinates roughly corresponding to easting and northing. Vertical coordinates (heights H) were given in NAP (Normaal Amsterdams Peil or Amsterdam Ordnance Datum). Strictly speaking the combined coordinates are not Cartesian, but for a small area we can consider them as such. Receiver coordinates were chosen at locations in the RD-NAP system (x_r, y_r, H_r).

In order to do LOS computations in RD-NAP, we computed pseudo-satellite coordinates close to the city model $(X_{\bar{s}}, Y_{\bar{s}}, Z_{\bar{s}})$ in such a way that they are still on the same LOS from receiver to satellite; compare (Taylor et al. 2007). For this purpose the chosen receiver coordinates were transformed to WGS-84 coordinates: $(X_r, Y_r, Z_r) \leftarrow (x_r, y_r, H_r)$. These WGS-84 coordinates were also used to compute elevation angles and to do DOP computations. Then pseudo-satellite coordinates where derived as $(X_{\bar{s}}, Y_{\bar{s}}, Z_{\bar{s}}) = (X_r, Y_r, Z_r) + (X_{rs}, Y_{rs}, Z_{rs}) \cdot \bar{\rho} / \rho$, where $\cdot_{rs} = \cdot_s - \cdot_r$, and ρ and $\bar{\rho}$ are the ranges from receiver to satellite and pseudo-satellite respectively. The range $\bar{\rho}$ is chosen such that the pseudo-satellite position lies just outside or on a bounding box around the city model. The pseudo-satellite coordinates are then

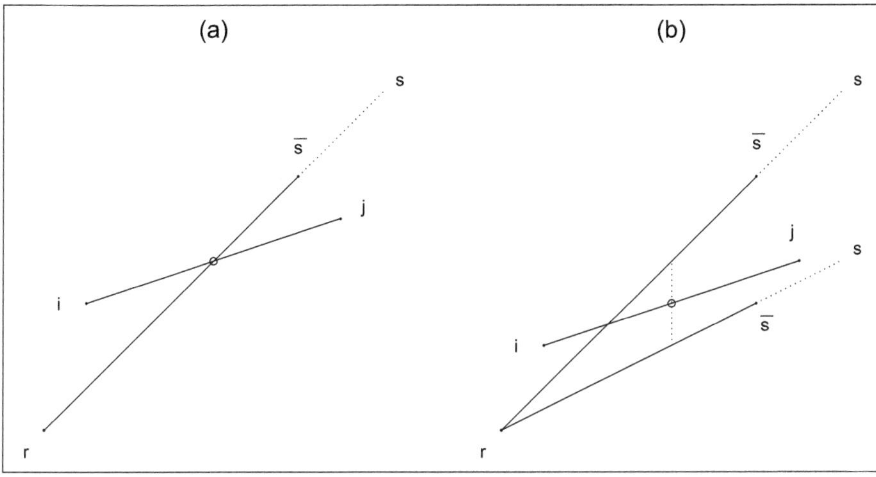

Fig. 23.1. Illustration of LOS computation for receiver r, satellite s and pseudo-satellite $\bar{s}$. (a) If in 2D the projection of the LOS (r, s) or $(r, \bar{s})$ crosses the projection of the rooftop edge (i,j) at the point indicated with 'o', the height of the LOS at that point is computed as well as the height of the line (i,j). (b) If in 3D the LOS (r, s) is above (i,j) the LOS is unblocked, else it is blocked.

transformed to RD-NAP: $(x_{\bar{s}}, y_{\bar{s}}, H_{\bar{s}}) \leftarrow (X_{\bar{s}}, Y_{\bar{s}}, Z_{\bar{s}})$. The transformations were done in two steps, using the equations by Bowring (1976) for transformation between ECEF coordinates and ellipsoidal coordinates (using the WGS-84 ellipsoid), and the equations by Schreutelkamp and Strang van Hees (2001) for transformation between ellipsoidal coordinates and RD; a geoid height of 43.1 m was assumed (De Min 1996). Although the latter transformation is an approximation (95% of the coordinates lie within 15 cm), it is considered sufficient for our purpose.

Once the pseudo-satellite coordinates in the RD-NAP system are obtained, they are used for the LOS computations in the following way (see *Figure 23.1*):

1. A 2D intersection is computed for each 2D rooftop polygon edge ij with coordinates (x_i, y_i, H_i) and (x_j, y_j, H_j) that potentially blocks a LOS $(x_r, y_r, H_r) - (x_{\bar{s}}, y_{\bar{s}}, H_{\bar{s}})$.
2. If there is an intersection, the height of the LOS at the intersection is compared with the height of the line ij at that intersection. If the line ij is higher than the LOS, the LOS is blocked, else it is not blocked by ij (but possibly by another rooftop line).

Although this is basically a very simple computation, computation costs will grow considerably with increasing amounts of rooftop edges, satellites, epochs, and receivers. Computation time was therefore limited by reducing the amount of potentially blocking rooftop lines: In a preprocessing step pointers were created pointing to all 2D rooftop edges in an artificial 2D grid cell, and in step 1 above only those grid cells were considered that cross the LOS.

23.3 GNSS Availability

Based on the GNSS fingerprints obtained from the LOS computations we can predict whether it is possible to compute the receiver coordinates. In a constellation with only GPS or Galileo satellites four satellites are needed for single-epoch (real-time) positioning because there are three coordinates and one receiver clock bias to be computed. In a constellation with both GNSS systems we assume there is an extra parameter to be determined: the time offset between GPS and Galileo (although in practise this might not be needed if this offset is sufficiently small or when information on this offset can be retrieved externally). In this case we need either two GPS satellites and three Galileo satellites, three GPS satellites and two Galileo satellites, or four satellites of either GNSS. The percentage of time that a sufficient number of satellites are visible we call availability. The availability of GPS, Galileo, and G2 is computed for both the urban canyon model and the Schiphol Airport data.

Bradbury et al. (2007) stressed that the urban environment is a multipath rich environment where one should take into account reflections and diffraction effects for computing availability. In our computations we did not take multipath into account, not only because of its complications, but since we are not particularly interested in specific time-location satellite availability but rather in availability expressed in percentages of time. It is unlikely that multipath is responsible for very different availability numbers as derived here for conventional (i.e. non high-sensitivity) GPS receivers since many receivers are mostly sensitive to short-delay multipath and many techniques are developed to suppress long-delay multipath; see, e.g. Sleewagen and Boon (2001). For high-sensitivity receivers however we see possible added value in using city models for predicting availability based on reflections as well. As shown by Bradbury (2007), because of diffraction, satellites may appear visible for a longer time (~5–10 minutes) than predicted based on direct LOS computations when setting behind a building or earlier when rising behind a building. Availability may therefore increase by diffraction. Considering a satellite that is being blocked twice per pass (when rising and setting) satellite visibility may then increase with ~5% of the total time for a satellite that is above the elevation cut-off angle for 5 hours. Availability may decrease however by line-of-sight blockage by cars, trees, and other objects.

23.3.1 The Urban Canyon Model

The urban canyon model is made up of twelve building blocks with streets in between. The four central building blocks are shown in *Figure 23.2.* In each of the cardinal directions two extra building blocks are located to represent a continuation of an infinitely large Manhattan-like city with equal building blocks. The model has three parameters: a street width w, a street length l, and a building block height

H (street heights are 0 m+NAP). Typical values for the street widths and building block heights are given in *Table 23.2.* As central location for the model we chose (arbitrary) coordinates from the Schiphol Airport area (x=113200 m, y=480250 m; corresponding to approximate latitude 52.31° N, and longitude 4.77° E). Availability numbers were computed for 65 locations in the model. Of these locations 23 are shown in *Figure 23.2.* Because the results were highly symmetrical (especially in east-west direction), only the results of 21 locations in the corner south-east of the central location 17 plus two additional locations are shown in *Figure 23.3.*

Table 23.2. Urban canyon model parameters used for the results of *Figures 23.3* and *23.4.*

height	8 m	regular house
	15 m	mansion
	30 m	low block of flats
	100 m	high block of flats
width	5 m	alley
	20 m	regular street
	80 m	open area

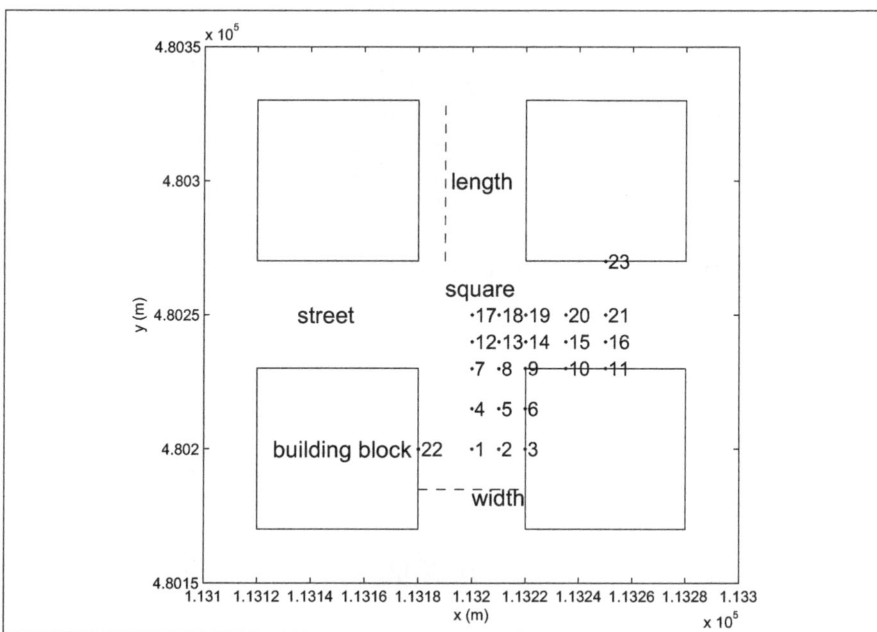

Fig. 23.2. Receiver locations used for the availability analyses in the urban canyon model. The locations 1, 4, 7, 12, and 17–21 are on the axes of the streets; locations 3, 6, 9–11, 22, and 23 are located 10 cm from the walls; locations 2, 5, 8, and 13–16 are at one (three) quarter of the street width; locations 1–3, 11, 16, and 21–23 are halfway the street length; locations 4–6, 10, 15, and 20 are at one (three) quarter of the street length.

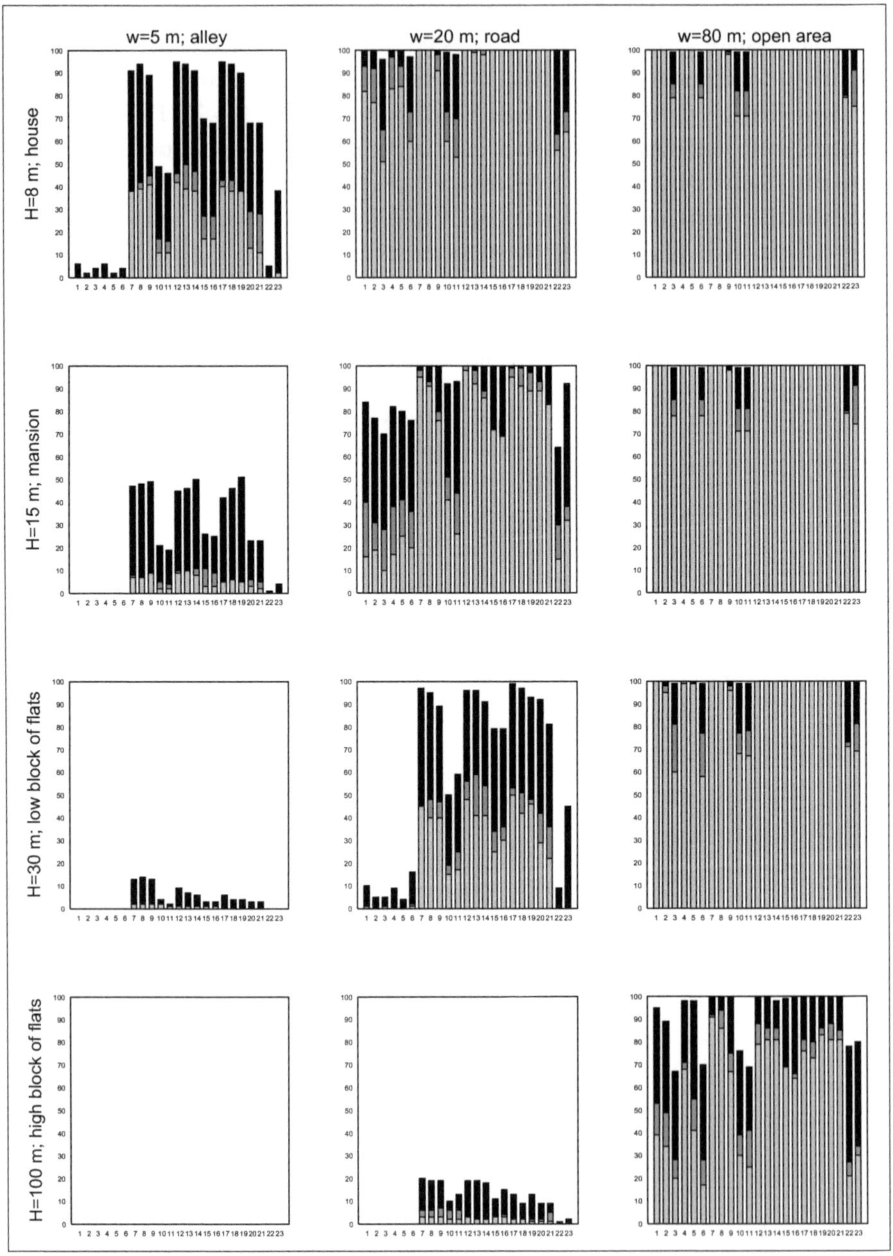

Fig. 23.3. Visibility percentages for the 23 locations of *Figure 23.2* in the 12 urban canyon scenarios (width $w = 5, 20, 80$ m; height $H = 8, 15, 30, 100$ m). Light grey: GPS only. Light grey plus dark grey: Galileo only. Total bar length: GPS + Galileo.

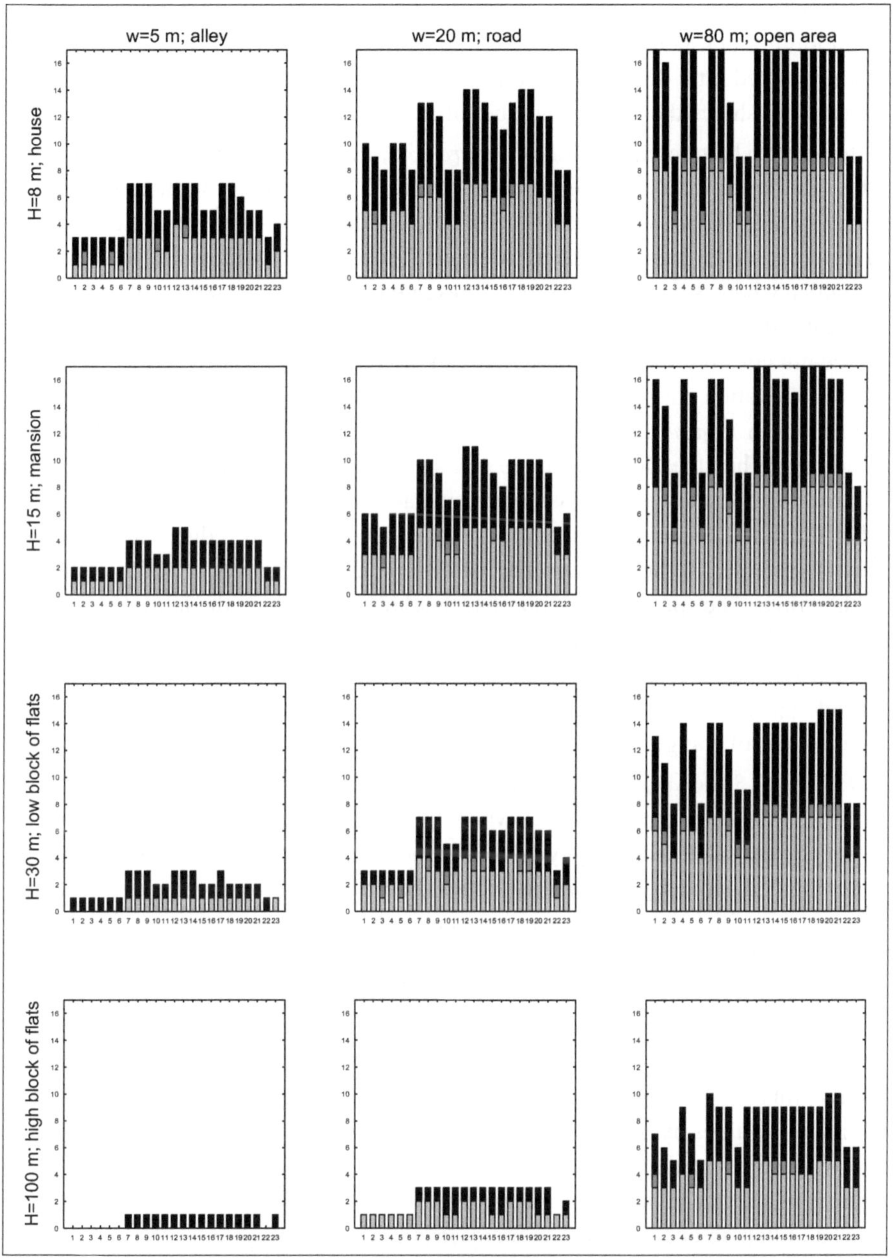

Fig. 23.4. Average number of visible satellites (rounded to integers) corresponding with the scenarios of *Figure 23.3*. Light grey: GPS only. Light grey plus dark grey: Galileo only. Total bar length: GPS + Galileo.

Figure 23.3 shows the results for twelve different models with each combination of building block height and street width as indicated in *Table 23.2*. Because the model showed little dependence on the street length we fixed it to one value: 50 m. *Figure 23.4* shows the average number of satellites on which *Figure 23.3* is based.

In order to help interpretation of *Figure 23.3, Figure 23.5* gives an impression of GPS satellite visibility for the central location 17 with model parameters $w = 20$ m and $H = 15$ m. *Figures 23.6* and *23.7* only show the rooftops to illustrate the effect of different model parameters and locations. These down-looking skyplots preserve azimuths, whereas zenith angles are represented by radial distances to the centre. In the centre the receiver position is shown by a + sign. The dashed lines represent the 80° zenith cut-off angle (the angle between the zenith and LOS) which corresponds with an elevation cut-off angle of 10°. All satellite positions in the 96 epochs used are shown in *Figure 23.5*. The solid lines represent the rooftops and walls of the buildings. When the elevation angle of a satellite is lower than the elevation angle of a rooftop the satellite is not visible. This translates in the skyplot as non-visibility when the radial distance of a satellite is larger than any rooftop edge. There is a clear lack of satellites in the northern part of *Figure 23.5* caused by the 55° inclination of the orbits. The specific location of such a hole depends on the latitude. The hole causes visibility and availability to be lower in north-south streets than in east-west streets. Also, a receiver location against a wall decreases visibility and availability considerately. This is especially so under mild conditions (low H/w) because the decrease in solid angle of clear sky visibility is more pronounced.

Figure 23.3 shows that Galileo has slightly higher availability numbers than GPS in most of the cases (the rare cases where this is not so are not shown). This may be caused by the satellite geometry as well as the additional satellite in the almanac. In the combined geometry availability values show a steep increase for those situations where availability for GPS or Galileo is already more than ~10%. An increase of 30%–50% of availability of G2 with respect to GPS-only can be seen for several locations. This increase agrees with the numbers given in (Swan et al. 2003) who claimed 56.3% availability for roads in Stuttgart and 99.7% for G2. Also it is confirmed that a higher value can be found for Galileo (78.3% in the Stuttgart case), although it would probably go too far as to say that Galileo would perform better in general. For availability values below ~10% the buildings are often blocking so much of the sky that extra satellites do not help much. A high increase can be seen, for example, for squares of alleys with houses or mansions, 20 m wide roads with mansions, and high blocks of flats in open areas. The increase is especially notable for those situations where the average number of satellites increases from below four to more than four.

In order to rely on GNSS, for many applications one would like to have at least 95% availability. An availability of 95% or higher for either GPS or Galileo can be seen for 20 m wide east-west roads with regular houses, but not against walls or in north-south streets, and for Galileo also at the centre of squares surrounded by mansions. For open areas surrounded by mansions also 95% can be reached, but not close to walls. At least about seven visible satellites on average are needed to reach this availability. In the G2 situation with 20 m wide roads and regular houses 95% is always reached. With mansions only at squares and east-west streets, but not against walls. With low blocks of flats this percentage is only reached at the centre of squares. G2 with open areas give 95% availability for mansions, even against walls, and for low blocks of flats if the receiver is not located against the wall. For alleys, only at the centre of squares and with regular houses, 95% can be reached.

A visibility of 5% or lower is seen for alleys with low blocks of flats, mansions if not at a square, or regular houses in streets in north-south direction. For regular streets this percentage is seen for high blocks of flats and streets surrounded by mansions in north-south streets.

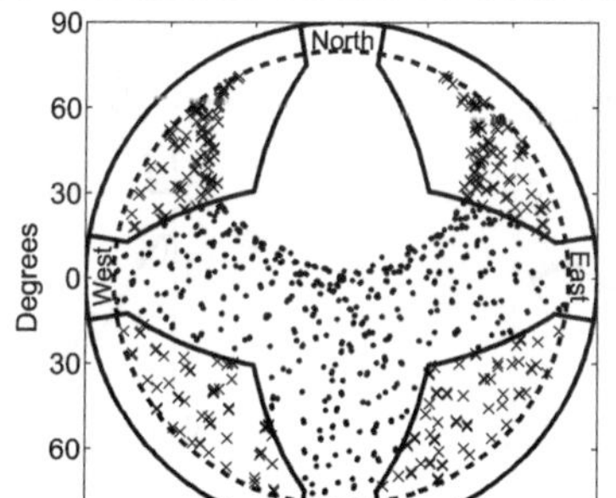

Fig. 23.5. GPS skyplot for the urban canyon model with parameters $H = 15$ m and $w = 20$ m, and for receiver location 17 (central location). Dots represent visible satellites; crosses represent non-visible satellites.

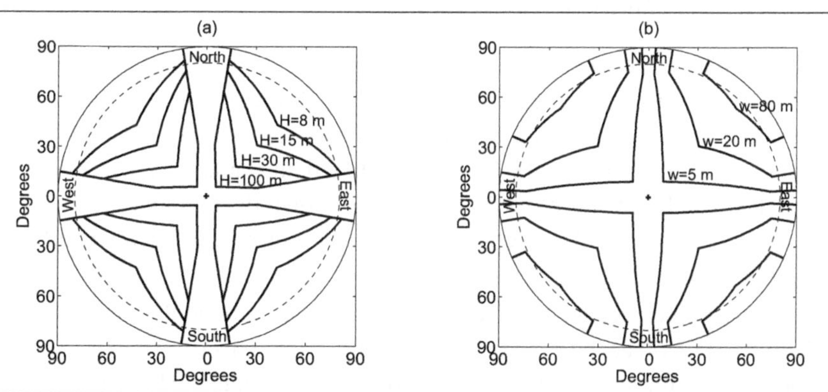

Fig. 23.6. Skyplots of rooftops in different versions of the urban canyon model. Receiver location 17 (central location). (a) Parameters: $w = 20$ m; $H = 8, 15, 30, 100$ m. (b) Parameters: $H = 15$ m; $w = 5, 20, 80$ m.

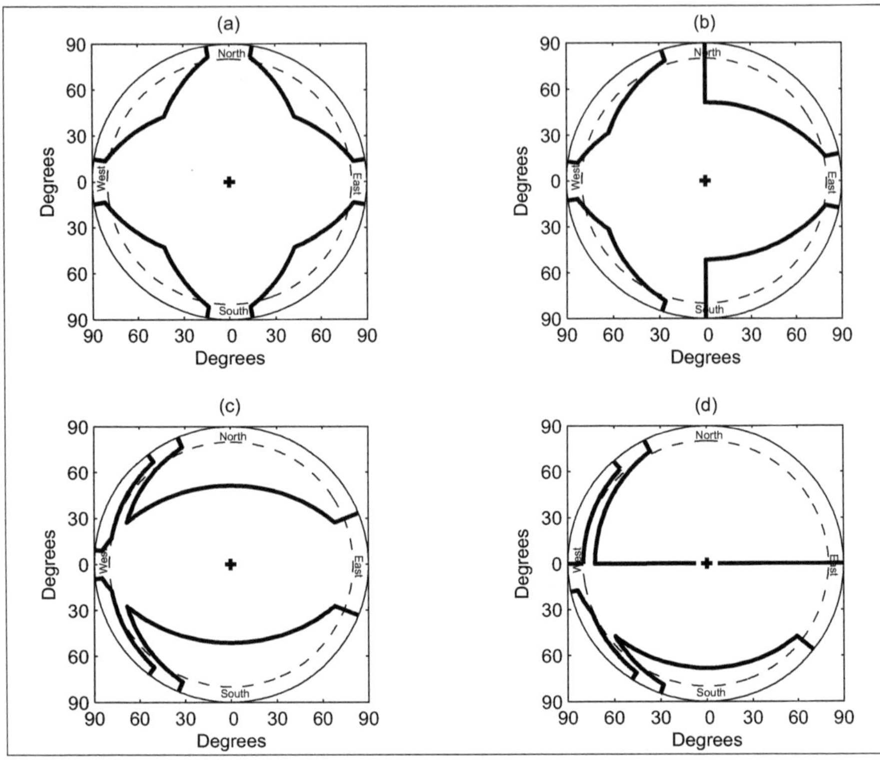

Fig. 23.7. Skyplots of rooftops in the urban canyon model for four different locations. Parameters: $H = 8\,\mathrm{m}$; $w = 20\,\mathrm{m}$. Receiver locations: 17 (a), 19 (b), 21 (c), and 23 (d).

23.3.2 Schiphol Airport

For the Schiphol dataset availability percentages were computed at a height of -3 m +NAP for every 10 m grid cell in the GPS-only, Galileo-only and G2 situation. *Figures 23.8* and *23.9* show the GPS and G2 cases respectively (the Galileo-only case was very similar to the GPS-only case). Only if an availability lower than 95% was reached a red area was plotted. In black are shown the building blocks as well as areas for which no availability was computed because the cell centres fell inside building polygons.

From these plots we can conclude that for the areas around the piers over 95% availability can be reached even in the GPS-only case. This can be expected since only on one side the signals are being blocked, which is comparable with the urban canyon model for open areas. Only around high buildings with more than one blocking side availability is lower. This is the case at some buildings of the Schiphol Group; but these are not the vital places for GNSS use.

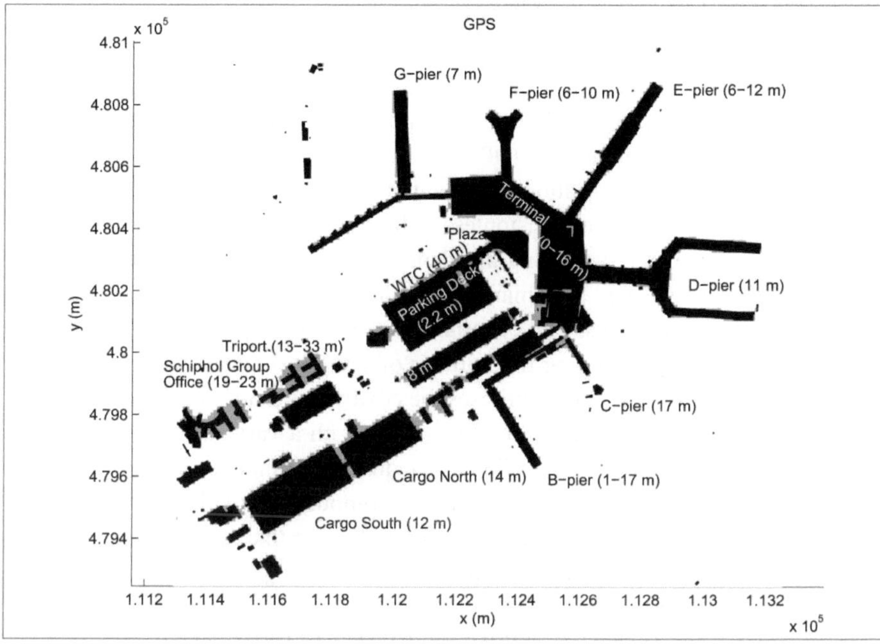

Fig. 23.8. GPS availability at Schiphol Airport. Gray: Less than 95%. White: More than 95%.

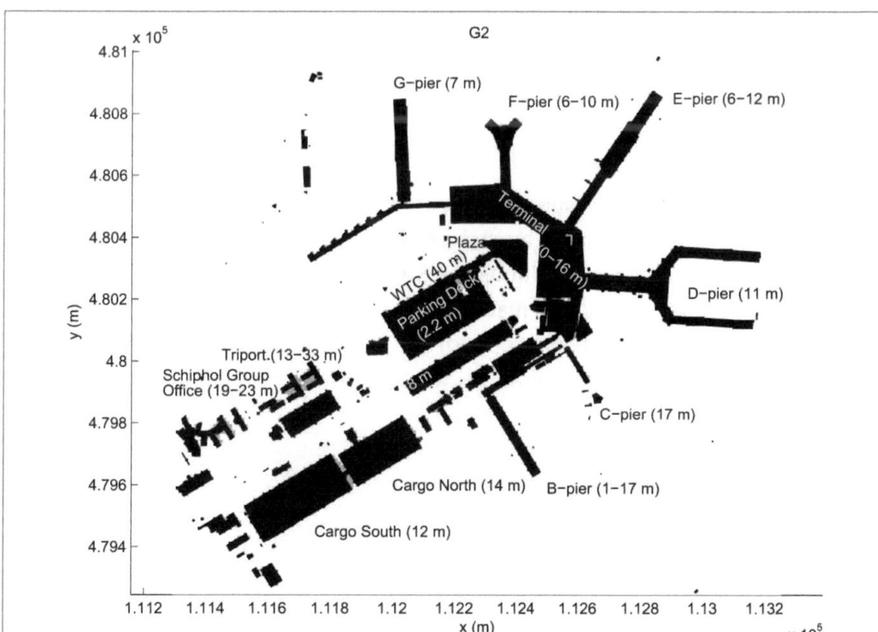

Fig. 23.9. G2 availability at Schiphol Airport. Gray: Less than 95%. White: More than 95%.

23.4 GNSS Dilution of Precision

In this section we deal with the accuracy that is achievable with one epoch of observations using either GPS, Galileo, or G2 under urban conditions. We do this by comparing dilution-of-precision (DOP) values. We can distinguish, amongst others, a vertical, horizontal, and position DOP, which are indicated by VDOP, HDOP and PDOP respectively. These dimensionless values are derived from the covariance matrix of the parameters assuming an identity matrix for a covariance matrix of the observations; see, e.g., (Hofmann-Wellenhof et al. 1992) for a discourse on how to do the computations. The covariance matrix of the parameters depends on the satellite geometry including the number of satellites, and the precision of the GPS observations. For practical purposes the covariance matrix of the observations is often assumed to be a diagonal matrix scaled with a variance scale factor. The square root of this scale factor is called user-equivalent range error (UERE), which depends on several error sources. A realistic computation of the UERE for GPS observations on one frequency is given in *Table 23.3*. In this table we also took into account the effect of multipath/diffraction.

Table 23.3. Error contribution to the UERE.

Error	st. dev.	variance
Ionosphere	4.0 m	16.00 m^2
Ephemeris	2.0 m	4.00 m^2
Satellite clock	2.0 m	4.00 m^2
Multipath	1.0 m	1.00 m^2
Troposphere	0.5 m	0.25 m^2
UERE	~5.0 m	25.25 m^2

In our computations we assumed four parameters for the GPS-only and Galileo-only case: three coordinates and a receiver-clock bias; see *Section 23.3*. In the G2 constellation we assumed an additional GPS-Galileo time offset as parameter. With the choice of a unit covariance matrix of the observations, the DOP values are computed from the diagonal elements of the covariance matrix of the estimated parameters transformed to a local north-east-up frame:

$$\text{VDOP (1D)} = (\sigma^2_{up})^{\frac{1}{2}};$$

$$\text{HDOP (2D)} = (\sigma^2_{north} + \sigma^2_{east})^{\frac{1}{2}};$$

$$\text{PDOP (3D)} = (\sigma^2_{north} + \sigma^2_{east} + \sigma^2_{up})^{\frac{1}{2}}.$$

An indication of radial accuracy for 1D vertical, 2D horizontal, and 3D position is then given by:

$$\sigma_{up} = \quad \text{UERE} \cdot \text{VDOP};$$

$$\sigma_{2D} = \frac{1}{\sqrt{2}} \text{UERE} \cdot \text{HDOP};$$

$$\sigma_{3D} = \frac{1}{\sqrt{3}} \text{UERE} \cdot \text{PDOP}.$$

Note that in the future GPS and Galileo measurements can be done on two civil frequencies. This means that the ionosphere will hardly contribute to the UERE any more, which will then reduce to about 2 m. However, strictly speaking, when we compute the DOP values in this case, we should include twice as many observations and introduce an ionospheric-delay parameter. This was not done in our computations. Also with EGNOS (European Geostationary Navigation Overlay Service) fully operational, the UERE should be reduced considerably as long as the signals of these satellites are not continuously blocked. *Figure 23.10* shows the distributions of PDOP, HDOP, and VDOP values under clear sky circumstances. Median values correspond to ratios of 0.5. The distribution of DOP values shows a long tail with a wider spread of DOP values higher than the median than DOP values smaller than the median. Especially for circumstances with high DOP values in the GPS-only and Galileo-only cases, the reduction of the DOPs in the G2 cases is large.

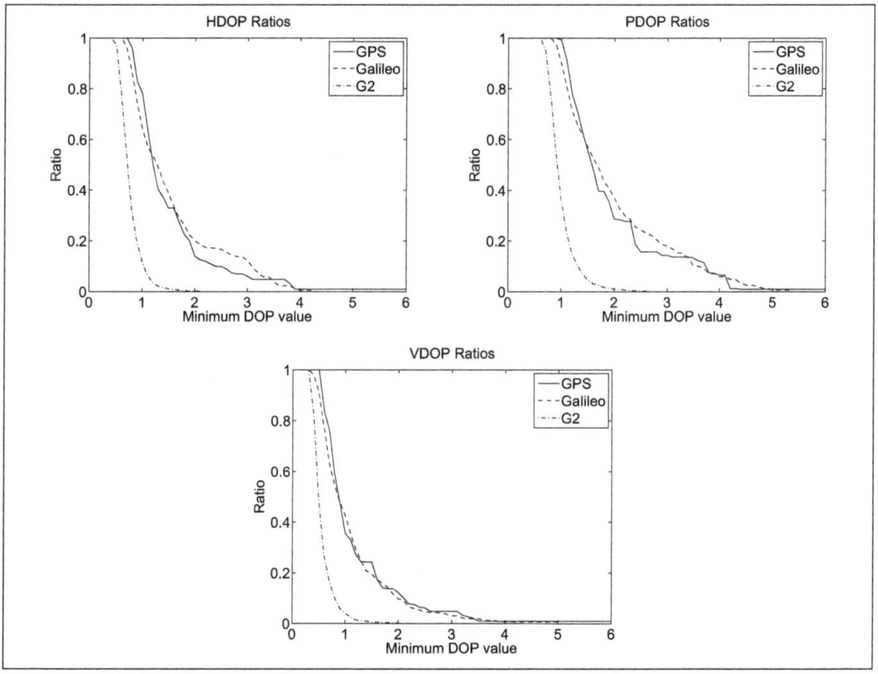

Fig. 23.10. DOP ratios for GPS, Galileo and G2 in case of a clear sky and an elevation cut-off angle of 10° for RD coordinates $x = 113400$, $y = 480250$. Based on 10 days of data, 1 epoch per minute.

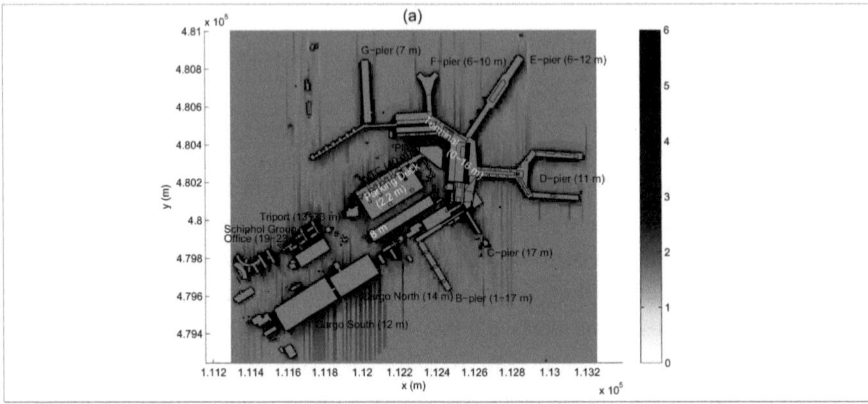

Fig. 23.11. Median GPS PDOPs at Schiphol Airport.

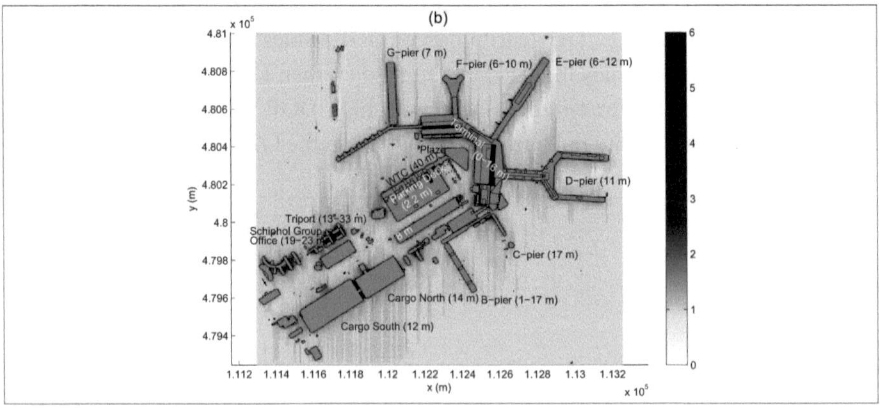

Fig. 23.12. Median G2 PDOPs at Schiphol Airport.

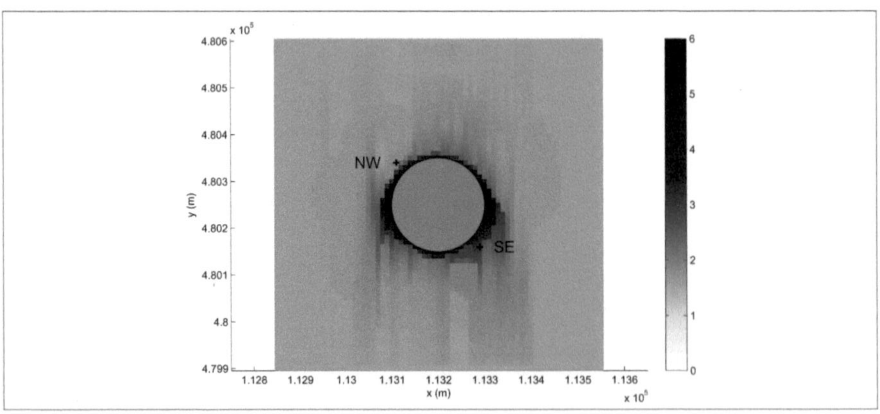

Fig. 23.13. Median GPS PDOPs around a (fictitious) 35-m high cylinder-shaped object of 200 m diameter at location Schiphol Airport. The north-west (NW) and south-east (SE) location are marked with +.

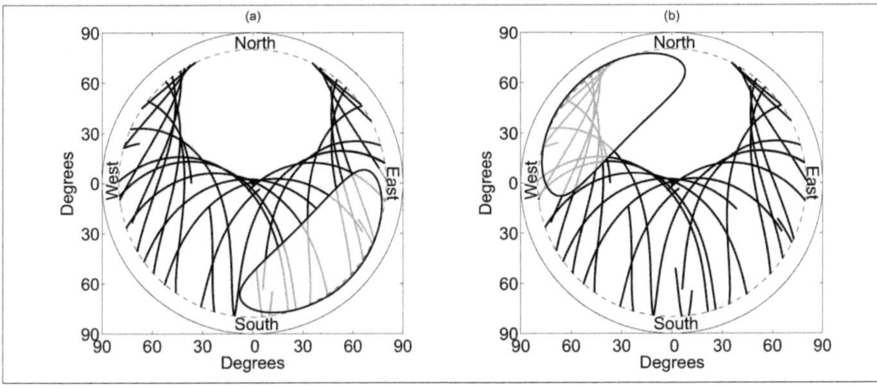

Fig. 23.14. GPS skyplots for the two locations indicated by + in *Figure 23.13*. Satellite positions are shown every minute for 12 hours. In blue are shown the satellites with unblocked LOS. In red the satellites blocked by the object. (a) the north-west (NW) location; (b) the south-east (SE) location.

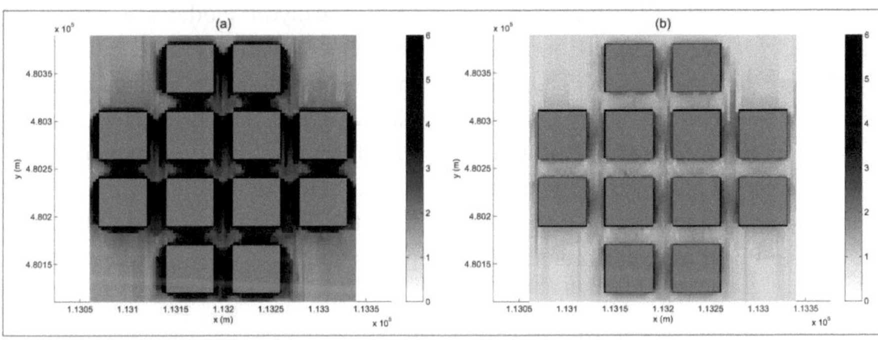

Fig. 23.15. Median PDOP values in the urban canyon model with parameters $l = 50$ m, $w = 20$ m, $H = 8$ m. (a) GPS only. (b) GPS and Galileo.

The median PDOPs over 96 epochs we computed at Schiphol Airport are given in *Figures 23.11* and *23.12*. Again the GPS and Galileo situation turned out to be very comparable (the latter is therefore not shown here), whereas the G2 situation gives a clear improvement. Also around the piers accuracy is clearly degraded in the GPS-only and Galileo-only situation.

Especially notable are the southern 'cast shadows' of the cargo buildings. Because of the hole in the satellite coverage in the north one would expect this shadow to be more clearly present in the north than in the south. Larger shadows on the south side than on the north side can also be seen for some of the piers, e.g. the E-pier. Because this is unexpected, we constructed a simple model of a 35 m high cylinder-shaped object and 200 m diameter in a further empty area at the Schiphol location. If there would be a symmetrical satellite coverage, the cast shadows should be equal in all directions. *Figure 23.13* shows the median GPS PDOPs for the same 96 epochs as used for the Schiphol computations. This figure tells that the shadow is not so much in the south direction but in the south-east and south-west direction. To rule out the possible effect of a heterogenous satellite distribution caused by a limited amount of epochs, we repeated the computation for two locations shown in *Figure 23.13* indicated by a + sign for every minute during 12 hours. *Figure 23.14* shows the skyplots for these locations. For the latter computation we found median GPS PDOPs of 1.90 and 2.95 instead of previously found values of 1.89 and 2.97 of the north-west and south-east locations respectively, thus ruling out that the limited amount of 96 epochs was an issue. The amount of LOSs blocked by the cylinder turned out to be 4724 and 5009 respectively. So we saw more satellites in the north-west than in the south-east causing the lower median PDOP values in the north-west, although that is also where more outliers of high PDOP values were actually found. Although there is a clear hole in the satellite coverage in the exact north the density of satellites in the north-west (and north-east) is higher than in the southern directions, thus causing lower median PDOPs in the north.

The median PDOPs computed for the urban canyon model are shown in *Figure 23.15*. We restricted ourselves to one set of relatively mild conditions. Notable are the high median PDOP values in north-south direction for the GPS-only case. Since availability under these circumstances is low (an unsufficient amount of satellites results in a PDOP of infinity) the median values are often larger than 6. For presentation purposes the locations with unsufficient availability are indicated with the same colour as a median PDOP of 6. In the G2 case in north-south streets the median PDOPs are at a much more acceptable level.

23.5 Conclusions

GNSS availability in urban areas is seriously degraded by buildings blocking LOSs from more than one side. Only under mild conditions 95% availability can be reached. For our mid-latitude computations this was the case for, e.g., 20 m wide streets surrounded by 8 m-high buildings, but only in east-west streets or on squares, but not close to buildings. In the G2 situation substantial improvements in availability (30%–50%) can be made for those situations where availability is above ~10% for GPS only. This means that for mild conditions, like 20 m wide streets surrounded by 8 m high buildings, 95% availability can be reached anywhere. But still, under more difficult conditions 95% cannot be reached everywhere and anytime in urban canyons. In a constellation with three GNSSs, including the Russian GLONASS, the availability improvement should be even higher. But it is unexpected that this will lead to 95% availability everywhere and anytime. In areas with buildings on one side, like the piers of Schiphol Airport,

95% availability can be reached with GPS (and Galileo) only. The (time) median accuracy of the determined coordinates obtained from GNSS single-epoch observations in those situations can however degrade by up to a factor three. In the G2 situation the DOPs will be about a factor two better, and, with two civil frequencies for both GPS and Galileo, the accuracy will improve even further. This means that even close to buildings an accuracy under the 10 m (a mismatch of 10 m or lower, 95% of the time), as is often required for LBS, can be achieved in the future, although one should realize that there can be other blocking objects with a deteriorating effect that we have not considered. Under more difficult conditions in urban canyons, GNSS positioning should at least involve using a history of measurements, e.g. by applying Kalman filtering techniques, or additional information from, e.g., inertial navigation or other observations.

Acknowledgements

This research was part of the Bsik project RGI-150 (Ruimte voor Geo-Informatie) *3D positioning infrastructure in the built-up environment.* We thank Dick Viveen of the Schiphol Group for his cooperation and for providing the Schiphol data set.

References

Bowring BR (1976) Transformation from spatial to geographical coordinates, Survey Review, 23, pp. 323–327.

Bradbury J (2007) Prediction of urban availability and signal degradation using virtual reality city models, ION GNSS 2007, Fort Worth, Texas.

Bradbury J, Ziebart M, Cross PA, Boulton P, Read A (2007) Code multipath modelling in the urban environment using large virtual reality city models: Determining the local environment, Journal of Navigation, 60, pp. 95–105.

Hofmann-Wellenhof B, Lichtenegger H, Collins J (1992) GPS: Theory and practice, Springer-Verlag, Wien, 3rd edn.

Liu G, Zhang K, Densley L, Wu F, Retscher G (2006) High fidelity 3D urban model-based signal performance simulation of the current and future GNSS in Australia, IGNSS Symposium, Holiday Inn Surfers Paradise, Queensland, Australia.

Min EJ de (1996) De geoïde van Nederland, PhD thesis, Delft University of Technology, the Netherlands.

Odijk D, Kleijer F (2007) Performance of high-sensitivity GPS for personal navigation at Schiphol Airport, the Netherlands, IGNSS Symposium, Sydney, Australia.

Odijk D, Kleijer F (2008) Can GPS be used for location-based services at Schiphol Airport, the Netherlands?, Proceedings of the 5th Workshop on Positioning, Navigation and Communication 2008, Hannover Germany, pp. 143–148.

Schreutelkamp FH, Strang van Hees GL (2001) Benaderingsformules voor de transformatie tussen RD- en WGS84-coördinaten, Geodesia, pp. 64–69.

Sleewaegen JM, Boon F (2001) Mitigating short-delay multipath: A promising new technique, ION GPS 2001, Salt lake City, Utah.

Swann J, Chatre E, Ludwig D (2003) Galileo: Benefits for Location Based Services, Journal of Global Positioning Systems, 1, pp. 58–66.

Taylor G, Li J, Kidner D, Brunsdon C, Ware M (2007) Modelling and prediction of GPS availability with digital photogrammetry and LiDAR, International Journal of Geographical Information Science, 21, pp. 1–20. DOI: 10.1080/13658810600816540.

Tiberius C, Verbree E (2004) GNSS positioning accuracy and availability within Location Based Services: The advantages of combined GPS-Galileo positioning. In: 2nd ESA/Estec workshop on Satellite Navigation User Equipment Technologies, G.S. Granados (Ed), ESA publications division, Noordwijk, pp. 1–12.

Verbree E, Tiberius C, Vosselman G (2004) Combined GPS-Galileo positioning for location based services in urban environment, In: Proceedings of the Location Based Services & Telegraphy Symposium 2004, Austria, Vienna University of Technology, pp. 99–107.

Verhagen S (2002) Innovation Studying the performance of Global Navigation Satellite Systems: A new software tool, GPS World, 13, pp. 60–65.

Walker JG (1984) Satellite constellations, Journal of the British Interplanetary Society, 37, pp. 559–571.

24 An Investigation of the Signal Performance of the Current and Future GNSS in Typical Urban Canyons in Australia Using a High Fidelity 3D Urban Model

Kefei Zhang[1], Gang-Jun Liu[1], Falin Wu[1], Liam Densley[2] and Günther Retscher[3]

[1] School of Mathematical and Geospatial Sciences, RMIT University, Australia
[2] Spatial Vision, Melbourne, Australia
[3] Institute of Geodesy and Geophysics, Vienna University of Technology, Austria

Abstract

With the rapid development of spatial information infrastructure in US, Europe, Japan, China and India, there is no doubt that the next generation Global Navigation Satellite Systems (GNSS) will improve the integrity, accuracy, reliability and availability of the position solution. GNSS is becoming an essential element of geospatial infrastructure and consequently part of our daily lives. However, the applicability of GPS in supporting a range of location-sensitive applications such as location based services (LBS) in an urban environment is severely curtailed by the interference of the 3D urban settings. No investigation has been carried out to accurately quantify and reliably evaluate the upcoming improvements like Galileo in typical 3D Australian urban environments.

A high-fidelity 3D urban model of Melbourne Central Business District is built using ArcGIS and large scale high-resolution spatial data sets to characterise and gain in-depth understanding of such interferences, and to enable an effective implementation of location-based optimisation alternatives. This model is used to support a comprehensive simulation study of current and future GNSS signal performance, in terms of signal continuity, availability, geometry, positioning accuracy and reliability based on a number of scenarios. The design, structure and major components of the simulator are first outlined. Useful time-stamped spatial patterns of the signal performance over the experimental urban areas have been revealed which are very valuable for supporting LBS applications, such as emergency responses, the optimisation of wireless communication infrastructures and vehicle navigation services.

Keywords: Global Navigation Satellite System, location based services, 3D urban model, numerical simulation

24.1 Introduction

Global Navigation Satellite Systems (GNSS) are a primary positioning technology for a large number of critical industry sectors such as location based services (LBS), maritime, aviation, agriculture, mining, surveying, military, road transport, and personal mobility (Misra & Enge 2006, Parkinson et al. 1996). GNSS is a critical space-bone geospatial infrastructure and plays an important part in our daily lives. However, current GNSS is still dominated by the US Navstar GPS and additional spatial infrastructures are being (or to be) built in Europe, Japan, China and India (Sisodia et al. 2003, Breeuwer et al. 2001, Zhang et al. 2005, Wu et al. 2004). For example, the European Union is currently developing and deploying its own GNSS, known as Galileo (European Commission 2001, Montenbruck et al. 2006). It is anticipated that the next generation GNSS will offer over one hundred satellites for positioning and navigation in the next decade.

The applicability of GNSS in supporting a range of location-sensitive applications such as LBS in an urban environment raises a number of problems (Swann et al. 2003, Chatre & Ludwig 2003). One of the dominating problems is the difficulties related to signal obstructions by features such as buildings, urban canyons, bridges and trees, as well as the effects of multipath caused by signal reflections from buildings and other surfaces (Bradbury 2007). There is no doubt that the combined use of different GNSS will improve the availability, accuracy, reliability and integrity of the position solution. It is anticipated that more GNSS constellations will undoubtedly give a user to view more satellites at a given location and time. This in turn leads to an increased accuracy with much improved availability and positioning continuity. Therefore, it is critical to accurately quantify and reliably evaluate these improvements in a typical urban context, where most human activities are taking place.

To evaluate the performance of the current and future GNSS in an urban environment, a 3D model of urban buildings in Melbourne Central Business District (CBD) is built first, a satellite visibility simulation system based on the 3D urban model is developed, and a comprehensive simulation is carried out. The GNSS satellite visibility tool considers direct line of sight (LOS) obstructions of satellite signals using the skeleton of urban settings to form surfaces of obstructions. To characterise and gain in-depth understanding of signal obstruction, and to enable an effective implementation of location-based optimisation alternatives, a high-fidelity three-dimensional (3D) model of urban buildings in Melbourne CBD is developed using ArcGIS and large scale high-resolution spatial data sets. The model is used to support a comprehensive simulation of current and future GNSS performance, in terms of signal continuity, availability, geometry, positioning accuracy and reliability. In addition, the system design, structure and major components of the simulator are outlined and a comprehensive simulation is carried out in Melbourne

CBD. Two types of simulations, i.e., spatial simulation and temporal simulation, are performed. Simulation results are presented using the number of visible satellites (NVS) as a measure of availability and Dilution of Precision (DOP) as a measure of accuracy with spatial and temporal variations.

24.2 The Development of a High Fidelity 3D Urban Model

The interest in developing 3D urban models has significantly increased recently due to advances in software, and increasing availability of high quality spatial datasets at reduced costs. Historically, the main focus of developing 3D urban models is on visualisation or virtual reality applications. The development of these urban models has therefore concentrated on more aesthetically appealing criteria for visual enhancement of the "true" urban settings for marketing benefits. Conversely, a high "fidelity" (in terms of accuracy and resolution) urban model is required to accurately evaluate the GNSS integrity, accuracy, reliability and availability in a typical 3D urban context.

24.2.1 Methodology

Recent research focus within photogrammetry and computer science has been directed at the automatic extraction of building features from aerial photography. The "cost" associated with manual interpretation and digitisation of urban scenes has limited the manual development of such models. Naturally, much research focuses on reducing the "cost" associated with 3D urban modelling by introducing automatic feature extraction techniques. These techniques were initially considered, although a more labour intensive method was adopted instead. The reasons for this are:

- To date there still remain many issues associated with automatic feature extraction methods. Extensive editing work is still required for the development of a high fidelity 3D urban model.
- Much of the labour intensive photogrammetrical work had already been performed.
- Manual interpretation of building features from ortho-rectified high-resolution imagery can be of high spatial accuracy.
- Any problems associated with the data supplied can be identified and rectified in the process.

24.2.2 Melbourne CBD Simulation Region

Melbourne, the capital of the state of Victoria and the second largest city in Australia, has been selected for the study. The Melbourne CBD precinct covers an area of approximately 1.88 km^2. The first survey plan for Melbourne was produced in 1837 by surveyor Robert Hoddle. Hoddle proposed that all streets should be ninety-five feet wide; this plan was objected at the time by the governor Sir Richard Burke. After outlining the benefits of wide open streets to the governor, and convincing him that wide streets were advantageous on the score of health and convenience to the future city of Victoria, the streets were laid out in their present position, known as the Hoddle Grid. The Hoddle grid is rotated 345° (approximately) form true north. Streets run East-North-East (e.g. Burke St) and North-North-West (e.g. Swanston St), the direction that the streets run and the width of the roads, will have a significant impact on the visibility of satellites. Melbourne CBD contains several large sky scrapers, the tallest building being the Rialto Tower at 248 m. The distribution of large buildings within the CBD is concentrated in the South-West region. Other tall buildings are located in the South East corner. Initial observation suggest that satellite visibility will be higher than in many other large cities world wide, due to our wide streets and relatively small building heights.

24.2.3 Data Set Used

The Melbourne City Council was approached for the development of the Melbourne CBD model. The following datasets were originally requested for the development of the 3D urban model.

- Aerial photography: Large scale, high resolution imagery of the CBD area. True ortho-rectified and mosaiced imagery was preferred to reduce manual photogrammetric workload.
- Cadastre dataset: Dataset containing cadastre boundaries.
- Building heights: Point or line datasets containing building elevation data.
- Feature line work: Building outlines, verandas, chimneys, air-conditioning units etc.
- Light Detecting And Ranging (LIDAR) data of the CBD.

The council initially provided some of the data required and indicated the possibility to provide the remaining data when sufficient progress is made, and the benefit to the council and the wider community is evident. The following datasets were made available for this research:

- Cadastre data sets in both dwg. and shp. formats.
- Large scale aerial photographs captured in October 2002 as true ortho-rectified georeferenced image with 0.063 m resolution in tiff. format.

- Feature datasets including polyline data showing building/feature outlines and heights, point data showing building/feature elevations and spot height ground data, in both dwg. and shp. formats.

The data is provided under a strict data license agreement with the Melbourne City Council.

24.2.4 Data Quality

Data quality and integrity was paramount in the development of a high fidelity 3D urban model. Comprehensive visual investigation was performed to identify data quality issues. The ortho-image supplied had a different creation date to that of the building feature data, creating issues in areas of significant building development. The true ortho-images also contained the following data reliability issues.

- *Ghost images:* The roofs of building are superimposed onto the bottom of the building and the roofs of the buildings appear twice. This phenomenon is called "ghost image" (Rau et al. 2002), see *Figure 24.1.*
- *Inaccuracy of digital surface model (DSM):* Many small objects weren't ortho-rectified due to inadequacy of DSM, boundaries of buildings were also inaccurately ortho-rectified due to DSM inadequacy, see *Figure 24.2.*
- *Mosaic alignment:* Feature alignment issues were observed along image seams.
- *Insufficient refilling to occlusion:* Insufficient slave images are in the refill of the occluded region on the master ortho-image (Rau et al. 2002).

Many of the data issues associated with the ortho-image are overcome through visual interpretation. Small features not ortho-rectified due to inadequacy of DSM are digitised by visual interpretation of their bases.

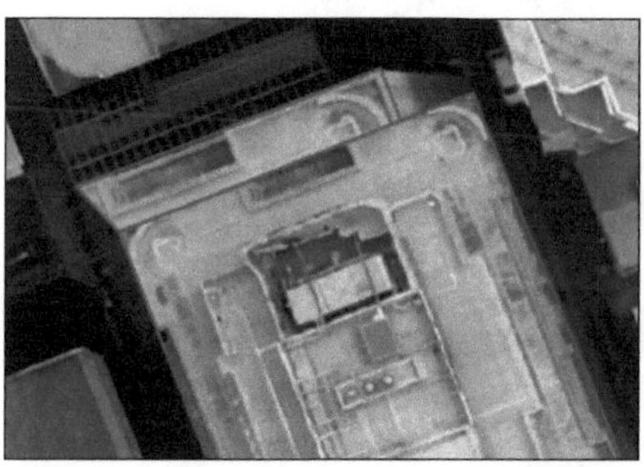

Fig. 24.1. The ghost image of the roof can be seen above the ortho-rectified location.

Fig. 24.2. Building roof top has been ortho-rectified although some minor rooftop structures have not been rectified.

24.2.5 Digitisation of Buildings and Features

The software used to perform the tasks of digitising the building features is Environmental Systems Research Institute's (ESRI) ArcGIS suite, in particular ArcMap (ESRI 2006). The ortho-image, feature dataset and cadastre data are used to digitise the buildings. The building boundaries are digitised by the following criteria:

- The highest weight is given to the building feature line work in the digitisation process. Relevant feature line work is used as a template for digitising when available.
- Where building feature line work is not available the ortho-image is used. Feature line work is often not available for smaller structures on the rooftop of buildings, e.g. air-conditioning towers.

The buildings are digitised as a polygon shapefile (see *Figure 24.3)*. The shapefile feature class contains the following attributes:

- Max_Height: Each polygon was assigned a maximum elevation. The feature data set was used to assign the corresponding heights.
- Comment: Polygons were assigned with a descriptive comment where relevant, e.g. poor quality data.
- Reg_Code: The city is broken into 9 regions; this enables at a later stage the separation of the shapefiles for editing within regions.

Approximately 4000 polygons are produced. *Figure 24.4* shows the 3D urban model of Melbourne CBD completed through the above process.

Fig. 24.3. Creating building polygons in the process of digitisation.

Fig. 24.4. A urban 3D model of Melbourne CBD developed with ArcGIS.

24.2.6 Integration of the Urban Model Developed with Simulation Software

With the issues associated with the data quality, the initial simulation testing is performed over a region of the city that contains no significant modelling problems. The regions used are the four south west city blocks shown on *Figures 24.4* and *24.6.* The ArcGIS shapefile data format the building polygons were created in was converted to the ASCII (American Standard Code for Information Interchange) format to be integrated into the simulation software it is required to be. The first step in the conversion was to reproject the building shapefile from the Australian Map Grid, Zone 66 (AMG66) to the GPS reference system WGS84. After reprojecting the shapefile it is then converted to the raster format. The final step is to convert it to the ASCII format. This process was carried out using ArcGIS functionality.

24.3 GNSS Performance Simulation

24.3.1 Simulation System Design

The simulation system consists of three packages, i.e., the three-dimensional model generation package, the satellite position estimation package and the GNSS performance evaluation package (see *Figure 24.5)* (Wu et al. 2007). The three-dimensional model generation package is designed to generate a high fidelity 3D urban model. The generation method and data source of a high fidelity 3D urban model have been given in *Section 24.2.* The satellite position estimation package is designed to estimate GNSS satellite position using Keplerian principle (Misra & Enge 2006). The package imports satellite orbit parameters from an almanac, such as YUMA almanac (USCG Navigation Center 2006), broadcast ephemeris, such as RINEX navigation file. Then the GNSS satellite positions are derived from satellite orbit parameters. The GNSS performance evaluation package is designed to simulate the signal propagation from satellite to receiver, and to evaluate GNSS performance in urban areas. The package primarily simulates direct GNSS signals that are not blocked by the 3D terrain and building objects. Only line of sight (LOS) propagation model is used in this research. More sophisticated propagation models, such as diffraction and reflection (Bradbury 2007), haven't been investigated in this research. The simulation results are presented using the NVS as a measure of availability and PDOP as a measure of accuracy with spatial and temporal variations.

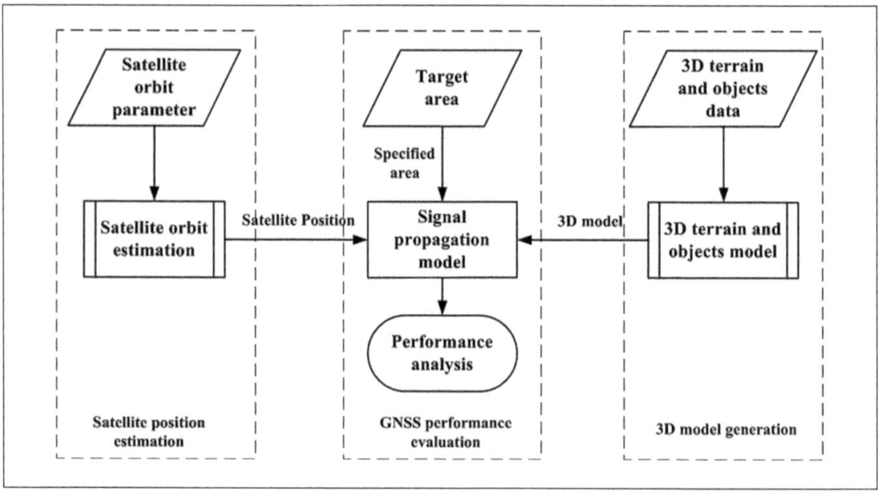

Fig. 24.5. An outline of the simulation system architecture.

24.3.2 Typical Simulation Results

Two types of simulations, i.e., spatial simulation and temporal simulation, are carried out. In these simulations, a GPS constellation with twenty-nine satellites is used via GPS YUMA almanac (USCG Navigation Center 2006), and a Galileo constellation with a total of 30 Mean Earth Orbiting (MEO) satellites configured as *Walker* constellation (i.e. distributed over three orbital planes), is used (Wu 2004). *Table 24.1* gives the almanac parameters for the Galileo constellation simulated.

Table 24.1. Parameters of the simulated Galileo constellation for this study.

Semi-major axis	a	29,994 km
Inclination	i	56°
Eccentricity	e	0.0
Right ascension	$\Omega_\circ$	-120°, 0°, 120°
Rate of right ascension	$\dot{\Omega}$	0.0°/day
Argument of perigee	ω	0.0°
Mean anomaly (1st orbit plan)	$M_\circ$	-160°, 120°, ..., 120°, 160°

In the spatial simulation, the receiver-satellite geometries are simulated at 00:00 on April 9, 2006 in the study area, with a sampling grid of 0.5m × 0.5m. In the temporal simulation, the receiver-satellite geometries are simulated at a cross point of King Street and Collins Street, Melbourne, Australia from 00:00 April 9, 2006 to 24:00 April 15, 2006. The experimental simulation area and point are shown in *Figure 24.6.* The point in the center of *Figure 24.6* is the cross point between King Street and Collins Street. *Figures 24.7* and *24.8* show the high fidelity 3D model and raster image of the specified areas respectively.

Fig. 24.6. Specified simulation areas and a street junction point in Melbourne CBD.

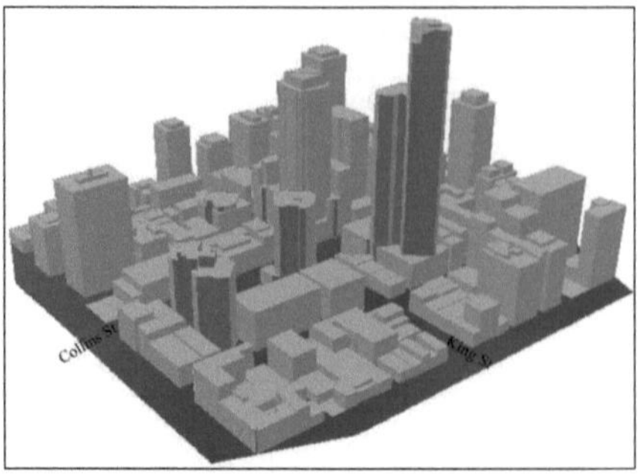

Fig. 24.7. A high fidelity 3D model of the specified areas.

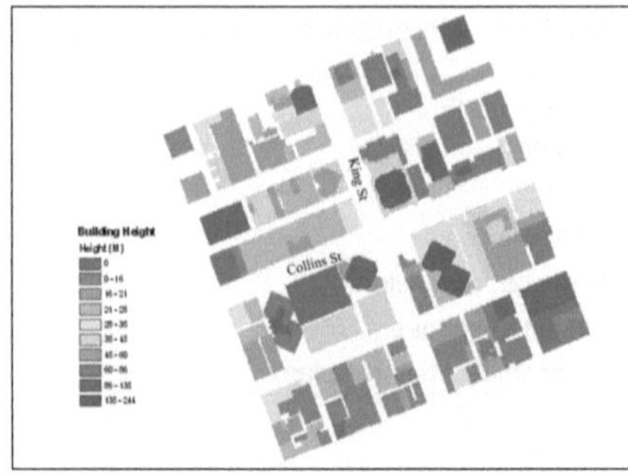

Fig. 24.8. The raster image of the specified areas with a scale of building heights.

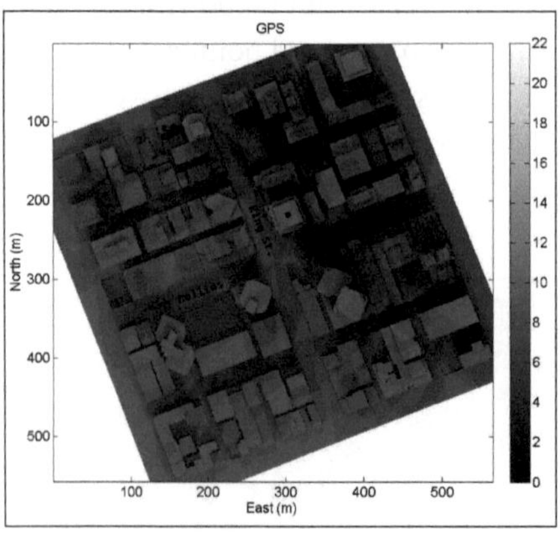

Fig. 24.9. Spatial variations of the number of visible satellites for GPS only.

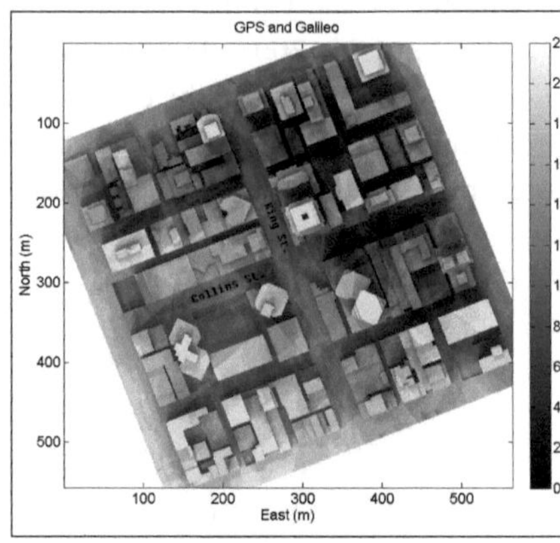

Fig. 24.10. Spatial variations of the number of visible satellites for a combination of GPS and Galileo systems.

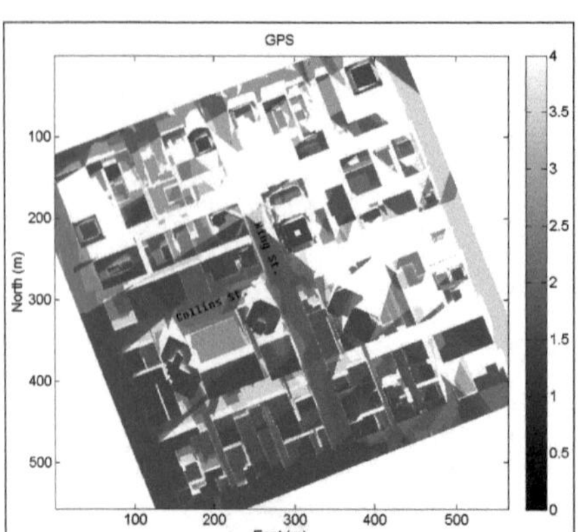

Fig. 24.11. Spatial variations of PDOP values for GPS only.

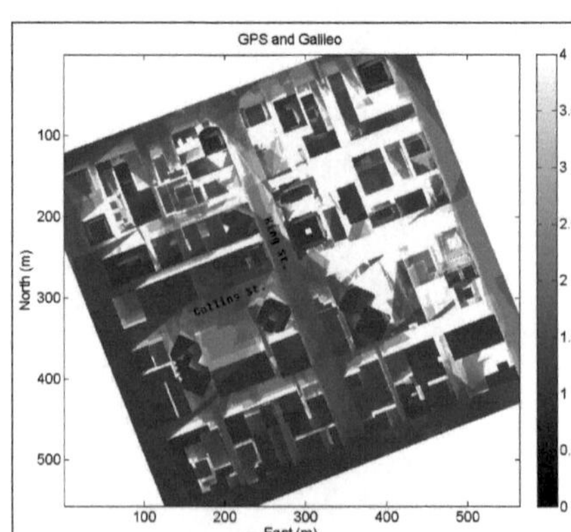

Fig. 24.12. Spatial variations of PDOP values for integrated GPS/Galileo..

Figures 24.9 and *24.10* show the spatial variations of NVS and PDOP of GPS and integrated GPS/Galileo systems in the region of interest. GNSS signals within the specific region are blocked by features, such as high buildings, bridges and trees. This is evident by the low NVS numbers. In some areas GNSS service becomes unavailable, due to unavailability of the required four NVS for GNSS positioning. It has been shown that the integrated GPS and Galileo systems can improve the satellite visibility and extend the positioning available area significantly.

Figures 24.11 and *24.12* show the spatial variations of the PDOP values of GPS and integrated GPS/Galileo systems in the test area. Correspondingly the PDOP values are also higher within urban environments of the simulation region. The results show that the integrated GPS and Galileo systems will offer better PDOP and improve positioning accuracy considerably.

Figures 24.13 and *24.14* compare the temporal variations of the NVS and PDOP values of GPS, and the integrated GPS and Galileo system at the cross point of King Street and Collins Street, Melbourne, with an open sky area. It is evident from *Figures 24.13* and *24.14,* that GNSS service performance is significantly affected by the features of the local urban environments, such as Melbourne CBD. The results show that the integrated GPS and Galileo system will not only improve the satellite visibility, extend the positioning available time, but also offer better DOP, improve positioning accuracy in urban environments.

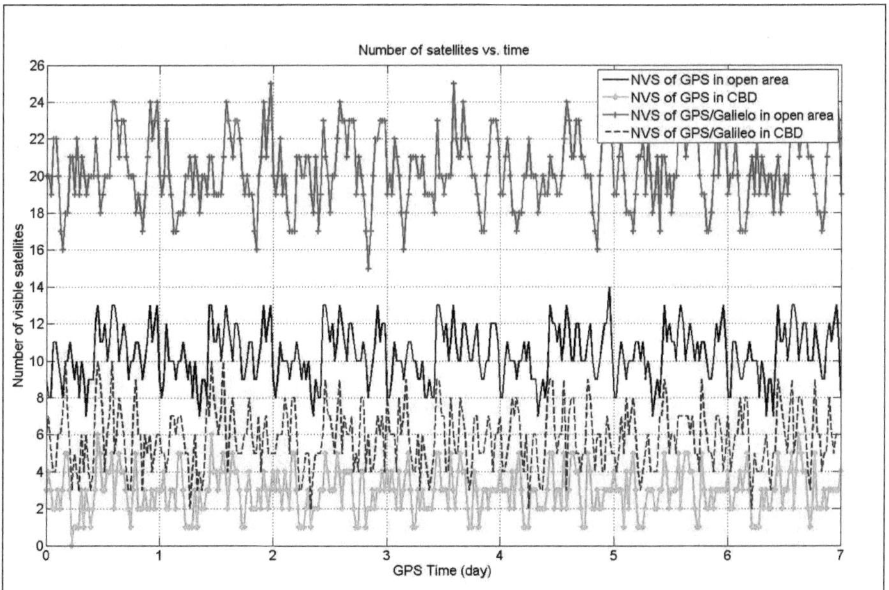

Fig. 24.13. Temporal variations of the number of visible satellites in four different scenarios (i.e. GPS only in open areas and Melbourne CBD, GPS/Galileo in open areas and Melbourne CBD).

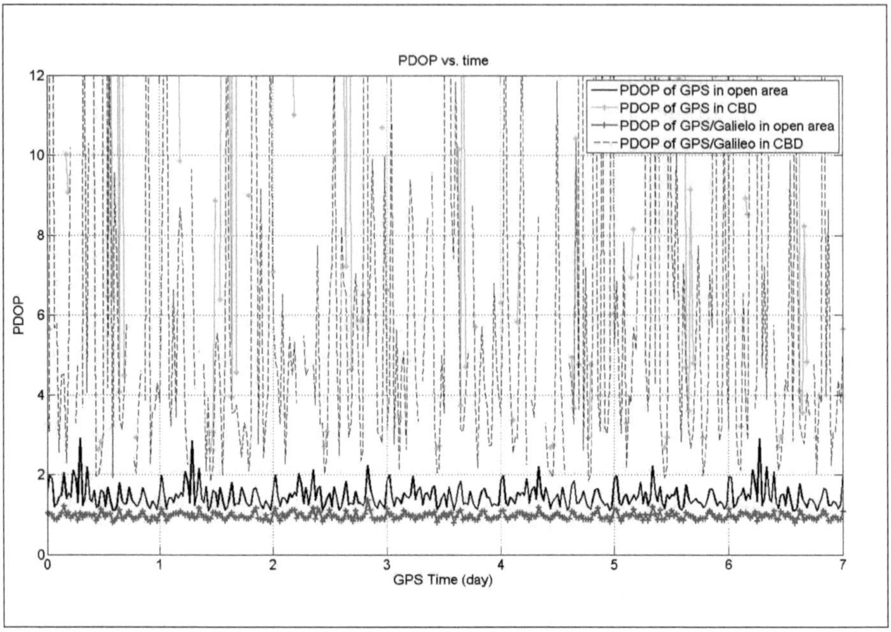

Fig. 24.14. Temporal variations of PDOP values in four different scenarios (i.e. GPS only in open areas and Melbourne CBD, GPS/Galileo in open areas and Melbourne CBD).

Figures 24.15 and *24.16* compare the sky plots at an open sky area, and in Melbourne CBD. The doted line in *Figure 24.16* indicates the margin of sky view, caused by obstruction blockage.

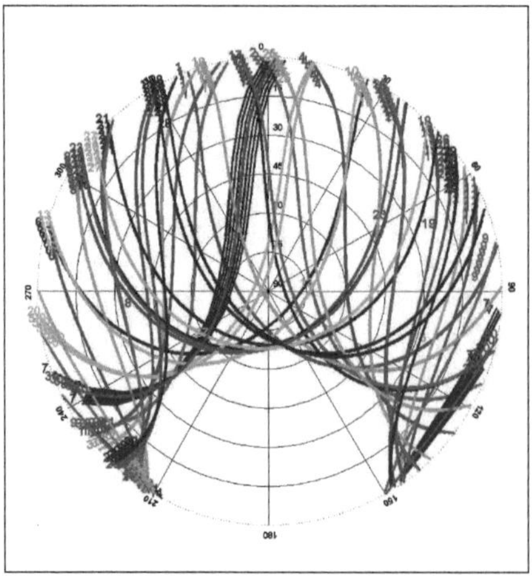

Fig. 24.15. A schematic sky plot in an open sky area.

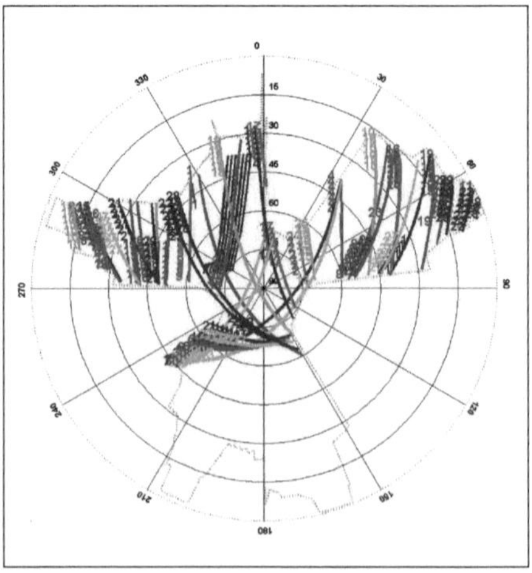

Fig. 24.16. A schematic sky plot in Melbourne CBD.

24.4 Conclusions

In this study, a high fidelity 3D model of urban buildings in Melbourne CBD has been built, a simulation system based on a variety of urban environment settings has been developed, and a number of comprehensive simulations have been carried out. Systematic analyses of the detailed results have revealed some useful time-stamped spatial patterns of the performance over the experimental urban area. These time-stamped spatial patterns can be very valuable for supporting applications such as car navigation, tourist information services, emergency responses and the optimisation of wireless communication infrastructures.

The performance of an integrated GPS and Galileo system in urban areas has been investigated by software simulations. The availability and accuracy of hybrid GPS and Galileo system have been analysed spatially and temporally. The integration of the two systems provides stronger geometry and better availability. These properties allow us to achieve a higher reliability of the overall system. The increased availability gives a better precision in the position domain and can provide an improved satellite geometry in the most difficult signal masking environments, such as urban canyons. Current research efforts are directed in validating simulation results with real GPS data, incorporating non-LOS propagation models, such as diffraction and reflection, and establishing an intelligent 3D model of urban buildings to support location-sensitive applications that require seamless navigation and positioning operations under both indoor and out-door settings in urban environments.

Acknowledgements

The Melbourne City Council is highly appreciated for providing Melbourne CBD data. Part of this research was presented in iGNSS conference, Sydney in 2006.

References

Bradbury J (2007) Prediction of Urban GNSS Availability and Signal Degradation Using Virtual Reality City Models. *Proceedings of the 20th International Technical Meeting of the Satellite Division of The Institute of Navigation.* Fort Worth, Texas, USA.

Breeuwer E, Farnworth R, Soley S (2001) EGNOS Operational Validation: A Status Update. *Proceedings of the 14th International Technical Meeting of the Satellite Division of the Institute of Navigation.* Salt Lake City, UT, USA.

Chatre J, Ludwig D (2003) Galileo: Benefits for Location Based Services. *Journal of Global Positioning Systems,* 1, 57–66.

ESRI (2006) ArcGIS – The Comple Geographic Information System. http://www.esri.com/software/arcgis/index.html.

European Commission (2001) European Transport Policy for 2010: Time to Decide. European Commission, http://ec.europa.eu/transport/white_paper/.

Misra P, Enge P (2006) *Global Positioning System – Signals, Measurements, and Performance,* Lincoln, Massachusetts, USA, Ganga-Jamuna Press.

Montenbruck O, Günther C, Graf S, Garcia-Fernandez M, Furthner J, Kuhlen H (2006) GIOVE – A Initial Signal Analysis. *GPS Solutions,* 10, 146–153.

Parkinson BW, Spilker JJ, Axelrad P, Enge P (1996) *Global Positioning System: Theory and Applications,* 370 L'Enfant Promenade, SW, Washington, DC 20024-2518, American Institute of Aeronautics and Astronautics, Inc.

Rau JY, Chen NY, Chen LC (2002) True Orthophoto Generation of Built-up Areas Using Multi-view Images. *Photogramm, Engm Remot Sens.,* 68, 581–588.

Sisodia AK, Hota JK, Vora NJ, Siddiqui A, Sanandiya HC, Bera SC (2003) Indian Plan for Satellite-Based Navigation System for Civil Aviation. *Proceedings of the 16th International Technical Meeting of the Satellite Division of The Institute of Navigation.* Portland, Oregon, USA.

Swann J, Chatre E, Ludwig D (2003) Galileo: Benefits for Location Based Services. *Proceedings of the 16th International Technical Meeting of the Satellite Division of The Institute of Navigation.* Portland, Oregon, USA.

USCG Navigation Center (2006) GPS Almanacs, NANUs, and Ops Advisories (Including archives). Navigation Center, United States Coast Gard, U.S. Department of Homeland Security, http://www.navcen.uscg.gov/gps/almanacs.htm.

Wu F (2004) A Study on High Precision Navigation Satellite System. *Laboratory of Communication Engineering.* Tokyo, Japan, Tokyo University of Marine Science and Technology.

Wu F, Kubo N, Yasuda A (2004) Performance Evaluation of GPS Augmentation Using Quasi-Zenith Satellite System. *IEEE Transactions on Aerospace and Electronic Systems,* 40, 1249–261.

Wu F, Zhang K, Yasuda A, Liu GJ (2007) Performance Improvement of Integrated Modernized GPS and Galileo System. *Journal of Geospatial Engineering,* 9, 12–21.

Zhang K, Seynat C, Mowlam A, Wu F, Kealy A (2005) An Investigation of Future GNSS in Support of Research and Development of Positioning Technology in Australia. *International Symposium on GPS/GNSS 2005.* Hong Kong, China.

25 Using RFID and INS for Indoor Positioning

Qing Fu and Günther Retscher

Institute of Geodesy and Geophysics
Vienna University of Technology, Austria

Abstract

This paper describes the current research work in the project UCPNAVI (Ubiquitous Cartograpy for Pedestrian Navigation) and outlines the methods by using active RFID (Radio Frequency Identification) in combination with INS (Inertial Navigation Systems) for positioning. In RFID positioning totally three different methods have been employed, i.e., cell-based positioning, trilateration using ranges to the surrounding RFID transponders (so-called tags) deduced from received signal strength measurements and RFID location fingerprinting. These technologies can be employed depending on the density of the RFID tags in the surrounding environment providing different levels of positioning accuracies. The positioning is restricted, however, to areas where at least one RFID signal can be detected. In order to overcome the lack of coverage of signals of the RFID tags we propose to integrate a low-cost INS in addition. INS measurements would be utilized to calculate the trajectory based on the method of strapdown mechanization. At the same time Kalman Filter would be used to correct the positions and velocity obtained. Since the INS components produce small measurement errors that accumulate over time and cause drift errors, the positions determined by RFID would be needed regularly to eliminate these errors. After a description of the principles and methods for positioning using active RFID the determination of pedestrian trajectories using INS and RFID is described briefly. The approach is verified by field tests and first test results are presented.

Keywords: indoor positioning, active RFID, INS, Kalman Filter, sensor fusion

25.1 Introduction

The technology of RFID can be used nowadays for positioning, where the location estimation is based on RSSI (received signal strength indication) which is a measurement of the power present in a received radio signal. The receiver can compute its position using various methods based on RSSI, e.g. a range-based positioning system based on trilateration or modelling of indoor positioning systems based on location fingerprinting. In the ongoing research project UCPNAVI the use of long range active RFID for positioning is investigated. Totally, three different methods have been employed, i.e., cell-based positioning, trilateration using ranges to the surrounding RFID tags deduced from received signal strength measurements and RFID location fingerprinting. In the case of a relative complex scenario, for example for the pedestrian path from a nearby underground station to one of the offices in a big building these three methods would be combined. Inside of the underground station and the building, the position would be determined by RFID location fingerprinting. At the entrance of the building the method of trilateration would be utilized to position the RFID reader. And outside of the building the user would be located by using cell-based positioning. The experiments showed that these approaches using RFID are suitable to navigate the user. However, the accuracy of the positioning is restricted in the area where RFID signals could be only detected from one or two tags. And in the area not covered by signals of RFID tags the position of the RFID reader could not be determined. In other words, if there is lack of the coverage of signals of the RFID tags, the RFID reader will lose its orientation. In order to overcome these shortages we propose to integrate INS with RFID.

Inertial navigation systems (INS) obtain measurements for the rate of turn using a gyroscope and acceleration using an accelerometer. These measurements need to be integrated over time to obtain orientation changes and velocity measurements. In the like way, the current position could be derived by means of integrating the obtained orientation changes and velocity measurements over time if the start position is given for the integration. In this way the trajectory of the user in the lack of the coverage between two RFID tags can be calculated. Inside of the signal coverage the position determined by RFID can be corrected regularly. However, the INS components produce small measurement errors that accumulate over time and cause drift errors. Therefore the sensor is accurate over short time intervals. In this regard, the INS needs position determined by RFID as a start point for the integration and for regular updates. This means that the RFID and the INS would be fused with each other, in order to position more accurately.

This paper is organized as follows; the second section describes the principles and methods of positioning using active RFID. The third section explains the determination of the trajectories using INS briefly. Section four presents and discusses the fusion of RFID and INS. The conclusions and an outlook are given in the fifth section and the acknowledgements will be expressed in the last section.

25.2 Positioning Using Active RFID

RFID stands for radio-frequency identification. Thanks to this technology, data can be transmitted from RFID tags to a reader via radio waves without line of sight contact. And the data might include the ID and the information of the position of the RFID tags. In the presented work only one reader and a large number of active tags are used. The moving object is equipped with a RFID reader and the unmoving objects in the surrounding environment are equipped with RFID tags whose positions are known and stored in the devices. If the reader moves inside the surrounding of the RFID tags its position can be calculated by using various methods based on RSSI. In the previous work three different methods have been employed, i.e., cell-based positioning, trilateration using ranges to the surrounding RFID tags deduced from received signal strength measurements and RFID location fingerprinting. These methods will be described in this section respectively.

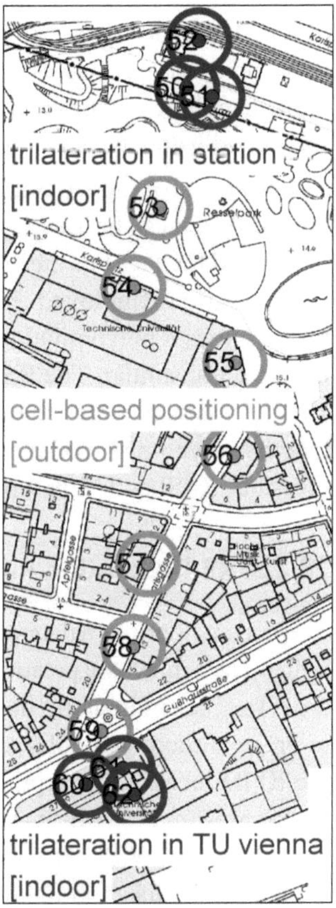

Fig. 25.1. An example of a cell-based positioning concept in outdoor areas of the city of Vienna in conjunction with the trilateration concept for indoor areas.

25.2.1 Cell-Based Positioning

The first method used in the research was cell-based positioning, which is an algorithm to determine the location of the user in a cell around the RFID tag with a size defined by the maximum range of the RFID signals. However, this method is only well suited for applications where accuracy is not that important. In the experiment for the path from a public transport stop (i.e., underground station 'Karlsplatz') to the university building cell-based positioning has been applied in outdoor areas as an alternative to GPS positioning (see *Figure 25.1*). Three RFID tags were installed at the entrance of the underground station 'Karlsplatz' (indoor area). Along a road between the underground station and the university building ('TU Vienna' in *Figure 25.1;* outdoor area) seven tags were installed. Additionally, three tags were installed at the building's entrance (indoor area). Each circle indicates a different cell.

As the accuracy of cell-based positioning generally depends on the size of the distinguishable cells, the achievable positioning accuracies might not be sufficient for indoor areas. Ranges from the RFID tags location have been achieved up to around 20 m. In the indoor area mostly higher positioning accuracies are required. Therefore several RFID tags have been installed in the transition zones between outdoor to indoor to be able to locate the user with a higher precision using trilateration.

25.2.2 Trilateration

Trilateration is the second method of determining the relative positions of moving objects (e.g. a user with a portable RFID reader) using the known location of two or more unmoving objects (tags), and the range between the moving object and the unmoving objects. To accurately and uniquely determine the relative position of a RFID reader on a 2-D plane two ranges are necessary. Furthermore, in order to obtain a very precise 3-D positioning at least three ranges are needed. Finally, if more than three ranges are available, the position fix can be calculated using a least squares adjustment.

The available ranges could be deduced from the received signal strength indication (RSSI). It can be assumed that the RSSI depends on the range between the RFID reader and the tag. For the conversion of the signal strength measurement different models can be employed. In the case of our study we used a polynomial model that assumes a polynomial relationship between the measured RSSI and the range from the transmitter. The RSSI decreases linear with the range (see Retscher & Fu 2007). In advance, the coefficients of the polynomial model must be determined that the polynomial model could be used. This step is also called calibration which measures RSSI at different places whose distances to the transmitter are known.

25.2.3 RFID Location Fingerprinting

The accuracy of the trilateration depends entirely on the accuracy of the measured range. The achievable positioning accuracies might not be sufficient for small indoor areas (e.g. location along the corridor at the centre of the office building). In order to create a more generally applicable localization method than using trilateration, the RFID location fingerprinting is used for positioning to provide more accurate location estimates. The principle of operation of RFID fingerprinting is similar to that used in WiFi networks (see e.g. Retscher et al. 2007, Kaemarungsi 2005). Generally, the location fingerprinting could be divided into two phases: an off-line or training phase and an on-line or positioning phase.

25.2.3.1 The Off-Line Phase

The first phase is called off-line or calibration phase. In this phase a database of signal strength values to the surrounding RFID tags at known locations is created.

The so-called location fingerprints are collected by performing a site-survey of the received signal strength indication (RSSI) from multiple RFID tags at every location for four different directions of the antenna of the RFID reader. The RSSI values at a particular location to a certain tag in certain direction are stored in a database in form of means of the total scans. The accuracy of position measuring can be improved by increasing the number of scans and increasing the number of points in the database. Therefore the database can be created not only with calibration points but also with interpolation points (see *Table 25.1).*

*Table 25.*1 shows the database with the calibration points (CP) which have values RSSI*(p,d,t)* of *C* calibration points in *D* directions to *T* tags in the off-line phase assigned; where *p* is the total number of points running from 1 to *C*, *d* is the total number of directions running from 1 to *D* and *t* is the total number of tags running from 1 to *T*. The rows in the table describe the measured RSSI for each tag from every calibration point CP. Usually the number of directions *D* is four in the calibration. For densification of the points in the database, a database with interpolation points can be created by using different interpolation methods between the calibration points, i.e., two-point interpolation and four-point interpolation. The purpose of the interpolation is to increase the number of points in the database so that the accuracy of position measurement can be improved.

Table 25.1. The database with RSSI*(p,d,t)* of C calibration points CP in D directions to T tags in the off-line phase.

CP RSSI(*p*,*d*,*t*)	Tag	Direction		
		1		*D*
1	1	RSSI(1,1,1)	...	RSSI(1,*D*,1)
	...	...	...	...
	T	RSSI(1,1,*T*)	...	RSSI(1,*D*,*T*)
...	...	...	...	...
C	1	RSSI(*C*,1,1)	...	RSSI(*C*,*D*,1)
	...	...	...	...
	T	RSSI(*C*,1,*T*)	...	RSSI(*C*,*D*,*T*)

25.2.3.2 The On-Line Phase

The second phase is the on-line phase which is also called positioning phase. In this phase a RFID reader will report a measurement of RSSI from different RFID tags at certain location where the reader should be positioned. To estimate the location an algorithm named "nearest neighbor in signal space" (NNSS) can be employed. In this algorithm the Euclidean distance between the measurement of the rssi*(p,d,t)* and each fingerprint in the database RSSI*(p,d,t)* is computed in signal strength space.

Table 25.2. The rssi*(p,d,t)* of *M* measurement points MP in *D* directions to *T* tags in the on-line phase.

MP rssi(*p*,*d*,*t*)	Tag	Direction		
		1		*D*
1	1	rssi(1,1,1)	...	rssi(1,*D*,1)
	...	...	...	...
	T	rssi(1,1,*T*)	...	rssi(1,*D*,*T*)
...	...	...	...	...
M	1	rssi(*M*,1,1)	...	rssi(*M*,*D*,1)
	...	...	...	...
	T	rssi(*M*,1,*T*)	...	rssi(*M*,*D*,*T*)

Four different approaches can be distinguished to assign the measured RSSI to the location in the on-line phase, namely, the direction-based approach (DBA), the tag-based approach (TBA), the direction-tag-based approach (DTBA) and the heading-based approach (HBA). In the experiment, DBA, DTBA use measurements in four directions and TBA and HBA use measurements in heading (one direction).

(1) Direction-Based Approach (DBA)

The direction-based approach (DBA) can be used to estimate the location of the RFID reader with measurements in four directions to one tag. DBA develops the NNSS algorithm using RSSI (circle with cross) and rssi (arrow) in four directions (D_1, D_2, D_3, and D_4) per tag per point (see *Figure 25.2)*.

P is the number of the points in the database; *D* is the number of directions. The *find* function can find the point number with minimum r_{DBA}. If there are several tags, the *hisfu* function can be used which is a frequency function that can count the number of occurrences of a point number (PNR) and select the point number with maximum frequency. Then the following MATLAB code was used to investigate the range r_{DBA} between the two vectors for positioning:

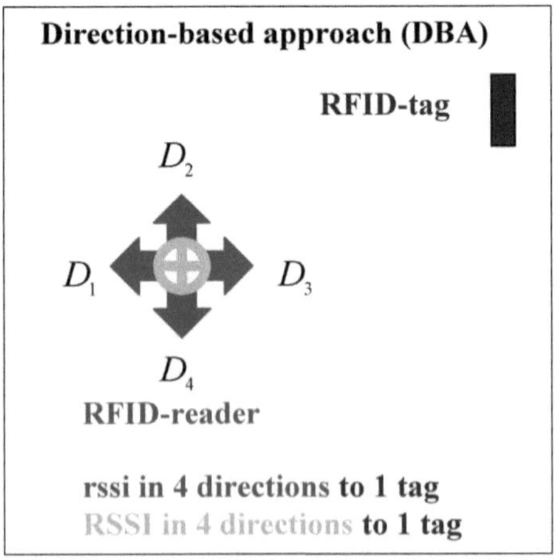

Fig. 25.2. Direction-based approach (DBA).

for $t = 1:T$ *% T is number of tags*
for $p = 1:M$ *% M is number of measurement point*

$$r_{DBA} = \sqrt{\sum_{d=1}^{D} (RSSI(1:P,d,t) - rssi(p,d,t))^2}$$

$PNR(p,t) = find(r_{DBA} = \min(r_{DBA}))$
end
end
for $p = 1:M$
$[PNR(p), frequency] = hisfu(PNR(p,1:T)')$
end

(2) Tag-Based Approach (TBA)

The tag-based approach (TBA) can be used to estimate the location of the RFID-reader with measurements in heading to several tags. TBA develops the NNSS algorithm using RSSI (circle without cross) and rssi (arrow) in one direction (i.e., D_4) to several tags (i.e., 4 tags) per point (see *Figure 25.3*). The following MATLAB code was used to investigate the range r_{TBA} between the two vectors for positioning:

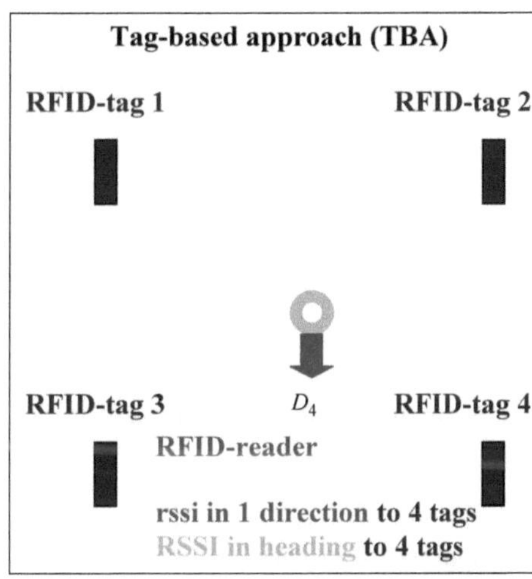

Fig. 25.3. Tag-based approach (TBA).

for $d = 1:D$ % D is number of directions
for $p = 1:M$ % M is number of measurement point

$$r_{TBA} = \sqrt{\sum_{t=1}^{T} (RSSI(1:P,d,t) - rssi(p,d,t))^2}$$

$PNR(p,d) = find(r_{TBA} = \min(r_{TBA}))$
end
end
for $p = 1:M$
$[PNR(p), frequencecount] = hisfu(PNR(p,1:D)')$
end

(3) Direction-Tag-Based Approach (DTBA)

The direction-tag-based approach (DTBA) can be used to estimate the location of the RFID-reader with measurements in four directions. DTBA develops the NNSS algorithm using RSSI (circle with cross) and rssi (arrow) in four directions (D_1, D_2, D_3, and D_4) to several tags (i.e., 4 tags) per point (see *Figure 25.4*). And this approach could be implemented by the following MATLAB code:

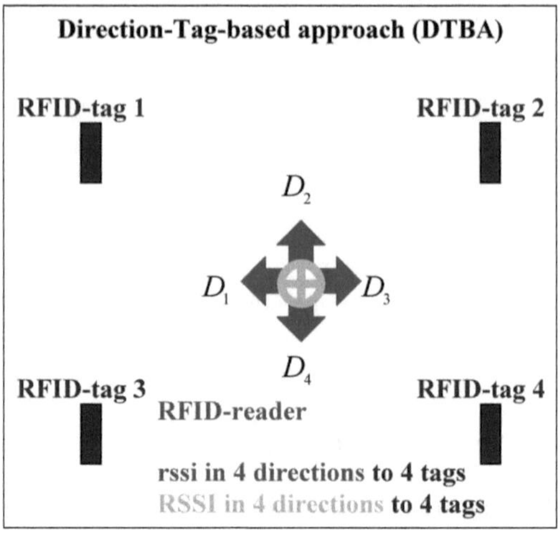

Fig. 25.4. Direction-Tag-based approach (DTBA).

for $= 1{:}M$

$$r_{DTBA} = \sqrt{\sum_{d=1}^{D}\sum_{t=1}^{T}(RSSI(1:P,d,t) - rssi(p,d,t))^2}$$

$$PNR(p) = find(r_{DTBA} = \min(r_{DTBA}))$$

end

(4) Heading-Based Approach (HBA)

The heading-based approach (HBA) can be used to quickly estimate the location of the RFID-reader with only measurements in heading. HBA develops the NNSS algorithm using RSSI (circle with cross) in four directions to several tags (i.e., 4 tags) and rssi (arrow) in one direction (i.e., D_4) to several tags (i.e., 4 tags) per point (see *Figure 25.4*).

Similar to the three approaches above, the range r_{HBA} between the two vectors for positioning in HBA can be investigated using MATLAB code:

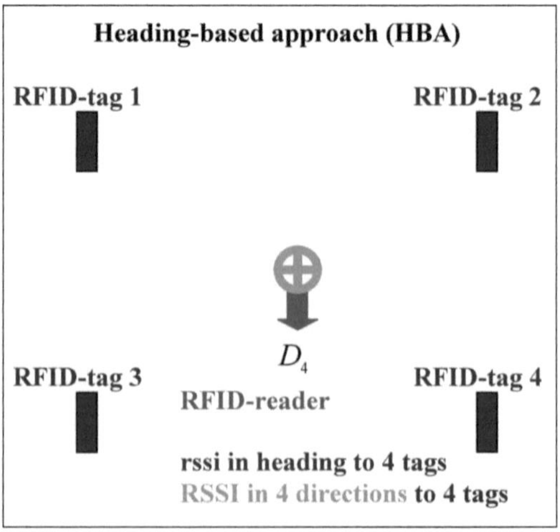

Fig. 25.5. Heading-based approach (HBA).

for $p = 1{:}M$

$$r_{HBA} = \sqrt{\sum_{d=1}^{D}\sum_{t=1}^{T}(RSSI(1:P,d,t) - rssi(p,h,t))^2}$$

$$PNR(p) = find(r_{HBA} = \min(r_{HBA}))$$

end

Table 25.3 shows an overview of these four approaches for the on-line phase of RFID location fingerprinting. The methods differ from each other in their approaches, in run time and their achievable accuracies. DBA and TBA are basic approaches and DTBA and HBA are more advanced approaches. DTBA is a combination of DBA and TBA to obtain the benefits of both. Then high accurate point positioning can be achieved, however, with the tradeoff of a slow run time. It also requires the measurement in four directions at every point. For this reason, HBA was developed based on TBA and DTBA to achieve middle to high accuracy with a fast run time.

Table 25.3. An overview of the four approaches.

Approach	DBA	TBA	DTBA = DBA+TBA	HBA = TBA+DTBA
Run time	slow	fast	slow	fast
Accuracy	low	middle	high	middle to high

25.3 Determination of the Trajectory Using INS

In our project we use mainly the three methods using RFID described in the *Section 25.2* for the positioning in indoor and urban environment. However, the accuracy of the positioning is restricted in the area where RFID signals could be only detected from one or two tags. And in the area not covered by signals of RFID tags the position of the RFID reader could not be determined. In other words, if there is lack of the coverage of signal of RFID tags, the RFID reader will lose its orientation. In order to overcome these shortages we propose to use INS (short for Inertial Navigation System) to determine the trajectory in the place where RFID signals are not available. The INS device used in our project is Xsens MTi manufactured by the Xsens Technologies B.V. in the Netherlands. This sensor is a low-cost MEMS based INS with a weight of only 50 g. It is therefore suitable to be carried by a pedestrian user and can be connected to a PDA or notebook computer.

INS obtains measurements for the rate of turn using a gyroscope and acceleration using an accelerometer. These measurements need to be integrated over time to obtain orientation changes and velocity measurements. Then the current position could be derived by means of integrating the obtained orientation changes and velocity measurements over time if start position can be given for the integration. For such a process as mentioned above there are lots of different approaches. In our project we use the strapdown mechanization.

The strapdown mechanization has several advantages compared to gimbaled or stabilized platform techniques. Such modern systems have removed most of the mechanical complexity of platform systems by having the sensors attached rigidly, or "strapped down", to the body of the host vehicle. The potential benefits of this approach are lower cost, reduced size, and greater reliability compared with equivalent platform systems (Tampere University of Technology 2008). In our approach the strapdown mechanization uses the orientation data from the INS (unit quaternions or Euler parameters $\mathbf{q}_0$, $\mathbf{q}_1$, $\mathbf{q}_2$, $\mathbf{q}_3$) and the calibration data (rate of turn from the gyroscope $\mathbf{gyr}_{x\_b}$, $\mathbf{gyr}_{y\_b}$, $\mathbf{gyr}_{z\_b}$ and acceleration from the accelerometer $\mathbf{acc}_x$, $\mathbf{acc}_y$, $\mathbf{acc}_z$) to obtain the trajectory of the INS. This process is divided into two steps which are presented in *Figure 25.6* and *Figure 25.7* respectively. The indices **b** or **n** for the different parameters denote either the body frame or navigation frame. A transformation from the body frame to the navigation frame is indicated with index **nb** and from the navigation frame to the body frame with **bn**.

In the first step the acceleration $\mathbf{acc}_{x\_b}$, $\mathbf{acc}_{y\_b}$, $\mathbf{acc}_{z\_b}$ and the rotation matrix $\mathbf{R}_{nb}$ are calculated from the measured orientation and calibration data. Thereby two kinds of acceleration values have to be distinguished, i.e., the original measured linear acceleration and the 'free' acceleration (i.e., 2nd derivative of the position). The original measured linear acceleration includes the acceleration due to gravity and centrifugal force. This is inherent to all accelerometers (Xsens 2007). Therefore, the gravity and centrifugal force must be first subtracted from the measured linear acceleration in order to obtain the 'free' acceleration. The second part of *Figure 25.6* illustrates the calculation of the 'free' acceleration from the original measured linear acceleration, whereby the rotation matrix results from the measured quaternion and rate of turn from the gyroscope as demonstrated in the first part of the *Figure 25.6*. Actually, *Figure 25.6* shows the calculation of the acceleration and rotation matrix from the measured orientation and calibration data clearly. The boxes with the heading 'Quaternion', 'Rate of trun (gyro)' and 'Acceleration (measured)' denote the data input of original measured data; the white boxes show the process of the calculation, and the boxes with the heading 'Rotation matrix' and 'Acceleration (free)' present the result output from the calculation.

Concerning that there are drifts in the measurements of rate of turn and acceleration, one more measurement was carried out by keeping the Xsens device unmoved for a short time period, in order to found out the drifts. And these drift corrections ($\mathbf{gyr}_{x0\_b}$, $\mathbf{gyr}_{y0\_b}$, $\mathbf{gyr}_{z0\_b}$ and $\mathbf{acc}_{x0}$, $\mathbf{acc}_{y0}$, $\mathbf{acc}_{z0}$) are subtracted from the measurements while the Xsens device was moved.

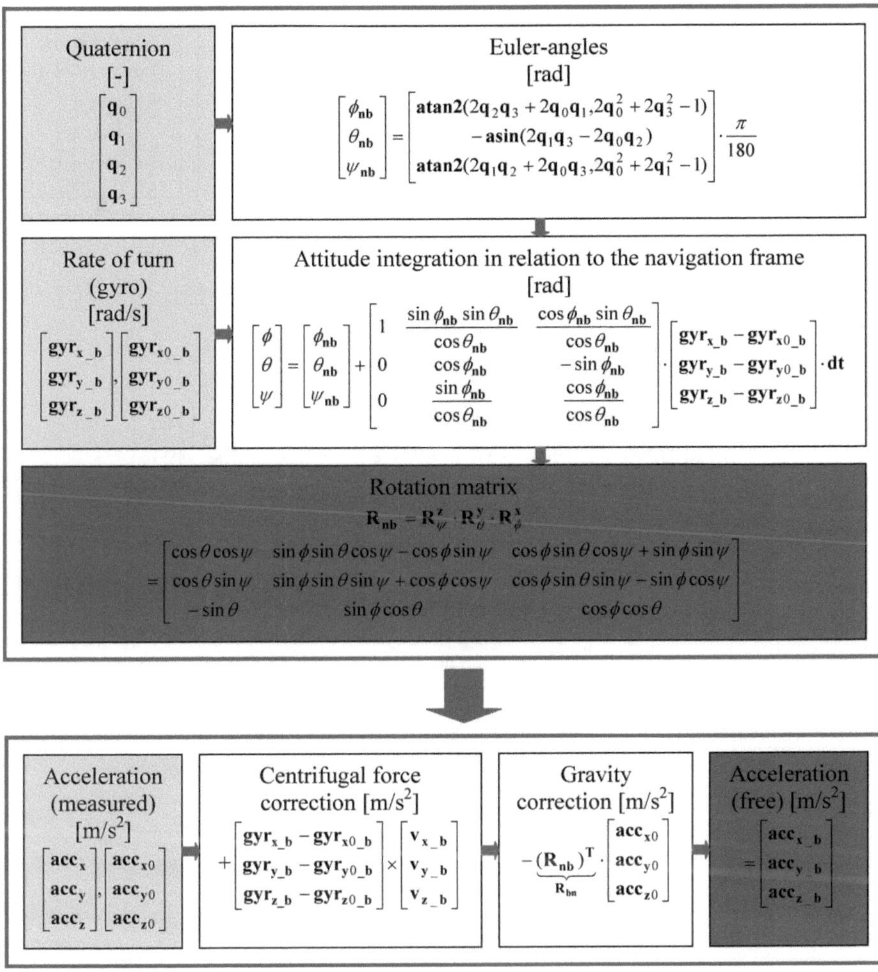

Fig. 25.6. Calculation of the 'free' acceleration and rotation matrix from the measured orientation and calibration data.

After the 'free' acceleration and rotation matrix have been calculated from the original measured orientation and calibration data they can be integrated over time to obtain the orientation changes and velocity measurements. Then the current position could be derived by means of integrating the obtained orientation changes and velocity measurements over time and using the position from the last calculation. These processes occur in the second step of the determination of the trajectory using INS and are presented in *Figure 25.7*.

In our experiments we have only used the strapdown mechanization for determining a trajectory of a straight line with a length of 22.2 meters (see *Figure 25.7)*. More complex cases will be tested in the future.

Rotation matrix

$$\mathbf{R_{nb}} = \mathbf{R}_{\psi}^{z} \cdot \mathbf{R}_{\theta}^{y} \cdot \mathbf{R}_{\phi}^{x} = \begin{bmatrix} \cos\theta\cos\psi & \sin\phi\sin\theta\cos\psi - \cos\phi\sin\psi & \cos\phi\sin\theta\cos\psi + \sin\phi\sin\psi \\ \cos\theta\sin\psi & \sin\phi\sin\theta\sin\psi + \cos\phi\cos\psi & \cos\phi\sin\theta\sin\psi - \sin\phi\cos\psi \\ -\sin\theta & \sin\phi\cos\theta & \cos\phi\cos\theta \end{bmatrix}$$

Acceleration (body) [m/s^2]

$$= \begin{bmatrix} \mathbf{acc_{x\_b}} \\ \mathbf{acc_{y\_b}} \\ \mathbf{acc_{z\_b}} \end{bmatrix}$$

Acceleration integration and velocity integration

$$\begin{bmatrix} \mathbf{v_{x0\_b}} \\ \mathbf{v_{y0\_b}} \\ \mathbf{v_{z0\_b}} \end{bmatrix} = \begin{bmatrix} \mathbf{p_{x0\_b}} \\ \mathbf{p_{y0\_b}} \\ \mathbf{p_{z0\_b}} \end{bmatrix} = \begin{bmatrix} 0 \\ 0 \\ 0 \end{bmatrix}$$

$$\begin{bmatrix} \mathbf{acc_{x0\_b}} \\ \mathbf{acc_{y0\_b}} \\ \mathbf{acc_{z0\_b}} \end{bmatrix} = \begin{bmatrix} \mathbf{acc_{x\_b}} \\ \mathbf{acc_{y\_b}} \\ \mathbf{acc_{z\_b}} \end{bmatrix}_1 - \mathbf{R_{bn}} \cdot \begin{bmatrix} \mathbf{acc_{x0}} \\ \mathbf{acc_{y0}} \\ \mathbf{acc_{z0}} \end{bmatrix}_1, \quad \begin{bmatrix} \mathbf{acc_{x\_b}} \\ \mathbf{acc_{y\_b}} \\ \mathbf{acc_{z\_b}} \end{bmatrix} = \begin{bmatrix} \mathbf{acc_{x\_b}} \\ \mathbf{acc_{y\_b}} \\ \mathbf{acc_{z\_b}} \end{bmatrix}_2 - \mathbf{R_{bn}} \cdot \begin{bmatrix} \mathbf{acc_{x0}} \\ \mathbf{acc_{y0}} \\ \mathbf{acc_{z0}} \end{bmatrix}_2$$

For i = 1 : length(t)

$$\mathbf{VB_i} = \begin{bmatrix} \mathbf{v_{x\_b}} \\ \mathbf{v_{y\_b}} \\ \mathbf{v_{z\_b}} \end{bmatrix} = \begin{bmatrix} \mathbf{v_{x0\_b}} \\ \mathbf{v_{y0\_b}} \\ \mathbf{v_{z0\_b}} \end{bmatrix} + \frac{1}{2} \cdot \begin{bmatrix} \mathbf{acc_{x0\_b}} + \mathbf{acc_{x\_b}} \\ \mathbf{acc_{y0\_b}} + \mathbf{acc_{y\_b}} \\ \mathbf{acc_{z0\_b}} + \mathbf{acc_{z\_b}} \end{bmatrix} \cdot \mathbf{dt}, \quad \mathbf{VN_i} = \begin{bmatrix} \mathbf{v_{x\_n}} \\ \mathbf{v_{y\_n}} \\ \mathbf{v_{z\_n}} \end{bmatrix} = \mathbf{R_{nb}} \cdot \begin{bmatrix} \mathbf{v_{x\_b}} \\ \mathbf{v_{y\_b}} \\ \mathbf{v_{z\_b}} \end{bmatrix}$$

$$\mathbf{PB_i} = \begin{bmatrix} \mathbf{p_{x\_b}} \\ \mathbf{p_{y\_b}} \\ \mathbf{p_{z\_b}} \end{bmatrix} = \begin{bmatrix} \mathbf{p_{x0\_b}} \\ \mathbf{p_{y0\_b}} \\ \mathbf{p_{z0\_b}} \end{bmatrix} + \frac{1}{2} \cdot \begin{bmatrix} \mathbf{v_{x0\_b}} + \mathbf{v_{x\_b}} \\ \mathbf{v_{y0\_b}} + \mathbf{v_{y\_b}} \\ \mathbf{v_{z0\_b}} + \mathbf{v_{z\_b}} \end{bmatrix} \cdot \mathbf{dt}, \quad \mathbf{PN_i} = \begin{bmatrix} \mathbf{p_{x\_n}} \\ \mathbf{p_{y\_n}} \\ \mathbf{p_{z\_n}} \end{bmatrix} = \mathbf{R_{nb}} \cdot \begin{bmatrix} \mathbf{p_{x\_b}} \\ \mathbf{p_{y\_b}} \\ \mathbf{p_{z\_b}} \end{bmatrix}$$

$$\begin{bmatrix} \mathbf{v_{x0\_b}} \\ \mathbf{v_{y0\_b}} \\ \mathbf{v_{z0\_b}} \end{bmatrix} = \begin{bmatrix} \mathbf{v_{x\_b}} \\ \mathbf{v_{y\_b}} \\ \mathbf{v_{z\_b}} \end{bmatrix}, \quad \begin{bmatrix} \mathbf{acc_{x0\_b}} \\ \mathbf{acc_{y0\_b}} \\ \mathbf{acc_{z0\_b}} \end{bmatrix} = \begin{bmatrix} \mathbf{acc_{x\_b}} \\ \mathbf{acc_{y\_b}} \\ \mathbf{acc_{z\_b}} \end{bmatrix}, \quad \begin{bmatrix} \mathbf{p_{x0\_b}} \\ \mathbf{p_{y0\_b}} \\ \mathbf{p_{z0\_b}} \end{bmatrix} = \begin{bmatrix} \mathbf{p_{x\_b}} \\ \mathbf{p_{y\_b}} \\ \mathbf{p_{z\_b}} \end{bmatrix}$$

end

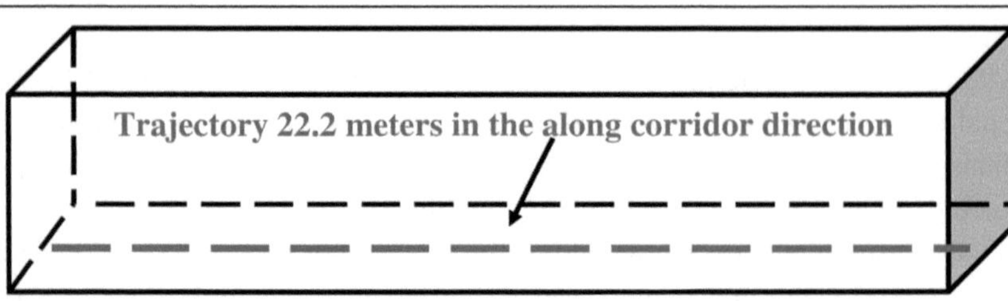

Fig. 25.7. Determination of the trajectory by integrating acceleration and rotation matrix.

The Kalman Filter was used to correct the position and velocity dynamics of the INS sensor. In our approach we used a so called Time-Varying Kalman Filter developed by Matlab which is a generalization of the steady-state filter for time-varying systems with nonstationary noise covariance and is given by the recursions (Matlab 2007).

All equations utilized in the Time-Varying Kalman Filter and the work flow of the calculation are shown in a flow-chart in *Figure 25.8* whereby the coefficients in the system noise covariance matrix were derived from the bias and drifts given by the manufacturer. The state equation shows a continual Kalman Filter state model (above) and a discrete state model (below). In the following the discrete model was employed. The epoch of the filter state is indicated with **k**.

25.4 Experimental Results

The experimental test run was performed along a corridor in an office building of the Vienna University of Technology along a 22.2 m straight line. The line runs along the x-axis on the xy-plan in a local coordinate system. At the start and end of the line four RFID tags where located to provide the initial position and an update determined by trilateration or fingerprinting. The Xsens MTi sensor bandwidth was 100 Hz. The measured data were processed according to the algorithms of the strapdown mechanisation which were presented in *Section 25.3*. At the same time, the obtained position and acceleration were filtered by using the Kalman Filter. *Figure 25.9* shows the measured response with and without drift correction. The end point has a deviation of 12.0 m for the measured response without drift correction and 9.1 m for the measured response with drift correction. *Figure 25.10* shows the Kalman Filter result of the INS trajectory of the measured response with drift correction. The dashed line represents the measured response and the solid line the filtered response. It can be seen that the Kalman Filter is able to reduce the measurement noise and produce a smooth trajectory. The deviation at the end, however, is quite large and we do not reach the true point. The maximum error in y-direction at the end is 6.4 m and in x-direction 6.5 m. An update of the drift can be obtained using RFID location fingerprinting or trilateration. Then the trajectory can be corrected towards the end point. The covariance plot can be used to prove whether the output covariance did indeed reach a steady state or not by using the Kalman Filter. *Figure 25.11* and *25.12* show the output covariance along the x-axis and the y-axis. It can be seen that the output covariance did reach a steady state. Further results can be found in Retscher and Fu (2008).

1. State equation (continual (above) and discrete (below))

$$\underbrace{\begin{bmatrix} \dot{p}_{x\_n\_kf} \\ \dot{p}_{y\_n\_kf} \\ \dot{v}_{x\_n\_kf} \\ \dot{v}_{y\_n\_kf} \\ \dot{\psi}_{kf} \end{bmatrix}_k}_{\substack{\dot{\mathbf{x}}[\mathbf{k}] \\ \text{state vector} \\ \text{at time k}}} = \underbrace{\begin{bmatrix} 0 & 0 & 1 & 0 & 0 \\ 0 & 0 & 0 & 1 & 0 \\ 0 & 0 & 0 & 0 & 0 \\ 0 & 0 & 0 & 0 & 0 \\ 0 & 0 & 0 & 0 & 0 \end{bmatrix}}_{\substack{\mathbf{A}^* \\ \text{state transition matrix}}} \cdot \underbrace{\begin{bmatrix} p_{x\_n\_kf} \\ p_{y\_n\_kf} \\ v_{x\_n\_kf} \\ v_{y\_n\_kf} \\ \psi_{kf} \end{bmatrix}_k}_{\substack{\mathbf{x}[\mathbf{k}] \\ \text{state vector} \\ \text{at time k}}} + \underbrace{\begin{bmatrix} 1 & 0 & 0 & 0 & 0 \\ 0 & 1 & 0 & 0 & 0 \\ 0 & 0 & \cos\psi & -\sin\psi & 0 \\ 0 & 0 & \sin\psi & \cos\psi & 0 \\ 0 & 0 & 0 & 0 & 1 \end{bmatrix}}_{\substack{\mathbf{B}^* \\ \text{input matrix}}} \cdot \underbrace{\begin{bmatrix} \mathbf{0} \\ \mathbf{0} \\ \mathbf{acc_{x\_b}} \\ \mathbf{acc_{y\_b}} \\ \mathbf{gyr_{z\_b}} \end{bmatrix}_k}_{\substack{\mathbf{u}[\mathbf{k}] \\ \text{input vector}}} + \mathbf{G}^* \cdot \underbrace{\boldsymbol{\omega}[\mathbf{k}]}_{\substack{\text{gaussian} \\ \text{process} \\ \text{noise with} \\ \text{covariance} \\ \text{Q at time k}}}$$

$$\underbrace{\begin{bmatrix} p_{x\_n\_kf} \\ p_{y\_n\_kf} \\ v_{x\_n\_kf} \\ v_{y\_n\_kf} \\ \psi_{kf} \end{bmatrix}_{k+1}}_{\substack{\mathbf{x}[\mathbf{k+1}] \\ \text{state vector} \\ \text{at time k+1}}} = \underbrace{\begin{bmatrix} 1 & 0 & \Delta t & 0 & 0 \\ 0 & 1 & 0 & \Delta t & 0 \\ 0 & 0 & 1 & 0 & 0 \\ 0 & 0 & 0 & 1 & 0 \\ 0 & 0 & 0 & 0 & 1 \end{bmatrix}}_{\substack{\mathbf{A} \\ \text{state transition matrix}}} \cdot \underbrace{\begin{bmatrix} p_{x\_n\_kf} \\ p_{y\_n\_kf} \\ v_{x\_n\_kf} \\ v_{y\_n\_kf} \\ \psi_{kf} \end{bmatrix}_k}_{\substack{\mathbf{x}[\mathbf{k}] \\ \text{state vector} \\ \text{at time k}}} + \underbrace{\begin{bmatrix} \Delta t & 0 & \frac{1}{2}\Delta t^2 \cdot \cos\psi & -\frac{1}{2}\Delta t^2 \cdot \sin\psi & 0 \\ 0 & \Delta t & \frac{1}{2}\Delta t^2 \cdot \sin\psi & \frac{1}{2}\Delta t^2 \cdot \cos\psi & 0 \\ 0 & 0 & \Delta t \cdot \cos\psi & -\Delta t \cdot \sin\psi & 0 \\ 0 & 0 & \Delta t \cdot \sin\psi & \Delta t \cdot \cos\psi & 0 \\ 0 & 0 & 0 & 0 & \Delta t \end{bmatrix}}_{\substack{\mathbf{B} \\ \text{input matrix}}} \cdot \underbrace{\begin{bmatrix} \mathbf{0} \\ \mathbf{0} \\ \mathbf{acc_{x\_b}} \\ \mathbf{acc_{y\_b}} \\ \mathbf{gyr_{z\_b}} \end{bmatrix}_k}_{\substack{\mathbf{u}[\mathbf{k}] \\ \text{input vector}}} + \mathbf{G} \cdot \underbrace{\omega[\mathbf{k}]}_{\substack{\text{gaussian} \\ \text{process} \\ \text{noise with} \\ \text{covariance} \\ \text{Q at time k}}}$$

2. Measurement equation

$$\underbrace{\begin{bmatrix} p_{x\_n} \\ p_{y\_n} \end{bmatrix}_k}_{\substack{\mathbf{y_V}[\mathbf{k}] \\ \text{observation vector} \\ \text{at time k}}} = \underbrace{\begin{bmatrix} 1 & 0 & 0 & 0 & 0 \\ 0 & 1 & 0 & 0 & 0 \end{bmatrix}}_{\substack{\mathbf{C} \\ \text{observation matrix}}} \cdot \underbrace{\begin{bmatrix} p_{x\_n\_kf} \\ p_{y\_n\_kf} \\ v_{x\_n\_kf} \\ v_{y\_n\_kf} \\ \psi_{kf} \end{bmatrix}_k}_{\substack{\mathbf{x}[\mathbf{k}] \\ \text{state vector} \\ \text{at time k}}} + \underbrace{\upsilon[\mathbf{k}]}_{\substack{\text{measurement noise} \\ \text{with covariance R} \\ \text{at time k}}}$$

Design a Kalman filter to estimate the output y based on the noisy measurement

The system noise covariance matrix

$$\mathbf{Q} = \begin{bmatrix} \sigma^2_{p\_x\_noise} & 0 & 0 & 0 & 0 \\ 0 & \sigma^2_{p\_y\_noise} & 0 & 0 & 0 \\ 0 & 0 & \sigma^2_{v\_x\_noise} & 0 & 0 \\ 0 & 0 & 0 & \sigma^2_{v\_y\_noise} & 0 \\ 0 & 0 & 0 & 0 & \sigma^2_{\psi\_noise} \end{bmatrix}$$

$\sigma_{p\_x\_noise} = 0.5 \cdot \mathbf{bias} \cdot \mathbf{t}^2$	$\sigma_{p\_y\_noise} = 0.2 \cdot \sigma_{p\_x\_noise}$	
$\sigma_{v\_x\_noise} = \mathbf{bias} \cdot \mathbf{t}$	$\sigma_{v\_y\_noise} = 0.1 \cdot \sigma_{v\_x\_noise}$	
$\mathbf{bias} = 0.012[\mathbf{m/s^2}]$	$\mathbf{drift} = 0.007[\mathbf{rad/s}]$	$\sigma_{\psi\_noise} = \mathbf{drift} \cdot \mathbf{t}$

The observation error covariance matrix

$$\mathbf{R} = \begin{bmatrix} \sigma^2_{p_{x\_n}} & 0 \\ 0 & \sigma^2_{p_{y\_n}} \end{bmatrix}$$

$$\sigma_{p_{x\_n}} = \sigma_{p_{y\_n}} = 1.97\,\text{m}$$

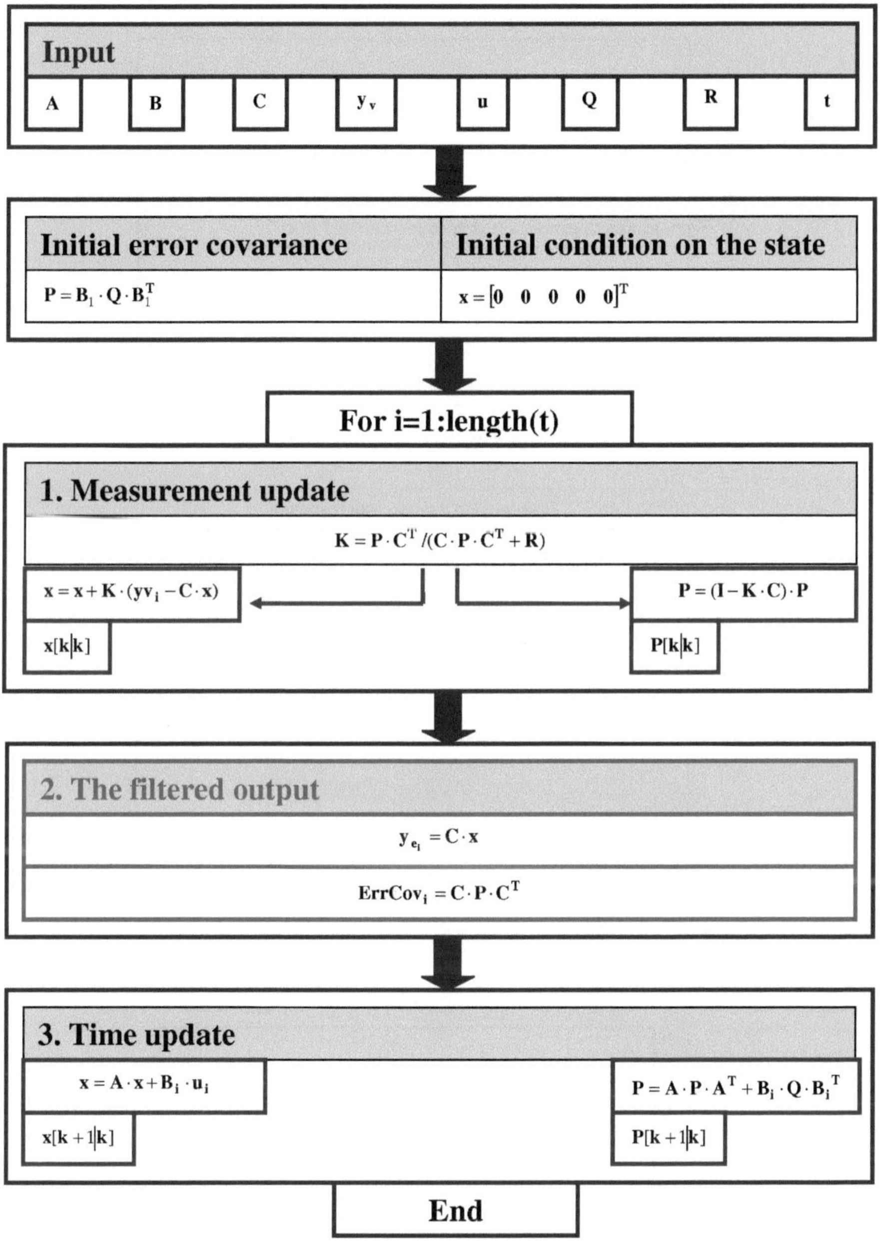

Fig. 25.8. Correction of position and velocity using a Kalman Filter.

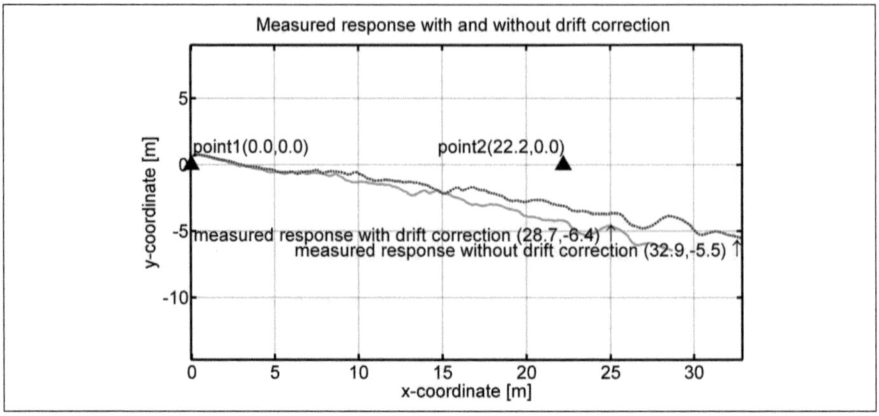

Fig. 25.9. Measured response with and without drift correction.

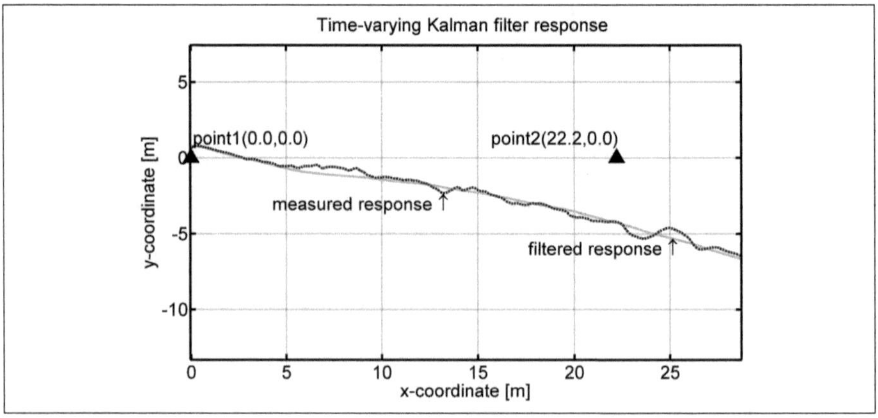

Fig. 25.10. Kalman Filter result of INS trajectory along a straight line.

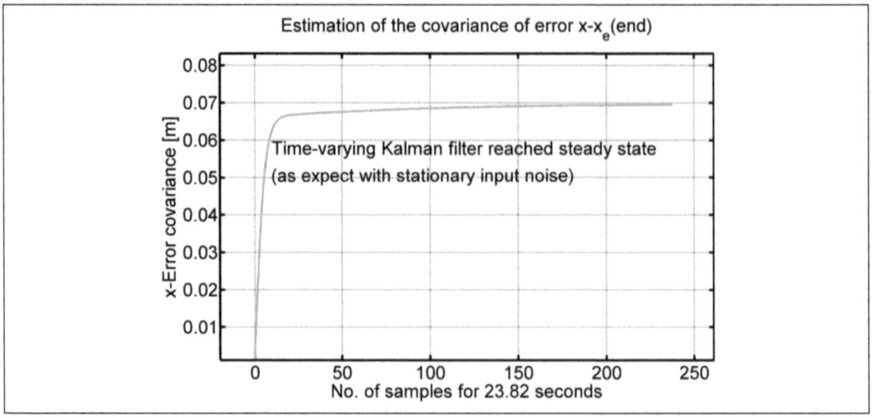

Fig. 25.11. Covariance plot along x-axis.

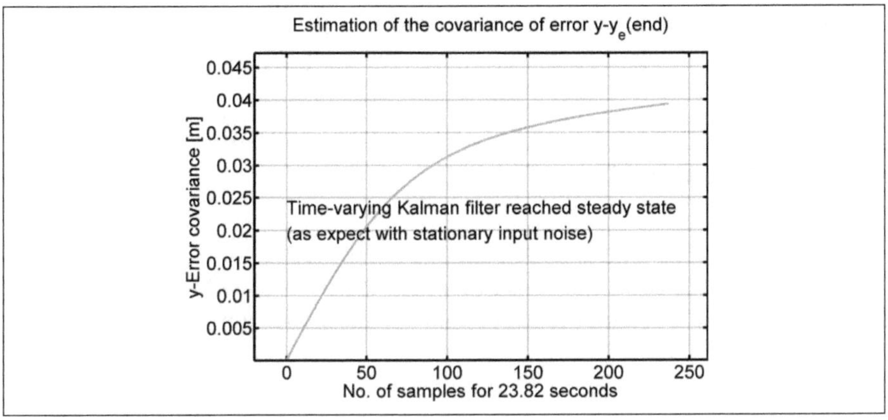

Fig. 25.12. Covariance plot along y-axis.

25.5 Summary

This paper addresses the use of long range active RFID for positioning in indoor environments. Three different methods have been investigated, i.e., cell-based positioning, trilateration using ranges to the surrounding RFID tags deduced from received signal strength measurements and RFID location fingerprinting. On this base the positioning method using interpolation has been introduced and developed for RFID location fingerprinting, in order to improve the positioning accuracy. For location fingerprinting four different approaches have been developed.

However, the accuracy of the positioning is restricted in the area where RFID signals could be only detected from one or two tags. And in the area not covered by signals of RFID tags the position of the RFID reader could not be determined. In other words, if there is lack of the coverage of signal of RFID tags, the RFID reader will lose its orientation. In order to overcome these shortages we propose to use a low-cost INS to determine the trajectory in the place where RFID signals are not available.

For calculating the positions from the measured data of the Xsens MTi sensor the strapdown mechanization was employed. Furthermore, the Kalman Filter was used to correct the position and acceleration resulted in the strapdown mechanization. The experiment showed that the determination of trajectory using Xsens MTi along a 22.2 m straight line achieved an accuracy of around 6.5 m in two directions on the xy-plan. The positioning accuracy can be improved by employing a Kalman Filter. A positioning accuracy of around 0.5–1.0 m can be achieved, if combined with RFID trilateration or fingerprinting. This accuracy is suitable for most positioning applications in indoor environments.

In the near future the combined RFID and INS positioning will be tested and improved in more complicated environments and along different trajectories.

Acknowledgements

The research work presented in this paper is fully supported by the FWF Project "Ubiquitous Cartograpy for Pedestrian Navigation UCPNAVI" (Project No. P19210-N15) of the Austrian Science Fund (FWF Fonds zur Förderung wissenschaftlicher Forschung).

References

Dorobantu R, Zebhauser B (1999) Field Evaluation of a Low-Cost Strapdown IMU by means GPS. Ortung und Navigation 1/1999, DGON, Bonn, 51-65, see also at http://www.imar-navigation.de/download/iapg_2.pdf (Last accessed: June 2008).

Fu Q (2008) Active RFID for Positioning Using Trilateration and Location Fingerprinting Based on RSSI. in: Papers presented at the ION GNSS 2008 Conference, September 16-19, 2008, Savannah, Georgia, USA, CD-Rom Proceedings.

Kaemarungsi K (2005). Design of Indoor Positioning Systems Based on Location Fingerprinting Technique, Dissertation, University of Pittsburgh, see http://etd.library.pitt.edu/ETD/available/etd-02232005-235903/unrestricted/dissertation28Feb05.pdf (Last accessed: June 2008).

Matlab (2007) Control System Toolbox™ 8.1, Kalman Filter Design. http://www.mathworks.com/products/control/demos.html?file=/products/demos/shipping/control/kalmdemo.html (Last accessed: June 2008).

Retscher G, Moser E, Vredeveld D, Heberling D, Pamp J (2007) Performance and Accuracy Test of a WiFi Indoor Positioning System, Journal of Applied Geodesy, Vol. 1, No. 2, pp. 103-110.

Retscher G, Fu Q (2007) Using Active RFID for Positioning in Navigation Systems, in: Papers presented at the 4th Symposium on Location Based Services and Telecartography, November 8-10, 2007, Hong Kong, PR China.

Retscher G, Fu Q (2008) GNSS, RFID and INS Integration for Pedestrian Navigation, in: Papers presented at the International Symposium on GPS/GNSS, November 11-14, 2008, Toyko, Japan.

Schiffer J (2007) Systemidentifikation mit dem Kalman Filter, Hauptseminar Technische Kybernetik. *www.kybs.de/daver/download.php?fileno=245* (Last accessed: September 2008).

Tampere Universty of Technology (2008) Basic Principles of Inertial Navigation, Seminar on Inertial Navigation Systems, http://www.tkt.cs.tut.fi/kurssit/9628/Intro.pdf (Last accessed: September 2008).

Xsens (2007) MTi and MTx User Manual and Technical Documentation, Product Manual, Xsens Technologies B.V., The Netherlands.

26 GPS/WiFi Real-Time Positioning Device: An Initial Outcome

Joon Wayn Cheong, Binghao Li, Andrew G. Dempster and Chris Rizos

School of Surveying and Spatial Information Systems, UNSW, Australia

Abstract

The Global Positioning System (GPS) has a typical outdoor positioning accuracy of up to 15m for civilian users. Thus, it has become a viable method for civilian to carry out coarse positioning. However, it has its shortcomings; it is available only in indoors with a clear view of the sky. Since WiFi has become another proven positioning technology that is capable of performing positioning in indoor environments and urban canyons, it is desirable to combine both of these technologies for ubiquitous positioning. Therefore, by means of integrating GPS positioning with a WiFi positioning system, indoor and outdoor positioning may be performed using only one device. The device can be implemented using FPGA embedded systems technology that allows easy reconfiguration of the device. Such a combination allows GPS and WiFi positioning technology to transition seamlessly.

Keywords: GPS/WiFi integration, GPS/WiFi positioning, localisation, orientation, Namuru

26.1 Introduction

Since the rapid growth of the user community of IEEE 802.11a/b/g wireless networks (WiFi), WiFi has become another viable positioning method (Bahl and Padmanabhan 2000) (Li et al. 2006). Depending on the method of positioning used, the WiFi positioning system may not require the MU to have a Line of Sight (LOS) path to the WiFi Access Points (AP) and is least affected by multipath. Implicitly, WiFi is capable of performing positioning in indoor environments and urban canyons. Therefore, by means of integrating GPS positioning with WiFi positioning system, indoor and outdoor real-time positioning may be performed using only one device. Such a combination allows GPS and WiFi positioning technology to transition seamlessly, such that WiFi pos. may compensate for areas where GPS unavailable, and vice-versa.

The application of such an independent device may be in the field of navigation for the Blind and Vision Impaired (BVI). Such an application allows the BVI to navigate not only in outdoor but also indoor environments. Another application of such a device will be in the field of location based services (LBS). The rapid growth of technology has allowed many useful technologies to be integrated together into mobile personal handheld devices such as handheld gaming devices, Personal Digital Assistant (PDA), mobile phones, and laptops. The integration of GPS and WiFi into these devices may allow mobile users to easily find the location of shops or businesses in both indoor and outdoor environments.

Preliminary investigation found little theory or implementation of GPS-WiFi integration. The only technique for such integration can only be found in (Singh et al. 2004). However, the WiFi positioning technique uses trilateration rather than fingerprinting. Hence, this research attempts to use the fingerprinting method for WiFi positioning and changed the structure of the GPS-WiFi integration mechanism as proposed in (Singh et al. 2004).

Taking advantage of the reconfigurable nature of the newly developed GPS receiver, the Namuru II (Mumford et al. 2006) will be adopted for use to provide GPS coordinates, and the soft-core NIOS II processor within the Namuru II will be used to process a fully tested WiFi positioning algorithm. Effort is also put into redeveloping the WiFi positioning algorithm by using MATLAB programming that resembles the C/C++ programming style. The successful algorithm is then appended into the Namuru II. The reason to use MATLAB is to ease the debugging process and verification of the developed system. The reason to implement the integration of the two positioning technologies rather than simply simulate it is to be able to demonstrate the practicality of such integration and to ensure real-time positioning is realizable.

26.2 Developing and Testing the WiFi Positioning Algorithm

The development and testing done in this section ensures that the implementation of the WiFi positioning algorithm in hardware will be smooth and easy to realize in the next section. In addition, the testing of the algorithm with various parameters allows the best algorithm implementation, with minimum computational requirements and least error to be chosen.

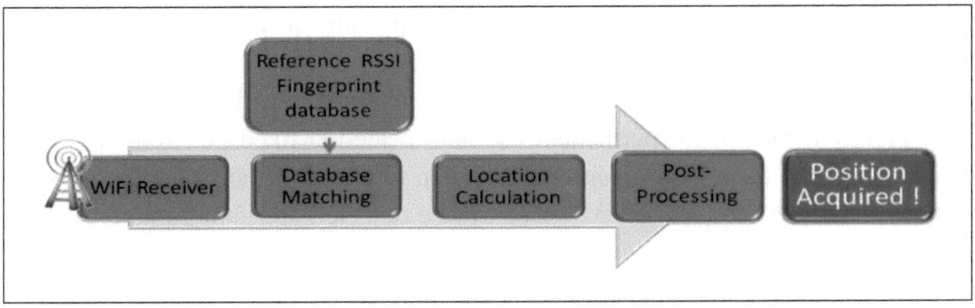

Fig. 26.1. Block diagram of the overall WiFi positioning system.

The empirical method of "Location Fingerprinting" can be described by two stages, namely the online stage and the offline stage. In the offline stage (training phase), a fingerprint database containing Radio Signal Strength Indicator (RSSI) measurements and location of Reference Points (RP) must be created. To generate the database, RPs must be carefully selected so that the entire test-bed is covered (Lee et al. 2005). Furthermore, the number of RPs to be used should be a balance between considerations of accuracy (the more RPs the higher the accuracy, because the granularity is lower) and the labour effort required to survey the patterns (or "fingerprints") of RSSIs at the RPs for the database (Li et al. 2007). In the online stage (positioning phase), the Mobile User (MU) measures the RSSIs whenever it requires its position to be determined. An appropriate algorithm is used to compare the measurements with the fingerprints in the database (Li et al. 2007). The outcome is the likeliest location of the MU.

Understanding the significant effects of MU's orientation on RSSI (Li et al. 2007), a directional approach is used to collect fingerprints of RPs. The non-directional method takes the RSSI of 4 cardinal directions and averages the RSSI into a single entry per Access Point (AP) per RP. In comparison, the directional method takes the RSSI of the 4 cardinal directions and stores the RSSI collected from each direction as a separate entry, resulting in 4 entries per AP per RP. The RSSI vector is defined as a set of RSSI arranged with respect to its AP in vector form. Thus, the directional approach will result in 4 RSSI vectors assigned to one RP while the non-directional approach results in only one RSSI vector assigned to one RP (Haeberlen. et al. 2004). The block diagram below illustrates the proposed implementation of the WiFi positioning algorithm. It is worth noting that for the scope of this discussion, the Post-Processing stage is not implemented.

26.2.1 Missing AP Resolution

Wireless technology has emerged to be an essential tool in communications. Almost every building in urban areas is equipped with Wireless Access Points (AP). For example, in many developed countries (and especially in the U.S.), the Central Business District (CBD) area of cities, university campuses, public hospitals, shopping malls, and other public spaces, WiFi coverage is widespread. These APs are typically deployed for permanent use and located at fixed positions. It would be advantageous for a WiFi positioning system to exploit its surrounding APs for better accuracy. Therefore, all APs that can be detected are incorporated into the system.

Since we attempt to use APs that we have no control over in both the Reference Database and Test data, we thus cannot ensure that all APs will cover the entire test-bed. Hence, the issue of Test Points (TP) collecting RSSI from an AP that is not listed in some RPs, or vice-versa, has to be addressed. This situation is referred to as the "missing AP" case.

The "missing AP" cases can be due to one of the following reasons:

1. Case 1: The TP is far away from the RP and hence cannot acquire any signal from some APs in the list of APs for that RP or vice-versa.

2. Case 2: The TP near a RP detects a newly deployed (turned on) AP near the test-bed that the RP did not detect or vice-versa (Cheng et al. 2005).

3. Case 3: The TP near a RP is unable to detect an AP the RP had detected earlier or vice-versa because the AP was turned off (Cheng et al. 2005).

To accommodate these situations, the APs that cannot be detected either by the Reference Point or Test Point is replaced with an RSSI of -100dBm (Li 2006). This seems to be an intuitive solution for case 1 because the weakest RSSI that we can practically acquire is approximately -90dBm while -100dBm appears to be an interpolated value to indicate that the AP is far away from the TP or RP. This method may also well accommodate for cases 2 and 3, where no other method can be used to predict whether an AP is turned off or newly deployed, the respective APs should not be penalised too much.

However, it would be almost impossible for a TP to be near a RP to have many "missing AP" cases occurring at once. Hence, the case where there are too many "missing AP" should be heavily penalised. There are 2 methods of detecting such cases:

1. Method 1: If "missing AP" cases are more than a pre-determined threshold (Cheng et al. 2005).

2. Method 2: If "missing AP" cases are more than a pre-determined percentage of APs being compared.

Since the number of APs being compared between a TP and a particular RP has a huge variation from one RP to another, method 2 should be used. It is experimentally found through testing herein that allowing 75% of APs being compared to have "missing AP" cases is optimal (Li et al. 2007).

26.2.2 Inverse Distance Weightage (IDW) Interpolation

In the directional Reference Database, each RP has four RSSI vectors with one RSSI vector representing each direction of the same RP, as opposed to the non-directional database which has only one RSSI vector per RP. Since the four NNs are chosen from a list of "4 x number_of_RP" (4 refers to the four cardinal directions, North, East South, West (NESW)), it is very possible that the case where "2 or 3 of the 4 NNs are from the same RP" may arise. In this case, the possible area that the calculated position could lie in is smaller than the normal case. (The normal case assumes the 4 NNs are from different RPs.) This case is illustrated in *Figure 26.2.*

Figure 26.2 illustrates the usage of 1-NN to 4-NN in the non-directional case. To find 4-NNs from the directional Reference Database, all four NNs must come from all four different RPs to span the geometry formed by the 4-NN case in *Figure 26.2.* Suppose if the directional Reference Database returned four NNs with two of the NNs originating from the same RP, the geometry will be spanned by only three RPs, resulting in the similar geometry of 3-NN in *Figure 26.2.* As well as, if only two RPs are returned by the directional Reference Database, in which three of the NNs come from the same RP, a geometry equivalent to the 2-NN case in *Figure 26.2* will be formed. Therefore, if the four NNs have three of the NNs from the same RP in the directional case, it would be equivalent to performing a 2-weighted-NN in the non-directional case, with most of the weights applied to the 3NN RP.

Even though we have lost spatial diversity by allowing this situation to occur, it is intuitively desirable since the correlation between the TP with a particular RP is so high, it is not necessary to have a large area of uncertainty, as we are very confident with this particular RP. The result of using this concept is a lesser computational burden.

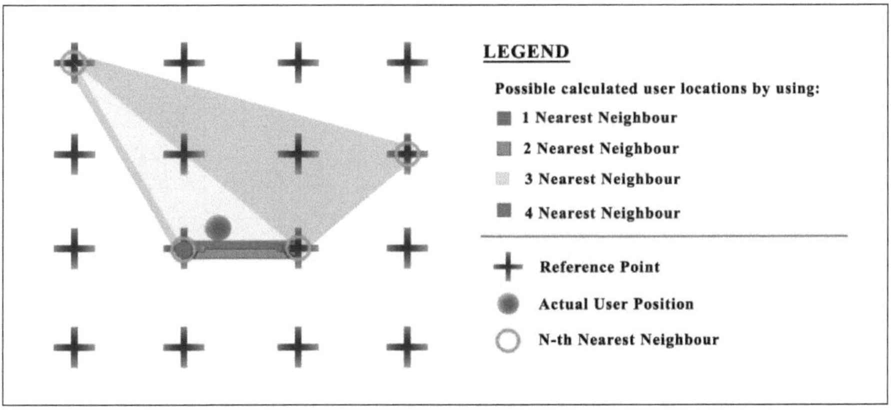

Fig. 26.2. Illustration of area of possible calculated positions using lower number of NNs.

26.2.3 Evaluation of the WiFi Positioning System

To test the proposed WiFi positioning algorithm, data collection was carried out on the test-bed located on levels 2, 3 and 4 of UNSW's Electrical Engineering Building (G17). The large stairwell (marked by GQ10, 3Q13 and 4Q7) allows most of the AP signals to propagate from one level to another more easily. There were a total of 7 APs that were self-deployed (1 on level 4, 3 on level 3, 3 on level 2). All other available APs detected by the WiFi receiver were also used. The data collection is done at times when human movement is minimal (i.e. at night, university holidays) due to the well known fact that human activity causes undesired fluctuations of RSSI (Cheng et al. 2005).

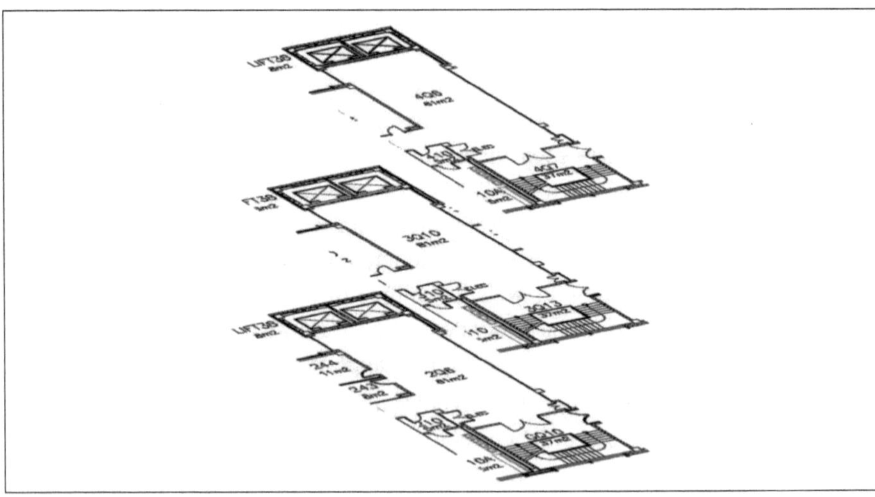

Fig. 26.3. Map of test-bed, level 2, 3 and 4 of Electrical Engineering Building, UNSW.

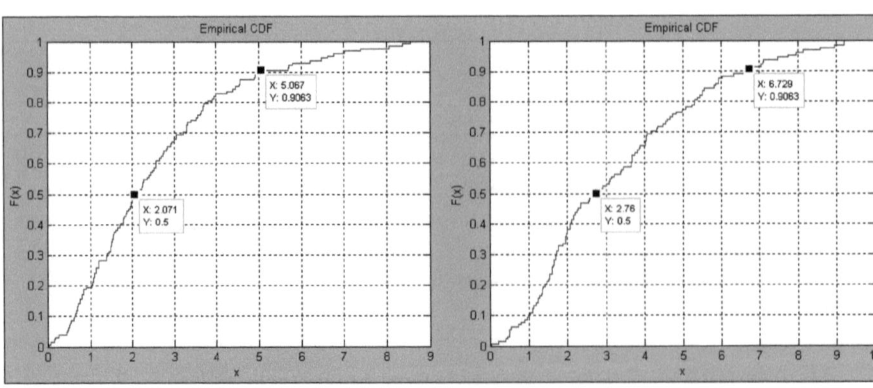

Fig. 26.4. CDF plot of the magnitude of error for all test points using directional method (left) and non-directional method (right). In the highlighted points for both plots, F(x) of 0.9 corresponds to the 90th percentile error, while F(x) of 0.5 corresponds to the median error.

From the TP and RP RSSI collection, the algorithm is tested with various parameters to investigate the effects of changing those parameters. There are 91 Reference Points and 32 Test Points used in the analyses. Each Reference Point has 4 sets of RSSI vectors, each corresponding to one of the NESW directions, and similarly in the case of the TPs. Hence there are 4x32 tests to calculate instead of 32, since the directional method is used. The Cumulative Distribution Function (CDF) plot is used to analyse the magnitude of errors caused by inaccurate positioning. The use of both Directional and Non-Directional (conventional) methods are compared.

The detailed horizontal error statistics generated based on the directional method shows 1-sigma error to be 3.25 m, median error to be 2.07 m and 90th percentile error is found to be approximately 5 m. As for the non-directional case, the 1-sigma error is found to be 3.98 m with a median error of 2.75 m and 90th percentile error of 7 m. Hence, the directional method shows significant improvement over the non-directional method, as indicated by the CDF graphs. Compared to the non-directional method, which has a 90th percentile error of about 7m, the directional database gives a significant improvement of 2 m.

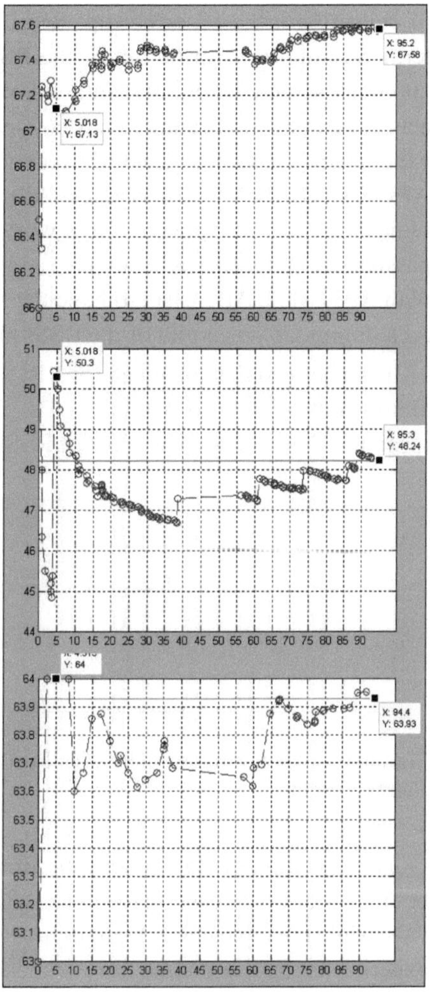

Fig. 26.5. Plot of variation of "RSSI (-dBm) mean over time (seconds)" (circle markers) and the "mean RSSI (-dBm) of 100 seconds" (red line) for AP1 (top), AP2 (middle), and AP3 (bottom).

In the Nearest Neighbour algorithm, the RSSI used for each AP at each reference point in the reference database is the mean of the RSSI signals for each AP over the collection time. The principle of taking the mean RSSI instead of an instantaneous RSSI is justified because the RSSI is corrupted by fast fading, resulting in the RSSI varying quickly as a function of time. However, the corrupted RSSI typically varies about the mean RSSI, resembling a Gaussian distribution (Haeberlen et al. 2004) (Cheng et al. 2005). Therefore taking the mean RSSI can mitigate the problem of collecting a biased or corrupt RSSI. Many literatures have discussed the long term variation or distribution of the RSSI signal (Wang et al. 2003, Cheng et al. 2005). However, the optimal duration of sampling of the RSSI signals such that the actual mean RSSI is captured has not yet been discussed. To examine this, a simple experiment consisting of a WiFi receiver/logger (Sony VAIO VGN-U70P) and 3 APs (SparkLAN 802.11b Wireless AP) in a corridor is conducted with all devices remaining in a static position. It is worth noting that the WiFi receiver is set to use passive scanning. AP1 is set to have approximately equal distance from the receiver with AP3, while AP2 is set to have the nearest distance to the receiver amongst all APs. *Figure 26.5* is a plot of the RSSI mean for each of the 3 APs over time.

It can be seen that the mean for AP1 and AP3 converges to the 100-second-mean almost immediately to within +/- 1%. The plot of AP2 shows large variations of mean RSSI over time. The plots of RSSI over time are shown in *Figure 26.6*.

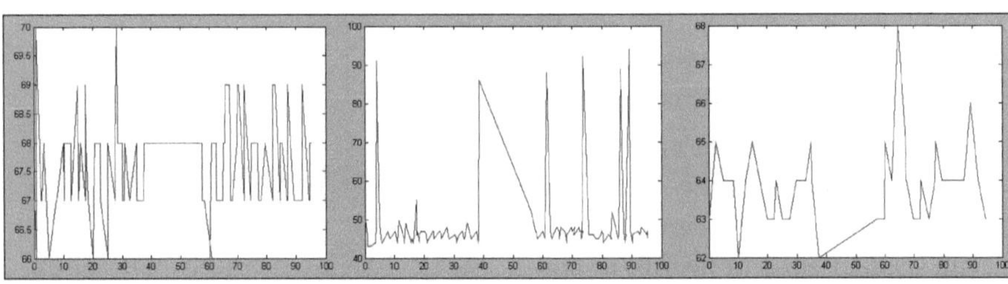

Fig. 26.6. Plot of variation of RSSI (-dBm) over time (seconds) for AP1 (left), AP2 (middle), and AP3 (right).

The deviation of RSSI over time from the mean is approximately +/- 3 dBm for both AP1 and AP3 cases. However, AP2 shows peaks (-90dBm) at irregular intervals, while the typical RSSI ranges in -40dBm to -50dBm. It has been found through experiments that cases like those indicated for AP2 are not uncommon in practical situations. This case may be explained by the presence of strong multipath signals or the behaviour of the passive scanning software/hardware, however the exact cause of this issue has not been exactly identified yet. Statistically, this suggests outliers that corrupt the estimation of the mean RSSI.

However, the Nearest Neighbour algorithm is trying to match the difference between the user collected mean RSSI and the RP's mean RSSI. This means that as long as the reference point captures the outlier that the user might capture whilst standing at the reference point, the difference in mean will be in the range of +/- 3dBm. This difference is relatively small compared to the possible range of mean RSSI from -35dBm to -90dBm, thus not affecting the final result. The final positioning accuracy will undoubtedly improve if this effect can be mitigated. Also from the RSSI plots of AP1 and AP3, it can be suggested that a RSSI collection time of less than 5 seconds is sufficient to approximate the RSSI mean from one AP.

26.3 GPS-WiFi Integration

The integration of GPS with WiFi for positioning is a useful yet immature domain. Preliminary literature studies have shown only one example of such integration (Singh et al. 2004). Although such a system is desirable, it is still at the state of development. As being discussed earlier, the unavailability of GPS in indoors and the availability of WiFi positioning in indoors are able to complement each other, thus providing a practical approach to realise a true indoor and outdoor ubiquitous positioning system.

The integration strategy originates from the method (Singh et al. 2004) used to overcome multipath in GPS by extracting information from WiFi RSSI readings and the local map. However, this method uses the trilateration-based WiFi positioning, which has been proven to be less accurate and less reliable than the fingerprinting method in heavy multipath environments, such as indoors (Cheng et al. 2005).

To improve the accuracy of indoor positioning, the fingerprinting method is proposed as a replacement for the trilateration method. Modifications are also made to the flowchart as proposed in (Singh et al. 2004) because it is highly dependent on the GPS parameters. It is possible to acquire GPS multipath signals in indoor environments, which may lead to a wrong determination of MU's position. Although the GPS signals are acquired indoors, they are most probably multipath signals that are reflected by nearby walls and buildings. Therefore, we suggest using both WiFi and GPS parameters to allow better detection of the user's location.

The implemented algorithm will require the entire Reference Database to be separated into sections/buildings/vicinities (referred to as Blocks in this paper). This method is more feasible if the system is implemented across a large area (e.g. a metropolitan area). Hence, the system will not search the entire database to calculate the user's position, but the vicinity of the user is first identified, then a Block of Reference Database (or two) is selected for subsequent computations. The benefits of using such an implementation are:

1. Huge reduction in computational burden.
2. Allows nearby Blocks to be loaded or unloaded from the system when necessary.
3. Allows different coordinate systems to be used in different situations (e.g. x, y, z for indoors, and LLA for outdoors).

The last benefit listed concerns the coordinate system of GPS which uses LLA. This is not an intuitive coordinate system for indoors and most maps of building have their own origin, and use the xyz Cartesian coordinate system. It would be more practical and less computationally demanding if the output coordinate system can be set to local x,y,level coordinates (NED) when the user is indoors, and Latitude, Longitude, Altitude (LLA) in the WGS84 datum when the user is outdoors. The reason being that outdoor users typically would acquire maps from sources (i.e. Google Earth) that use LLA coordinates, while indoor users typically use a map of a building which has its own origin of x,y,z coordinate. Consider if the output coordinate of the system is to be matched with a number of indoor building maps and a large road map, the output coordinates can easily adapt to these maps without any conversion. If the system only outputs LLA coordinate, a conversion process (LLA to xyz) will have to take place in order to match the user to the indoor maps when the MU is indoors, hence increasing computational burden. Hence, it is proposed to have a 'block' of reference database for each indoor building and each area of pre-surveyed outdoor vicinities.

26.3.1 Integration Hardware

The implementation stage uses two main hardware components. The WiFi signals are acquired using the G2 Microsystems's G2C501-HDK Hardware Development Kit, while the GPS signal acquisition, correlation, and receiver processing are executed on the UNSW SNAP lab's Namuru II GPS receiver board. The WiFi positioning algorithm and the WiFi/GPS integration algorithm are implemented on the Namuru II processor as well.

Firstly, the G2C501-HDK Hardware Development Kit is a development board produced by G2 Microsystems to support 2.4GHz WiFi communication and serial UART communication, along with many other functions. Thus, we exploit the 2.4GHz WiFi communication capability of the device to acquire RSSI readings and send it to the Namuru II through the UART for further processing. This device will be simply referred to as the "G2 receiver".

The Namuru II is an open-source GPS receiver research platform based on an L1 GPS RF frontend and is implemented on an Altera® Cyclone II™ Field Programmable Gate Array (FPGA) chip. The FPGA is composed of an array of configurable logic cells, and each cell can be programmed to perform one of a variety of simple digital logic functions. This is used to implement the correlators and a soft-core processor known as NIOS II™. In addition, the Namuru II incorporates a dual UART, 3 axis accelerometer, 2 axis gyro, and ample of memory. *Table 26.1* summarises the technologies available and the technologies utilised for this implementation.

Altera has provided a PC-based development environment known as Quartus II, for configuring the logic elements using Verilog and VHDL. The NIOS II C/C++ IDE is used to develop, compile and load the compiled binary onto Namuru II. The NIOS II IDE (on the PC) also has a console that maintains connection with the NIOS II CPU (on the Namuru II) after the binary is loaded onto it. This is known as the NIOS II console which is used to display debug information during runtime.

The Namuru II acquires the RSSI readings through the UART and performs the WiFi positioning and GPS/WiFi integration utilising the NIOS II soft-core processor. GPS coordinates are acquired from the GPS Architect software. All software processes are developed in NIOS II C/C++ Integrated Development Environment (IDE).

Figure 26.7 illustrates the block diagram of the overall system. The GPS Architect block *(Figure 26.6.1)* is the Mitel® firmware which is already ported to the Altera® NIOS II soft-core processor for the required processing of the GPS correlators and timing signals. These functions are implemented as real-time interrupts that runs as a higher priority task.

The blocks of WiFi positioning algorithms are similar to those described earlier. However, additional functions were added to accommodate the separated blocks of reference database. *(Section 26.2* implemented only a single reference database for all reference points.)

Table 26.1. Comparison of capabilities between the G2 Development Kit (used as a WiFi receiver) and Namuru II Development Kit (used as a GPS receiver and GPS/WiFi integration module). Shaded boxes show the technologies used.

	G2 Development Kit	Namuru II Development Kit
Microprocessor	LEON 2 SPARC V8 32-bit CPU, clocked at 44 MHz	Altera Cyclone II, up to 260Mhz
Wireless Communication	External 2.4 GHz and EPC antennas	N/A
Short Range Communication	3 axis 125 kHz magnetic interface	N/A
Wired Communication	Dual RS-232, SPI	Dual UART, USB 2.0
Positioning	N/A	2x Zarlink GP2015 GPS L1/L2 front ends
Sensors	temperature, security seal, motion, 3 axis shock	3 axis accelerometers, 2 axis gyro
Memory	320kb ROM 80kb RAM (2kb NVM)	32MB ROM, 64MB SDRAM, 256K SRAM

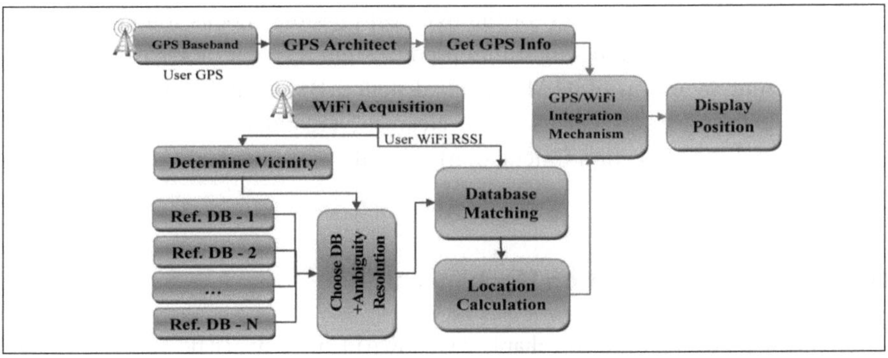

Fig. 26.7. Block diagram of overall integrated system ported into the NIOS II soft-core processor.

All of the implemented functions runs in/from the main() task, which is the lowest priority task (i.e. last executed when the processor is freed by other Interrupt Service Routines (ISR)). However, the UART receive function is an exception, which uses the Altera ISR to retrieve data from the Namuru II UART.

26.3.2 Separated Reference Database

As explained earlier, the entire reference database has to be separated into several smaller Blocks, which is clearly visible from *Figure 26.7.* To determine which Block to be used, a search for the best matching Block(s) of Reference Database has to be done. This is called the FindBlock function. This technique uses the Media Access Control (MAC) address of the APs that the user acquires and attempts to match it with the list of MAC addresses of the APs in each Block of Reference Database. A non-matching AP, where the user acquires a AP that is not in a particular Block of Reference Database, is a case of "missing AP". The number of "missing AP" cases in each Reference Database is recorded, and the Block with the least "missing AP" cases will be chosen. This is implemented in findBlock() as follows:

1. For each Block "k", attempt to match the user's AP MACs with the list of MAC addresses of the AP's in each Block of Reference Database. The counts of "missing AP" cases will be stored in missing_APs[k].
2. Sort the missing_APs[] array in ascending order.
3. The k-th Block with the least "missing AP" cases is chosen.
4. The "wifi_mode" is set to INDOOR or OUTDOOR depending on the environment (outdoor/indoor) of the chosen Block of Reference Database.

FindBlock may produce the same number of "missing AP" for two or more Blocks. This condition may occur when the user is at the border between one Block of Reference Database and another (i.e. between Indoor and Outdoor). Hence, if more than one Block of Reference Database suggested by FindBlock to use, then Database Matching is executed for the two best possible Blocks found instead of one. The Database Matching algorithm produces the residual of k-Nearest-Neighbours matching by means of its Manhattan distance. Based on this metric, the most possible Block is chosen and its set of Nearest-Neighbours are used to perform interpolation (a.k.a. Location Calculation) using the IDW method as described and tested in *Section 26.2.*

26.3.3 GPS/WiFi Integration Mechanism

The implemented integration mechanism is shown in the form of a flowchart in *Figure 26.8.* The flowchart is implemented in the form of if-else statements. "wifi_mode" is a variable determined in the online phase by a function that calculates the user's

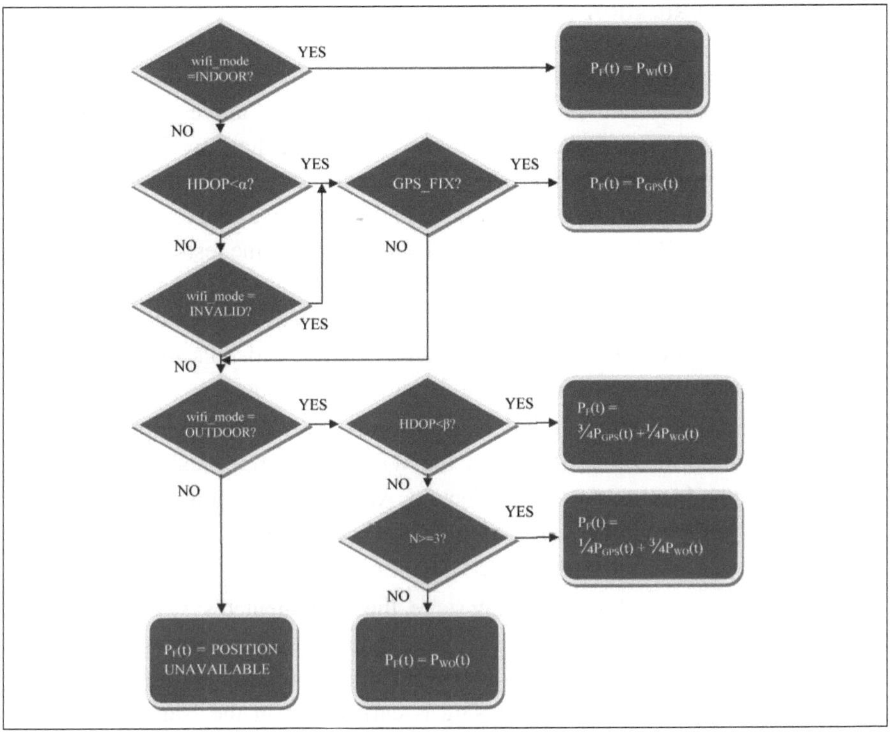

Fig. 26.8. Flowchart of GPS/WiFi Integration Mechanism

vicinity (“INDOOR” / ”OUTDOOR”). If the user’s vicinity is undetermined due to APs not matching those contained in the Reference Database (i.e. user in an area not mapped by the reference database), “INVALID” will be assigned to “wifi_mode”. Horizontal Dilution of Precision (HDOP) is a function of how good the geometry of the tracked satellite signals is. The lower the HDOP, the more precise the MU’s calculated horizontal position is. More details are explained in (Kaplan et al. 2006). GPS_FIX is a Boolean variable determined by the Geometric Dilution of Precision (GDOP) and the number of satellite signals acquired and the Code to Noise Ratio. This variable is acquired from GPS Architect. N is the number of satellite signals tracked. PGPS(t) is the GPS calculated position in the Latitude, Longitude, Altitude (LLA) system. PWO(t) is the outdoor WiFi calculated position in the LLA system. PWI(t) is the indoor WiFi calculated position in the x,y,z system, with reference to the origin of the local building map. PF(t) is the final calculated position of the overall system.

The basic concept of mapping the flowchart is to first take the WiFi parameters as a priority to determine the indoor/outdoor status of the user, then using the validity of GPS parameters to determine which technology (GPS or WiFi) to perform positioning for the user.

26.3.4 Initial Evaluation

The location chosen as the test-bed *(Figure 26.9)* is level G of the EE building, UNSW, where a rear entrance allows the test-device to move from indoors to outdoors, and vice-versa. This location is chosen because there are fewer passersby at non-peak hours. There are three steps to set up the system and the test-bed. The first is to ensure sufficient APs are available in the vicinity of the test-bed. Then the fingerprints of each Reference Point are collected. Finally, the test-device is used to collect test data for analysis. In the first step, a WiFi pre-survey in the vicinity of the test-bed is carried out to identify areas where insufficient APs are detected. Then, an adequate number of APs are deployed in/near the test-bed. APs are also deployed outdoors to ensure WiFi positioning also is possible outdoors.

For the second step, Kismet (a device customised for WiFi fingerprint database collection) is used to collect RSSIs for the WiFi Fingerprint database. Again, the directional method is used for data collection. After the collection, the data is transferred to a PC and pre-processed by a MATLAB program into C/C++ initializations before it is loaded into the Namuru II.

In the third step, which is the testing stage, the implemented system as described in *Section 26.3.2* is brought up to test. The equipment set-up is shown in *Figure 26.10*. The implemented system collects WiFi signals through the integrated hardware and all subsequent data processing is carried out in the Namuru II, which includes WiFi positioning, GPS/WiFi integration, etc. Then, the Namuru II sends the output and debug information to a PC, and displays it on the NIOS II console.

Analysis of the results is done only for 2D. All altitude values are assumed to be 0. 6 Test Points are strategically chosen in the test-bed *(Figure 26.9)*, in both indoor

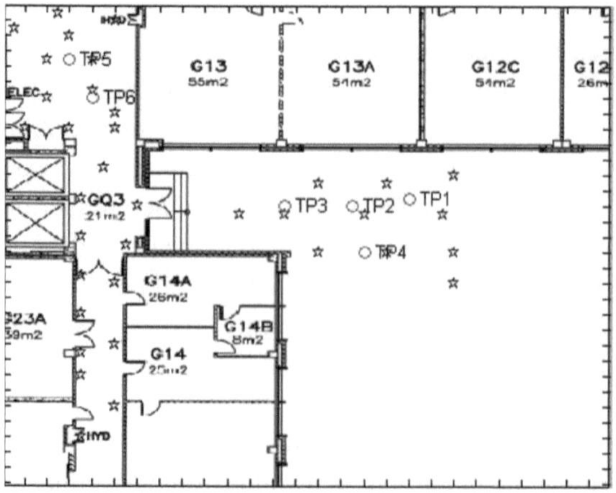

Fig. 26.9. Map of test-bed with reference points (stars) and Test Points (circles) embedded (each tick in x and y axis indicates a meter).

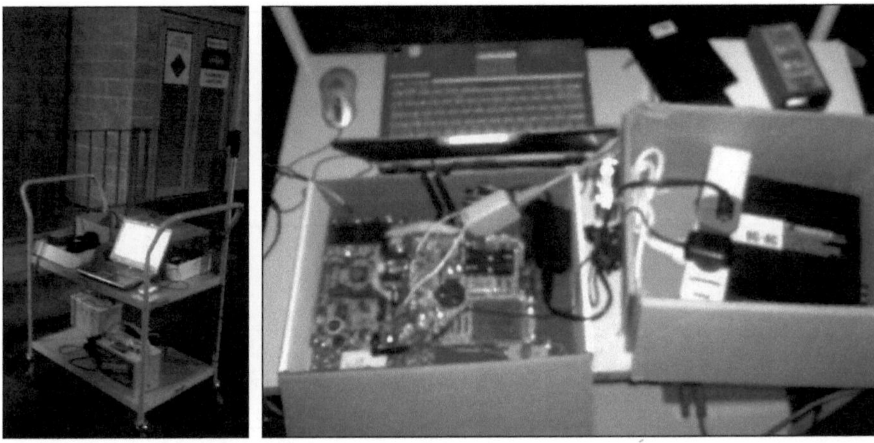

Fig. 26.10. Left: equipments set up for testing, right: A laptop connected to the Namuru II, the UART cable connecting Namuru II to G2 Dev Board, and inverter connected to a battery to power up the boards (right).

and outdoor environments. A number of observations (ranging from 10 to 20, over a 2 minute interval) are collected to analyse the statistical positioning accuracy of each point.

Combining all the error vectors for each test point, the overall accuracy behaviour of the system can be summarised by the CDF plot in *Figure 26.11.* 60% of the points can be positioned within 4.5m of accuracy, while a large probability of misdetection (approximately 40%) due to the inaccurate indoor/outdoor detection of the user causes an error of more than 10m.

The test-bed has been intentionally chosen to be in the vicinity of tall buildings, which resemble urban canyons, so that the importance of WiFi positioning can be

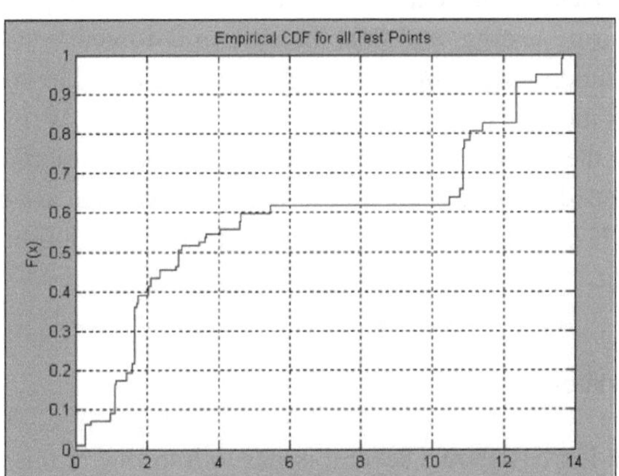

Fig. 26.11. CDF plot for all test points (x in meters).

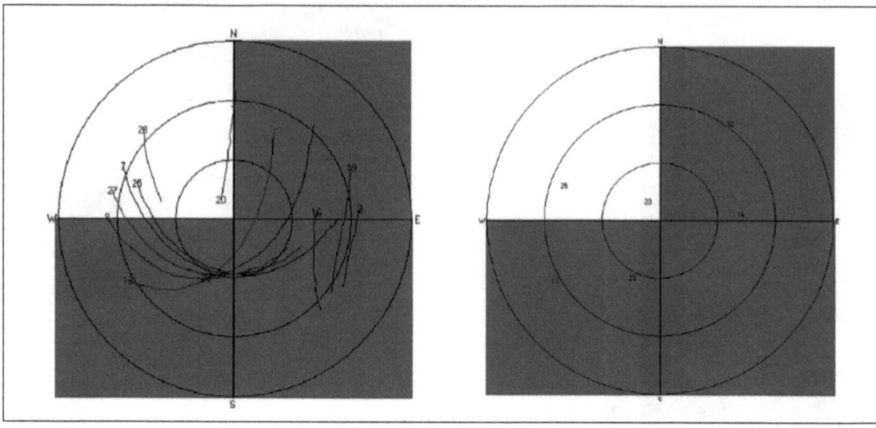

Fig. 26.12. Possible Satellites that can be seen. Shaded region shows the part of sky blocked by tall buildings.

demonstrated. To analyse the availability of GPS, an investigation of the possible satellites visible during the test period was conducted using SkyplotTM.

Figure 26.12 shows the possible satellites that are visible above the horizon on the night of the experiment. The EE building in the test-bed is aligned North in the downward direction in the map *(Figure 26.9)*. Thus, only the Northwest direction of the sky is clear, while the other sky sectors are mostly blocked by the building. The shaded region in *Figure 26.12* indicates the possibly blocked area of the sky. Thus, only satellites 27, 7, 25, 28 and 20 are likely to be detected. *Figure 26.12 (right)* shows the satellites that can be detected at one time (satellites 25 and 20), and also suggests most probably only two satellites can be detected at any one time from the test-bed.

Relating this back to the experiment, this situation compromises the availability of GPS during testing, and thus the system is forced to use WiFi for positioning even in outdoors since only 2–3 GPS satellites could be tracked. However, when the system moves further away from the test-bed, a WIFI INVALID case is observed, and the system is forced to use GPS for positioning whenever the conditions mentioned in *Section 26.3.3* are satisfied (i.e. there is a good view of the sky). A POSITION UNAVAILABLE case is observed when a user is indoors of an unknown building.

26.3.5 Further Work

The initial evaluation has shown several issues to improve in this implementation. The different hardware used for RSSI collection has identified an issue in the transformation of RSSI signals from one device to another. Haeberlen et al. (2004) and

Cheng et al. (2005) has suggested the mapping of received RSSI from one device to another is a linear transformation and will be implemented in the future.

Also, the WiFi/GPS Integration Algorithm is not sufficiently tested yet as GPS was unavailable on the test-bed. Another test-bed with clear view of the sky should be tested with the system. Also, the positioning accuracy of a moving test device has not been addressed due to the fragile prototype, which is susceptible to physical movements and jitter during operation.

26.4 Conclusions

In short, the research has undertaken a simulate-develop-test approach to achieve the outlined objectives. The simulate stage includes the development and testing of the k-wNN fingerprinting algorithm for the WiFi Positioning, in which the parameters chosen attest to the expected accuracy of the system. In the development stage, the system is modified to accommodate for practical considerations, and successfully ported into Namuru II to be integrated with the already available GPS system. In the testing stage, the integrated system is put to test in a test-bed resembling challenged indoor environments and urban canyons, such that WiFi is capable of assisting GPS during its unavailability. The implementation has also suggested a solution to search a large-sized WiFi Reference Database efficiently. The separation of the entire Reference Database into several Blocks is an effective method to reduce computational burden and positioning delay.

The final result has achieved its main objective, which is to be able to seamlessly transition between indoor and outdoor environments by using the newly proposed algorithm. However, more testing are still required to prove the transition between the 2 different technologies, namely the GPS and WiFi positioning system, is feasible. A few simple static testing has suggested improvements to be made to the current real-time positioning system implementation.

For a feasible realisation of the system, the entire Reference Database (e.g. the Reference Database of a metropolitan city) may not be ported into the system due to memory limitations. Hence, Blocks of Reference Database in the user's surrounding can be acquired through an internet server and 'cached'. Opposing to the method of sending MU's RSSI vectors to the internet server where the server then executes the positioning process and returns the position to the MU (e.g. Skyhook Wireless http://www.skyhook.com), this method reduces the computational burden of the internet server. Also, this method allows the MU to perform positioning without having continuous internet connectivity, since intermittent internet connectivity is sufficient for the MU to download the necessary Reference Database. Hence, this method is proposed to be implemented in later stage to support positioning while internet connectivity is unavailable. The probabilistic algorithm for WiFi positioning as introduced in (Roos et al. 2002) is a more accurate method due to its ability to account for the probability distribution of the RSSI signals. However, in depth analysis on the computational capability of the embedded system should be performed before undertaking this task. The system should also be extended to accommodate building level or altitude information, such that a fully operational 3D positioning device is achieved.

References

Bahl P, Padmanabhan VN (2000) RADAR: An in-building RF-based user location and tracking system (Vol. 2). Tel-Aviv, Israel: Proc. IEEE INFOCOM.

Cheng Y, Chawathe Y, LaMarca A, Krumm J (2005) Accuracy Characterization for Metropolitan-scale Wi-Fi Localization. Proceedings MobiSys 2005.

Haeberlen A, Flannery E, Ladd A, Rudys A (2004) Practical Robust Localization over Large-Scale 802.11 Wireless Networks. Philadelphia, PA: Proc. of the 10th ACM International Conference on Mobile Computing and Networking (MOBICOM).

Kaplan ED, Hegarty CJ (2006) Understanding GPS : principles and applications. Boston : Artech House.

Lee HK, Li B, Rizos C (2005) Implementation procedure of wireless signal map matching for location-based services. Proc. IEEE Int. Symposium Signal Processing and Information Tech.

Li B, Kam J, Lui J, Dempster A (2007) Use of Directional Information in Wireless LAN based indoor positioning. Sydney, Australia: Symp. on GPS/GNSS (IGNSS2007).

Li B (2006) Terrestrial mobile user positioning using TDOA and Fingerprinting techniques, PhD thesis, University of New South Wales.

Li B, Salter J, Dempster A, Rizos C (2006) Indoor Positioning Techniques Based on Wireless LAN. Sydney, Astralia: First IEEE International Conference on Wireless Broadband and Ultra Wideband Communications.

Mumford P, Parkinson K, Dempster A (2006) An Open GNSS Receiver. Surfers Paradise, Australia: Symp. on GPS/GNSS 2006.

Retscher G (2007) Performance and accuracy test of a WiFi indoor positioning system. Journal of Applied Geodesy 2007.

Roos T, Myllymaki P, Tirri H, Misikangas P, Sievanenen J (2002) A Probabilistic Approach to WLAN User Location Estimation. Int. Journal of Wireless Information Networks 2002.

Singh R, GuainazzoM, Regazzoni CS (2004) Location determination using WLAN in conjunction with GPS Network. Proc. IEEE 59th VTC—Spring.

Wang Y, Jia X, Lee H (2003) An Indoors Wireless Positioning System Based on Wireless Local Area Network Infrastructure. Proceedings of the 6th International Symposium on Satellite Navigation Technology Including Mobile Positioning & Location Services.

MIX
Papier aus verantwortungsvollen Quellen
Paper from responsible sources
FSC® C105338

If you have any concerns about our products,
you can contact us on
ProductSafety@springernature.com

In case Publisher is established outside the EU,
the EU authorized representative is:
Springer Nature Customer Service Center GmbH
Europaplatz 3, 69115 Heidelberg, Germany

Printed by Libri Plureos GmbH
in Hamburg, Germany